STRUCTURAL STEELWORK
Analysis and Design

STRUCTURAL STEELWORK
Analysis and Design

S.S. RAY
BE (Cal), CEng, FICE, MBGS

Blackwell
Science

© 1998 by
Blackwell Science Ltd
Editorial Offices:
Osney Mead, Oxford OX2 0EL
25 John Street, London WC1N 2BL
23 Ainslie Place, Edinburgh EH3 6AJ
350 Main Street, Malden
  MA 02148 5018, USA
54 University Street, Carlton
  Victoria 3053, Australia
10, rue Casimir Delavigne
  75006 Paris, France

Other Editorial Offices:

Blackwell Wissenschafts-Verlag GmbH
Kurfürstendamm 57
10707 Berlin, Germany

Blackwell Science KK
MG Kodenmacho Building
7–10 Kodenmacho Nihombashi
Chuo-ku, Tokyo 104, Japan

The right of the Author to be identified as the Author of this Work has been asserted in accordance with the Copyright, Designs and Patents Act 1988.

All rights reserved. No part of this publication may be reproduced, stored in a retrieval system, or transmitted, in any form or by any means, electronic, mechanical, photocopying, recording or otherwise, except as permitted by the UK Copyright, Designs and Patents Act 1988, without the prior permission of the publisher.

First published 1998

Set in 10½/12 pt Times
by Aarontype Limited, Bristol
Printed and bound in Great Britain by
MPG Books Ltd, Bodmin, Cornwall

The Blackwell Science logo is a trade mark of Blackwell Science Ltd, registered at the United Kingdom Trade Marks Registry

DISTRIBUTORS

Marston Book Services Ltd
PO Box 269
Abingdon
Oxon OX14 4YN
(*Orders*: Tel: 01235 465500
         Fax: 01235 465555)

USA
Blackwell Science, Inc.
Commerce Place
350 Main Street
Malden, MA 02148 5018
(*Orders*: Tel: 800 759 6102
              781 388 8250
         Fax: 781 388 8255)

Canada
Login Brothers Book Company
324 Saulteaux Crescent
Winnipeg, Manitoba R3J 3T2
(*Orders*: Tel: 204 837-2987
         Fax: 204 837-3116)

Australia
Blackwell Science Pty Ltd
54 University Street
Carlton, Victoria 3053
(*Orders*: Tel: 03 9347 0300
         Fax: 03 9347 5001)

A catalogue record for this title is available from the British Library

ISBN 0-632-03857-8

Library of Congress
Cataloging-in-Publication Data

Ray, S.S
  Structural steelwork: analysis and design   /
S.S. Ray.
      p. cm.
  Includes bibliographical references and index.
  ISBN 0-632-03857-8
  1. Building, Iron and Steel.   2. Steel, Structural.
I. Title.
TA684.R37   1998
624.1′821–dc21   97-34497
                                                CIP

For further information on
Blackwell Science, visit our website:
www.blackwell-science.com

Contents

Preface	xi
Notation	xiii

Chapter 1 Strength of Materials — 1
1.1 Bending stress in beams — 1
    1.1.1 Assumptions — 1
    1.1.2 Position of neutral axis — 3
    1.1.3 Bending stress in asymmetrical sections — 4
    1.1.4 Product inertia of a rectangle about orthogonal axes — 5
1.2 Shear stress in beams — 8
    1.2.1 Shear centre for thin-walled asymmetrical section — 9
1.3 Torsional shear stress — 11
    1.3.1 Torsion of thin rectangular members — 12
    1.3.2 Torsion of thin open sections — 13
1.4 Strain energy in axial load, bending, torsion and shear — 14

Chapter 2 Theory of Structures — 19
2.1 Polygon of forces — 19
2.2 Equations of equilibrium — 20
2.3 Internal forces — 21
    2.3.1 Compound trusses: the method of section — 23
    2.3.2 Maxwell diagram for simple trusses — 24
    2.3.3 The method of section by Maxwell diagram — 25
    2.3.4 Deflection of pin-jointed structures — 27
2.4 Bending moment and shear force — 28
    2.4.1 Simply supported beams — 29
    2.4.2 Slope–deflection equations — 29
    2.4.3 Area moment theorems — 31
    2.4.4 Generalised slope–deflection equations — 33
    2.4.5 Fixed-end beams — 35
    2.4.6 Theorem of three moments — 35
    2.4.7 Method of moment distribution — 37
    2.4.8 Arches — 39
    2.4.9 Symmetrical two-hinged arches — 40
    2.4.10 Hingeless symmetrical arches — 43
2.5 Influence lines — 46
    2.5.1 Influence lines of cantilever beams — 46
    2.5.2 Influence lines of a simply supported beam — 47
    2.5.3 Influence lines of three-hinged arches — 48
    2.5.4 Influence lines of simple trusses — 51

2.6	Matrix method of structural analysis	52
	2.6.1 Truss analysis by the displacement method	52
	2.6.2 Continuous beam analysis by the displacement method	55
	2.6.3 Continuous frame analysis by the displacement method	57
2.7	Structural dynamics	61
	2.7.1 Single degree of freedom (SDOF) systems	62
	2.7.2 Effect of viscous damping on free vibration	65
	2.7.3 Forced vibration of undamped SDOF systems due to harmonic excitation	65
	2.7.4 Forced vibration of undamped SDOF system by triangular impulse	67
	2.7.5 Free vibration of a simply supported beam with distributed mass	68
	2.7.6 Response of an SDOF system to a very short impulse	71
	2.7.7 Response of an SDOF system to any arbitrary time-dependent loading	71
	2.7.8 Seismic response of a damped SDOF system	72
	2.7.9 Response spectra	73
2.8	Analysis of plates	74
	2.8.1 Elastic analysis of plates	74
	2.8.2 Yield-line analysis of plates	76
2.9	Methods of plastic analysis	79
	2.9.1 Definition of plastic behaviour	79
	2.9.2 Plastic behaviour of a section of a beam in pure bending	79
	2.9.3 Plastic behaviour of a section of a beam with bending and axial load	80
	2.9.4 Plastic behaviour of a section of a beam with bending and shear	81
	2.9.5 Load combinations in plastic analysis	82
	2.9.6 Plastic analysis of structures	82

Chapter 3 Analysis of Structures: Worked Examples   85

3.1	Example 3.1: Roof truss	85
	3.1.1 Type of truss	85
	3.1.2 Loading on the truss	86
	3.1.3 Analysis by method of section	90
	3.1.4 Method of section using Maxwell diagram	91
3.2	Example 3.2: Continuous beam	95
	3.2.1 Analysis of continuous beam by the three moment theorem	98
	3.2.2 Analysis of continuous beam by the method of moment distribution	101
	3.2.3 Analysis of continuous beam by the matrix method	103

3.3	Example 3.3: Frame structure	105
	3.3.1 Analysis of a rigid frame by the method of moment distribution	105
	3.3.2 Analysis of a rigid frame by the matrix method	110
3.4	Example 3.4: Analysis of a hingeless arch	114
3.5	Example 3.5: Yield-line analysis of a rectangular plate	118
3.6	Example 3.6: Seismic analysis of a tall cantilever structure	120
3.7	Example 3.7: Plastic analysis of a pitched portal frame	127
	3.7.1 Haunched pitched portal frame	133

Chapter 4   Design of Structures — 137

4.1	Principal issues	137
4.2	Material grade selection and section type selection	138
4.3	Manufacturing process	138
4.4	Connection design	139
	4.4.1 Simple connections	140
	4.4.2 Moment connections	140
	4.4.3 Trusses and open-web girders	141
4.5	Check list of actions and design considerations	142
4.6	Ultimate limit state design	144
	4.6.1 Step-by-step method of semi-rigid design	145
	4.6.2 Load combinations at the ultimate limit state as per BS 5950: Part 1	146
	4.6.3 Patterned loading	147
	4.6.4 Structural stability against lateral loads	148
	4.6.5 Stability of multistorey rigid frames	148

Chapter 5   Design of Struts — 155

5.1	Axial capacity of a column or a strut	155
5.2	Types of failure of a column or strut	155
5.3	Design basis of columns and struts	156
	5.3.1 Compressive strength (p_c)	156
	5.3.2 Boundary conditions	156
	5.3.3 Combined axial compression and bending moment	157
5.4	Step-by-step design procedure of columns/struts	159
5.5	Worked examples	183
	5.5.1 Example 5.1a: Design the internal compression member of a roof truss	183
	5.5.2 Example 5.1b: Same member as in Example 5.1a but use circular hollow section Grade 50	185
	5.5.3 Example 5.2a: Design the rafter of a roof truss	185
	5.5.4 Example 5.2b: Design the same rafter as in Example 5.2a using a rectangular hollow section	189
	5.5.5 Example 5.3: Design the vertical leg of a portal frame	190
	5.5.6 Example 5.4: Design the corner column of a multistorey building	195
	5.5.7 Example 5.5: Design the compound column of an industrial building with a heavy-duty crane	200

viii Contents

Chapter 6	**Design of Ties**	211
6.1	Principal issues	211
6.2	Design basis	211
6.3	Combined axial tension and bending moment	211
	6.3.1 Principal issues	211
	6.3.2 Design basis	212
6.4	Step-by-step design of members in tension	213
6.5	Worked examples	220
	6.5.1 Example 6.1: Design the main tie of the roof truss in Example 3.1	220
	6.5.2 Example 6.2: Design the tie of a 75 m span latticed girder for the roof of an aircraft hangar	222
Chapter 7	**Design of Beams**	227
7.1	Principal issues	227
7.2	Design basis	227
	7.2.1 Local capacity	227
	7.2.2 Lateral torsional buckling	229
	7.2.3 Buckling of web	230
	7.2.4 Tension-field action in thin webs	231
7.3	Step-by-step design of beams	232
7.4	Worked examples	269
	7.4.1 Example 7.1: Beam supporting the floor of a workshop	269
	7.4.2 Example 7.2: Main beam in a multistorey building using simple construction	273
	7.4.3 Example 7.3: Main beam with full end fixity to concrete wall	278
	7.4.4 Example 7.4: Design of a stiffened plate girder	284
	7.4.5 Example 7.5: Design of a gantry girder for an electric overhead travelling crane	300
7.5	Beams subject to torsion	317
	7.5.1 Torsional resistance	317
	7.5.2 Stresses in closed sections	318
	7.5.3 Stresses in open sections	318
	7.5.4 Checks for capacity	320
7.6	Example 7.6: Design of a beam with concentrated applied torque at the centre of span	322
Chapter 8	**Design of Composite Beams and Columns**	329
8.1	Composite beams	329
	8.1.1 Principal issues	329
	8.1.2 Design basis	330
	8.1.3 Effective breadth	330
	8.1.4 Modular ratio	331
	8.1.5 Transformed section: elastic section properties	331
	8.1.6 Second moment of area for elastic analysis	333
	8.1.7 Plastic section properties	335
	8.1.8 Redistribution of support moments	340
	8.1.9 Shear connection	341

		8.1.10	Longitudinal shear transfer	342
		8.1.11	Reduction of plastic moment capacity due to high shear	342
		8.1.12	Deflections	343
		8.1.13	Effect of dead load	343
		8.1.14	Effect of shrinkage and creep	344
		8.1.15	Effect of imposed loads	344
		8.1.16	Step-by-step design procedure for composite beams	345
	8.2	Example 8.1: Composite beam		365
	8.3	Composite columns		383
		8.3.1	Step-by-step design of an encased composite column	384
		8.3.2	Step-by-step design of concrete filled hollow circular section	389
	8.4	Example 8.2: Composite column		390

Chapter 9 Connections in Steelwork — 397

9.1	Bolted connections		397
	9.1.1	Types of bolts	397
	9.1.2	Capacity of bolts	398
	9.1.3	In-plane loading of a group of bolts	400
	9.1.4	Out-of-plane loading of a group of bolts	401
	9.1.5	Out-of-plane bending moment, direct load and shear on a group of bolts	407
	9.1.6	Local capacity check of connected elements in a moment connection	411
	9.1.7	Design of stiffeners and haunched ends of beams	414
	9.1.8	Bolted splices	421
9.2	Welds and welding		427
	9.2.1	Types of weld	428
	9.2.2	Defects in welds	431
	9.2.3	Weld inspection methods: non-destructive testing (NDT)	431
	9.2.4	Design of fillet welds	434
	9.2.5	Design of butt welds	434
	9.2.6	Analysis of weld group	435
	9.2.7	Welded beam-to-column connection	436
9.3	Notched beams		440
	9.3.1	Plain shear check at notched end of beam	441
	9.3.2	Block shear check at notched end of beam	441
9.4	Beam-to-beam connection: shear capacity		442
9.5	Column bases		443
	9.5.1	General rules	443
	9.5.2	Empirical design of base plates for concentric forces	443
	9.5.3	Analysis of column-to-foundation connection	444
	9.5.4	Analysis of base plate	447
	9.5.5	Base plate with stiffeners	449
	9.5.6	Holding-down bolts	451

	9.5.7	Shear transfer from column to concrete	453
	9.5.8	Rules to determine trial size of base plate	454
9.6	Worked examples		454
	9.6.1	Example 9.1: Rigid bolted connection of column-to-roof truss	454
	9.6.2	Example 9.2: Welded connections of members in a roof truss	457
	9.6.3	Example 9.3: Beam-to-beam connection using fin plates	458
	9.6.4	Example 9.4: Beam-to-beam connection using double angle web cleats	465
	9.6.5	Example 9.5: Beam-to-beam connection using end plates	469
	9.6.6	Example 9.6: Portal frame eaves haunch connection	473
	9.6.7	Example 9.7: Bolted splice of a beam	485
	9.6.8	Example 9.8: Bolted splice of a column	491
	9.6.9	Example 9.9: Design of a column base	496

Chapter 10	**Corrosion Protection**	501
10.1	Process of corrosion in steel	501
10.2	Protective system	501
10.3	Types of coating	502
10.4	Components of a corrosion protection system	502
10.5	Specification of a protective system	502
10.6	Sample corrosion protection systems	503

Chapter 11	**Material Properties**	505
Table 11.1	Influence coefficients for continuous beams	506
Table 11.2	Bolt data and capacities	508
Table 11.3	Capacity of welds	508
Table 11.4	Dimensions and properties of steel sections	510
Table 11.5	Dimensions for detailing and surface areas	530
Table 11.6	Hp/A values	535
Table 11.7	Back marks in channel flanges and angles	544
Table 11.8	Back marks in flanges of joists, UBs and UCs	545
Table 11.9	Bending moments, shear forces and deflections	546
Table 11.10	Fire protection methods for steelwork	554

Index 555

Preface

There are a number of excellent books available which deal with the design of structural steel elements but, in the author's view, they lack certain important features. The design of structural steel members requires many checks in a systematic, structured manner: this book has been written to provide a step-by-step approach to this task with a view to achieving completeness of the design process in all respects.

This highly prescriptive approach provides a comprehensive book which is at the same time user friendly. The task of quality assurance becomes less arduous and the output of a design office becomes fully standardised if the approach in this book is strictly followed. For students, too, the book should prove to be invaluable because the essential elements of the theory of structures and analysis are discussed, followed by a structured approach to the design of all elements in a building, including connections. The numerous worked examples will be very useful to practitioners, students and lecturers alike.

The design principles are illustrated at each stage by sketches. The extensive use of illustrations is a particular feature of the book and is intended to clarify ambiguities in the codes of practice.

ACKNOWLEDGEMENTS

Extracts from BS 5950: Part 1: 1990 are reproduced with the permission of BSI. Complete editions of the standards can be obtained by post from BSI Customer Services, 389 Chiswick High Road, London W4 4AL.

The author wishes to thank the British Standards Institution for their kind permission to reproduce some of the essential charts from their Codes of Practice, and Mr Malcolm Smith for helping to produce the sketches in this book.

S.S. Ray
Great Bookham
Surrey

Notation

α	Angle of inclination of line AB to horizontal
α	Coefficient of thermal expansion of materials
α_e	Modular ratio
α_l	The modular ratio for long-term loading taking into account creep of concrete
α_s	The modular ratio for short-term loading
γ_m	Material factor
γ_{mc}	Material factor for concrete
γ_t	Factor to allow for thread stripping effects
δ	Corrected final deflection of a composite beam allowing for partial shear connection
δ	Deflection at the centre of the column
δ_0	The deflection of the span due to loading as a simply supported beam
δ_c	Deflection of a composite beam with full shear connection
δ_c	The deflection of a continuous composite beam at the middle of one of the spans due to imposed loading
δ_i	Axial extension of member i of a truss
δ_n	Deflection at the point of application of P_n in the direction of the force
δ_s	Deflection of steel beam alone subjected to loading
δ_{st}	Static deflection of a single degree of freedom system due to gravity load Mg
Δ_i	Displacement of joint i
Δ_s	Altitude of site above mean sea level
ε	Axial strain
ε_{AB}	Strain at a distance y from the neutral axis on a circular arc AB
ε_s	Effective shrinkage strain in concrete
η	Perry factor
η_{LT}	Perry coefficient
θ	Angle to the horizontal made by the diagonal stiffener
θ	Angle of inclination between X and U axes
θ	Angle of rotation when the length of the member is l
θ	Angle subtended at the centre of the circular arc AB
θ	Angle between the X–X and U–U axes measured anticlockwise from the X–X axis
θ	Beam span-end rotation
θ	Torsional rotation of a circular shaft about the longitudinal axis
θ_A	Slope of beam at end A
θ_B	Slope of beam at end B
Θ_{AV}	Rigid body rotation of beam AB due to settlement of support B
Θ_i	Rotation of joint i
λ	Slenderness ratio
λ_c	Minor axis slenderness ratio corresponding to L_c in bracing system
λ_{LO}	Limiting equivalent slenderness ratio
λ_{LT}	Equivalent slenderness ratio for a column or beam

xiv Notation

λ_w	Web slenderness factor
λ_x	Slenderness ratio of the effective section of bearing stiffener
λ_x	Slenderness ratio corresponding to the effective length about the X–X axis
λ_y	Slenderness ratio corresponding to the effective length about the Y–Y axis
μ_1	Snow load shape coefficient
μ_f	Average load factor for ultimate limit state
ρ_l	The proportion of total loading which is long term
σ	Bending stress on the section compression positive
σ	Direct axial stress
σ_{AB}	Stress at a distance y from the neutral axis on a circular arc AB
σ_c	Direct stress in tension
σ_w	Bending stress due to warping torsion
σ_x	Stress due to applied moment M_x
σ_y	Stress due to applied moment M_y
τ	Maximum shear stress in the circular shaft at a distance of radius r from the centre
τ	Shear stress at a distance y_0 from neutral axis
τ	Shear stress in thin rectangular section along the long edge
τ	Torsional shear stress
τ_t	Shear stress in a closed section subjected to torque
τ_w	Warping shear stress in the flanges
ν	Poisson's ratio for the material of the plate
ϕ	Angle of the centre-line of the arch rib with respect to the X-axis
ϕ	Angle of inclination of the lacing bars to the axis of the member
ϕ	Diameter of steel tube
ϕ	Haunch angle
ϕ	Inclination of the tangent at point D of the arch to the horizontal
ϕ	Rotation of the member about the longitudinal X-axis
ϕ	Shear strain
ϕ	Total angle of twist
$\psi(x)$	Shape function
ω_d	Damped natural frequency
ω_n	Undamped natural frequency (radians per second)
Ω	Frequency of excitation
a	Diagonal dimension
a	Distance from the centroid of bolts in tension to the Y–Y axis
a	Distance from the innermost vertical line of fastener holes to the edge of the web
a	Distance of hinge C from support A of arch
a	Distance from the steel beam centre to the concrete centre
a	Length of greater projection of the plate beyond the column
a	Length of run of weld in the X–X direction
a	Robertson constant
a	Spacing of transverse stiffeners
a	Torsional bending constant
a'	Distance of unit load from support A of the arch
a_h	Minimum hook approach
a_w	Wheel spacing in end carriage
A	Amplitude of motion
A	Area of bending moment diagram

A	Area of cross-section of rib of arch
A	Area of the section cut off by a plane parallel to neutral axis at a distance of from the neutral axis
A	Area of segment bordered by the yield lines (m^2)
A	Area of steel beam or column section
A	Area of stiffener in contact with flange
A	Net area of plate allowing for holes
A_0	Area of the cross-section of the arch rib at the crown.
A_0	Area of the rectilinear element of the section which has the largest dimension parallel to the direction of the load
A_1	Area of free-span bending moment diagram for span 1 (AB)
A_2	Area of free-span bending moment diagram for span 2 (BC)
A_c	Area of concrete in the steel tube
A_c	Net area of concrete in the encasement after deducting the area of the steel section and any steel reinforcement used
A_{cv}	Mean area of the concrete shear surface per unit length of beam
A_e	Effective area of batten after deduction for holes
A_e	Effective area of flange plates allowing for bolt holes
A_e	Effective area in tension
A_{eff}	Effective area of stiffener considering 20 times web thickness on either side of the stiffener
A_f	Area of flange
A_g	Gross cross-sectional area
A_g	Gross cross-sectional area (lacing bar)
A_{gf}	Gross area of flange
A_{go}	Gross area of unconnected leg
A_{gs}	Gross area of stem
A_h	Area enclosed by the mean perimeter of the closed section
A_h	Tensile area of web from the innermost vertical line of fastener holes to the edge of the web along a horizontal line
A_{he}	Effective tensile area of web from the innermost vertical line of fastener holes to the edge of the web along a horizontal line
A_i	Area of section of member i of truss
A_n	Net area
A_{nc}	Net area of connected leg
A_{nf}	Net area of flange
A_{nw}	Net area of web
A_r	Area of any steel reinforcement in the concrete
A_r	Area of steel reinforcement in the concrete encasement
A_r	Area of tensile reinforcement in the concrete slab in the effective section
A_s	Area of shank where threads do not appear in the shear plane
A_s	Area of steel in the steel tube cross-section
A_s	Area of steel section encased
A_s	Effective area of bolt in shear
A_{sc}	Equivalent area of web stiffener
A_{sc}	Equivalent area of stiffeners including 20 times thickness of web
A_{sg}	Area of the stiffener assembly
A_{sg}	Gross area of stiffener assuming a pair of plates symmetrically placed about the web
A_{sv}	Cross-sectional area of the total steel reinforcement crossing the potential shear failure plane
A_t	Area of the bolt at the bottom of the threads
A_t	Effective area of bolt in tension

A_v	Effective shear area
A_v	Shear area of column section
A_v	Total shear area of the web with web plates
A'_v	Vertical shear area of web from the notch to the bottom-most fastener hole
A_{ve}	Effective shear area at the notch
A'_{ve}	Effective vertical shear area of web from the notch to the bottom-most fastener hole
A_{vg}	Gross shear area at the notch
A'_{vg}	Gross vertical shear area of web from the notch to the bottom-most fastener hole
A_{vn}	Net shear area at the notch after deduction of holes
A'_{vn}	Net vertical shear area of web from the notch to the bottom-most fastener hole
AB	Length of the circular arc along the line AB
b	Actual breadth of flange on each side of the beam centre-line measured as half the distance to the adjacent beam
b	Distance from the centre of the web to the tip of the flange
b	Distance of hinge C from arch support B
b	Length of run of weld in the $Y-Y$ direction
b	Width of individual panels
b	Width of rectangular section
b	Width of the section at a distance y_0 from neutral axis
b	Width of smaller flange
b'	Distance of unit load from support B of the arch
b_1	Stiff bearing length determined by assuming 45° dispersion of load
b_c	Breadth of concrete encasement
b_e	Width of end post
b_1	Stiff bearing length which cannot deform appreciably in bending
b_r	Breadth of the concrete rib
b_s	Width of bearing stiffener
b_s	Width of stiffener plates
b_s	Width of web plate
b'_s	Width of stiffener in contact with the flange
b_{sg}	Gross outstand of the stiffener
b_{sg}	Width of stiffener on each side of the web
b_{sn}	Net outstand in contact with the flange of the column allowing for corner snipe
B	Building plan width
B	Length of side of square anchor plate
B	Overall breadth of box section
B	Width of the base plate
B	Width of base plate along the $Y-Y$ axis
B	Width of flange (steel beam or column section)
B_b	Width of flange of the beam
B_c	Width of flange of the column
B_e	Effective width of the concrete flange
B_{eff}	Effective width of the column flange for load transfer
c	Average cover to the two orthogonal layers of reinforcement at the top of a foundation
c	Centroid of the area A (loaded by r_u) from the line of rotation of the segment
c	Damping coefficient

c_{cr}	Critical damping coefficient
C	Compressive internal force in the section
C	Torsional modulus constant
C	Total upward contact reaction of a concrete foundation on the base plate
C_a	Size effect factor
C_p	Standard pressure coefficient
C_r	Dynamic augmentation factor
d	Clear depth of web of the steel beam
d	Depth between fillets of column section
d	Depth of rectangle
d	Depth of girder
d	Depth of web
d	Depth of web between fillets
d	Diameter of the bolt
d	Diameter of holes in a row perpendicular to the direction of tensile force
d	Effective depth of embedment
d	Length of each constituent thin rectangle of the section
d	Length of rectangular section
d	Nominal diameter of the bolt
d	Nominal shank diameter of the studs
d'	Total depth of web remaining after notches, without deduction of holes for fasteners
d''	Depth of web from the notch to the bottom-most fastener hole
d_p	Depth of the plate stiffener at the point of maximum moment in the base plate
d_w	Depth of web
dA	Elemental area of a section
dS	Elementary length along rib of arch
D	Depth of the base plate
D	Diameter of circular shaft
D	Diameter of the column
D	Diameter of holes for bolts
D	Diameter of holes for fasteners
D	Flexural rigidity of plate
D	Length of base plate along the X–X axis
D	Overall depth of box or column section
D	Total depth of embedment of the holding-down bolt
D_1	Overall depth of beam B_1
D_2	Overall depth of beam B_2
D_b	Overall depth of beam
D_c	Overall depth of the column
D_f	Distance between the centre of flanges
D_p	Length of a side of the square base plate
D_p	Overall depth of profile sheet
D_r	Distance between the centre of the tensile and compressive reinforcement
D_r	Distance from the top of the steel beam to the centroid of the tensile reinforcement
D_s	Overall depth of the concrete slab
e	Eccentricity of in-plane loading on a group of bolts
e	Eccentricity of reaction R_L from the centre of gravity of effective section along the centre-line of web

e	End distance in the direction of load
e	Shear centre of a thin-walled section
e	Spacing of the end posts
$e(t)$	Time-dependent vertical displacement
E	Modulus of elasticity
E_i	Elastic modulus of member i of a truss
E_P	Elastic modulus of the panel material
E_s	Elastic modulus of steel
EI	Flexural rigidity of section
f	Mean longitudinal stress in the smaller flange due to moment and axial load on the beam at the point of interest
f	Number of rotationally and translationally fixed supports (fixed)
f	Rise of arch rib
f_c	Stress in compression due to axial load
f_c	Stress in concrete fibre
f_{cu}	Characteristic 28 day cube strength of concrete
f_{ed}	Compressive stress in web at the compression edge of the web
f_m	Longitudinal fibre stress due to bending moment
f_m	Maximum shear stress in the weld group due to the applied moment M
$f_{m,x}$	Shear in the weld in the X–X direction due to moment on the weld group
$f_{m,y}$	Shear in the weld in the Y–Y direction due to moment on the weld group
f_n	Natural frequency of a single degree of freedom system
f_n	Undamped natural frequency of vibration $= 1/T_n = \omega_n/2\pi$ (cycles per second, Hz)
f_r	Resultant shear stress in the weld due to combined moment and shear
f_s	Stress in the steel beam
f_v	Punching shear stress at the effective perimeter
f_v	Shear stress due to applied shear force V
f_v	Shear stress due to direct shear F_v in the Y–Y direction
f_y	Yield strength of steel reinforcement
f_y	Yield stress
F	Applied direct tension
F	Axial compression in a column
F	Lateral load at the bearing
F	Lateral load on the stiffener applied at the beam compression flange
F	Magnitude of the impulsive force
F	Ultimate axial compression load at a section in a member
F	Ultimate maximum axial tensile force in a compound member
$F_{1,b}$	Flange force in the bottom flange of beam B_1
$F_{1,t}$	Flange force in the top flange of beam B_1
$F_{2,b}$	Flange force in the bottom flange of beam B_2
$F_{2,t}$	Flange force in the top flange of Beam B_2
F_c	Compressive force in the concrete flange
F_c	Horizontal compressive force for connection equilibrium
F_c	Total compressive force acting at the centre of compression
F_c	Ultimate compressive load
F_f	Flange forces
F_n	The longitudinal tensile force in the reinforcement at the negative moment region
F_P	The longitudinal compressive force in the concrete slab due to maximum positive moment

F_q	Stiffener force
F_r	Tensile force in steel reinforcement
F_{ri}	Actual tensile load in the bolt at the ith row
F_{rj}	Actual load in the bolts in row j
F_s	Applied shear force
F_t	Applied direct tension
F_v	Applied load through flange
F_v	Applied maximum shear force in the panel under consideration
F_v	Applied shear force
F_v	Applied shear on the weld group
F_v	Column web panel shear
F_v	Shear in the stiffeners
$F_{v,max}$	Applied maximum shear
F_{vx}	Applied shear about the X–X axis
F_{vy}	Applied shear about the Y–Y axis
F_x	External load on the stiffener
FEM	Applied joint moment or algebraic sum of fixed-end moments at joint
g	Distance between holes at right angles to the direction of the tensile load
g	Horizontal spacing of bolts at the connection
G	Modulus of rigidity
G	Shear modulus
h	Distance between centroids of flanges
h	Number of translationally fixed supports (hinged)
h	Overall height of stud
h	Thickness of plate
h_b	Distance from neutral axis to centroid of bottom flange
h_c	Depth of concrete encasement
h_i	Distance of the ith bolt hole in the plate from the centre-line parallel to the axis of the beam or column
h_i	The distance of the bolt in the ith row from the centre of compression
h_n	Distance from line of action of direct load N to the centre of compression
h_S	Distance between centres of the flanges of the steel beam
h_t	Distance from neutral axis to centroid of top flange
H	Building height
H	Horizontal component of the support reaction H'
H	Horizontal reaction at supports
H	Warping constant
H'	Reaction at arch support in the direction AB
H_0	Height of average roof-top level around the building
H_A	Horizontal reaction at end A
H_e	Effective height
H_q	Anchorage force parallel to the longitudinal horizontal axis of a beam
H_{max}	Maximum horizontal base shear
H_r	Reference height of building
ΣH	Summation of all the resolved forces in the horizontal direction
I	Moment of inertia of the cross-section of a beam about its neutral axis
I	Moment of inertia of a section of arch rib
I	Moment of inertia of a section of beam or column
I	Second moment of area about the neutral axis

I_0	Moment of inertia of a section of arch rib at the crown
I_1	Moment of inertia of section of beam in span 1
I_2	Moment of inertia of section of beam in span 2
I_{cf}	Second moment of inertia of the compression flange about the minor axis
I_{eff}	Effective moment of inertia
I_g	Second moment of area of an uncracked composite section
I_n	Second moment of area of a cracked composite section for negative moments
I_p	Second moment of area of a cracked composite section for positive moments
I_P	Polar moment of inertia of the group of bolts
I_P	Polar moment of inertia of the weld group
I_s	Minimum stiffness of transverse web stiffener
I_s	Second moment of area of stiffeners only about web centre-line
I_{sc}	Equivalent moment of inertia including 20 times web thickness
I_{tf}	Second moment of inertia of the tension flange about the minor axis
I_U	Moment of inertia of the section about the principal axis U–U
I_{UV}	Product inertia of the section about the U and V axes
I_V	Moment of inertia of the section about the principal axis V–V
I_x	Moment of inertia about the major axis
I_x	Second moment of area about the major axis through the neutral axis
I_x	Second moment of area of the steel beam about the major axis
I_X	Moment of inertia of the section about the X-axis
I_{xx}	Moment of inertia of the weld group about X–X axis
I_{XY}	Product inertia of the section about the X and Y axes
I_y	Second moment of area about the minor axis through the neutral axis
I_Y	Moment of inertia of the section about the Y-axis
I_{yy}	Moment of inertia of the weld group about Y–Y axis
ΣI	Second moment of area of all the components in a composite column about the minor axis expressed in steel units
j	Number of joints in a truss
j	Total number of joints in the structure
J	Number of lacing bars cut by a plane perpendicular to the axis of the member
J	Number of shear connectors in a row perpendicular to the axis of the beam
J	Torsion constant
k	Stiffness
k	Torsional stiffness of a circular shaft
k^*	Generalised stiffness of a system
k_1	The reduction factor for ribs perpendicular to the beam
k_2	The reduction factor for ribs parallel to the beam
k_b	Beam stiffness
k_c	Column stiffness
k_i	Joint restraint coefficient of joint i
k_G^*	Generalised geometric stiffness
K	Stiffness of a single degree of freedom system (N/m)
l	Horizontal distance between supports A and B of arch
l	Horizontal span of an arch
l	Length of member

l	Length of a side of segment or projected length of a side of a segment
l	Length of span of beam
l	Storey height
l_1	Length of span 1
l_2	Length of span 2
l_i	Length of member i of a truss
L	Actual length of strut
L	Building plan length
L	Length along the beam or column between adjacent points of restraint
L	Length of rafter between points of intersection with web members
L	Length of web plate
L_{aa}	Effective length about the A–A axis
L_{bb}	Effective length about the B–B axis
L_c	Actual overhang of base plate on the compression side
L_c	Effective length between points of intersection of lacing with main members
L_c	Length of end plate
L_c	Span of crane bridge
L_e	Effective length of beam
L_{eff}	Effective length of flange or end plate gone beyond yield
L_{eff}	Effective length of the stiffener
L_E	Effective length of beam or column between restraints
L_E	Effective length of strut
L_{EX}	Effective length about the X–X axis
L_{EY}	Effective length about the Y–Y axis
L_j	Distance between the first row of bolts and the last row of bolts in the direction of loading in a long joint
L_s	Length of the stiffener
L_s	Length of web plate
L_t	Actual overhang of base plate
L_t	Effective length of web resisting tension assuming 60° spread of load
L_{vv}	Effective length about the V–V axis
L_x	Length along the beam or column to a point of interest from a point of restraint
L_Z	Effective span L for simply supported beam
m	Distance from the centre of bolt to 20% of the distance into the root of rolled section or the fillet weld
m	Equivalent uniform moment factor
m	Mass per unit length of beam
m	Number of members in a frame
m	Number of members in a truss
m^*	Generalised mass of a system
m_c	Effective overhang of base plate on the compression side
m_t	Effective overhang of base plate from the centre of tension of the holding-down bolts
M	Applied bending moment
M	Applied bending moment about an axis perpendicular to D
M	Applied in-plane moment on the weld group
M	Applied moment at the connection
M	Beam span-end moments
M	Bending moment at any point along the length of a beam along the X-axis

M	Bending moment in an arch due to applied loading
M	Constant applied moment on column section
M	Mass of a single degree of freedom system (kg)
M	Mass of system
M	Maximum ultimate bending moment
M	Numerically larger end moment of the beam in the unrestrained length
M	Resultant moment on a circular section
M'	Bending moment due to external loading
M'	Bending moment in an arch rib when H is zero
M'	Revised plastic moment of resistance with shear
M_0	Bending moment at midspan of unrestrained length of beam
M_0	Maximum span bending moment in span due to loading as a simply supported beam
M_1	Bending moment in beam B_1 clockwise positive
M_2	Bending moment in beam B_2 clockwise positive
M_a	Bending moment at support A of continuous beam
M_A	Maximum moment in the length of the member
M_{ab}	Applied moment at end A of beam AB
M_b	Bending moment at support B of continuous beam
M_b	Buckling resistance moment
M_{ba}	Applied moment at end B of beam AB
M_{bs}	Buckling moment of resistance of a simple column
M_c	Bending moment in arch rib at the crown of an arch
M_c	Bending moment in the cantilever base plate overhang on the compression side
M_c	Elastic moment of resistance of flange plates about the plate centre-line parallel to the axis of the beam or column, allowing for bolt holes
M_c	Moment capacity of the connection
M_c	Plastic moment capacity of a composite section
M_c	Plastic moment of resistance of the section
M_c	Section moment capacity
M_{cx}	Plastic moment capacity of a section about the X–X (major) axis
M_{cx}	Plastic moment capacity of a section in the absence of axial compressive load about the X–X (major) axis
M_{cy}	Plastic moment capacity of the section about the minor axis
M_{cy}	Plastic moment capacity of a section in the absence of axial compressive load about the Y–Y (minor) axis
M_d	Bending moment at point D of arch
M_E	Elastic critical moment
M_f	Flange plate bending moment due to minor axis bending moment M_{yy}
M_f	Plastic moment capacity of the section remaining after deduction of shear area A_v
M_{local}	Local bending moment in a rafter due to concentrated load of purlin
M_m	Equivalent bending moment
M_{max}	Maximum moment in a column section due to moment and direct load
M_{max}	Minor axis bending moment due to strut action
$M_{max,x}$	Minor axis bending moment due to strut action at any point in the beam, assuming sinusoidal distribution
M_p	Ultimate plastic moment capacity of a plate section per unit length (Nm/m)
M_P	Plastic moment capacity
M_P	Plastic moment of resistance of the beam section

M_P		Plastic moment of resistance of flange or end plate per unit length
M_{pf}		Plastic moment capacity of the smaller flange about its own equal area axis parallel to the flange
M_{pw}		Plastic moment capacity of the web about its own equal area axis perpendicular to the web
M_{rc}		Maximum moment of resistance of base plate
M_{rx}		Reduced local section moment capacity about the X–X axis in the presence of axial load
M_{ry}		Reduced local section moment capacity about the Y–Y axis in the presence of axial load
M_s		Applied moment due to transverse eccentricity of R_L
M_s		Moment on a stiffener due to eccentrically applied load
M_s		Moment on a stiffener due to eccentricity of transverse load
M_s		Plastic moment capacity of the steel beam
M_t		Bending moment in the base plate overhang on the tension side
M_T		In-plane moment on a group of bolts
M_u		Resolved moment on the section about the U–U axis
M_v		Resolved moment on the section about the V–V axis
M_x		Applied moment about the major axis at the splice
M_x		Applied moment about the X–X axis
M_x		Applied ultimate moment about the major axis
M_x		Bending moment per unit length acting on the edges parallel to the Y-axis
M_x		Ultimate nominal bending moment about the X–X axis
M_{xy}, M_{yx}		Torsional moment per unit length on the edges of a plate
M_y		Applied moment about the Y–Y axis
M_y		Applied ultimate moment about the minor axis
M_y		Bending moment per unit length acting on the edges parallel to the X-axis
M_y		Ultimate nominal bending moment about the Y–Y axis
M_{ys}		Moment capacity of the stiffener based on its elastic modulus
M_{yy}		Minor axis bending moment due to applied minor axis moment and additional minor axis bending moment due to strut action
ΣM		Summation of all moments of the forces (or the resolved forces) about any point in the system
n		Depth of the plastic neutral axis from the top of the concrete flange
n		Effective edge distance
n		Number of bolts in the group
n		Number of fastener holes in any single vertical line
n		Number of holding bolts in the tension zone
n		Number of plate stiffeners resisting the compression side moment M_c and the tension side moment M_t in the base plate
n		Slenderness correction factor
n'		Number of fastener holes along the horizontal line from the innermost line of fastener to the edge of the web
n_1		The length obtained by dispersion at 45° through half the depth of the section
n_2		The length obtained by dispersion through the flange to the flange-to-web connection at a slope of 1:2.5 to the plane of the flange
n_c		Number of bolts in the compression zone
n_p		Number of web plates
n_t		Number of bolts in the tension zone
N		Actual number of shear connectors for positive or negative moment as relevant

N	Applied direct axial load, compression positive
N	Applied direct tension
N	Axial direct load from column
N	Direct axial compression in the column applied at the intersection of the X–X and the Y–Y axis
N	Internal axial load due to applied loading
N	Number of shear connectors in a group
N	Total number of shear connectors in a spacing S
N	Total number of wheels
N	Vertical load
N'	Axial thrust due to external loading only
N'	Axial compression in the arch rib when H is zero
N_1	Direct tension or compression in beam B_1, positive in the direction of positive x
N_2	Direct tension or compression in beam B_2, positive in the direction of positive x
N_a	Actual number of shear connectors used for positive moment
N_c	Axial thrust in arch rib at the crown of an arch
N_d	Shear in the bolt due to axial load
N_{dx}	Direct load shear in each bolt along the X–X axis
N_i	Internal force in member i due to external loading
N_L	Axial tension or compression in lacing bar
N_m	Shear in the ith bolt due to moment
N_{mix}	Shear in the ith bolt in the X–X direction due to moment M_f
N_{miy}	Shear in the ith bolt in the Y–Y direction due to moment M_f
N_n	Number of shear connectors due to negative moment
N_p	Number of shear connectors due to positive moment
N_p	Number of shear connectors required to give full shear connection and for positive moment capacity of the section to develop
N_s	Shear in bolt due to direct shear
NA	Length of the circular arc along the line NA
p_0	Amplitude of force of excitation
p_b	Bending strength
p_{bb}	Design strength of bolt in bearing
p_{bs}	Design strength of connected ply in bearing
p_c	Compressive strength
p_c	Compressive strength of column acting as strut
p_c	Compressive strength of column web
p_{cx}	Compressive strength corresponding to slenderness about the X–X axis
p_{cy}	Compressive strength corresponding to slenderness about the Y–Y axis
p_{ed}	Compressive strength due to edge loading
$p_{\text{eff}}^*(t)$	Generalised effective load
p_E	Euler strength
p_s	Design strength of bolts in shear
p_t	Design strength of bolts in tension
p_w	Design strength of fillet weld made using electrodes complying with BS 639
p_y	Design strength of material (flange plates, beam and web)
p_y'	Design strength of backing plate
p_{yb}	Design strength of beam
p_{yc}	Design strength of column
p_{yf}	Design strength of the flange
p_{yg}	Design strength of plate stiffener

p_{yp}	Design strength of base plate
p_{yp}	Design strength of end plate
p_{yp}	Design strength of profiled steel sheet
p_{ys}	Design strength of the stiffener
p_{ys}	Design strength of web plate
p_{yw}	Design strength of the web
$p(t)$	Applied time-dependent force
P	Applied external force
P	Applied point load between stiffeners
P	Generalised form of notation of external loads
P	In-plane load on a group of bolts
P_0	Minimum shank tension as per BS 4604
P_{bb}	Bearing capacity of bolts
P_{bg}	Bearing capacity of friction-grip bolts
P_{bs}	Bearing capacity of connected ply
P_{BC}	Capacity of the batten in compression
P_{BT}	Capacity of the batten in tension
P_c	Bearing resistance of the beam flange by compression
P_c	Compressive strength of strut
P_c	Force in concrete in compression
P_{cr}	Elastic critical load
P_{crx}	Elastic critical load corresponding to effective length L_x
P_{cry}	Elastic critical load corresponding to effective length L_y
P_{cy}	Flexural buckling failure axial load
P_{fc}	Force in steel flange in compression
P_{ft}	Force in steel flange in tension
P_{LC}	Capacity of the lacing bar in compression
P_{LT}	Capacity of the lacing bar in tension
P_{LW}	Local capacity of web at bearing or under concentrated load
P_q	The buckling resistance of the intermediate web stiffener
P_{rc}	Force in reinforcement in compression
P_{ri}	Maximum allowable tensile load in the bolt at the ith row
P_{rt}	Force in reinforcement in tension
P_s	Reduced bolt shear capacity
P_s	Resistance of horizontal shear of an inclined diagonal stiffener
P_s	Shear capacity of bolts
P_{sl}	Slip resistance
P_{sl}	Slip resistance of waisted shank friction-grip bolts
P_{st}	Tension capacity of the tension stiffener
P_t	Local tension capacity of column web
P_t	Resistance of web in tension
P_t	Tension capacity of flange at the connection
P_t	Tensile capacity
P_t	Tension capacity of friction-grip bolts
P_t	Ultimate maximum tensile load in the encased column
P_t'	Enhanced bolt tensile capacity without allowance for prying
P_u	Composite column section capacity in axial compression
P_u	Ultimate maximum compressive load in a 'stocky' column
P_v	Block shear capacity
P_v	Plain shear capacity of notched beam
P_v	Resistance of the web in panel shear
P_v	Shear capacity
P_{vx}	Section shear capacity about the X–X axis

P_{vy}	Section shear capacity about the Y–Y axis
P_w	Buckling resistance of unstiffened web
P_{wc}	Force in web in compression
P_{wt}	Force in web in tension
P_x	Buckling resistance of a load carrying stiffener
ΔP_n	Increment of external applied load
q	Loading per unit area on plate
q	Ratio of the tapered length to the total length between effective torsional restraints
q_b	Basic shear strength
q_b	Buckling shear strength
q_c	Elastic critical shear stress
q_{cr}	Critical shear strength
q_f	Flange-dependent shear strength
q_s	Dynamic pressure
Q	Capacity of shear connectors for positive or negative moment as relevant
Q	First moment of area about the neutral axis
Q_k	Characteristic strength of headed studs
Q_n	Capacity of shear connector in the negative moment region
Q_p	Capacity of shear connector in the positive moment region
Q_{xb}	First moment of area of bottom flange about neutral axis
Q_{xt}	First moment of area of top flange about neutral axis
r	Number of translationally fixed support in one orthogonal direction (roller)
r	Radius of circular shaft
r	Web stress ratio
r_{aa}	Radius of gyration of strut section about the A–A axis
r_{bb}	Radius of gyration of strut section about the B–B axis
r_c	Root radius of column section
r_{eff}	Effective radius of gyration
r_i	Distance of the line from origin O to the bolt i
r_{max}	Maximum distance to a point on the weld from the centroid of the weld group
r_{sc}	Radius of gyration about the minor axis of the effective section
r_{sc}	Radius of gyration of the equivalent core section of the transverse web stiffener
r_u	Ultimate resistance (N/m^2)
r_{vv}	Radius of gyration of strut section about the V–V axis
r_x	Radius of gyration about the major axis
r_x	Radius of gyration of strut section about the X–X axis
r_y	Radius of gyration of the section about the minor axis
r_y	Radius of gyration of strut section about the Y–Y axis
R	Radius of curvature
R	Ratio of the greater depth to the lesser depth of effective torsional restraint
R_a	Vertical reaction at support A of three-hinged arch
R_A	Vertical reaction at end A
R_b	Vertical reaction at support B of three-hinged arch
R_B	Vertical reaction at end B
R_c	Resistance of concrete flange
R_f	Resistance of flange of the steel beam
R_L	End reaction of beam at the bearing

R_L	Maximum ultimate reaction or maximum applied load on the flange
R_n	Resistance of slender steel beam
R_o	Resistance of slender web of the steel beam
R_q	Resistance of shear connection
R_r	Resistance of steel tensile reinforcement
R_s	Resistance of steel beam
R_v	Resistance of clear web depth
R_w	Resistance of overall web depth
s	Leg length of fillet weld
s	Length along rib of arch
s	Longitudinal spacing of groups of shear connectors
s	Number of degrees of freedom of sidesway
s_b	Basic snow load
s_d	Snow load on roof
S	Leg length of fillet weld
S	Minor axis plastic modulus of the smaller section at the splice.
S	Plastic modulus of section
S_a	Altitude factor
S_A	Spectral acceleration
S_d	Direction factor
S_D	Spectral displacement
S_f	Plastic modulus of flanges only
S_i	Internal force in member i of a truss
S_p	Probability factor
S_p	Spacing of holes along the direction of load
S_s	Seasonal factor
S_t	Throat thickness of fillet weld
S_v	Plastic modulus of shear area A_v
S_{vx}	Plastic modulus of shear area only about the X–X axis
S_{vy}	Plastic modulus of shear area only about the Y–Y axis
S_V	Pseudo-spectral velocity
S_{ws}	Warping statical moment
S_x	Plastic modulus about the X-axis (major axis)
S_y	Plastic modulus about the minor axis
t	Minimum web thickness required for spacing of stiffeners equal to a using tension field action
t	Rise in temperature of an arch rib (°C)
t	Thickness of base plate
t	Thickness of connected ply
t	Thickness of each constituent thin rectangle of the section
t	Thickness of the element in which shear stress is determined
t	Thickness of flange or end plate
t	Thickness of each panel
t	Thickness of the steel tube
t	Thickness of web of steel beam
t	Thickness of web of steel section
t'	Thickness of backing plate
t_c	Thickness of column web
t_d	Duration time of impulsive loading
t_e	Thickness of end post
t_{eff}	Effective thickness of web
t_p	Thickness of base plate
t_p	Thickness of flange plates

t_p	Thickness of plate stiffener
t_p	Thickness of profiled steel sheet
t_p	Thickness of web plate
t_s	Thickness of bearing stiffener
t_s	Thickness of stiffener plate
t_s	Thickness of web plate
t_w	Thickness of web
T	Applied tensile force at the connection
T	Applied torque
T	Flange tension
T	Kinetic energy of a system
T	Tensile internal force in section
T	Tension in a single bolt or in a group of bolts
T	Thickness of flange (steel beam and steel section)
T	Thickness of flanges of symmetrical steel section
T	Thickness of the smaller flange
T_1	Tension in holding-down bolt no.1
T_1	Thickness of flange of beam B_1
T_2	Tension in holding-down bolt no.2
T_2	thickness of flange of beam B_2
T_b	Thickness of flange of beam
T_c	Thickness of flange of column
T_g	The total thickness of all the plies joined together by a bolt
T_i	Tension in ith holding-down bolt
T_n	Undamped natural frequency of vibration $= 2\pi/\omega_n$ (seconds)
T_P	Resistance by uniform torsion
T_q	Applied torque
T_r	Total torsional resistance at any point in a beam
T_w	Tension on the fusion face
T_W	Resistance by warping
ΣT	Sum of all tensile forces in the bolts in the tension zone
u	Buckling parameter
u	Co-ordinates along the U-axis
u	Time-dependent transverse deflection of beam
$u(t)$	Time dependent displacement
U	Effective perimeter for punching shear stress
U	Initial strain energy of a system
U	Strain energy
U	Strain energy stored in the arch
U_f	Ultimate tensile strength of bolt
U_s	Specified minimum ultimate tensile strength
U_s	Ultimate tensile strength of material of the beam
UTS	Ultimate tensile strength
ΔU	Increment of strain energy
v	Co-ordinates along the V-axis
v	Slenderness factor
v_b	Basic tension field strength of web
v_c	Design concrete shear stress
v_p	Contribution of the profiled steel sheeting
V	Applied shear force on the section or in a beam
V	Maximum shear force adjacent to the stiffener
V	Potential energy of a system
V	Transverse shear force acting parallel to the planes of lacing or batten

V	Ultimate shear at the section under consideration
V_b	Basic wind speed
V_b	Shear buckling resistance of a stiffened panel of web
V_{cr}	Shear buckling resistance of web
V_d	Shear force at point D of an arch
V_e	Standard effective wind speed
V_f	Strain energy due to flexure
V_{ht}	Vertical shear per unit length due to wheel load in gantry girder
V_N	Potential energy of load N
V_s	Maximum ultimate transverse shear in a direction parallel to the planes of lacing or batten
V_s	Shear buckling resistance of the web panel without tension field action
V_s	Transverse shear in the member at any section
V_S	Site wind speed
V_w	Shear on the fusion face
V_x	Shear force per unit length of plate due to M_x
V_{xb}	Horizontal shear per unit length at the connection between the bottom flange and the web
V_{xt}	Horizontal shear per unit length at the connection between the top flange and the web
V_y	Shear force per unit length of plate due to M_y
ΣV	Summation of all of the forces in the vertical direction
w	Deflection of plate in the out-of-plane direction
w	Distributed load per unit length over the whole panel a
w	Intensity of uniformly distributed load on a beam
w	Pressure on the underside of the plate assuming uniform distribution
w	Uniformly distributed vertical load intensity
W	Total distributed load over shorter-than-panel dimension a
W	Ultimate point load from purlin
W	Weight of mass $= mg$
W_c	Weight of crane bridge
W_{cap}	Crane live load including hook load
W_{cb}	Weight of crab
W_{ns}	The normalised warping constant
W_s	Moment due to unfactored service load
W_u	Moment due to factored ultimate load
W_v	Wheel load in kN
x	Co-ordinates along the X-axis
x	Length along the longitudinal axis of the member
x	Torsional index
$\bar{x}$	Location of the neutral axis parallel to the major axis
$\bar{x}_1$	Centre of gravity of area A_1 from left-hand support A
$\bar{x}_2$	Centre of gravity of area A_2 from right-hand support C
X	Co-ordinate of the centroid of the rectangle along the global X–X axis
X	Distance of the plastic neutral axis from the edge of web
X	Upwind spacing of the building from existing obstructions
X_1	The larger side of the assumed trapezoidal contact pressure diagram under the base plate
X_2	The smaller side of the assumed trapezoidal contact pressure diagram under the base plate
X_i	Distance along the X–X axis of the ith bolt from the centroid of the bolt group

X_i	Generalised expression for all types of displacement at the joints
X_P	Distance of the plastic neutral axis from the edge of the compressive face of the concrete
ΣX	Summation of all of the forces in the X–X direction
y	Co-ordinates along the Y-axis
y	Distance from the neutral axis to the arc AB
$\bar{y}$	Distance of centroid of area A from the neutral axis
$\bar{y}$	Location of the neutral axis parallel to the minor axis
y_b	Basic tension field strength
Y	Co-ordinate of the centroid of the rectangle along the global Y–Y axis
Y	Depth of web in compression from the compression flange of the steel section
Y_i	distance along the Y–Y axis of the ith bolt from the centroid of the bolt group
Y_s	Specified minimum yield strength
ΣY	Summation of all of the forces in the Y–Y direction
Z	Elastic modulus of base plate along the axis of moment M_c
Z	Elastic modulus of the plate stiffener
Z	Elastic section modulus about relevant axis
Z_c	Elastic section modulus of compression flange
Z_f	Elastic modulus of flanges only
Z_{fc}	Elastic section modulus of flanges only relative to the compression flange
Z_{ft}	Elastic section modulus of flanges only relative to the tension flange
Z_P	Polar modulus of the weld group
Z_t	Elastic section modulus of tension flange
Z_x	Elastic section modulus about the X–X (major) axis
Z_{xb}	Elastic modulus about major axis to bottom fibre
Z_{xt}	Elastic modulus about major axis to top fibre
Z_y	Elastic section modulus about the Y–Y (minor) axis

Chapter 1
Strength of Materials

1.1 BENDING STRESS IN BEAMS

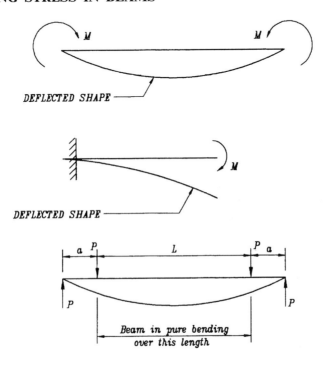

SK 1/1 Beam in pure bending.

A beam is in pure bending where shear force is zero and hence bending moment is constant over a length of the beam. A plane section is assumed to remain plane before and after bending. When shear force is present, the plane section is warped. But for all practical purposes this warping due to shear force may be ignored and the theory of pure bending may be extended to cover the cases of bending where the bending moment is not constant.

1.1.1 Assumptions

(1) Plane section remains plane.
(2) Material is homogeneous and isotropic.
(3) Lateral strains (anticlastic curvature) are neglected.

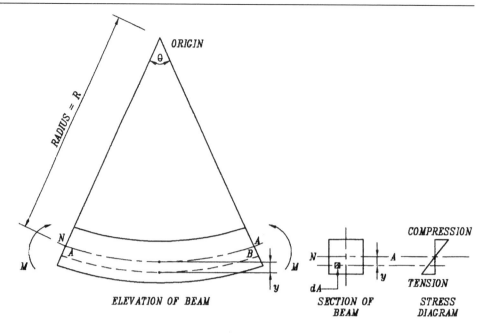

SK 1/2 Bending strains and stresses.

(4) Local effects under loads are neglected.
(5) Beam bends in a circular arc of radius R.
(6) The radius of curvature is large compared to the cross-sectional dimension of the beam.
(7) Modulus of elasticity, E, has the same value in compression and tension.

The neutral axis, where stress is zero, is denoted on the circular arc by NA. The stress σ_{AB} at a distance y from the neutral axis is on the circular arc AB.

$$\text{Strain} = \varepsilon_{AB} = \frac{(AB - NA)}{NA} = \frac{(R+y)\theta - R\theta}{R\theta} = \frac{y}{R}$$

$$\varepsilon_{AB} = \frac{\sigma_{AB}}{E} \quad \therefore \quad \frac{\sigma_{AB}}{E} = \frac{y}{R}$$

or $\quad \dfrac{\sigma}{y} = \dfrac{E}{R}$

where $E =$ modulus of elasticity
   $R =$ radius of curvature
   $\sigma_{AB} =$ stress at a distance y from the neutral axis on a circular arc AB
   $\varepsilon_{AB} =$ strain at a distance y from the neutral axis on a circular arc AB
   $\theta =$ angle subtended at the centre of the circular arc AB
   $AB =$ length of the circular arc along the line AB

NA = length of the circular arc along the line NA
y = distance from the neutral axis to the arc AB

1.1.2 Position of neutral axis

dA is an element of area at a distance y from NA. Net normal force on the cross-section must be zero for pure bending.

$$\therefore \int_A \sigma \, dA = 0$$

or $$\frac{E}{R} \int_A y \, dA = 0$$

$$\frac{E}{R} \neq 0$$

The location of the neutral axis should satisfy:

$$\int_A y \, dA = 0$$

The external applied moment, M, is balanced by internal forces. Taking moment of the internal forces about NA:

$$M = \int_A \sigma y \, dA = \frac{E}{R} \int_A y^2 \, dA = \frac{EI}{R}$$

where $I = \int_A y^2 \, dA$
I = moment of inertia of the section of the beam about the neutral axis

$$\therefore \quad \frac{\sigma}{y} = \frac{M}{I} = \frac{E}{R}$$

If the moment about an axis perpendicular to the axis about which M is applied is zero, then that axis is called the principal axis of the section.

$$\therefore \int_A \sigma x \, dA = 0$$

or $$\int_A xy \, dA = 0$$

because $$\sigma = \frac{E}{R} y$$

The axes about which the product of inertia $\int xy \, dA$ is zero are called the principal axes. An axis of symmetry in a section is a principal axis.

1.1.3 Bending stress in asymmetrical sections

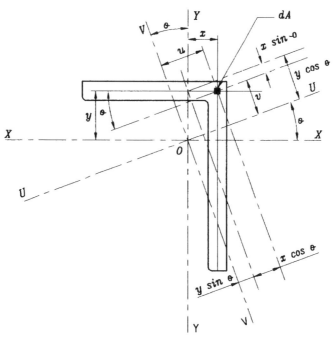

SK 1/3 Principal axes through centroid of a section.

dA is the elemental area of a section; x and y are co-ordinates about axes X–X and Y–Y; u and v are co-ordinates about axes U–U and V–V; and U–U and V–V are principal axes about which the product of inertia is zero.

$$u = x\cos\theta + y\sin\theta$$
$$v = y\cos\theta - x\sin\theta$$

Product of inertia $\quad I_{UV} = \int_A uv\, dA$

where dA = elemental area of a section
x = co-ordinate along X-axis
y = co-ordinate along Y-axis
u = co-ordinate along U-axis
v = co-ordinate along V-axis
θ = angle of inclination between X and U axes
I_{UV} = product inertia of the section about U and V axes
I_X = moment of inertia of section about the X-axis
I_Y = moment of inertia of section about the Y-axis
I_{XY} = product inertia of section about X and Y axes

$$I_{UV} = \int_A (x\cos\theta + y\sin\theta)(y\cos\theta - x\sin\theta)\, dA$$

$$= (I_X - I_Y)\frac{\sin 2\theta}{2} + I_{XY}\cos 2\theta$$

when $I_{UV} = 0$, $\tan 2\theta = 2I_{XY}/(I_Y - I_X)$

$$I_X = \int_A y^2 \, dA$$

$$I_Y = \int_A x^2 \, dA$$

$$I_{XY} = \int_A xy \, dA$$

$$I_U = \int_A v^2 \, dA = I_X \cos^2\theta - I_{XY} \sin 2\theta + I_Y \sin^2\theta$$

Similarly,

$$I_V = I_Y \cos^2\theta + I_{XY} \sin 2\theta + I_X \sin^2\theta$$

or $I_X + I_Y = I_U + I_V$

1.1.4 Product inertia of a rectangle about orthogonal axes

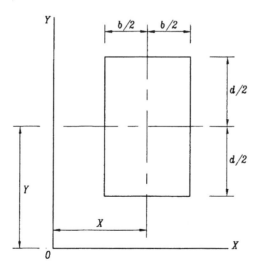

SK 1/4 Rectangular element product inertia.

To find product inertia of any section about a set of orthogonal axes X–X and Y–Y, divide the section into rectangles and find the summation of product inertia for each rectangle about the orthogonal axes. I_{XY_i} is the product inertia of each rectangle about the X–X and Y–Y axes.

$$I_{XY_i} = \iint xy\,dx\,dy = \left[\frac{x^2}{2}\right]_{x-b/2}^{x+b/2} \times \left[\frac{y^2}{2}\right]_{y-d/2}^{y+d/2}$$

$$= b\,d\,XY$$

where X = co-ordinate of the centroid of the rectangle along the global X–X axis
Y = co-ordinate of the centroid of the rectangle along the global Y–Y axis
b = width of rectangle
d = depth of rectangle
I_{XY} = product inertia of the section about the global X–X and Y–Y axes = $\sum I_{XY_i}$

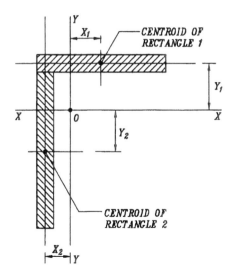

SK 1/5 Product inertia of an angle section.

Product inertia of an angle section about the centroidal axes X–X and Y–Y = product inertia of rectangle 1 about X–X and Y–Y + product inertia of rectangle 2 about X–X and Y–Y.

For sections which are asymmetrical, the orthogonal applied moments M_x and M_y about any set of X–X and Y–Y axes should be resolved about the principal axes U–U and V–V of the section to obtain bending stresses in the section due to the applied moments.

Bending stresses in an angular section
Find principal axes U–U and V–V of the section and resolve bending moments M_X and M_Y about the principal axes.

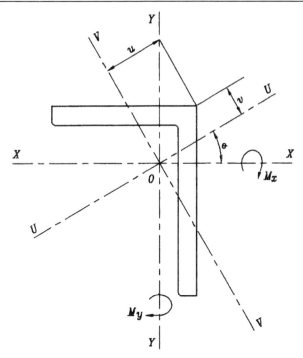

SK 1/6 Principal axes of an angle section.

$$M_U = M_X \cos\theta + M_Y \sin\theta$$

$$M_V = M_X \sin\theta + M_Y \cos\theta$$

$$\sigma = \frac{M_U v}{I_U} + \frac{M_V u}{I_V}$$

where σ is the bending stress at any point on the angle section and u and v are co-ordinates of the point along the principal axes U–U and V–V. Always observe a sign convention which is consistent throughout.

Sign convention
Bending moments producing compression in the positive zone are taken as positive.

Positive M_x produces compressive stresses in the zone where y is positive, and similarly, positive M_y produces compressive stresses in the zone where x is positive.

x and u are co-ordinates which are positive to the right of the origin along the X and U axes respectively.

y and v are co-ordinates along the Y and V axes respectively, with positive values in the same direction as x and u but at an angle of 90 degrees in the anticlockwise direction from the X and U axes respectively.

8 Structural Steelwork

θ = angle between X–X and U–U axes measured anticlockwise from the X–X axis

M_x = applied moment on the section about X–X axis

M_y = applied moment on the section about Y–Y axis

M_u = resolved moment on the section about U–U axis

M_v = resolved moment on the section about V–V axis

I_U = moment of inertia of the section about the principal axis U–U

I_V = moment of inertia of the section about the principal axis V–V

σ = bending stress on the section compression positive

1.2 SHEAR STRESS IN BEAMS

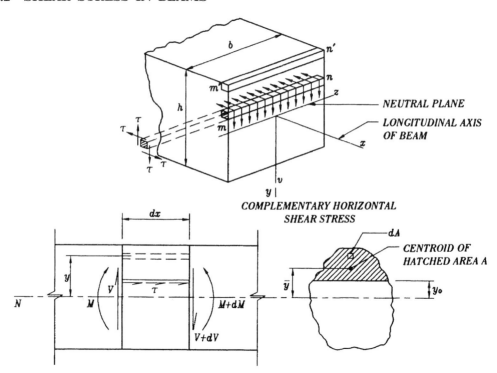

SK 1/7 Shear stresses in beam section.

The concept of transverse vertical shear stress in a section of a beam with complementary horizontal shear stress in layers of the beam parallel to the neutral plane is illustrated in SK 1/7 by the use of a rectangular cross-section.

Consider an elemental length dx of a beam. The shear stress in the transverse plane will give rise to a complementary shear stress in the longitudinal plane. The bending moments at two transverse sections spaced dx apart are M and $M + dM$. The corresponding shear forces are V and $V + dV$. The shear stress on a plane bounded by b and dx is τ. The normal stresses on an elemental area dA are σ and $\sigma + d\sigma$, corresponding to M and $M + dM$. For longitudinal force equilibrium:

$$\tau b \, dx = \int d\sigma \, dA$$

Substituting $\sigma = My/I$ and $d\sigma = dMy/I$:

$$\tau b \, dx = \frac{dM}{I} \int_{y_0} y \, dA$$

or $\quad \tau = \left(\dfrac{dM}{dx}\right) \dfrac{A\bar{y}}{bI} = \dfrac{VQ}{Ib}$

where $Q = A\bar{y}$
$\quad A$ = area of the section cut off by a plane parallel to the neutral axis at a distance of y_0 from the neutral axis
$\quad \bar{y}$ = distance of centroid of area A from neutral axis
$\quad \tau$ = shear stress at a distance y_0 from neutral axis
$\quad b$ = width of the section at a distance y_0 from neutral axis

1.2.1 Shear centre of thin-walled asymmetrical section

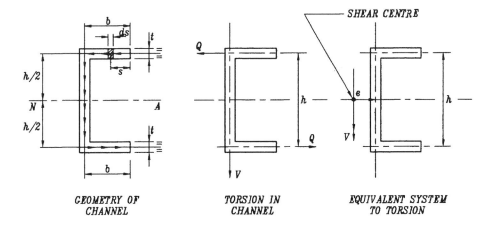

GEOMETRY OF CHANNEL   TORSION IN CHANNEL   EQUIVALENT SYSTEM TO TORSION

SK 1/8 Shear centre of a channel section.

Each elemental area $t \, ds$ of the flange is at a distance of $h/2$ from the neutral axis. The complementary shear stress at a distance s from the tip of the channel is given by:

$$\tau_s = \frac{Vh}{2It} \int_0^s t \, ds = \frac{Vhs}{2I}$$

where I = moment of inertia of section about the neutral axis

Elemental force in the flange due to shear flow $= \tau_s t \, ds$

Table 1.1 Shear centres of asymmetrical structural shapes.

Shape of section	$e =$ distance of shear centre
	Shear centre at the corner of angle
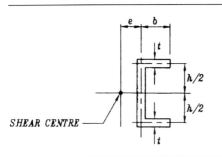	$e = \dfrac{b^2 h^2 t}{4 I_{xx}}$
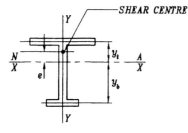	$e = \dfrac{y_t I_{ytf} - y_b I_{ybf}}{I_{yy}}$ $I_{ytf} =$ moment of inertia of top flange about Y–Y axis $I_{ybf} =$ moment of inertia of bottom flange about Y–Y axis
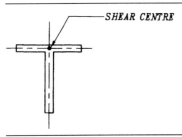	Shear centre at the intersection of flange and web
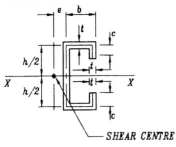	$e = \dfrac{b^2 h^2 t}{4 I_{xx}} \left[1 + \dfrac{2c}{b} - \dfrac{8c^3}{3bh^2} \right]$

The force Q in the flange is given by:

$$Q = \int_0^b \tau_s t\, ds = \frac{Vht}{2I} \int_0^b s\, ds = \frac{Vhtb^2}{4I}$$

The forces Q in the flanges constitute a couple $= Qh$

e = shear centre of the thin-walled section
V = applied shear force on the section
t = thickness of the flange of the asymmetrical section
b = distance from the centre of web to the tip of flange
h = distance of centre-to-centre of flanges

Assume that the shear force V acts at the shear centre e of the section, then the resultant torsion on the section is $Ve - Qh$.

If the resultant torsion on the thin-walled section is zero, then:

$$Ve = Qh \quad \text{or} \quad e = \frac{Qh}{V} = \frac{b^2 h^2 t}{4I}$$

1.3 TORSIONAL SHEAR STRESS

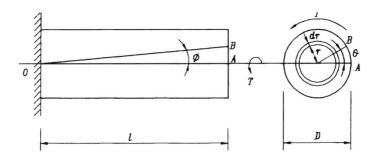

SK 1/9 Circular shaft subjected to torsion.

A line on the surface of the circular shaft is OA, which moves to OB when the torsion T is applied to the free end of the shaft of length l. The shear strain ϕ is the angle AOB. In the circular section the point A moves to B subtending an angle θ at the centre. For small angle ϕ it can be said that:

$$l\phi = \text{arc } AB = R\theta$$

R = radius of circular shaft

$$\phi = \text{shear strain} = \frac{\text{stress}}{\text{modulus}} = \frac{\tau}{G}$$

or $\quad \dfrac{\tau_R}{R} = \dfrac{G\theta}{l} = \dfrac{\tau}{r}$

where G = modulus of rigidity
τ_R = shear stress at radius R
τ = shear stress at radius r from centre of shaft

12 Structural Steelwork

At a distance r from the centre of the circular section, the shear stress is τ acting on an elemental area of $2\pi r\, dr$. Moment of the elemental shear load about the centre of section is the applied torque T.

$$T = \int \tau(2\pi r\, dr)r = \frac{G\theta}{l}\int (2\pi r\, dr)r^2 = \left(\frac{G\theta}{l}\right)J$$

where J = polar moment of inertia

$$= \frac{\pi D^4}{32} \text{ for a solid circular shaft}$$

$$= \frac{\pi(D^4 - d^4)}{32} \text{ for a hollow circular section with outer diameter } D \text{ and inner diameter } d$$

$$k = \text{torsional stiffness of a circular shaft} = \frac{T}{\theta} = \frac{GJ}{l}$$

and $\dfrac{T}{J} = \dfrac{\tau}{r} = \dfrac{G\theta}{l}$

Maximum shear stress $= \tau_{max} = \dfrac{TD}{2J} = \dfrac{16T}{\pi D^3}$ for a circular shaft of diameter D

1.3.1 Torsion of thin rectangular members

Maximum shear stress in the rectangular section is assumed to be τ along the long edge and τ' along the short edge. The variation of the

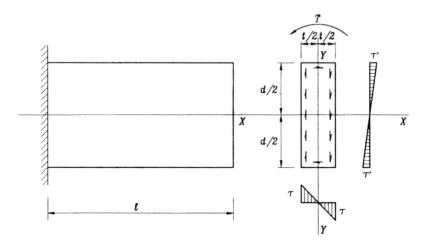

SK 1/10 Torsional shear stress in rectangular thin-walled section.

shear stress across the section is assumed linear. It is found experimentally that:

$$\frac{\tau'}{\tau} = \frac{t}{d}$$

Shear stress at a distance of x from the Y–Y axis $= 2\tau x/t$ on an elemental area equal to $d \cdot dx$.

$$\text{The elemental force} = \frac{2\tau x}{t} d \cdot dx$$

The torsion T is equal to the sum of the moments of these elemental forces about the Y–Y and X–X axes.

$$T = 2 \int_0^{t/2} \frac{2\tau x}{t} d \cdot x \cdot dx + 2 \int_0^{d/2} \frac{2\tau' y}{d} t \cdot y \cdot dy$$

$$= \frac{1}{6} \tau \cdot d \cdot t^2 + \frac{1}{6} \tau' \cdot d^2 \cdot t$$

Substituting $\tau' = \tau \cdot t/d$

$$T = \frac{1}{3} \tau \cdot d \cdot t^2$$

where $T =$ applied torque on thin rectangular section
 $\tau =$ shear stress in thin rectangular section along the long edge
 $d =$ length of thin rectangular section
 $t =$ thickness of thin rectangular section

1.3.2 Torsion of thin open sections

For I-, channel-, angle- and open-curved sections the same principle as above will apply.

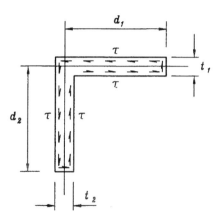

SK 1/11 Torsional shear stress in an angular section.

The angle section can be broken down into two rectangles.

Assume that each rectangular section carries a part of the total applied torsion T in proportion to the stiffness factor dt^2 of the rectangle and that τ_1 and τ_2 are torsional shear stresses in each rectangle.

$$T = T_1 + T_2$$

$$T_1 = \frac{Td_1 t_1^2}{\sum dt^2} = \frac{1}{3}\tau_1 d_1 t_1^2$$

$$T_2 = \frac{Td_2 t_2^2}{\sum dt^2} = \frac{1}{3}\tau_2 d_2 t_2^2$$

$$\frac{T_1}{T_2} = \frac{\dfrac{Td_1 t_1^2}{\sum dt^2}}{\dfrac{Td_2 t_2^2}{\sum dt^2}} = \frac{\dfrac{1}{3}\tau_1 d_1 t_1^2}{\dfrac{1}{3}\tau_2 d_2 t_2^2}$$

This proves $\dfrac{\tau_1}{\tau_2} = 1$ because the assumption is $\dfrac{T_1}{T_2} = \dfrac{d_1 t_1^2}{d_2 t_2^2}$

or $\quad \tau_1 = \tau_2 = \tau$

$$T = \frac{1}{3}\tau \sum dt^2$$

$$\theta = \frac{3Tl}{G \sum dt^3}$$

where θ = angle of rotation when the length of the member is l
 l = length of member
 T = applied torque
 d = length of each constituent thin rectangle of the section
 t = thickness of each constituent thin rectangle of the section
 τ = torsional shear stress

1.4 STRAIN ENERGY IN AXIAL LOAD, BENDING, TORSION AND SHEAR

Strain energy in axial load

For a load gradually applied, P, to produce deflection x, the strain energy stored in the system is the work done by the load.

Strain energy $= U = 0.5 Px$

which is the area of the load–displacement curve.

$$U = \frac{\sigma^2}{2E} A \cdot l \text{ because } x = \varepsilon \cdot l, \ \varepsilon = \frac{\sigma}{E} \text{ and } P = \sigma \cdot A$$

Strength of Materials 15

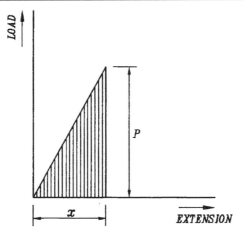

SK 1/12 Strain energy in axial load.

where σ = direct axial stress
 ε = axial strain
 E = Young's modulus of the material
 A = area of cross-section
 l = length of the member
 U = strain energy

$A.l$ is the volume of the member and the term $\sigma^2/2E$ is called the strain energy per unit volume.

Strain energy in bending

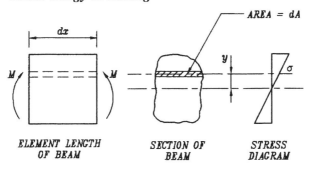

SK 1/13 Strain energy in bending.

Assume that over a very short length dx of a beam the bending moment M remains constant. σ is the bending stress on an element of area dA at a distance y from the neutral axis. The strain energy in the beam of length dx subjected to a moment M is dU.

$$dU = \int \frac{\sigma^2}{2E} dx \cdot dA \quad \text{(as in strain energy for axial load)}$$

$$= \frac{dx}{2E} \int \sigma^2 dA = \frac{dx}{2E} \int \frac{M^2 y^2 dA}{I^2} = \frac{M^2}{2EI} dx$$

16 Structural Steelwork

because

$$\sigma = \frac{My}{I} \quad \text{and} \quad \int y^2 \, dA = I$$

Hence for the entire length of the beam:

$$U = \int \frac{M^2}{2EI} \, dx$$

where M = bending moment at any point on the beam along its length along X axis
E = modulus of elasticity of the material of the beam
I = moment of inertia of the cross-section of the beam about its neutral axis

Strain energy due to shear

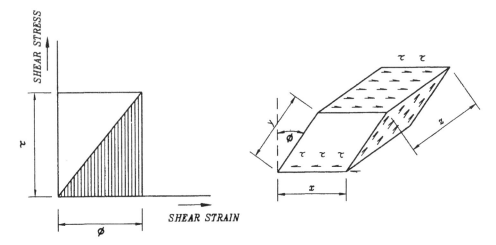

SK 1/14 Strain energy due to shear.

Strain energy due to shear is found by equating the energy to the work done by the moment of the internal shear stresses causing the angular strain ($= 0.5 \times$ moment $\times$ rotation):

$$U = \frac{1}{2} (\tau \cdot y \cdot z) \cdot x \cdot \phi$$

where $\phi = \dfrac{\tau}{G}$
G = modulus of rigidity
ϕ = shear strain
τ = shear stress

$$\therefore \quad U = \left(\frac{\tau^2}{2G}\right) x \cdot y \cdot z$$

$$= \left(\frac{\tau^2}{2G}\right) \text{volume}$$

Strain energy of a circular shaft in torsion

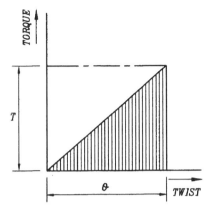

SK 1/15 Strain energy of a circular shaft in torsion.

For a circular shaft of length l subjected to a torsion T, the strain energy U is given by:

$$U = \frac{1}{2} T\theta$$

$$\theta = \frac{\tau \cdot l}{G \cdot r} = \frac{2\tau \cdot l}{G \cdot D}$$

and $\quad \tau = \dfrac{16T}{\pi D^3}$

$$T = \frac{\pi D^3 \tau}{16}$$

$$\therefore \quad U = \frac{1}{2}\left(\frac{\pi D^3 \tau}{16}\right)\left(\frac{2\tau \cdot l}{G \cdot D}\right)$$

$$= \left(\frac{\tau^2}{4G}\right) \text{volume}$$

where τ = maximum shear stress in the circular shaft at a distance of radius r from the centre
T = applied torsion
G = modulus of rigidity
D = diameter of circular shaft
r = radius of circular shaft

Strain energy of a thin rectangular section in torsion

For a thin rectangular section with sides of length d and t, shear stresses are τ on the long side and τ' on the short side as maximum values. The shear stress is equal to $2\tau x/t$ acting on an elemental area $dx \cdot d$. The volume of the elemental bar is equal to $d \cdot l \cdot dx$.

$$U = \frac{1}{2} T\theta = \left(\frac{\tau^2}{2G}\right) \text{volume}$$

$$= 2 \int_0^{t/2} \left(\frac{2\tau x}{t}\right)^2 \frac{dl}{2G} dx + 2 \int_0^{d/2} \left(\frac{2\tau' y}{d}\right)^2 \frac{tl}{2G} dy$$

$$= \frac{\tau^2 dtl}{6G}\left(1 + \frac{t^2}{d^2}\right)$$

assuming $\tau'/\tau = t/d$ approximately

$$\theta = \frac{\tau l}{tG} = \frac{3Tl}{dt^3 G}$$

ignoring t^2/d^2 and substituting $T = 1/3\tau \, dt^2$

where d = length of thin rectangular section
t = thickness of thin rectangular section
l = length of the member

Chapter 2
Theory of Structures

2.1 POLYGON OF FORCES

Several forces acting in a plane at a point O can be reduced to one single force, which is called their resultant. Forces can be represented by vectors. Vectors have magnitude and direction. If two forces P_1 and P_2 are represented by their vectors $\overline{OA}$ and $\overline{OB}$, then their resultant is $\overline{OC}$, which can be found using the triangle of forces. In the triangle of forces $\overline{AB}$ represents P_1 in direction and magnitude, $\overline{BC}$ represents P_2 in direction and magnitude and $\overline{AC}$ is the closing side of the triangle representing the resultant R. When this resultant vanishes, the forces are said to be in equilibrium.

Similarly, any number of forces acting at a point can be represented by vectors and the resultant R of these forces can be found using the polygon of forces. When several forces acting at a point are in equilibrium, the polygon of forces will close, meaning the resultant is zero.

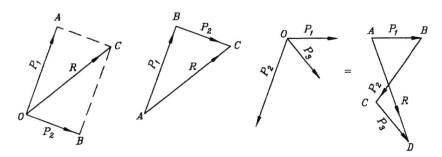

SK 2/1 Polygon of forces.

Any force can be resolved into two forces acting at right angles to each other parallel to a chosen orthogonal axis system X–X and Y–Y. The force P can be resolved into $P\cos\theta$ and $P\sin\theta$ parallel to the X–X and Y–Y axes respectively. If a system of forces acting at point O is in equilibrium, then:

$\sum X =$ summation of all resolved forces in the X–X direction $= 0$

$\sum Y =$ summation of all resolved forces in the Y–Y direction $= 0$

19

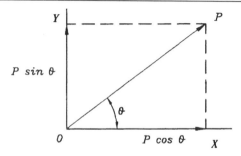

SK 2/2 Forces resolved into two orthogonal axes.

2.2 EQUATIONS OF EQUILIBRIUM

A system of coplanar forces may not all intersect at a point. If the forces do not intersect at one point, they may reduce to a resultant force, a resultant couple or be in equilibrium. A couple is a set of equal and opposite forces separated by a distance. The polygon of forces will close, indicating that there is no resultant force in the system, but a resultant couple may remain.

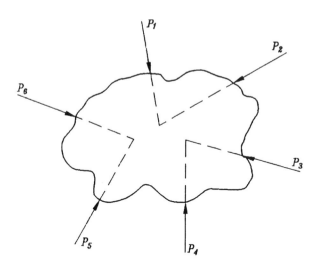

SK 2/3 Coplanar system of forces.

The following equations of equilibrium must be satisfied for a system of coplanar forces to be in equilibrium:

$\sum X =$ summation of all resolved forces in the $X\!-\!X$ direction $= 0$

$\sum Y =$ summation of all resolved forces in the $Y\!-\!Y$ direction $= 0$

$\sum M =$ summation of all moments of the forces (or the resolved forces) about any point in the system $= 0$

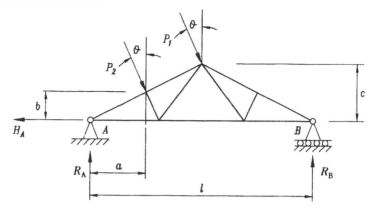

SK 2/4 Equilibrium of coplanar forces on a simple truss.

For a simple truss the reactions at the supports can be determined as follows. Assuming that there is no frictional resistance at the rollers at support B, the horizontal reaction at that support is zero. For equilibrium the equations are:

$\sum H =$ summation of all the forces in the horizontal direction $= 0$

or $\quad H_A = (P_1 + P_2) \sin \theta$

$\sum V =$ summation of all the forces in the vertical direction $= 0$

or $\quad R_A + R_B = (P_1 + P_2) \cos \theta$

$\sum M =$ summation of moments of all the forces about $A = 0$

$$R_B l = (P_1 \sin \theta)c + (P_1 \cos \theta)\frac{1}{2} + (P_2 \sin \theta)b + (P_2 \cos \theta)a$$

Knowing the geometry of the truss and the loading directions and magnitudes, the unknown reactions R_A, R_B and H_A can be found from the three equations of equilibrium.

$R_A =$ vertical reaction at end A
$R_B =$ vertical reaction at end B
$H_A =$ horizontal reaction at end A

2.3 INTERNAL FORCES

The truss in SK 2/5 has nodes marked 1, 2, 3, 4 etc. Members are numbered as follows: member 1 is between nodes 1 and 2, member 2 is between nodes 2 and 3 etc. The internal forces in members are N_1, N_2, etc., corresponding to the number of members.

At any node in the truss two equations of equilibrium can be written taking into account the internal and external forces at that node. Unknown internal forces in the two members of the truss joining at that node can be found from the two equations.

22 Structural Steelwork

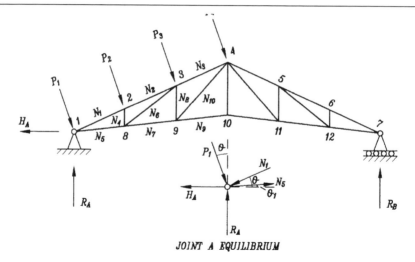

SK 2/5 Internal forces in a truss.

The first step is to find the end reactions of the truss at supports A and B using the methods described in Section 2.2. The joint A is in equilibrium under the action of H_A, R_A, N_1 and N_5. Applying equations of equilibrium:

$\sum H =$ summation of all the forces in the horizontal direction $= 0$

or $H_A + N_1 \cos\theta - N_5 \cos\theta_1 - P_1 \sin\theta = 0$

$\sum V =$ summation of all the forces in the vertical direction $= 0$

or $R_A - N_1 \sin\theta + N_5 \sin\theta_1 - P_1 \cos\theta = 0$

where $H_A =$ horizontal reaction at end A
$N_i =$ internal force in member i
$R_A =$ vertical reaction at end A
$R_B =$ vertical reaction at end B

Knowing H_A, R_A and θ, the unknown internal forces N_1 and N_5 can be found from the two equations of equilibrium. Having found N_1, at node 2 there are two unknown member forces, N_2 and N_4. They can be found by the two equations of equilibrium at node 2. Similarly, progressing a node at a time, the internal forces in the rest of the members can be found.

For a determinate truss the relationship between the number of members m and the number of joints j is given by:

$m = 2j - 3$

where $m =$ number of members in a truss
$j =$ number of joints in a truss

The truss becomes indeterminate if the number of members exceeds that given by the above equation. For indeterminate trusses, considerations of deformations and strain energy have to be made to solve the internal forces. If the number of members is less than that given by the equation, then the truss is not a rigid structure and becomes a mechanism. For a cantilever truss, however, the joints fixing the truss to the foundation are not taken into consideration, and the equation for a determinate cantilever truss reduces to:

$$m = 2j$$

2.3.1 Compound trusses: the method of section

A compound truss may be defined as a truss where all joints cannot be solved by the equations of equilibrium at the joints, because the number of unknown member forces at a joint is more than two. The truss is divided by a section and a convenient moment centre helps to determine unknown member forces.

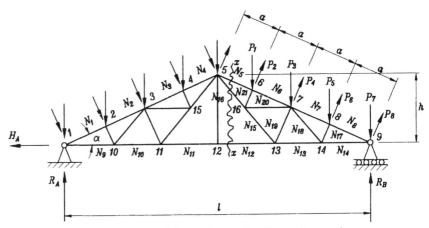

SK 2/6 Determination of internal member forces by section.

Take a section x–x through the truss as shown in SK 2/6. Consider the equilibrium of the right side of the truss acted on by applied external loads, the vertical reaction R_B and the internal forces N_5, N_{12} and N_{16} in members 5, 12 and 16 respectively acting along the line of the members. Select node 5 as the centre of moment which eliminates internal forces N_5 and N_{16} because the line of action of these two forces meets at that point. Taking moments about node 5 and equating to zero for equilibrium of the truss on the right side of the section:

$$R_B \frac{1}{2} - N_{12}h - P_1 a\cos\alpha + P_2 a - P_3 2a\cos\alpha + P_4 2a$$
$$- P_5 3a\cos\alpha + P_6 3a - P_7 4a\cos\alpha + P_8 4a = 0$$

From this equation the only unknown member force, N_{12}, can be found. Having determined N_{12} the rest of the member forces can be found by the two equations of equilibrium at every joint. P is the applied external force.

2.3.2 Maxwell diagram for simple trusses

Solving for all the joints in a truss with numerous members can be very tedious using the equations of equilibrium at every joint. A graphical method of solution of internal forces in the truss could be easier – this method is called the Maxwell diagram.

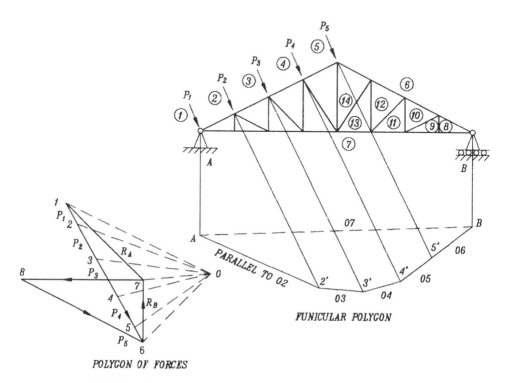

SK 2/7 Maxwell diagram and the funicular polygon of a truss.

The roof truss is supported and loaded as shown in SK 2/7. The polygon of applied known forces P_1 to P_5 is drawn in direction and magnitude as a straight line from 1 to 6 with 1–2 representing P_1, 2–3 representing P_2 etc. A pole O is selected and rays from the pole are drawn to the points 1, 2, 3 etc. on the polygon of forces. Effectively the force P_1 is replaced by two forces 1O and O2, and P_2 is replaced by two forces 2O and O3 etc. Starting at support A, the line of action of the reaction R_A must pass through the fixed point A. The funicular polygon is started from this fixed point A by drawing a line parallel to the ray O2 in the force diagram. This ray meets the line of action of load P_2 at point $2'$ on the funicular polygon. The ray O1 for load P_1 is not taken into account because it passes through the fixed point A. From point $2'$ a line is drawn parallel to the ray O3 and it meets the line of action of load P_3 at $3'$. By using the same process the line parallel to the ray O6 from point $5'$ meets the line of action of reaction R_B at B. The line of action of reaction R_B is vertical because of the roller supports. The direction of the closing line AB in the funicular polygon should be parallel to a ray in the force polygon which is common between reactions R_A and R_B. A line parallel to this closing line is drawn from the pole O to meet a vertical line from point 6 at point 7 on the force polygon. The magnitude and direction of the reaction at B is the vector 6–7. Similarly, the vector 7–1 represents the reaction at support A in magnitude and direction.

Having found the reactions R_A and R_B, the internal member forces in the simple truss can be found by the solution of the polygon of forces in equilibrium graphically at the joints in a progressive manner. The triangular spaces inside the truss are marked 8 to 14 as shown. This helps to create the force diagram. The members of the truss are at the boundary between two spaces. For example, the member 13–7 is the central tie. Starting at joint B a line is drawn from point 7 parallel to member 7–8, which is horizontal, and another line parallel to member 8–6 is drawn from point 6 in the force diagram. These two lines meet at point 8. The magnitude of internal forces in members 7–8 and 8–6 can be determined by measurement from the force diagram. At each joint a clockwise direction of designation of member forces through the space numbers in the truss will give the actual directions of applied forces at the joint. Similarly, all the joints in the simple truss can be solved for internal forces. The points in the force diagram are determined in a progressive manner from 8 to 14. The force in a member acting towards a joint means that the member is in compression. If the member force acts away from the joint, then the member is in tension.

2.3.3 The method of section by Maxwell diagram

A compound truss may be defined as a truss where all joints cannot be solved by the equations of equilibrium at the joints because the number of unknown member forces at a joint is more than two. In these cases a method of section has to be applied. The method of section can be applied graphically by the use of Maxwell diagrams.

The joints are numbered from 1 to 11 in the truss in SK 2/8. The members are designated by the number of joints at the two ends.

The force diagram can be constructed for joint 1 only, because beyond that joint, at joints 2 and 8, there are more than two unknown member forces. If the section m–n is taken through members 4–5, 4–11 and 8–10, as shown in SK 2/8, the internal forces in these members along with the external forces P_5, P_6, P_7 and R_B on the right-hand side of the section must be in equilibrium.

The two rays parallel to O9 and O5 in the funicular polygon meet at a point r where a vertical force R establishes equilibrium on the right-hand side of the section m–n. The line of action of the internal force in member 8–10 meets the vertical line of action of the force R at r'. The resultant of the remaining two internal member forces in 4–5 and 4–11 must pass through this point r' for equilibrium to exist. Hence the line of action of the resultant of internal forces in

26 Structural Steelwork

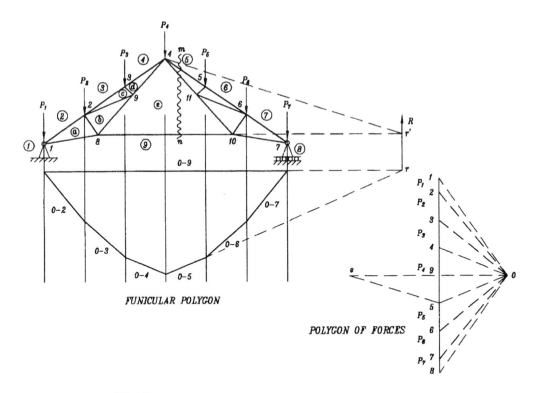

SK 2/8 Maxwell diagram of a truss using method of section.

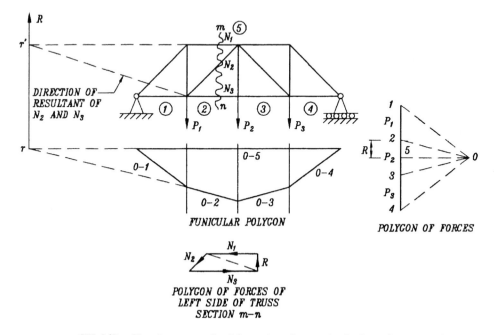

SK 2/9 Simple truss solved by using the method of section graphically.

members 4–5 and 4–11 must be on the line 4–r'. Draw a line in the polygon of forces diagram from point 5 parallel to the direction 4–r'. Draw another line from point 9 parallel to the direction of the member force in 8–10. These two lines meet at point e in the force diagram. The internal force in the member 8–10 is e–9 in the force diagram. Having solved for joint 8, the rest of the joints can be easily solved by using the force diagram because at all other joints there are no more than two unknown member forces.

Consider another example which is not a compound truss. The principle of section is explained again by using the funicular polygon and Maxwell diagram. The diagrams in SK 2/9 are self-explanatory.

2.3.4 Deflection of pin-jointed structures

Castigliano's second theorem states that a partial derivative of the strain energy of a loaded elastic system with respect to a force in the system is the deflection of the structure at the point of application of that force in the direction of that force.

Suppose an elastic structure is acted on by forces P_1 to P_n. If P_n is increased slightly by an amount ΔP_n, then the strain energy of the elastic system increases by an amount ΔU.

$$\Delta U = \frac{\partial U}{\partial P_n} \Delta P_n$$

where U = initial strain energy of system
 ΔU = increment of strain energy
 ΔP_n = increment of external applied load

The increased strain energy becomes:

$$U + \Delta U = U + \frac{\partial U}{\partial P_n} \Delta P_n$$

Suppose ΔP_n is applied first to the structure at the point of application of load P_n. The deflection produced by this very small load may be ignored. Forces P_1 to P_n are then applied to the elastic structure. The deflection at the point of application of P_n in the direction of the force becomes δ_n. The initially applied force ΔP_n goes through a deflection of δ_n, and the work done by this load is $\Delta P_n \delta_n$. Then the strain energy of the structure may be written as $U + \Delta P_n \delta_n$. Equating the previous expression of increased strain energy with this expression:

$$U + \frac{\partial U}{\partial P_n} \Delta P_n = U + \Delta P_n \delta_n$$

or $\delta_n = \dfrac{\partial U}{\partial P_n}$ which is the proof of Castigliano's second theorem

δ_n = deflection at the point of application of P_n in the direction of the force

28 Structural Steelwork

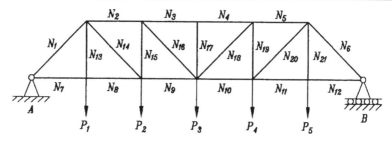

SK 2/10 Deflection of a pin-jointed structure.

The deflection under the load P_3 in the truss in SK 2/10 is given by:

$$\delta_3 = \frac{\partial U}{\partial P_3} = \sum \frac{N_i l_i}{A_i E} \frac{\partial N_i}{\partial P_3}$$

where $\partial N_i/\partial P_3$ is the internal force in each member of the truss due to unit load at the point of application and in the direction of P_3. This is a dimensionless quantity and is a pure number. The expression $N_i l_i/A_i E$ is the elongation of each member of the truss due to the internal force as a result of the external loading P_1 to P_5. A consistent system of sign convention must be adhered to for the computation of deflections. Compressive internal forces in the members of the pin-jointed structures may be designated as negative forces.

The unit load can be applied in any direction at any joint and the deflection of the pin-jointed structure due to a system of applied external loads can be found in the direction of the unit load at the joint by summation of the product of the individual member extensions due to the application of the external system of loads and the internal forces in the members due to the unit load.

N_i = internal force in the member i due to external loading
l_i = length of member i
A_i = sectional area of member i
E = modulus of elasticity

2.4 BENDING MOMENT AND SHEAR FORCE

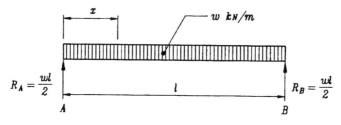

SK 2/11 A simply supported beam with a uniform loading.

2.4.1 Simply supported beams

A simply supported beam of span length l is loaded with uniformly distributed load w per unit length. The total load on the beam is wl. The reaction at each support is $wl/2$. The bending moment at a distance x from A is given by:

$$M_x = R_A x - wx\frac{x}{2} = \frac{wlx}{2} - \frac{wx^2}{2} = \frac{wx}{2}(l-x)$$

Shear force $= V_x = \dfrac{wl}{2} - wx = \dfrac{w}{2}(l-2x)$

2.4.2 Slope–deflection equations

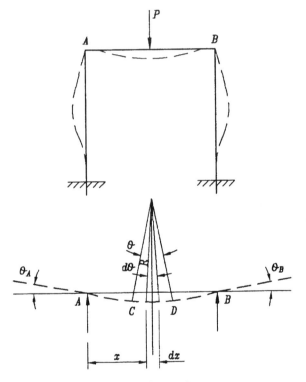

SK 2/12 Shape of a deflected beam.

Neglecting shear deformation, the curvature of the beam in SK 2/12 may be written as:

$$\frac{1}{R} = \frac{M}{EI} \quad \text{(see Section 1.1.2)}$$

where $M =$ bending moment
$EI =$ flexural rigidity of section
$I =$ moment of inertia of section
$R =$ radius of curvature of beam

Assume $d\theta$ is the angle subtended at the centre of curvature for a length dx of the beam:

$$\therefore \quad d\theta = \frac{dx}{R} = \frac{M\,dx}{EI}$$

Change of slope between two points C and D in the beam is given by:

$$\theta_D - \theta_C = \theta = \int_C^D \frac{M\,dx}{EI}$$

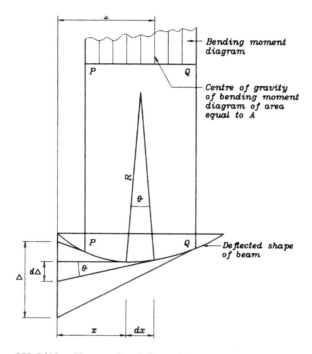

SK 2/13 Shape of a deflected beam.

Assume R is the radius of curvature at a point on the beam between sections P and Q. The angle between the tangents at the ends of a short length dx is $d\theta$, where $dx = R\,d\theta$. The intercept of these tangents on the orthogonal Y-axis is $d\Delta$, where $d\Delta = x\,d\theta$, because $d\theta$ is very small.

$$d\Delta = x\,d\theta = \frac{x\,dx}{R} = \frac{Mx\,dx}{EI}$$

$$\Delta = \int_Q^P \frac{Mx\,dx}{EI} = \frac{A\bar{x}}{EI}$$

where EI is constant. Otherwise, EI is expressed as a function of x and the integration is carried out either mathematically or numerically.

A = area of bending moment diagram between P and Q
$\bar{x}$ = distance of centre of gravity of bending moment diagram from origin of x

2.4.3 Area moment theorems

The area moment theorems follow from the slope–deflection equations.

Theorem 1: The change in slope in a beam between any two points is the area of the bending moment diagram divided by EI between those two points.

Theorem 2: The intercept on the orthogonal Y-axis of the tangents to the deflected shape of the beam between any two points is the moment taken about the Y-axis of the bending moment diagram between those two points divided by EI.

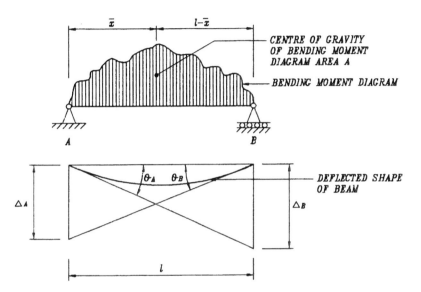

SK 2/14 Simply supported beam with an arbitrary loading.

The bending moment diagram of the simply supported beam with an arbitrary loading has an area equal to A and the centre of gravity of this area is at $\bar{x}$ from support A. The second area moment theorem for small angles of θ_A and θ_B gives:

$$-\theta_B l = \Delta_A = \frac{A\bar{x}}{EI} = \int_0^l \frac{Mx\,dx}{EI}$$

$$\theta_A l = \Delta_B = \frac{A(l-\bar{x})}{EI} = \int_0^l \frac{M(l-x)\,dx}{EI}$$

or $\quad \theta_A = \int_0^l \frac{l-x}{l} \frac{M\,dx}{EI}$

and $\quad \theta_B = -\int_0^l \frac{x}{l} \frac{M\,dx}{EI}$

Angles measured in the clockwise direction from the beam are assumed to be positive.

32 Structural Steelwork

From the above expressions it is clear that if a fictitious loading on the beam is assumed to be the bending moment diagram of the beam, then the end slopes are the shearing forces at the supports due to the fictitious loading divided by EI, where EI remains constant.

For a uniformly distributed loading of intensity w on the beam the bending moment diagram is a parabola with area equal to $wl^3/12$ and $\bar{x}$ equal to $l/2$.

$$\therefore \quad \theta_A = -\theta_B = \frac{A\bar{x}}{EIl} = \frac{wl^3}{24EI}$$

where θ_A = slope of beam at end A
θ_B = slope of beam at end B
w = intensity of uniformly distributed load on beam
l = span of beam
E = modulus of elasticity
I = moment of inertia of section of beam

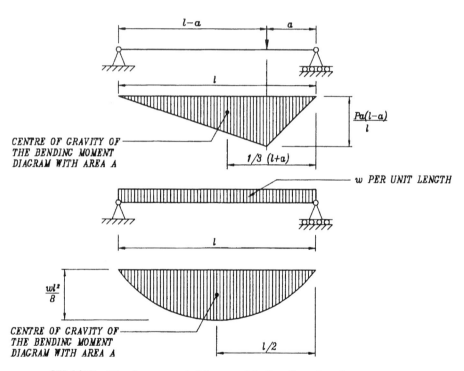

SK 2/15 Simply supported beam with distributed and concentrated loads.

For the case of a simply supported beam with a concentrated load P at a distance a from the right-hand support B, the angles of rotations at A and B are given by:

$$\theta_A = \frac{Pa(l^2 - a^2)}{6lEI}$$

$$\theta_B = -\frac{Pa(l - a)(2l - a)}{6lEI}$$

2.4.4 Generalised slope–deflection equations

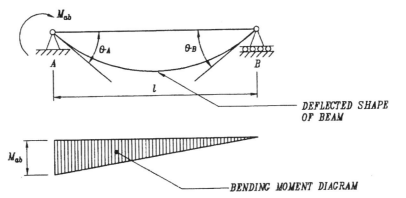

SK 2/16 Beam with applied moment at one end.

Consider a beam AB of length l subjected to an applied moment M_{ab} at A when the bending moment is zero at B. The flexural rigidity EI is assumed constant.

$$A = \text{area of bending moment diagram} = \frac{M_{ab}l}{2}$$

The centre of gravity of the area $A = \bar{x} = \frac{1}{3}l$

$$\theta_A = \frac{A(l-\bar{x})}{EIl} = \frac{M_{ab}l}{3EI}$$

$$\theta_B = -\frac{M_{ab}l}{6EI}$$

When both M_{ab} and M_{ba} (clockwise positive at both ends of the beam) are acting together, then the end rotations are given by:

$$\theta_A = \frac{M_{ab}l}{3EI} - \frac{M_{ba}l}{6EI}$$

$$\theta_B = \frac{M_{ba}l}{3EI} - \frac{M_{ab}l}{6EI}$$

Expressing the end moments in terms of the end rotations gives:

$$M_{ab} = \frac{2EI}{l}(2\theta_A + \theta_B)$$

$$M_{ba} = \frac{2EI}{l}(2\theta_B + \theta_A)$$

If an arbitrary loading is applied to this beam which produces a free span bending moment diagram of area A with a centre of gravity at $\bar{x}$

from the left-hand support A, then the end rotations are given by:

$$\theta_A = \frac{M_{ab}l}{3EI} - \frac{M_{ba}l}{6EI} + \frac{A(l-\bar{x})}{EIl}$$

$$\theta_B = \frac{M_{ba}l}{3EI} - \frac{M_{ab}l}{6EI} - \frac{A\bar{x}}{EIl}$$

Expressing the end moments in terms of the end rotations gives:

$$M_{ab} = \frac{2EI}{l}(2\theta_A + \theta_B) - \frac{2A}{l^2}(2l - 3\bar{x})$$

$$M_{ba} = \frac{2EI}{l}(2\theta_B + \theta_A) + \frac{2A}{l^2}(3\bar{x} - l)$$

where M_{ab} = applied moment at end A of beam AB
M_{ba} = applied moment at end B of beam AB
θ_A = slope of beam at end A
θ_B = slope of beam at end B
l = span of beam
E = modulus of elasticity
I = moment of inertia of section of beam

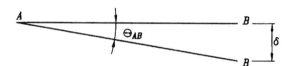

SK 2/17 Beam rotating as a rigid body due to settlement of support.

For the most generalised case of a beam, if it also rotates as a rigid body due to settlement of support B and the rotation Θ_{AB} is clockwise positive, then the end rotations are given by:

$$\theta_A = \frac{M_{ab}l}{3EI} - \frac{M_{ba}l}{6EI} + \frac{A(l-\bar{x})}{EIl} + \Theta_{AB}$$

$$\theta_B = \frac{M_{ba}l}{3EI} - \frac{M_{ab}l}{6EI} - \frac{A\bar{x}}{EIl} + \Theta_{AB}$$

Solving for the end moments in terms of end rotations, the generalised equations are:

$$M_{ab} = \frac{2EI}{l}(2\theta_A + \theta_B) - \frac{2A}{l^2}(2l - 3\bar{x}) - \frac{6\Theta_{AB}EI}{l}$$

$$M_{ba} = \frac{2EI}{l}(2\theta_B + \theta_A) + \frac{2A}{l^2}(3\bar{x} - l) - \frac{6\Theta_{AB}EI}{l}$$

Θ_{AB} = rigid body rotation of beam AB due to settlement of support B.

2.4.5 Fixed-end beams

For any beam with both ends fixed against rotation and displacement, the terms θ_A and θ_B vanish and the generalised slope–deflection equations become:

$$FEM_{ab} = -\frac{2A}{l^2}(2l - 3\bar{x})$$

$$FEM_{ba} = \frac{2A}{l^2}(3\bar{x} - l)$$

FEM_{ab} is the fixed-end moment at end A of a beam with both ends fixed against rotation and displacement and loaded with any arbitrary loading which produces a free span bending moment diagram of area equal to A with a centre of gravity at $\bar{x}$ from the left-hand support.

Using a notation $k = EI/l$, which is termed the stiffness factor, the slope–deflection equations can be written as:

$$M_{ab} = 2k(2\theta_A + \theta_B) + FEM_{ab} - 6k\Theta_{AB}$$

$$M_{ba} = 2k(2\theta_B + \theta_A) + FEM_{ba} - 6k\Theta_{AB}$$

The fixed-end moments FEM_{ab} and FEM_{ba} for beams with different loading are given in Table 11.9.

2.4.6 Theorem of three moments

Consider span AB of length l_1 which has applied bending moments M_{ab} and M_{ba} at ends A and B respectively due to the effects of continuity or boundary conditions, and an arbitrary loading on span AB

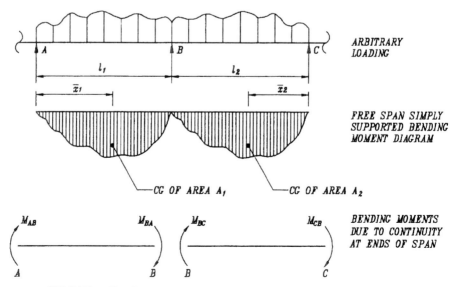

SK 2/18 Continuous beam subjected to arbitrary loading.

which produces a free span bending moment diagram of area equal to A_1 with a centre of gravity at $\bar{x}_1$ from the left-hand support A:

$$\theta_{BA} = \frac{M_{ba}l_1}{3EI_1} - \frac{M_{ab}l_1}{6EI_1} - \frac{A_1\bar{x}_1}{EI_1l_1}$$

Similarly, considering span BC, where the area of the free span bending moment diagram is A_2 and the centre of gravity of the area A_2 is at $\bar{x}_2$ from the right-hand support:

$$\theta_{BC} = \frac{M_{bc}l_2}{3EI_2} - \frac{M_{cb}l_2}{6EI_2} + \frac{A_2\bar{x}_2}{EI_2l_2}$$

Using the conventional sign for moments in beams (i.e. positive moments produce bottom tension and negative moments produce top tension), the bending moments in the slope deflection equations may be changed as follows:

$$M_{ba} = -M_b; \quad M_{bc} = M_b; \quad M_{ab} = M_a; \quad M_{cb} = -M_c$$

Equating θ_{BA} and θ_{BC} and cancelling E from both sides:

$$M_a\frac{l_1}{I_1} + 2M_b\left(\frac{l_1}{I_1} + \frac{l_2}{I_2}\right) + M_c\frac{l_2}{I_2} = -\frac{6A_1\bar{x}_1}{I_1l_1} - \frac{6A_2\bar{x}_2}{I_2l_2}$$

This is known as the theorem of three moments. The application of this theorem results in the creation of a number of equations matching the number of unknown support moments in a continuous beam. At the simply supported end of the beam the bending moment is zero. If the beam is built-in at one end, say at support A, then the expression of rotation θ_{AB} at end A is equated to zero, giving the necessary number of equations to solve the problem.

$$\theta_{AB} = \frac{M_a l_1}{3EI_1} + \frac{M_b l_1}{6EI_1} + \frac{A_1(l_1 - \bar{x}_1)}{EI_1 l_1} = 0$$

or $\quad 2M_a + M_b = -\dfrac{6A_1(l_1 - \bar{x}_1)}{l_1^2}$

where $M_a =$ bending moment at support A of continuous beam
$M_b =$ bending moment at support B of continuous beam
$A_1 =$ area of free span bending moment diagram for span 1 (AB)
$A_2 =$ area of free span bending moment diagram for span 2 (BC)
$\bar{x}_1 =$ centre of gravity of area A_1 from left-hand support A
$\bar{x}_2 =$ centre of gravity of area A_2 from right-hand support C
$l_1 =$ length of span 1 (AB)
$l_2 =$ length of span 2 (BC)
$I_1 =$ moment of inertia of the section of beam in span 1
$I_2 =$ moment of inertia of the section of beam in span 2
$E =$ modulus of elasticity

2.4.7 Method of moment distribution

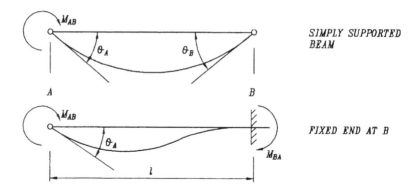

SK 2/19 Determination of carry-over factors.

Carry-over factor

It was established earlier that when a bending moment M_{ab} is applied at end A of a simply supported beam, then the end rotations are given by:

$$\theta_A = \frac{M_{ab}l}{3EI}; \quad \theta_B = -\frac{M_{ab}l}{6EI}$$

EI remains constant over the length l of the beam. If a moment M_{ba} is applied at the end B of the beam to reduce θ_B to zero, then the end rotation at B can be written as:

$$\theta_B = \frac{M_{ba}l}{3EI} - \frac{M_{ab}l}{6EI} = 0 \quad \text{or} \quad M_{ba} = \frac{1}{2} M_{ab}$$

$$\therefore \quad \theta_A = \frac{M_{ab}l}{3EI} - \frac{M_{ba}l}{6EI}$$

$$= \frac{M_{ab}l}{3EI} - \frac{M_{ab}l}{12EI} = \frac{M_{ab}l}{4EI}$$

$$\therefore \quad M_{ab} = \frac{4EI\theta_A}{l} \quad \text{and} \quad M_{ba} = \frac{2EI\theta_A}{l}$$

The carry-over factor for a beam with constant EI is $+\frac{1}{2}$. When a moment M is applied at one end of a beam then a moment $\frac{1}{2}M$ is carried over to the other end of the beam if that end is fixed.

Distribution factor

Four prismatic members of uniform cross-section and fixed ends are connected at joint A as shown in SK 2/20. The members AB, AC, AD and AE have stiffness factors k_1, k_2, k_3 and k_4 respectively, where $k_i = EI_i/l_i$.

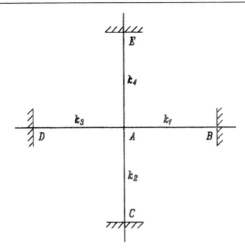

SK 2/20 Determination of distribution factors.

Assume that due to loading on all four members joint A rotates by an angle θ_A. Since the other ends of the members are fixed with zero slope, the slope–deflection equations can be written as follows (see section 2.4.5):

$$M_{ab} = 4k_1\theta_A + FEM_{ab}$$

$$M_{ac} = 4k_2\theta_A + FEM_{ac}$$

$$M_{ad} = 4k_3\theta_A + FEM_{ad}$$

$$M_{ae} = 4k_4\theta_A + FEM_{ae}$$

The algebraic sum of all the moments at joint A must be zero because there is no applied moment at the joint:

$$M_{ab} + M_{ac} + M_{ad} + M_{ae} = 0$$

$$\therefore \quad 4\theta_A(k_1 + k_2 + k_3 + k_4)$$
$$= -(FEM_{ab} + FEM_{ac} + FEM_{ad} + FEM_{ae}) = M_a$$

The term M_a is defined as the unbalanced moment at the joint which has the opposite sign to the algebraic sum of the applied moments at the joint.

$$\theta_A = \frac{M_a}{4(k_1 + k_2 + k_3 + k_4)}$$

Substituting this expression for θ_A in the slope–deflection equations:

$$M_{ab} = FEM_{ab} + \frac{k_1}{k_1 + k_2 + k_3 + k_4} M_a$$

$$M_{ac} = FEM_{ac} + \frac{k_2}{k_1 + k_2 + k_3 + k_4} M_a$$

$$M_{ad} = FEM_{ad} + \frac{k_3}{k_1 + k_2 + k_3 + k_4} M_a$$

$$M_{ae} = FEM_{ae} + \frac{k_4}{k_1 + k_2 + k_3 + k_4} M_a$$

The ratios $k_i / \sum k$ are called the distribution factors.

After distribution at joint A the far ends of the beams will see an additional moment due to the carry-over equal to:

$$\frac{1}{2} \frac{k_i}{\sum k} M_a$$

The fixed-end moments may be due to transverse loading on the beam or due to displacement or rotation of the end joint. Fixed-end moments due to displacement of joints have a significant effect in sway frames. For sway frames the initial moment distribution is carried out ignoring sway and assuming supports at the joints in the direction of sway. After the initial moment distribution the reactions at these temporary supports are determined. Further moment distributions are carried out by releasing these temporary props one at a time and allowing the frame to sway in the direction opposite to the calculated reaction. The force required to cause a unit displacement at the joint of the frame is computed along with its effect on the other props in the system. Finally, a set of simultaneous equations matching the number of temporary props is created by equating the reactions at the props from the first moment distribution to the reactions due to unit sway multiplied by a constant at each point of sway. Solution of these constants from the equations gives the sway moments in the frame. These moments are then combined with the initial moments after first distribution.

2.4.8 Arches

The definition and assumptions of arches are as follows:

- bar curved in one plane only
- supported at both ends
- points of support immovable
- loading in the same plane as the curvature
- plane of curvature is the plane of symmetry of the cross-section
- depth of cross-section is small compared to the radius of curvature.

Three-hinged arches are statically determinate whereas two-hinged arches and hingeless arches are statically indeterminate. A two-hinged arch has four unknown reactions at the supports, viz. two vertical and two horizontal reactions. The three equations of equilibrium are not sufficient to find four unknown reactions. The degree of redundancy in this case is one. The hingeless arch has six unknown reactions at the supports and hence the degree of redundancy is three.

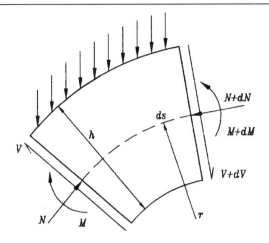

SK 2/21   Element of an arch rib.

Neglecting strain energy due to shearing forces, the strain energy of any arch may be written as:

$$U = \int_0^s \frac{M^2\,ds}{2EI} + \int_0^s \frac{N^2\,ds}{2AE}$$

where U = strain energy stored in arch
 M = bending moment in arch due to applied loading
 N = internal axial load due to applied loading
 ds = elementary length along rib of arch
 A = cross-sectional area of rib of arch
 I = moment of inertia of the section of rib of arch
 s = length along rib of arch

Generally for most arches the cross-sectional area A and the moment of inertia I vary along the length of the arch rib.

2.4.9 Symmetrical two-hinged arches

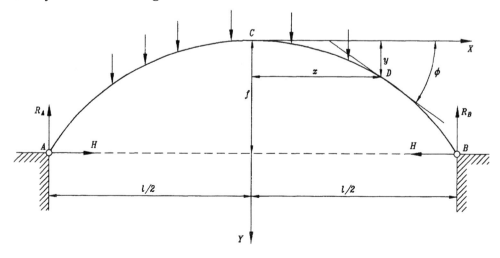

SK 2/22   Two-hinged symmetrical arch with vertical loading.

To find the redundant reaction H at both supports A and B due to applied vertical loading, apply the theorem of least work:

$$\frac{\partial U}{\partial H} = \int_0^s \frac{M}{EI}\frac{\partial M}{\partial H}\,ds + \int_0^s \frac{N}{AE}\frac{\partial N}{\partial H}\,ds = 0$$

Assume

$$M = M' - H(f-y)$$

$$\therefore \quad \frac{\partial M}{\partial H} = -(f-y)$$

and

$$N = N' + H\cos\phi$$

$$\therefore \quad \frac{\partial N}{\partial H} = \cos\phi$$

where M' = Bending moment in the arch rib when H is zero
N' = Axial compression in the arch rib when H is zero
f = rise of arch rib
ϕ = angle of the centre-line of the arch rib with respect to the X-axis
H = horizontal reaction at supports

$$\therefore \quad -\int_0^s \frac{M' - H(f-y)}{EI}(f-y)\,ds + \int_0^s \frac{N' + H\cos\phi}{AE}\cos\phi\,ds = 0$$

or

$$H = \frac{\int_0^s \frac{M'}{EI}(f-y)\,ds - \int_0^s \frac{N'}{AE}\cos\phi\,ds}{\int_0^s \frac{(f-y)^2}{EI}\,ds + \int_0^s \frac{\cos^2\phi}{AE}\,ds}$$

This equation is valid for any symmetrical two-hinged arch subjected to vertical loads and may be used for numerical integration. An arch with variable cross-sectional geometry may be divided up into discrete segments along the length of its rib and the numerator and the denominator of the expression for H evaluated by summation of the components of each segmental value using the segment length as ds in the equation.

Mathematically H can be calculated if the arch rib is assumed to follow a geometric profile and the geometry of the cross-section is expressed in terms of its slope to the X-axis. Assume:

$$y = \frac{4fx^2}{l^2}$$

$$\therefore \quad dy = \frac{8fx}{l^2}\,dx$$

$$ds = dx\left[\left(\frac{8fx}{l^2}\right)^2 + 1\right]^{\frac{1}{2}}$$

$$\cos\phi = \cfrac{1}{\sqrt{1+\left(\cfrac{8fx}{l^2}\right)^2}}$$

$$A = \frac{A_0}{\cos\phi}$$

$$I = \frac{I_0}{\cos\phi}$$

where A_0 = area of the cross-section of the arch rib at the crown
I_0 = moment of inertia of section of arch rib at crown

The arch when loaded with a uniformly distributed vertical load of intensity w per unit length along the X-axis satisfies the following expressions:

$$M' = \frac{wl^2}{8}\left(1 - \frac{4x^2}{l^2}\right)$$

$$N' = wx\sin\phi$$

Solving mathematically the horizontal reaction at the support may be written as:

$$H = \frac{wl^2}{8f}\frac{1}{(1+\beta)}$$

where $\beta = \dfrac{15}{32}\dfrac{l}{f^3}\dfrac{I_0}{A_0}\tan^{-1}\dfrac{4f}{l}$

l = horizontal span of arch
w = uniformly distributed vertical load intensity

Thermal stresses in symmetrical two-hinged arches

Assume a rise of temperature of the arch rib by $t°C$. If the support at B is free to slide then the support will move horizontally by a distance equal to $l\alpha t$, where α is the coefficient of thermal expansion for the material of the arch. Since the support is restrained against horizontal movement, this is equivalent to applying a force H at the support which will produce an opposite displacement equal to $l\alpha t$.

$$\therefore \frac{\partial U}{\partial H} = l\alpha t$$

$$M' = 0$$

$$N' = 0$$

$$H = \frac{l\alpha t}{\displaystyle\int_0^s \frac{(f-y)^2}{EI}\,ds + \int_0^s \frac{\cos^2\phi}{AE}\,ds}$$

This expression for H may be used for numerical integration.

2.4.10 Hingeless symmetrical arches

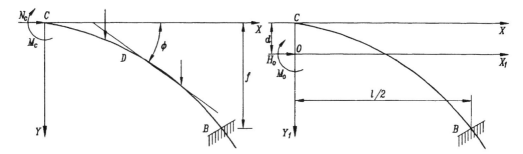

SK 2/23 Hingeless symmetrical arch under symmetrical vertical load.

Because the loading is symmetrical, the shear force must vanish at the crown. Only two redundant internal forces are present at the crown, viz. M_c and N_c. Assume s to be half the length of the arch rib. At the crown, owing to symmetry, the rotation of the arch rib and the horizontal displacement must be zero. Applying Castigliano's second theorem it can be said:

$$\frac{\partial U}{\partial M_c} = 2 \int_0^s \frac{M}{EI} \frac{\partial M}{\partial M_c} \, ds = 0$$

$$\frac{\partial U}{\partial N_c} = 2 \int_0^s \frac{M}{EI} \frac{\partial M}{\partial N_c} \, ds + 2 \int_0^s \frac{N}{AE} \frac{\partial N}{\partial N_c} \, ds = 0$$

Assume M' and N' are the bending moments and axial thrusts in the arch due to applied external loading only.

$$M = M_c + N_c y + M'; \qquad N = N_c \cos\phi + N'$$

From the above expressions:

$$\frac{\partial M}{\partial M_c} = 1; \qquad \frac{\partial M}{\partial N_c} = y; \qquad \frac{\partial N}{\partial N_c} = \cos\phi$$

Substituting these expressions in the equations of least work:

$$M_c \int_0^s \frac{ds}{EI} + N_c \int_0^s \frac{y \, ds}{EI} = -\int_0^s \frac{M'}{EI} \, ds$$

$$M_c \int_0^s \frac{y \, ds}{EI} + N_c \int_0^s \frac{y^2 \, ds}{EI} + N_c \int_0^s \frac{\cos^2\phi}{AE} \, ds$$
$$= -\int_0^s \frac{M' y}{EI} \, ds - \int_0^s \frac{N' \cos\phi}{AE} \, ds$$

Using these two equations the two unknown internal forces M_c and N_c can be determined knowing the geometry of the arch and the loading.

M' = bending moment due to external loading only
N' = axial thrust due to external loading only
M_c = bending moment in arch rib at crown of arch
N_c = axial thrust in arch rib at crown of arch

Elastic centre of the arch

If the origin is shifted to O as shown in SK 2/23 the following simplifications can be made:

$$y_1 = y - d$$

where y_1 = vertical ordinate with respect to the origin at O

$$d = \frac{\int_0^s \dfrac{y\,ds}{EI}}{\int_0^s \dfrac{ds}{EI}}$$

$$M_0 = M_c + H_0 d = \frac{\int_0^s \dfrac{M'\,ds}{EI}}{\int_0^s \dfrac{ds}{EI}}$$

$$H_0 = N_c = -\frac{\int_0^s \dfrac{M' y_1}{EI}\,ds + \int_0^s \dfrac{M'\cos\phi}{AE}\,ds}{\int_0^s \dfrac{y_1^2}{EI}\,ds + \int_0^s \dfrac{\cos^2\phi}{AE}\,ds}$$

The shifted origin at d is called the elastic centre of the arch. The expression for H_0 and N_c may be used for numerical integration.

The choice of the origin at the elastic centre is equivalent to attaching an infinitely stiff bracket at the crown. The forces H_0 and M_0 act at the end of this bracket at a depth d from the crown of the arch. The rigid bracket assumption validates that there is no rotation and horizontal displacement at either O or C of the hingeless symmetrical arch under symmetrical loading.

Thermal stresses in hingeless symmetrical arch

A symmetrical hingeless arch subjected to a uniform rise of temperature $t°C$ will try to expand by $\tfrac{1}{2}\alpha t l$ at the crown C. Because the arch is symmetrical, the rotation at the crown must be zero and the horizontal displacement is $\tfrac{1}{2}\alpha t l$. Applying the theorem of least work it can be said:

$$\frac{\partial U}{\partial M_0} = \int_0^s \frac{M}{EI}\frac{\partial M}{\partial M_0}\,ds = 0$$

$$\frac{\partial U}{\partial H_0} = \int_0^s \frac{M}{EI}\frac{\partial M}{\partial H_0}\,ds + \int_0^s \frac{N}{AE}\frac{\partial N}{\partial H_0}\,ds = \frac{\alpha t l}{2}$$

$$M = M_0 + H_0 y_1$$

$$N = H_0 \cos\phi$$

$$\therefore \quad \frac{\partial M}{\partial M_0} = 1; \quad \frac{\partial N}{\partial H_0} = \cos\phi; \quad \frac{\partial M}{\partial H_0} = y_1$$

$$M_0 = 0$$

$$H_0 = \frac{0.5\alpha t l}{\int_0^s \frac{y_1^2}{EI}\,ds + \int_0^s \frac{\cos^2\phi}{AE}\,ds}$$

This expression for H_0 may be used for numerical integration.

Antimetric loading on a hingeless arch

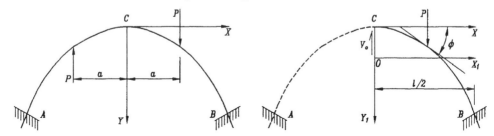

SK 2/24 Antimetric loading on a symmetrical hingeless arch.

The antimetric loading on a symmetrical hingeless arch is applied by two equal and opposite forces acting at an equal spacing on either side of the crown. From symmetry it can be concluded that each load P acting in the opposite direction at an equal distance from the crown will produce an equal and opposite internal bending moment and direct thrust at the crown. The moment and the axial thrust will vanish at the crown as a result of such a loading and only shearing force will remain. If the shearing force is V_0, then by using the theorem of least work it can be said:

$$\frac{\partial U}{\partial V_0} = 2\int_0^s \frac{M}{EI}\frac{\partial M}{\partial V_0}\,ds + 2\int_0^s \frac{N}{AE}\frac{\partial N}{\partial V_0}\,ds = 0$$

$$M = V_0 x_1 + M'; \qquad N = -V_0 \sin\phi + N'$$

Substituting in the expressions for least work:

$$\int_0^s \frac{V_0 x_1 + M'}{EI}x_1\,ds - \int_0^s \frac{-V_0 \sin\phi + N'}{AE}\sin\phi\,ds = 0$$

$$\text{or}\quad V_0 = \frac{-\int_0^s \frac{M' x_1}{EI}\,ds + \int_0^s \frac{N' \sin\phi}{AE}\,ds}{\int_0^s \frac{x_1^2}{EI}\,ds + \int_0^s \frac{\sin^2\phi}{AE}\,ds}$$

At any cross-section M' and N' are the bending moment and axial thrust for external loading only.

Having found V_0, the rest of the internal forces and reactions can be found easily. The expression for V_0 can be used for numerical integration. A combination of symmetric and antimetric loading can be used to represent any asymmetric loading on the arch as shown in SK 2/25.

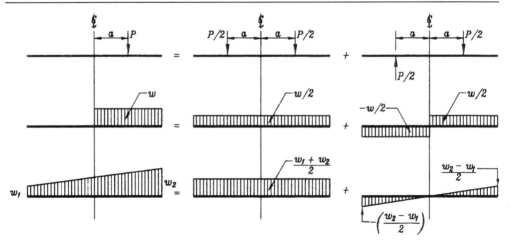

SK 2/25 Symmetric and antimetric loading.

2.5 INFLUENCE LINES

In any structure there may be several critical points where the internal forces have to be evaluated. Loading on any part of the structure gives rise to the internal forces at the critical points of interest. The influence coefficient at any point in a structure for any internal force at any critical point in the structure is the value of the internal force due to unit load at that point. Suppose the point of interest for bending moment in a simply supported beam is the centre of span. The influence coefficient at a quarter point is the measure of bending moment at the centre of span when the unit load is at the quarter point on the span. The influence line is the graphical representation of all the influence coefficients on a structure for an internal force at a critical point of interest. First select an internal force, next select a point of interest and then apply unit load at any point on the structure to find the influence coefficient at that point for the internal force at the chosen point of interest. The influence coefficients are graphically represented as ordinates on the structure and when joined up by a line produce what is called the influence line. The completed diagram is called the influence diagram for an internal force at a critical point of interest in the structure.

The benefit of using influence lines becomes evident when several moving loads are applied to the structure and the task is to find the worst effect of these loads. The effects of individual loads are added to find the combined effect of several loads.

2.5.1 Influence lines of cantilever beams

When a unit load is applied at the end of a cantilever beam of length l then the bending moment at the support is $-l$, which is represented as an ordinate $-l$ at the end of the cantilever beam. Similarly, if the unit load is at a distance x from the built-in end of the cantilever,

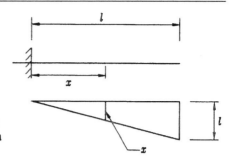

SK 2/26 Influence diagram for being moment at the support of a cantilever.

then the bending moment at the built-in support is $-x$. The influence line for bending moment at the built-in end of a cantilever beam is a straight line with an ordinate of $-l$ at the tip of the cantilever and zero ordinate at the built-in end. A load P applied at x from the built-in end of the cantilever will produce a bending moment $-Px$ at the built-in end. For a uniformly distributed load of w per unit length over the whole length of the cantilever beam the bending moment at the built-in end will be the area of the influence diagram multiplied by the intensity of loading.

$$M = \frac{1}{2} llw = \frac{wl^2}{2}$$

The influence line for shearing force at the built-in end of the cantilever is a constant ordinate line of coefficient equal to 1. The influence line for deflection at the tip of the cantilever is given by the expression:

$$\delta = \left[\frac{x^3}{3EI} + \frac{x^2}{2EI}(l-x) \right]$$

where x is the distance from the built-in end.

2.5.2 Influence lines of a simply supported beam

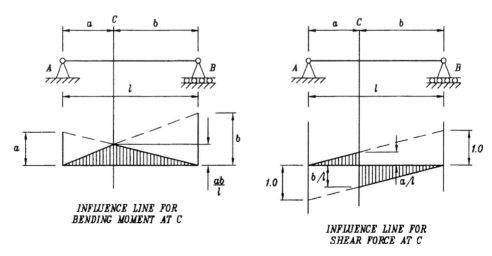

SK 2/27 Influence diagrams for bending moments and shearing force at point C of a simply supported beam.

A simply supported beam AB is under the action of a unit load. Consider the influence line for bending moment of a point C on the beam which is at a distance a from A and b from B. The bending moment at C is the reaction at $B(R_b)$ multiplied by the distance b when the load is at the left of C. Similarly, the bending moment at C is the reaction at $A(R_a)$ multiplied by the distance a when the load is at the right of C. The influence coefficients for $R_a a$ and $R_b b$ are the ordinates a and b at A and B respectively. The influence line for bending moment at C is the influence line for reaction R_a times a to the right of C and R_b times b to the left of C. This gives the triangular influence diagram with a maximum ordinate ab/l at C.

Similarly, the shearing force influence line at C is the reaction influence line for R_b when the load is to the left of C and the reaction influence line for R_a when the load is to the right of C.

For a train of concentrated moving loads on a simply supported beam the bending moment will be maximum under a load where the shearing force changes sign. The *absolute maximum bending moment* in that beam occurs under this load when the load is so positioned that the resultant of the train of loads and the load itself are equally spaced about the centre-line of the beam, or, in other words, when the midpoint of the span bisects the distance between the load and the resultant of the train of loads.

2.5.3 Influence lines of three-hinged arches

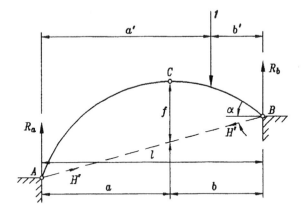

SK 2/28 Unit load on a three-hinged arch.

The three-hinged arch shown in SK 2/28 is subjected to a unit vertical load. Assume that the reactions at support A are R_a and H', and at support B they are R_b and H', where H' is acting in the direction of the line joining A and B. Taking moments of all external forces about A:

$$R_b = \frac{a'}{l}; \quad R_a = \frac{b'}{l}$$

R_a = vertical reaction at support A of three-hinged arch
R_b = vertical reaction at support B of three-hinged arch
a' = distance of unit load from support A of arch
b' = distance of unit load from support B of arch
l = horizontal distance between supports A and B of arch
H' = reaction at support of arch in the direction AB
H = horizontal component of H'
a = distance of hinge C from support A of arch
b = distance of hinge C from support B of arch
α = angle of inclination of line AB to horizontal

This means that the vertical reactions R_a and R_b are the same as in a simply supported beam of span l. Taking moments about the hinge C and equating to zero:

$$R_a a - H'f \cos \alpha = 0$$

or $\quad H = H' \cos \alpha = R_a \dfrac{a}{f}$

when the load is to the right of C, and:

$$H = H' \cos \alpha = R_b \dfrac{b}{f}$$

when the load is to the left of C.

It is concluded from the above expressions that the vertical simple span reaction influence lines when multiplied by a/f for the right part of C and b/f for the left part of C give the influence lines for H.

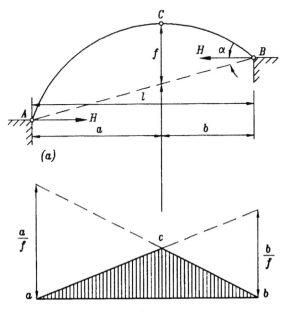

SK 2/29 Influence line for horizontal reaction of a three-hinged arch.

Bending moment at point D at a distance c from A (as shown in SK 2/30) may be expressed as follows:

$$M_d = R_a c - H' y_d \cos\alpha = R_a c - H y_d$$

The term $R_a c$ represents the bending moment influence line for point D of a simply supported beam of length l, and the term $H y_d$ represents the influence line of horizontal thrust multiplied by y_d. If the ordinates of these two influence lines are subtracted from one another, the result is the influence line of bending moment for point D in a three-hinged arch. This is illustrated in SK 2/30.

Similarly, the shearing force V_d at point D may be expressed as follows when the load is to the right of D:

$$V_d = R_a \cos\phi - H' \sin(\phi - \alpha) = R_a \cos\phi - H \frac{\sin(\phi - \alpha)}{\cos\alpha}$$

M_d = bending moment at point D of arch
V_d = shear force at point D of arch
ϕ = inclination of tangent at point D of arch to horizontal

The tangent at D is inclined to the horizontal by an angle ϕ and the angle of inclination of line AB to the horizontal is α. The first term in the expression corresponds to the influence coefficients for shearing force at point D of a simply supported beam of span l multiplied by $\cos\phi$, and the second term expresses the influence coefficients of horizontal thrust H multiplied by $\sin(\phi - \alpha)/\cos\alpha$. The influence

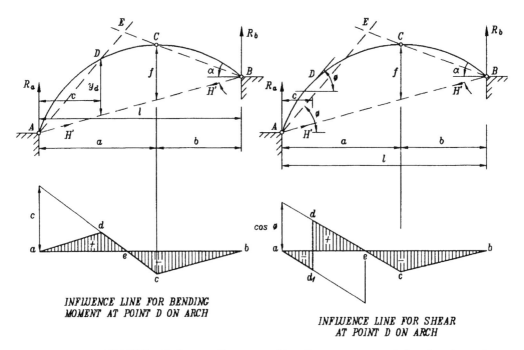

INFLUENCE LINE FOR BENDING MOMENT AT POINT D ON ARCH

INFLUENCE LINE FOR SHEAR AT POINT D ON ARCH

SK 2/30 Influence diagrams of bending moment and shearing force for a three-hinged arch.

diagram of shearing force at point D in the three-hinged arch is obtained by subtracting the ordinates of the second term from those of the first term. This is illustrated in SK 2/30.

2.5.4 Influence lines of simple trusses

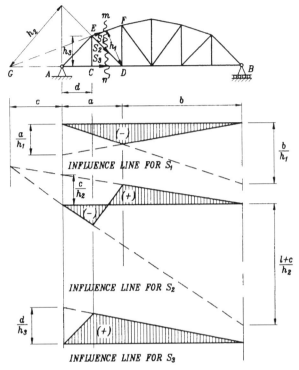

SK 2/31 Influence diagrams of a simple truss.

R_a is the vertical reaction at A and the load is right of point D in SK 2/31.

Consider the free body diagram to the left of section mn of the truss illustrated in SK 2/31. The part truss is in equilibrium under the action of R_a, S_1, S_2 and S_3. The lines of action of S_2 and S_3 meet at D, S_1 and S_3 meet at G and S_1 and S_2 meet at E.

Considering the equilibrium of the part truss on the left-hand side of section mn, taking moments of all the external forces about point D and equating the sum to zero:

$$R_a a + S_1 h_1 = 0$$

$$\therefore S_1 = -\frac{R_a a}{h_1}$$

Therefore the influence line ordinates for S_1 are the influence line ordinates of bending moment for point D of a simply supported beam divided by h_1. Similarly:

$$S_3 = \frac{R_a d}{h_3}$$

Therefore, the influence line ordinates for S_3 are the influence line ordinates for bending moment at point E for a simple beam divided by h_3.

By taking moments about point G it can be stated:

$$S_2 = \frac{R_a c}{h_2}$$

This expression is true when the load is to the right of D. Considering the equilibrium of the right-hand side of section mn of the truss when the load is to the left of C and taking moments about point G:

$$S_2 = -\frac{R_b(l+c)}{h_2}$$

The influence line ordinates for S_2 are the influence line ordinates of R_a or R_b (depending on the location of the load with respect to points D and C) multiplied by the appropriate factor as illustrated in SK 2/31.

Note: For a parallel chord truss the factors c/h_2 and $(l+c)/h_2$ are taken to be equal to 1.0 because the distances c and h_2 tend to infinity.

2.6 MATRIX METHOD OF STRUCTURAL ANALYSIS

It is assumed that the reader has a working knowledge of matrix algebra and all the notations. The basic principles of the matrix method of structural analysis are discussed in this section. The most powerful and popular method of structural analysis with the help of matrix algebra is the displacement method. This section deals with the application of the displacement method to trusses, continuous beams and rigid frames.

2.6.1 Truss analysis by the displacement method

The method of displacement is explained with reference to the truss shown in SK 2/32. Firstly the load equilibrium relationship between the external and internal forces is established at each joint, which are both free to move. Secondly, the joint deformation with member deformation relationship is established. The applied loads and the joint deformations are assumed to be positive corresponding to the direction of positive X and Y co-ordinates.

Applied external loads are P_1 to P_4 and the internal axial forces in the members of the truss are S_1 to S_5 for members 1 to 5 respectively.

Considering the load equilibrium relationship at all joints of the truss which are free to displace it can be stated:

$$S_1 \cos \alpha_1 - S_3 \cos \alpha_3 - S_5 = P_1$$
$$S_1 \sin \alpha_1 + S_3 \sin \alpha_3 = P_2$$
$$S_2 \cos \alpha_2 + S_4 \cos \alpha_4 + S_5 = P_3$$
$$S_2 \sin \alpha_2 - S_4 \sin \alpha_4 = P_4$$

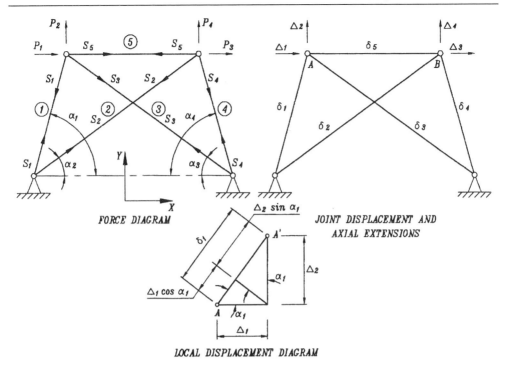

SK 2/32 Force and displacement diagrams of a truss.

Expressed in matrix form it becomes:

$$\begin{vmatrix} S_1 \\ S_2 \\ S_3 \\ S_4 \\ S_5 \end{vmatrix} \times \begin{vmatrix} +\cos\alpha_1 & 0 & -\cos\alpha_3 & 0 & -1 \\ +\sin\alpha_1 & 0 & +\sin\alpha_3 & 0 & 0 \\ 0 & +\cos\alpha_2 & 0 & -\cos\alpha_4 & +1 \\ 0 & +\sin\alpha_2 & 0 & +\sin\alpha_4 & 0 \end{vmatrix} = \begin{vmatrix} P_1 \\ P_2 \\ P_3 \\ P_4 \end{vmatrix}$$

which when expressed in a generalised matrix form gives:

$$\{S\}[A] = \{P\}$$

The matrix $[A]$ is called the geometry matrix or the static matrix.

Assume that the joints A and B displace by Δ_1, Δ_2, Δ_3 and Δ_4 along the X–X and Y–Y axes as shown in SK 2/32. The axial increases in the length of the truss elements due to these joint displacements are δ_1, δ_2, δ_3, δ_4 and δ_5 for members 1 to 5 respectively. The relationship between the joint displacements and the member extensions may be expressed as follows:

$$+\Delta_1 \cos\alpha_1 + \Delta_2 \sin\alpha_1 = \delta_1$$

$$+\Delta_3 \cos\alpha_2 + \Delta_4 \sin\alpha_2 = \delta_2$$

$$-\Delta_1 \cos\alpha_3 + \Delta_2 \sin\alpha_3 = \delta_3$$

$$-\Delta_3 \cos\alpha_4 + \Delta_4 \sin\alpha_4 = \delta_4$$

$$-\Delta_1 + \Delta_3 = \delta_5$$

Expressed in a matrix form:

$$\begin{vmatrix} \Delta_1 \\ \Delta_2 \\ \Delta_3 \\ \Delta_4 \end{vmatrix} \times \begin{vmatrix} +\cos\alpha_1 & +\sin\alpha_1 & 0 & 0 \\ 0 & 0 & +\cos\alpha_2 & +\sin\alpha_2 \\ -\cos\alpha_3 & +\sin\alpha_3 & 0 & 0 \\ 0 & 0 & -\cos\alpha_4 & +\sin\alpha_4 \\ -1 & 0 & +1 & 0 \end{vmatrix} = \begin{vmatrix} \delta_1 \\ \delta_2 \\ \delta_3 \\ \delta_3 \\ \delta_5 \end{vmatrix}$$

or $\{\Delta\}[A^T] = \{\delta\}$

The matrix $[A^T]$ is the transpose of the geometry matrix $[A]$ and is called the deformation matrix. The compatibility of deformations at all the joints is established by this relationship.

The relationship between the extension of a truss element and the internal force in the element can be expressed as follows:

$$S_i = \left(\frac{A_i E_i}{l_i}\right)\delta_i$$

This relationship expressed in matrix notation gives:

$$[S] = [K][\delta]$$

where $[K]$ is called the stiffness matrix. For the truss in question the stiffness matrix can be written as follows:

$$[K] = \begin{vmatrix} \dfrac{E_1 A_1}{l_1} & 0 & 0 & 0 & 0 \\ 0 & \dfrac{E_2 A_2}{l_2} & 0 & 0 & 0 \\ 0 & 0 & \dfrac{E_3 A_3}{l_3} & 0 & 0 \\ 0 & 0 & 0 & \dfrac{E_4 A_4}{l_4} & 0 \\ 0 & 0 & 0 & 0 & \dfrac{E_5 A_5}{l_5} \end{vmatrix}$$

This matrix is a diagonal matrix. By substitution of matrices we can get the following:

$$[S] = [K][\delta] = [KA^T][\Delta]$$

$$[P] = [S][A] = [AKA^T][\Delta]$$

$$\therefore \quad [\Delta] = [AKA^T]^{-1}[P]$$

The joint displacement matrix of the structure $[\Delta]$ can be solved by inversion of the $[AKA^T]$ matrix and multiplying it by the load matrix. Having found the joint displacements, the internal forces in the members of the truss can be computed with the help of the geometry matrix and the stiffness matrix.

P = external load
S_i = internal force in member i of truss
Δ_i = displacement of joint i
δ_i = axial extension of member i of truss
A_i = area of section of member i of truss
l_i = length of member i of truss
E_i = elastic modulus of member i of truss

2.6.2 Continuous beam analysis by the displacement method

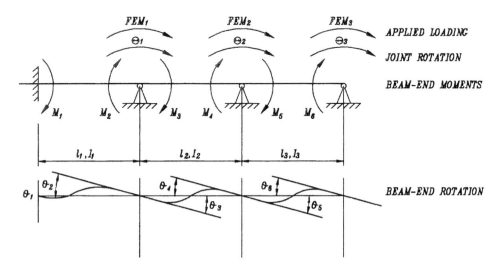

SK 2/33 Moment–rotation diagrams of a continuous beam.

The method of displacement applied to a continuous beam is best explained with the help of a three-span continuous beam as illustrated in SK 2/33. Assume that an external moment is applied at each joint. They are termed FEM_1, FEM_2 and FEM_3 at joints 1, 2 and 3 respectively. The applied joint moments are the algebraic sum of the fixed-end moments of the members connecting at the joint due to loading on the members. These applied moments make the joints rotate by angles Θ_1, Θ_2 and Θ_3 at joints 1, 2 and 3 respectively. Considering the equilibrium of moments at the joints it can be written:

$FEM_1 = M_2 + M_3$

$FEM_2 = M_4 + M_5$

$FEM_3 = M_6$

In matrix notation this can be expressed as:

$[FEM] = [A][M]$

where the geometry matrix $[A]$ is given by:

$$[A] = \begin{vmatrix} 0 & +1 & +1 & 0 & 0 & 0 \\ 0 & 0 & 0 & +1 & +1 & 0 \\ 0 & 0 & 0 & 0 & 0 & +1 \end{vmatrix}$$

Similarly, if the joint rotations at each joint are equated to the beam-end rotations for each beam connecting at the joint, the displacement matrix can be obtained. The joint and beam-end rotations are expressed as follows:

$$\theta_1 = 0$$

$$\theta_2 = \theta_3 = \Theta_1$$

$$\theta_4 = \theta_5 = \Theta_2$$

$$\theta_6 = \Theta_3$$

In matrix notation this can be expressed as:

$$[\theta] = [A^T][\Theta]$$

where $[A^T]$ is given by the following expression:

$$[A^T] = \begin{vmatrix} 0 & 0 & 0 \\ +1 & 0 & 0 \\ +1 & 0 & 0 \\ 0 & +1 & 0 \\ 0 & +1 & 0 \\ 0 & 0 & +1 \end{vmatrix}$$

The matrix $[A^T]$ is the transpose of the geometry matrix and may be called the deformation matrix.

From the slope–deflection equations derived in earlier parts of this chapter it is known that the end rotations and the end moments in prismatic members are related to one another. For a prismatic member with joint notations ij the following expressions can be written:

$$M_i = \left(\frac{4EI}{l}\right)\theta_i + \left(\frac{2EI}{l}\right)\theta_j$$

$$M_j = \left(\frac{4EI}{l}\right)\theta_j + \left(\frac{2EI}{l}\right)\theta_i$$

These equations expressed in matrix notation give:

$$[M] = [K][\theta]$$

where [K] is called the stiffness matrix and is written as follows:

$$[K] = \begin{vmatrix} \frac{4EI_1}{l_1} & \frac{2EI_1}{l_1} & 0 & 0 & 0 & 0 \\ \frac{2EI_1}{l_1} & \frac{4EI_1}{l_1} & 0 & 0 & 0 & 0 \\ 0 & 0 & \frac{4EI_2}{l_2} & \frac{2EI_2}{l_2} & 0 & 0 \\ 0 & 0 & \frac{2EI_2}{l_2} & \frac{4EI_2}{l_2} & 0 & 0 \\ 0 & 0 & 0 & 0 & \frac{4EI_3}{l_3} & \frac{2EI_3}{l_3} \\ 0 & 0 & 0 & 0 & \frac{2EI_3}{l_3} & \frac{4EI_3}{l_3} \end{vmatrix}$$

By substitution of matrices we can derive the following:

$$[M] = [K][\theta] = [KA^T][\Theta]$$

$$[FEM] = [A][M] = [AKA^T][\Theta]$$

$$\therefore \quad [\Theta] = [AKA^T]^{-1}[FEM]$$

The joint rotation matrix is computed from the geometry matrix, the displacement matrix, the stiffness matrix and the external joint moment matrix. The inversion of the matrix $[AKA^T]$ multiplied by the matrix $[FEM]$ gives the rotations at all the joints. When the joint rotations are known, the member internal forces can be calculated by the use of the geometry matrix and the stiffness matrix.

M_i = beam span-end moments at end i of beam
FEM_i = applied joint moment or algebraic sum of fixed-end moments at joint i
Θ_i = rotation of joint i
θ = beam span-end rotation at end i of beam
I = moment of inertia of section of beam number i
l = length of span of beam number i
E = elastic modulus

2.6.3 Continuous frame analysis by the displacement method

The rigid frame chosen to illustrate this method has a single degree of freedom in the direction of sidesway. To determine the number of degrees of freedom in the direction of sidesway the following equation may be used:

$$s = 2j - (2f + 2h + r + m)$$

where s = number of degrees of freedom of sidesway
j = total number of joints in the structure

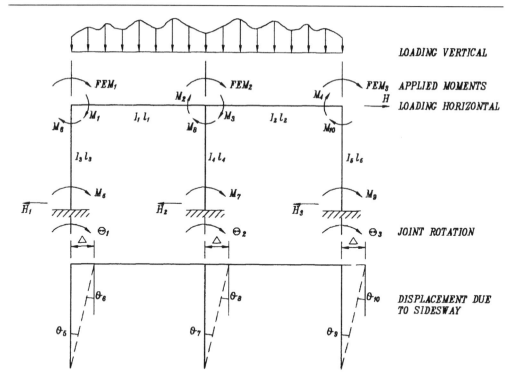

SK 2/34 Rotation–displacement diagrams of a continuous frame.

f = number of rotationally and translationally fixed supports (fixed)
h = number of translationally fixed supports (hinged)
r = number of translationally fixed supports in one orthogonal direction (roller)
m = number of members in the frame

Applying this equation to the problem illustrated in SK 2/34:

$$s = 2 \times 6 - [(2 \times 3) + (2 \times 0) + (0 + 5)] = 1$$

The number of degrees of freedom for sidesway is important in the analyses because in the displacement equations for the joints these unknown parameters have to be included. It may be noted that for the rigid frame illustrated the sidesway can occur at the level of the beam.

As in the case of the continuous beam discussed earlier, we can make the following assumptions:

- FEM_1, FEM_2 and FEM_3 are three algebraic sums of fixed-end moments at joints 1, 2 and 3 respectively. All clockwise moments are assumed to be positive. There is a net external applied force at the level of the beam in the direction of the sway displacement, and this force may be called H. These four elements form the external force matrix.

- The members 1 to 5 as shown in SK 2/34 have geometric properties defined by $(I_1 l_1)$, $(I_2 l_2)$, $(I_3 l_3)$, $(I_4 l_4)$ and $(I_5 l_5)$ respectively, where I_i is the moment of inertia and l_i is the length of the prismatic member.
- The beam-end moments are, for member 1 M_1 and M_2, for member 2 M_3 and M_4, for member 3 M_5 and M_6, for member 4 M_7 and M_8 and for member 5 M_9 and M_{10}. The beam-end rotations are marked exactly in the same fashion from θ_1 to θ_{10}.
- Assume the joint rotations are Θ_1, Θ_2 and Θ_3 at joints 1, 2 and 3 respectively. Also assume that the horizontal displacement at the level of the beam due to sidesway is Δ.

Considering the equilibrium of moments at the joints, the external load and internal force relationships may be written as follows, where P is the generalised form of notation of external loads:

$$P_1 = FEM_1 = M_1 + M_6$$

$$P_2 = FEM_2 = M_2 + M_3 + M_8$$

$$P_3 = FEM_3 = M_4 + M_{10}$$

$$P_4 = H = -(H_1 + H_2 + H_3)$$

$$H_1 = \frac{M_5 + M_6}{l_1}$$

$$H_2 = \frac{M_7 + M_8}{l_2}$$

$$H_3 = \frac{M_9 + M_{10}}{l_3}$$

External loads are termed P_i in a generalised fashion where P_i can be applied moments or direct loads. Expressed in matrix notation the above equations become:

$$[P] = [A][M]$$

where $[A]$ is given by the following expression:

$$[A] = \begin{vmatrix} +1 & 0 & 0 & 0 & 0 & +1 & 0 & 0 & 0 & 0 \\ 0 & +1 & +1 & 0 & 0 & 0 & 0 & +1 & 0 & 0 \\ 0 & 0 & 0 & +1 & 0 & 0 & 0 & 0 & 0 & +1 \\ 0 & 0 & 0 & 0 & -1/l_1 & -1/l_1 & -1/l_2 & -1/l_2 & -1/l_3 & -1/l_3 \end{vmatrix}$$

$[A]$ is termed the geometry matrix or the statics matrix.

The deformations of the joints are equated to the deformations of the ends of the prismatic members connecting at the joints. The effect of the sway of the frame at the level of the beam by an amount equal to Δ is equivalent to angular displacements of members 3, 4 and 5 when the following relationship is true for small displacements:

$$\theta_5 = \theta_6 = -\frac{\Delta}{l_1} = -\frac{X_4}{l_1}$$

$$\theta_7 = \theta_8 = -\frac{\Delta}{l_2} = -\frac{X_4}{l_2}$$

$$\theta_9 = \theta_{10} = -\frac{\Delta}{l_3} = -\frac{X_4}{l_3}$$

X_i is the generalised expression for all types of displacement at the joints.

The compatibility of deformations at all other joints will also give the following:

$$X_1 = \Theta_1 = \theta_1 = \theta_6$$
$$X_2 = \Theta_2 = \theta_2 = \theta_8$$
$$X_3 = \Theta_3 = \theta_4 = \theta_{10}$$

Expressed in matrix notation it gives:

$$[\theta] = [A^T][X]$$

where the matrix $[A^T]$ is given by:

$$[A^T] = \begin{vmatrix} +1 & 0 & 0 & 0 \\ 0 & +1 & 0 & 0 \\ 0 & +1 & 0 & 0 \\ 0 & 0 & +1 & 0 \\ 0 & 0 & 0 & -1/l_1 \\ +1 & 0 & 0 & -1/l_1 \\ 0 & 0 & 0 & -1/l_2 \\ 0 & +1 & 0 & -1/l_2 \\ 0 & 0 & 0 & -1/l_3 \\ 0 & 0 & +1 & -1/l_3 \end{vmatrix}$$

Exactly as in the case of a continuous beam we may proceed to find the stiffness matrix. The slope–deflection equations give the relationship between the end moments M_i and M_j and the end rotations θ_i and θ_j of any prismatic member with ends ij:

$$M_i = \left(\frac{4EI}{l}\right)\theta_i + \left(\frac{2EI}{l}\right)\theta_j$$

$$M_j = \left(\frac{4EI}{l}\right)\theta_j + \left(\frac{2EI}{l}\right)\theta_i$$

Expressed in matrix notation this gives:

$$[M] = [K][\theta]$$

where $[K]$ is called the stiffness matrix and is given by:

$$[K] = \begin{vmatrix} A & B & 0 & 0 & 0 & 0 & 0 & 0 & 0 & 0 \\ B & A & 0 & 0 & 0 & 0 & 0 & 0 & 0 & 0 \\ 0 & 0 & C & D & 0 & 0 & 0 & 0 & 0 & 0 \\ 0 & 0 & D & C & 0 & 0 & 0 & 0 & 0 & 0 \\ 0 & 0 & 0 & 0 & E & F & 0 & 0 & 0 & 0 \\ 0 & 0 & 0 & 0 & F & E & 0 & 0 & 0 & 0 \\ 0 & 0 & 0 & 0 & 0 & 0 & G & H & 0 & 0 \\ 0 & 0 & 0 & 0 & 0 & 0 & H & G & 0 & 0 \\ 0 & 0 & 0 & 0 & 0 & 0 & 0 & 0 & I & J \\ 0 & 0 & 0 & 0 & 0 & 0 & 0 & 0 & J & I \end{vmatrix}$$

$$A = \frac{4EI_1}{l_1}; \quad B = \frac{2EI_1}{l_1}; \quad C = \frac{4EI_2}{l_2}; \quad D = \frac{2EI_2}{l_2}; \quad E = \frac{4EI_3}{l_3};$$

$$F = \frac{2EI_3}{l_3}; \quad G = \frac{4EI_4}{l_4}; \quad H = \frac{2EI_4}{l_4}; \quad I = \frac{4EI_5}{l_5}; \quad J = \frac{2EI_5}{l_5}$$

By substitution of matrices the following is obtained:

$$[M] = [K][\theta] = [KA^T][X]$$

$$[P] = [A][M] = [AKA^T][X]$$

$$\therefore \quad [X] = [AKA^T]^{-1}[P]$$

The joint displacement matrix $[X]$ is first solved by the inversion of the $[AKA^T]$ matrix multiplied by the load matrix $[P]$. When the joint displacements are known, the member internal forces can be computed by the use of the geometry matrix and the stiffness matrix.

It is of interest to note that the matrix $[AKA^T]^{-1}$ is entirely dependent on the geometric and material properties of the frame structure and its boundary conditions. The external load matrix does not influence this structural property within elastic limits and small displacements.

2.7 STRUCTURAL DYNAMICS

This section deals with the fundamentals of the vibration of structures, free or forced. Most structures can be idealised into a mass–spring–dashpot system. Vibration maintained only by the spring force of the structure is called free vibration. An external time-dependent force applied to a structure causes a forced vibration.

The idealised spring represents the stiffness of the structure in any given direction and the dashpot is a measure of viscous damping, signifying the dissipating energy of the spring–mass system due to vibration. The linear viscous dashpot model of a structure represents a damping resistance which is velocity dependent.

2.7.1 Single degree of freedom (SDOF) systems

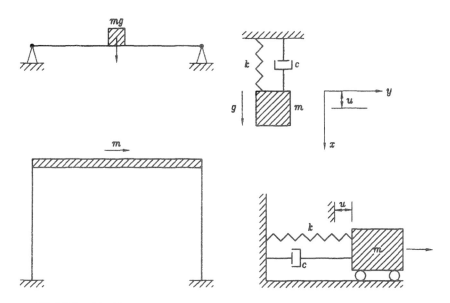

SK 2/35 Idealised dynamic models of structures.

The structures illustrated in SK 2/35 are idealised to dynamic models with a single degree of freedom (i.e. the lumped masses are allowed to move in one given linear direction only). It is assumed that the spring representing the structural stiffness is massless and has a stiffness equal to k (N/m). The viscous dashpot has a damping coefficient c (N sec/m).

For the case of the load supported by the spring vertically, assume the X-axis is along the line of motion and that the origin ($x=0$) is at the location where the spring is unstretched. Assume a force $p(t)$ is acting on the mass m, and causes a displacement $u(t)$ of the mass after an interval of time t.

Inertial force $= m\ddot{u}$ acting against the direction of motion
Damping force $= c\dot{u}$ acting against the direction of motion
Spring force $= ku$ acting against the direction of motion
$W =$ weight of mass $= mg$
$m =$ mass
$c =$ damping coefficient
$k =$ stiffness
$p(t) =$ applied time dependent force
$u(t) =$ time dependent displacement

The equation of dynamic equilibrium is given by:

$$p(t) - m\ddot{u} - c\dot{u} - ku + W = 0$$

$$\text{or} \quad m\ddot{u} + c\dot{u} + ku = W + p(t)$$

The static displacement of the spring due to load W is u_{st}, given by:

$$u_{st} = \frac{W}{k}$$

Assume $u = u_r + u_{st}$, where u_r is the relative displacement with respect to the static equilibrium configuration and u_{st} is a constant.

$$\therefore \quad m\ddot{u}_r + c\dot{u}_r + ku_r = p(t)$$

Similarly for the portal frame the idealised SDOF dynamic model will give:

$$m\ddot{u} + c\dot{u} + ku = p(t)$$

This is the fundamental equation of motion. Dividing both sides of the equation by m and using the notation $m = k/\omega_n^2$ the equation of motion may be rewritten as:

$$\ddot{u} + 2\xi\omega_n\dot{u} + \omega_n^2 u = \left(\frac{\omega_n^2}{k}\right)p(t)$$

where $\xi = c/c_{cr}$ and $c_{cr} = 2m\omega_n = 2k/\omega_n$.

The term ω_n (radians/sec) is called the undamped natural frequency of the structure and c_{cr} is the critical damping coefficient. The magnitude of the damping factor ξ is used to determine three cases of damping, viz. $0 < \xi < 1 =$ underdamped, $\xi = 1 =$ critically damped and $\xi > 1 =$ overdamped. For overdamped and critically damped structures there is no oscillation. Also:

$$T_n = \text{undamped natural period of vibration} = \frac{2\pi}{\omega_n} \text{ (seconds)}$$

$f_n =$ undamped natural frequency of vibration

$$= \frac{1}{T_n} = \frac{\omega_n}{2\pi} \text{ (cycles per second - Hz)}$$

Assuming that the system is underdamped and the forcing function is zero, the equation of motion for free vibration is given by:

$$\ddot{u} + \omega_n^2 u = 0$$

The solution of this second-order linear differential equation is:

$$u = C_1 \cos\omega_n t + C_2 \sin\omega_n t$$

where C_1 and C_2 are constants to be determined from the initial conditions of motion when time $t=0$. Assume that at time $t=0$, $u=u_0$ and $\dot{u}=\dot{u}_0$. Substituting these values in the equation of motion:

$$C_1 = u_0 \quad \text{and} \quad C_2 = \frac{\dot{u}_0}{\omega_n}$$

Therefore the equation of motion of a single degree of freedom system is:

$$u = u_0 \cos\omega_n t + \left(\frac{\dot{u}_0}{\omega_n}\right) \sin\omega_n t$$

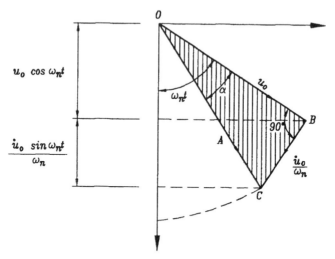

SK 2/36 Vectorial representation of a free vibration response.

The equation of motion has two parts. The first part is dependent on the initial displacement and the second part is dependent on the initial velocity of motion. Vectorial representation of these two parts is illustrated in SK 2/36. The rotating vector $\overline{OB}$ with a length u_0 is rotating anticlockwise with time t at an angular velocity ω_n. Vector $\overline{BC}$ with a length equal to $\dot{u}_0/\omega_n$ is at $90°$ to the vector $\overline{OB}$. The resultant vector is $\overline{OC}$ which has a magnitude equal to A given by:

$$A = \sqrt{u_0^2 + \left(\frac{\dot{u}_0}{\omega_n}\right)^2}$$

A is called the amplitude of the motion, which is the maximum value of u. The equation of motion can also be written as $u = A\cos(\omega_n t - \alpha)$, as shown in the vector diagram, where α is called the phase angle and is given by:

$$\alpha = \tan^{-1}\frac{\dot{u}_0}{\omega_n u_0}$$

The equation of motion may also be written as:

$$u = A\cos\omega_n\left(t - \frac{\alpha}{\omega_n}\right)$$

From this expression it can be concluded that a time equal to α/ω_n must pass after initial time $t = 0$ before u reaches the maximum value A equal to the amplitude of motion. This time is called the time lag.

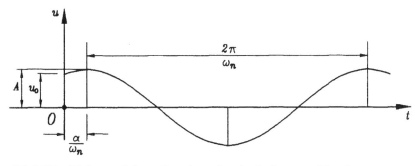

SK 2/37 Undamped free vibration of a single degree of freedom system.

2.7.2 Effect of viscous damping on free vibration

The general equation of motion is given by:

$$\ddot{u} + 2\xi\omega_n\dot{u} + \omega_n^2 u = 0$$

The underdamped case $\xi < 1$ is considered for the vibration analysis. Assuming the damped natural frequency $\omega_d = \omega_n\sqrt{(1-\xi^2)}$, solution of the differential equation results in:

$$u = e^{-\xi\omega_n t}(C_1 \cos\omega_d t + C_2 \sin\omega_d t)$$

The constants C_1 and C_2 are determined from the initial conditions as before. This will give:

$$u = e^{-\xi\omega_n t}\left[u_0 \cos\omega_d t + \left(\frac{\dot{u}_0 + \xi\omega_n u_0}{\omega_d}\right)\sin\omega_d t\right]$$

$$= A e^{-\xi\omega_n t}\cos(\omega_d t - \alpha)$$

The amplitude of the motion, A, is given by:

$$A = \sqrt{u_0^2 + \left(\frac{\dot{u}_0 + \xi\omega_n u_0}{\omega_d}\right)^2}$$

and $\quad \alpha = \tan^{-1}\left(\dfrac{\dot{u}_0 + \xi\omega_n u_0}{\omega_d u_0}\right)$

The term $e^{-\xi\omega_n t}$ is responsible for the dying out of the response over a period of time.

2.7.3 Forced vibration of undamped SDOF systems due to harmonic excitation

This section deals with the forced vibration of undamped single degree of freedom systems when the forcing function is harmonic. The solution to the general equation of motion is in two parts – i.e. the particular solution (u_p) and the complementary solution (u_c). The total response of the system is the superposition of the forced vibration and the natural vibration.

The equation of the forced motion is given by:

$$m\ddot{u} + ku = p_0 \cos \Omega t$$

where p_0 = amplitude of force of excitation
Ω = frequency of excitation

In the above equation the amplitude of the force of excitation p_0 and the frequency of excitation Ω remain constant. The solution of the differential equation may be written as $u_p = U \cos \Omega t$, where U is the amplitude of the steady-state response found by substitution of this solution in the equation of motion, and is given by:

$$U = \frac{p_0}{k - m\Omega^2} \quad \text{where } k - m\Omega^2 \neq 0$$

Assume $U_0 = p_0/k$, which is the deflection of the spring due to the amplitude p_0 of the harmonic excitation. Assume also that the ratio of frequencies (forced upon natural) is equal to r.

$r = \Omega/\omega_n$ and the response may also be expressed as $u_p = U_0/(1 - r^2) \cos \Omega t$ when $r \neq 1$ and $m/k = 1/\omega_n^2$.

It is observed from the above response equation that when r approaches 1, the amplitude of the response approaches a very high value. This is called resonance. The response is in phase when r is less than 1, and is 180° out of phase when r is greater than 1. This phenomenon is illustrated in SK 2/38. U/U_0 is called the steady-state dynamic magnification factor.

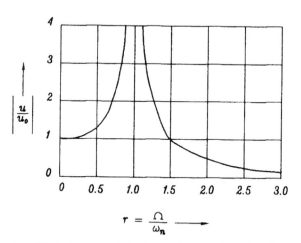

SK 2/38 **Steady-state dynamic magnification factor.**

The total response including the free vibration will be written as:

$$u = \left(\frac{U_0}{1 - r^2}\right) \cos \Omega t + C_1 \cos \omega_n t + C_2 \sin \omega_n t$$

The constants C_1 and C_2 may be determined from the initial conditions of motion.

2.7.4 Forced vibration of undamped SDOF system by triangular impulse

The spring–mass SDOF system is subjected to a triangular pulse load as in the case of a blast loading. The response of the system is calculated in two stages, viz. $t \leq t_d$ and $t > t_d$, where t_d is the time of duration of the impulsive loading.

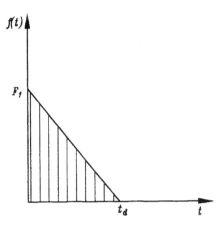

SK 2/39 Triangular impulse load-time history.

The equation of the forced motion due to the triangular pulse when $t \leq t_d$ is given as:

$$m\ddot{u} + ku = f(t) = F_1\left(1 - \frac{t}{t_d}\right)$$

The solution of this differential equation is given by:

$$u = C_1 \sin \omega_n t + C_2 \cos \omega_n t + \frac{F_1}{k}\left(1 - \frac{t}{t_d}\right)$$

At $t = 0$, $u = 0$ and $\ddot{u} = 0$, $\therefore C_1 = F_1/\omega_n k t_d$ and $C_2 = -F_1/k$.

$$\therefore u = \frac{F_1}{\omega_n k t_d} \sin \omega_n t - \frac{F_1}{k} \cos \omega_n t + \frac{F_1}{k}\left(1 - \frac{t}{t_d}\right)$$

The static displacement due to the maximum value of the impulsive load is $u_{st} = F_1/k$.

$$\text{Dynamic magnification factor} = \frac{u}{u_{st}} = \frac{\sin \omega_n t}{t_d \omega_n} - \cos \omega_n t + 1 - \frac{t}{t_d}$$

The equation of the forced motion due to a triangular impulse when $t > t_d$ is given by $m\ddot{u} + ku = 0$ because the external force is zero and a free vibration ensues. The solution to this equation is given by:

$$u = C_1 \sin \omega_n t + C_2 \cos \omega_n t$$

The constants C_1 and C_2 are determined from the initial condition of motion. The starting point of this motion is the end of the previous motion when t was less than t_d. Therefore the initial condition of this motion is found by substituting $t = t_d$ in the response equation of the previous motion when t was less than or equal to t_d. The response of this free vibration stage is therefore given by:

$$u = \frac{F_1}{k\omega_n t_d}[\sin \omega_n t - \sin \omega_n (t - t_d)] - \frac{F_1}{k}\cos \omega_n t$$

$$\therefore \frac{u}{u_{st}} = \frac{1}{\omega_n t_d}[\sin \omega_n t - \sin \omega_n (t - t_d)] - \cos \omega_n t$$

The dynamic magnification factor u/u_{st} depends entirely on $\omega_n t_d$ or the ratio t_d/T_n, where T_n is the natural period of vibration.

2.7.5 Free vibration of a simply supported beam with distributed mass

Consider a prismatic member with a constant flexural stiffness EI, a constant mass per unit length equal to m, and a span length l over simple supports. The inertial forces are acting against the flexural stiffness of the beam.

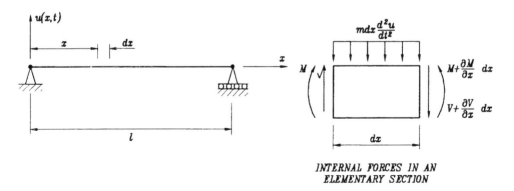

INTERNAL FORCES IN AN ELEMENTARY SECTION

SK 2/40 Free vibration of a simply supported beam with distributed mass.

Assume $u(x, t)$ is the transverse displacement of the beam, which is a function of x and t, where x is the distance along the length of the beam and t is the time elapsed in seconds. This is a case of free vibration and there is no applied external time-dependent force function on the system.

Consider the elementary section dx of the beam at a distance x from the left-hand support. The inertial force due to the elementary mass of the beam is $m\,dx\,a$, where a is the acceleration on the elementary mass given by d^2u/dt^2. On the elementary section dx the internal forces will act as shown in SK 2/40. The vertical force balance for dynamic equilibrium is given by:

$$V - \left(V + \frac{\partial V}{\partial x} dx\right) - m\, dx\, \frac{\partial^2 u}{\partial t^2} = 0$$

or $\quad \dfrac{\partial V}{\partial x} = -m \dfrac{\partial^2 u}{\partial t^2}$

The following relationships are true in any beam:

$$\frac{\partial M}{\partial x} = V; \quad \frac{\partial^2 M}{\partial x^2} = \frac{\partial V}{\partial x}; \quad M = EI \frac{\partial^2 u}{\partial x^2}$$

Substituting these expressions in the equation of vertical force equilibrium:

$$\frac{\partial^2 M}{\partial x^2} + m \frac{\partial^2 u}{\partial t^2} = 0$$

or $\quad \dfrac{\partial^2}{\partial x^2}\left(EI \dfrac{\partial^2 u}{\partial x^2}\right) + m \dfrac{\partial^2 u}{\partial t^2} = 0$

or $\quad u^{iv} + \dfrac{m}{EI} \ddot{u} = 0$

$$u^{iv} = \frac{\partial^4 u}{\partial x^4}; \quad \ddot{u} = \frac{\partial^2 u}{\partial t^2}$$

where V = shear force in beam
M = bending moment in beam
m = mass per unit length of beam
u = time-dependent transverse deflection of beam

This is the general equation of motion of the beam, where u is a function of time as well as distance x. By separating the variables one could use the following solution of the equation:

$$u(x, t) = \phi(x) y(t)$$

This assumption in physical terms means that the beam will have a defined constant flexural shape $\phi(x)$ and the amplitude of this shape varies with time according to the function $y(t)$. The function $\phi(x)$ is called the mode shape.

Substituting the above expression in the general equation of motion of the beam:

$$\phi^{iv}(x) y(t) + \frac{m}{EI} \phi(x) \ddot{y}(t) = 0$$

where $\phi^{iv}(x)$ means fourth-order differentiation with respect to x and $\ddot{y}(t)$ means second-order differentiation with respect to t. Dividing both sides of the equation by $\phi(x) y(t)$ we get:

$$\frac{\phi^{iv}(x)}{\phi(x)} + \frac{m \ddot{y}(t)}{EI y(t)} = 0$$

x and t are independent variables. The first term in the equation is a function of x only, whereas the second term is a function of t only. Therefore each part of the equation must be a constant with the same magnitude but of opposite sign to satisfy this equation for all arbitrary values of x and t.

Assume that $a^4 = \omega^2 m/EI$ is the magnitude of the constant which will satisfy the equations.

$$\therefore \quad \phi^{iv}(x) - a^4 \phi(x) = 0 \quad \text{and} \quad \ddot{y}(t) + \omega^2 y(t) = 0$$

Solution of the second equation is the free vibration response of an SDOF system:

$$y(t) = \frac{\dot{y}_0}{\omega_n} \sin \omega_n t + y_0 \cos \omega_n t$$

A solution of the first equation is of the form given by $\phi(x) = C e^{sx}$. This when substituted in the first differential equation results in $(s^4 - a^4) C e^{sx} = 0$, from which it can be derived that $s = \pm a, \pm ia$.

$$\therefore \quad \phi(x) = C_1 e^{iax} + C_2 e^{-iax} + C_3 e^{ax} + C_4 e^{-ax}$$

or $\phi(x) = A_1 \sin ax + A_2 \cos ax + A_3 \sinh ax + A_4 \cosh ax$

A_1, A_2, A_3 and A_4 can be determined from the beam boundary conditions.

At $x = 0$, $\phi(x)_0 = 0$ and $\phi(x)_0^{ii} = 0$ because $M = 0$

At $x = l$, $\phi(x)_l = 0$ and $\phi(x)_l^{ii} = 0$ because $M = 0$

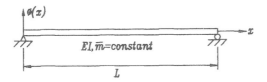

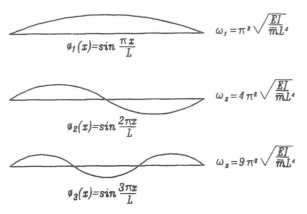

SK 2/41 Modes of vibration of a simply supported beam.

Substituting these conditions in the general equation of mode shape it is observed:

$$A_2 = A_4 = A_3 = 0 \quad \text{and} \quad \phi(x)_1 = A_1 \sin al = 0$$

This gives $\sin al = 0$, or $al = n\pi$ for $n = 0, 1, 2, 3, 4$, etc. which gives the following:

$$a = \frac{n\pi}{l}$$

$$\omega_n = n^2\pi^2 \sqrt{\frac{EI}{ml^4}}$$

$$\phi(x) = A_1 \sin \frac{n\pi}{l} x$$

2.7.6 Response of SDOF systems to a very short impulse

When the time of duration t_d of the impulse is very small compared to the natural period of vibration T_n of the system, the initial displacement u_0 in the response equation of free vibration may be taken as equal to zero and the initial velocity $\dot{u}_0$ may be given by the theorem of conservation of momentum as follows:

$$m\dot{u}_0 = \int F\,dt \quad \text{or} \quad \dot{u}_0 = \frac{\int F\,dt}{m}$$

Consider free vibration after this very short impulse, and the equation of response may be written as:

$$u \approx \frac{\int F\,dt}{m\omega_n} \sin \omega_n t$$

F = magnitude of the impulsive force

2.7.7 Response of SDOF systems to any arbitrary time-dependent loading

The arbitrary time-dependent loading may be approximated as a series of very short impulses of magnitude $F(\tau)\,d\tau$ occurring at elapsed time τ. This is illustrated in SK 2/42.

Assume that the total time elapsed is t, the time elapsed after the impulse is t' and the elapsed time when the impulse occurred is τ.

$$t = t' + \tau$$

The individual responses due to the very short impulses can be summed up to give the total response. From the previous section on

72 Structural Steelwork

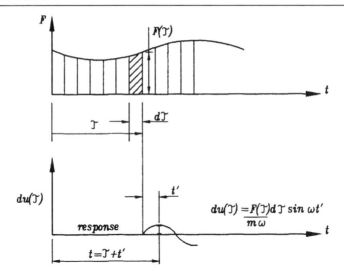

SK 2/42 Response of an SDOF system due to any arbitrary loading.

very short impulsive loads it can be said:

$$du = \frac{F(\tau)}{m\omega_n} \sin \omega_n t' \, d\tau = \frac{F(\tau)}{m\omega_n} \sin \omega_n (t - \tau) \, d\tau$$

$$\therefore u = \int_0^t \frac{F(\tau)}{m\omega_n} \sin \omega_n (t - \tau) \, d\tau$$

Following the same argument, the total response of a damped SDOF system due to any time-dependent arbitrary loading will be given by:

$$u = \int_0^t \frac{F(\tau)}{m\omega_D} e^{-\xi\omega_n(t-\tau)} \sin \omega_D(t - \tau) \, d\tau$$

2.7.8 Seismic response of a damped SDOF system

The very short impulsive loads $F(\tau)/m$ could be the equivalent seismic accelerations from the acceleration time-history of an earthquake given by the expression $\ddot{u}_g(\tau)$. Assuming $\omega_D = \omega_n$, the response equation due to seismic excitation may be written as:

$$u = \frac{1}{\omega_n} \int_0^t \ddot{u}_g(\tau) e^{-\xi\omega_n(t-\tau)} \sin \omega_n(t - \tau) \, d\tau$$

This is the earthquake displacement equation of a damped single degree of freedom system. The effective acceleration on the mass m of the structure is $\omega_n^2 u$ and the effective earthquake base shear is given by $m\omega_n^2 u$.

2.7.9 Response spectra

The spectral velocity function is $\omega_n u$ and may be expressed as:

$$V = \int_0^t \ddot{u}_g(\tau) e^{-\xi \omega_n (t-\tau)} \sin \omega_n (t - \tau) \, d\tau$$

The maximum value of this velocity response for a given damping coefficient and a given natural frequency of a single degree of freedom system can be computed from a given earthquake acceleration time-history, and this is called the pseudo-spectral velocity S_V given by:

$$S_V = \left\{ \int_0^t \ddot{u}_g(\tau) e^{-\xi \omega_n (t-\tau)} \sin \omega_n (t - \tau) \, d\tau \right\}_{\max}$$

A curve can be drawn for S_V against a series of SDOF systems with varying ω_n and fixed damping coefficient ξ. A series of these curves can also be drawn with different damping coefficients. These curves are called the response spectra for pseudo velocity.

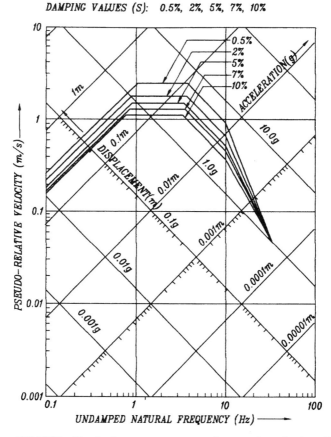

SK 2/43 Typical response spectra for pseudo velocity, displacement and acceleration.

Similarly response spectra curves can be drawn for the following:

Spectral displacement $S_D = S_V/\omega_n$

Spectral acceleration $S_A = \omega_n S_V$

From the spectral acceleration response spectra one could calculate the maximum effective base shear by multiplying the mass of the system by the appropriate acceleration from the spectra corresponding to the natural frequency of the system and an appropriate value of the damping coefficient.

Maximum horizontal base shear $H_{max} = MS_A$

where M = mass of system

The response spectra found from a given ground motion time-history will show large local peaks due to resonance, and for practical applications these peaks are broadened out.

2.8 ANALYSIS OF PLATES

2.8.1 Elastic analysis of plates

Assume that the load acting on the plate is normal to the surface of the plate, and the deflection of the plate due to the applied loading is small compared to the thickness of the plate. There is no in-plane loading on the plate and the plate is not restrained in the in-plane direction. Assume that during bending the middle plane of the plate has no strain.

The plane xy lies in the middle plane of the plate and the axis z is in the perpendicular out-of-plane direction. An element of the plate of sides dx and dy, cut out from the plate by two pairs of planes parallel to planes xz and yz, is shown in SK 2/44.

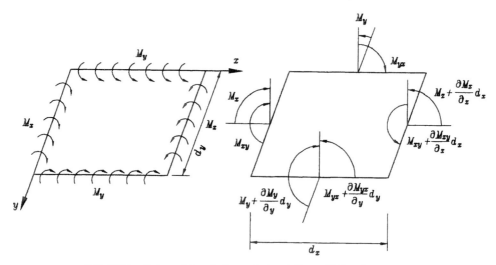

SK 2/44 Internal loads in a plate at the middle plane.

Assume that q is the loading per unit area on the plate and $q\,dx\,dy$ is the loading on the plate element of sides dx and dy. M_x is the bending moment per unit length acting on the edges of the plate element parallel to the y-axis and M_y is the bending moment per unit length acting on the edges parallel to the x-axis. M_{xy} and M_{yx} are the torsional moment per unit length on the edges of the plate element parallel to the y-axis and x-axis respectively. The out-of-plane shear forces V_x and V_y per unit length of the plate element correspond to bending moments M_x and M_y respectively.

Considering the force diagram in SK 2/44 the vertical force balance gives:

$$\frac{\partial V_x}{\partial x} dx\,dy + \frac{\partial V_y}{\partial y} dy\,dx + q\,dx\,dy = 0$$

$$\frac{\partial V_x}{\partial x} + \frac{\partial V_y}{\partial y} + q = 0$$

Taking moments about the x-axis gives:

$$\frac{\partial M_{xy}}{\partial x} dx\,dy - \frac{\partial M_y}{\partial y} dy\,dx + V_y\,dx\,dy = 0$$

$$\frac{\partial M_{xy}}{\partial x} - \frac{\partial M_y}{\partial y} + V_y = 0$$

Similarly, $$\frac{\partial M_{yx}}{\partial y} + \frac{\partial M_x}{\partial x} - V_x = 0$$

Observing that $M_{yx} = -M_{xy}$ and eliminating V_x and V_y from the equations gives:

$$\frac{\partial^2 M_x}{\partial x^2} + \frac{\partial^2 M_y}{\partial y^2} - 2\frac{\partial^2 M_{xy}}{\partial x \partial y} = -q$$

From pure bending of plates it is known that:

$$M_x = -D\left(\frac{\partial^2 w}{\partial x^2} + \nu \frac{\partial^2 w}{\partial y^2}\right)$$

$$M_y = -D\left(\frac{\partial^2 w}{\partial y^2} + \nu \frac{\partial^2 w}{\partial x^2}\right)$$

$$M_{xy} = -M_{yx} = D(1-\nu)\frac{\partial^2 w}{\partial x\,\partial y}$$

$$D = \frac{Eh^3}{12(1-\nu^2)}$$

where $D =$ *flexural rigidity* of plate
$h =$ thickness of plate
$w =$ deflection of plate in the out-of-plane direction
$\nu =$ Poisson's ratio for material of plate

M_x = bending moment per unit length acting on the edges parallel to the y-axis
M_y = bending moment per unit length acting on the edges parallel to the x-axis
M_{xy} and M_{yx} = torsional moment per unit length on the edges of the plate
q = loading per unit area on plate
V_x = shear force per unit length of plate due to M_x
V_y = shear force per unit length of plate due to M_y

By substitution of expressions for M_x, M_y and M_{xy} the differential equation in terms of w only becomes:

$$\frac{\partial^4 w}{\partial x^4} + 2\frac{\partial^4 w}{\partial x^2 \partial y^2} + \frac{\partial^4 w}{\partial y^4} = \frac{q}{D}$$

The shearing forces in the plate are expressed as:

$$V_x = \frac{\partial M_{yx}}{\partial y} + \frac{\partial M_x}{\partial x} = -D\frac{\partial}{\partial x}\left(\frac{\partial w}{\partial x^2} + \frac{\partial^2 w}{\partial y^2}\right)$$

$$V_y = \frac{\partial M_y}{\partial y} - \frac{\partial M_{xy}}{\partial x} = -D\frac{\partial}{\partial y}\left(\frac{\partial^2 w}{\partial x^2} + \frac{\partial^2 w}{\partial y^2}\right)$$

These are the classical differential equations for any plate with small deflections. The solution will depend on the loading and boundary conditions. Solutions for several loading and boundary conditions of different geometric shapes of plates are available in *Theory of Plates and Shells*, second edition, by Timoshenko and Woinowsky-Kreiger, published by McGraw-Hill Book Company, 1959.

2.8.2 Yield-line analysis of plates

In the ultimate limit state design of two-way elements it is not necessary to define the stress distribution in the element. Instead, assume plastic hinges in the regions of highest stresses. The yield-line method assumes that at the point of maximum moments plastic hinges will form and yielding will take place until the full plastic moment capacity is developed along the length of a yield line. Several yield lines will form in a plate until failure takes place by loss of flexural rigidity. Yield lines are idealised plastic moment formation lines in a two-way element. The idealised yield-line patterns in two-way elements are given in SK 2/45.

Plates are divided into segments along assumed yield lines. Yield lines must be straight lines forming axes of rotation of these segments. It has been shown by experiments that the use of idealised yield lines as shown in SK 2/45 results in a slight error on the side of conservatism.

The corner sections of two-way elements are stiffer in comparison to the internal parts of the elements. In order to model this additional stiffness, which will result in less rotation along the idealised yield

Theory of Structures 77

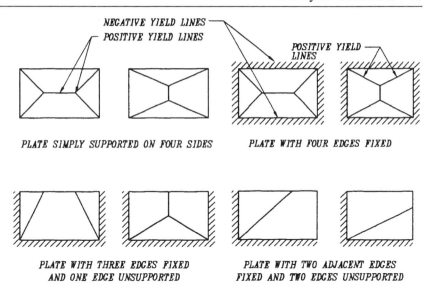

SK 2/45 Idealised yield-line patterns of two-way elements.

lines, it is recommended that the plastic moment of resistance is reduced to $\frac{2}{3}M_P$ over the corner region according to the diagrams in SK 2/46.

The ultimate resistance of a plate subjected to a uniformly distributed load is defined as the load per unit area that will cause

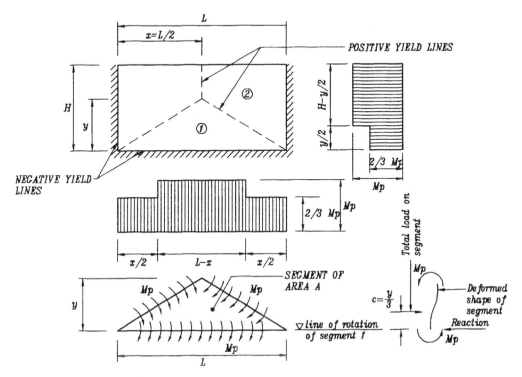

SK 2/46 Determination of ultimate unit resistance by yield lines.

failure along the idealised yield lines. The method of equilibrium may be used to determine the ultimate resistance. The equilibrium method requires shear forces on both sides of the yield lines to be equal and opposite. For simplicity, this method only checks the equilibrium of moments along a set of positive yield lines, assuming that shear force is zero along a positive yield line. Sagging moments are positive and hogging moments are negative. The ultimate moment of resistance may also be determined by the theory of virtual work, which requires a complex differentiation to find the minimum value of unit resistance.

The method of equilibrium involves the determination, for a segment, of the sum of the resisting moments along the yield lines and equating this sum to the moment of the external loading on the segment about the axis of rotation of the segment that is the support of the segment. This is expressed in mathematical terms as follows:

$$\sum M_P l = r_u A c$$

or $r_u = \dfrac{\sum M_P l}{A c}$

where M_P = ultimate plastic moment capacity of the plate section per unit length (Nm/m)
  l = length of a side of segment or projected length of a side of a segment over which M_P is acting to resist the rotation due to the applied loading
  r_u = ultimate resistance in N/m²
  A = area of the segment bordered by the yield lines (m²)
  c = centroid of the area A (loaded by r_u) from the line of rotation of the segment.

Consider the two-way plate element shown in SK 2/46. L and H are the dimensions of the plate. Locations of the yield lines are denoted by x and y. For symmetrical yield lines x is assumed equal to $L/2$. The only unknown parameter in the analysis is the dimension y. The segments are marked 1 and 2. The free body diagrams of the segments show the actions of M_P resisting the rotation about the segment support. Consider the free body diagram of segment 1:

$$\sum M_P l = 2\left[\dfrac{2}{3} M_P \left(\dfrac{L}{4}+\dfrac{L}{4}\right)+M_P\left(\dfrac{L}{2}\right)\right] = \dfrac{5}{3} M_P L$$

Assuming projected lengths for the inclined sides of the triangular segment:

$$c_1 = \dfrac{y}{3}; \quad A_1 = \dfrac{Ly}{2}$$

$$r_u A_1 c_1 = \sum M_P l = \dfrac{5}{3} M_P L$$

$$\therefore \ r_u = 10 \dfrac{M_P}{y^2}$$

Similarly, for segment 2 the moment balance will give:

$$\sum M_P l = 2\left[\frac{2}{3} M_P\left(\frac{y}{2}\right) + M_P\left(H - \frac{y}{2}\right)\right] = 2M_P\left(H - \frac{y}{6}\right)$$

$$c_2 = \frac{L(3H - 2y)}{6(2H - y)}; \qquad A_2 = \frac{L(2H - y)}{4}$$

$$A_2 c_2 = \frac{L^2(3H - 2y)}{24}$$

$$\therefore \quad r_u = \frac{8M_P(6H - y)}{L^2(3H - 2y)}$$

To find y equate the two expressions for r_u:

$$\frac{10M_P}{y^2} = \frac{8M_P(6H - y)}{L^2(3H - 2y)}$$

Having found y from the above equation r_u can be found.

2.9 METHODS OF PLASTIC ANALYSIS

2.9.1 Definition of plastic behaviour

The stress–strain relationship of mild steel in pure tension shows considerable ductility. Ductility is defined as the ultimate strain divided by the strain at yield. This property is generally found for all grades of carbon steel in tension and compression.

If the strain goes beyond yield at every fibre of a critical section in a loaded element made of structural steel, the section becomes fully plastic. Its internal resistance at that critical section becomes constant at the level of full plasticity. The strain can increase beyond this fully yielded strain to a maximum ultimate strain based on the available ductility. Externally applied loads on the element will work against this constant internal resistance to produce permanent deformation in the structure. Plastic strain is defined as the residual strain after all loads have been removed.

2.9.2 Plastic behaviour of a section of a beam in pure bending

Consider a beam with symmetrical section in pure bending. The tensile and compressive extreme fibre stresses in the critical section of the beam increase with the increase in loading. When the yield stress f_y is reached in the extreme fibres, the stress does not go beyond yield but remains constant. The strains in the extreme fibres can go beyond yield strain. With the increase in load beyond the first onset of yield in the extreme fibres, progressively from the outer fibres more and more fibres closer to the neutral axis go beyond yield strain. At some point in the incremental loading a resistance moment is reached in the section and all of the material in the section goes beyond yield strain. A plastic hinge forms in the section and an ultimate maximum plastic moment of resistance is offered by the section at that point.

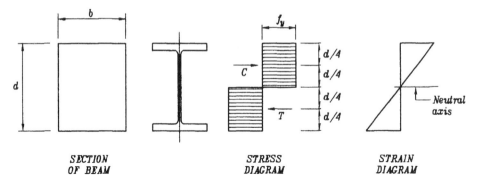

SECTION OF BEAM   STRESS DIAGRAM   STRAIN DIAGRAM

SK 2/47 Distribution of stress and strain in fully plastic sections.

Consider a rectangular section of width b and depth d. Assume that all the tensile and compressive fibres in the section have gone beyond yield strain. The stress blocks will be two rectangles on either side of the neutral axis as shown in SK 2/47. The tensile force T and the compressive force C of the stress blocks are given by:

$$T = C = \frac{1}{2} db f_y$$

b = width of rectangular section
d = depth of rectangular section
T = tensile internal force in section
C = compressive internal force in section
f_y = yield stress of material

The lever arm between T and C is $d/2$.

$$\therefore \quad M_P = T\frac{d}{2} = \left(\frac{bd^2}{4}\right) f_y = S f_y$$

M_P = plastic moment of resistance of section
S = plastic modulus of section

The ratio of the plastic moment of resistance to the elastic moment of resistance is called the shape factor. The shape factor of two flanges only is 1, and the shape factor of a universal beam is about 1.15. The design strength p_y should be used in the determination of M_P where the thickness exceeds 16 mm. The design strength is obtained from Table 6 of BS 5950: Part 1.

2.9.3 Plastic behaviour of a section of a beam with bending and axial load

Axial compressive load on a section along with bending has the effect of shifting the neutral axis towards the tension zone. The tensile stress block becomes smaller, signifying a reduction in the plastic moment capacity of the section.

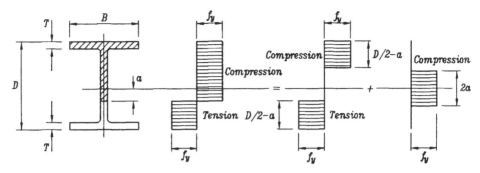

SK 2/48 Stress blocks in a section with bending and axial load.

The effect of compressive load is shown in SK 2/48. Assume that the tensile stress block is shifted by a distance a from the neutral axis and A is the total area of the section.

Axial load resistance $= 2atf_y$

Axial capacity of section $= Af_y$

$$n = \frac{\text{Load}}{\text{Capacity}} = \frac{2at}{A}$$

The revised moment capacity of the section is the full section pure bending plastic moment of resistance M_P less the moment capacity lost due to the axial load stress block at the centre of the section. The revised moment capacity is given by:

$$M'_P = M_P - taf_y a = M_P - t\frac{n^2 A^2}{4t^2} f_y = \left(S - \frac{n^2 A^2}{4t}\right) f_y$$

The ratio n is obtained from the applied compressive load. The design strength p_y should be used in place of f_y where the thickness exceeds 16 mm.

2.9.4 Plastic behaviour of a section of a beam with bending and shear

Longitudinal fibre stresses due to bending moment when combined with shear stresses will give rise to a yield criterion which will take place before the longitudinal bending stresses reach yield. This is the Von Mises yield criterion and is given by:

$$f_y^2 = f_m^2 + 3f_v^2$$

where $f_m =$ longitudinal fibre stress due to bending moment
$f_v =$ shear stress due to applied shear force V

In an I-section one could assume that the flanges are allowed to go up to yield stress f_y, but the web carries all the shear force and hence the bending stress in the web is limited to f_m. If the plastic moment of resistance of the web only is M_{pw}, then the revised plastic moment of resistance of the section with shear is M' given by:

$$M' = M_P - \left(\frac{f_y - f_m}{f_y}\right) M_{pw}$$

where $f_v = V/t_w d_w$ and $f_m = \sqrt{(f_y^2 - 3f_v^2)}$
M_{pw} = plastic moment of resistance of web only
t_w = thickness of web
d_w = depth of web

The reduced value of longitudinal fibre stress should be used to find a further reduction of plastic moment capacity if an axial compressive load is also present. Assuming that the axial load carrying capacity of the web only is not exceeded, the following relationship of axial load N can be written:

$$2a = \frac{N}{f_m t_w}$$

where a is the distance on either side of the central axis of the section over which the axial load is resisted. As in the case of bending and axial load, the plastic moment of resistance is further reduced to:

$$M' = M_P - \left(\frac{f_y - f_m}{f_y}\right) M_{pw} - t_w a^2 f_m$$

The design strength p_y should be used in place of f_y where the thickness exceeds 16 mm.

2.9.5 Load combinations in plastic analysis

The method of superposition cannot be used for structures where yielding has taken place because structural stiffness changes with yielding. Effects of different loads cannot be added together after any part of the structure has strained beyond the yield limit. The behaviour of the structure beyond this point becomes non-linear. Non-linear analysis requires the application of loads in increments till failure is reached by turning the structure into a mechanism.

2.9.6 Plastic analysis of structures

Plastic analysis of indeterminate structures should be carried out using the incremental load method. This is possible only when the ultimate plastic moment resistance is known for all parts of the structure. Before commencement of the analysis section properties must be available. The difficulty arises in the determination of ultimate moments which cause plastic hinges to appear in the critical parts of the structure. The ultimate moments are dependent on several factors, such as coacting shear and direct load and elastic instability.

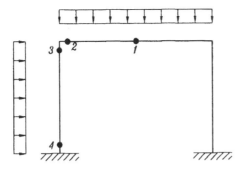

SK 2/49 Plastic analysis of a portal frame.

Consider a simple single span portal frame as shown in SK 2/49. The frame is subjected to a uniformly distributed load as shown. The plastic hinge at the centre of the beam (1) is governed by the plastic moment capacity of the section. The plastic hinge at the beam/column interface on the beam (2) is governed by the reduced plastic moment capacity of the section due to the presence of shear. The hinge on the column at the beam/column junction is governed by a reduced plastic moment capacity due to the combined bending moment, shear and direct load. The same is true for the hinge at 4. In a single span fixed-base portal frame the degree of redundancy is 3, and it requires four plastic hinges to form before the structure reduces to a mechanism. The analysis becomes iterative and time consuming because the value of the ultimate plastic moment changes with shear and direct load. Further complication arises if biaxial bending with torsion is present. The elastic analysis with factored loads does not take into account the redistribution of loads due to rotation of plastic hinges, and as such may be regarded as conservative. But non-linear plastic analysis with load-dependent plastic hinges can be very complicated and is not normally carried out.

Total factored and combined basic loads for an ultimate load scenario should be determined before the commencement of the plastic analysis. This combined loading should then be applied to the indeterminate structure fractionally and incrementally. At any stage of the incremental loading, if a plastic hinge appears in the structure, the structural configuration should be changed to incorporate the hinge and the additional incremental loading applied on the new structure with the hinge. At all remaining elastic parts of the structure the stresses due to incremental loading on the new structure should be additive. This process should continue till all the loads are used up or the structure turns into a mechanism due to the formation of several hinges.

The method of incremental loading is illustrated by the simple problem of a fixed-end beam with uniformly distributed load.

The fixed-end beam of span length l is subjected to a uniformly distributed load. The plastic moment of resistance of the beam section

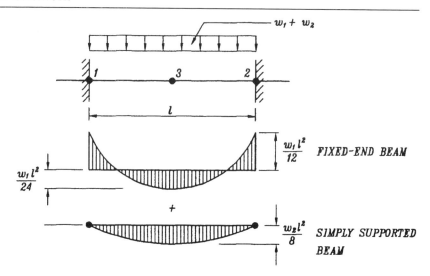

SK 2/50 Plastic analysis of a fixed-end beam.

is M_P. With incremental loading, when the loading becomes equal to w_1 per unit length then the plastic hinges at 1 and 2 form.

$$\therefore \quad M_P = \frac{w_1 l^2}{12}$$

The beam now behaves as simply supported till the hinge at 3 is formed due to further incremental loading. Then the system, with three hinges, turns into a mechanism. The incremental loading required to form the third hinge is w_2.

$$\therefore \quad \frac{w_1 l^2}{24} + \frac{w_2 l^2}{8} = M_P = \frac{w_1 l^2}{12}$$

$$\therefore \quad w_2 = \frac{w_1}{3}$$

w is the total ultimate uniformly distributed load on the span given by:

$$w = w_1 + w_2 = \frac{4}{3} w_1$$

$$w_1 = \frac{3}{4} w$$

$$\therefore \quad M_P = \frac{w l^2}{16}$$

Note: This analysis ignores the effect of shear on the plastic moment of resistance of the beam section.

Chapter 3
Analysis of Structures: Worked Examples

3.1 EXAMPLE 3.1: ROOF TRUSS

Design a truss to be used in a pitched roof for an industrial building. The span of the truss is 15 metres and the height of the building at the eaves is 7.5 metres.

3.1.1 Type of truss

Step 1 Select appropriate truss type
The table below is useful for the selection of the type of truss suitable for the intended span.

Table 3.1 Types of truss.

Type of truss	Sketch of truss	Span range	Span-to-depth ratio	Spacing of truss
Pratt		6 m to 12 m	4 to 5	3 m to 4 m
Howe		6 m to 12 m	4 to 5	3 m to 4 m
Fink		6 m to 15 m	5 to 7	3 m to 4 m
Mansard		15 m to 30 m	7 to 8	4 m to 6 m
Pratt		30 m to 50 m	15 to 25	6 m to 10 m
Warren		30 m to 50 m	15 to 25	6 m to 10 m

For a 15 metre span select a Fink truss with a height-to-apex of 3 metres. Assume a truss spacing equal to 4 metres.

Step 2 Determine the geometry of the truss
Length of rafter $= \sqrt{(7.5^2 + 3^2)} = 8.078$ m
Distance between nodes on rafter $= 8.078 \div 4 = 2.019$ m
Angle of inclination of rafter to horizontal $= \tan^{-1} \dfrac{3}{7.5} = 21.80°$

86 Structural Steelwork

3.1.2 Loading on the truss

Step 3 *Determine wind loading as per BS 6399: Part 2*
Basic geometry of the building:

H = building height = 10.5 m
L = building plan length = 30.0 m
B = building plan width = 15.0 m

Stage 1: C_r = dynamic augmentation factor
Type of building = portal shed
From Table 1 of BS 6399: Part 2 find $K_b = 2$
From Figure 3 of BS 6399: Part 2 for $K_b = 2$ and $H = 10.5$ find $C_r = 0.04$

Stage 2: Check $C_r < 0.25$ and $H < 300$ m (OK)

Stage 3: V_b = basic wind speed = 21 m/s from Figure 6 of BS 6399: Part 2 for the site.

Stage 4: V_s = site wind speed
Topography is not significant.

Δ_s = altitude of site above mean sea level = 100 m
S_a = altitude factor = $1 + 0.001\Delta_s = 1.1$
S_d = direction factor = 1 taken for all directions
S_s = seasonal factor = 1 for permanent works
S_p = probability factor = 1 for standard value of risk $Q = 0.02$
$V_s = V_b \times S_a \times S_d \times S_s \times S_p = 21 \times 1.1 = 23.1$ m/s

Stage 5: Determine effective height of building

$H = H_r$ = reference height of building = 10.5 m
H_0 = average level of height of roof tops around this building = 10 m
X = upwind spacing of this building from existing obstruction = 10 m
$X \leq 2H_0$

$\therefore H_e$ = effective height = $H_r - 0.8H_0 = 2.5$ m or $H_e = 0.4H_r = 4.2$ m (whichever is greater). $\therefore H_e = 4.2$ m

Stage 6: Choice of method: Use standard orthogonal wind method.

Stage 7: V_e = standard effective wind speed
Site is 10 km from sea.
Find from Table 4 of BS 6399: Part 2 $S_b = 1.51$ for $H_e = 4.2$ m.

$V_e = V_s \times S_b = 23.1 \times 1.51 = 34.88$ m/s

Stage 8: q_s = dynamic pressure

$q_s = 0.613; V_e^2 = 745$ N/m^2

Stage 9: C_p = standard pressure coefficient
For Duopitch roof with pitch angle 21.8° and the direction of wind angle $\theta = 0$, the external pressure coefficients in different zones of the roof are obtained from Table 10 of BS 6399: Part 2.

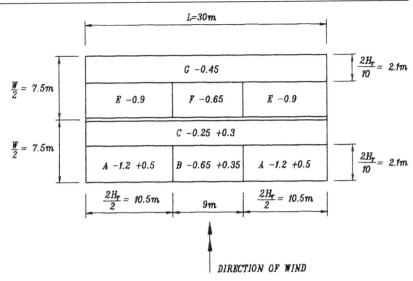

SK 3/1 Plan of roof showing pressure coefficients as per Figure 20 of BS 6399: Part 2.

Internal pressure coefficient is $C_{pi} = +0.2$ or -0.3, as per Table 16 of BS 6399: Part 2.

Stage 10: Determine wind loads
A roof truss in the zone ACEG will be most critically loaded by wind at $\theta = 0°$ according to the loading intensity plan in SK 3/1.

$$p = p_e + p_i = q_s \times C_{pe} \times C_{a(ext)} + q_s \times C_{pi} \times C_{a(int)}$$

a = diagonal dimension

For external pressure as per Figure 5 of BS 6399: Part 2 a is given by:

$$a = [(\text{rafter length})^2 + (\text{building length})^2]^{\frac{1}{2}}$$

$$= \sqrt{8.078^2 + 30^2} = 31.1\,\text{m}$$

For internal pressure as per Clause 2.6 of BS 6399: Part 2:

$$a = 10\,(\text{internal volume of building})^{\frac{1}{3}}$$

$$= 10\sqrt[3]{15 \times 30 \times 7.5 + 15 \times 30 \times 3 \times 0.5}$$

$$= 159.4\,\text{m}$$

C_a = size effect factor

Site is in town and 10 km from sea.
$H_e = 4.2$ m found in Stage 5.
$\therefore$ Graph C should be used in Figure 4 of BS 6399: Part 2. From this:

$C_{a(ext)} = 0.84$ for $a = 31.1$ m

$C_{a(int)} = 0.70$ for $a = 159.4$ m

Net surface pressures
Wind load case 1:

Zone A: $p = -q_s[C_{a(ext)}(1.2) + C_{a(int)}(0.2)]$

$= -745(0.84 \times 1.2 + 0.7 \times 0.2) = -855 \, \text{N/m}^2$

Zone C: $p = -q_s[C_{a(ext)}(0.25) + C_{a(int)}(0.2)]$

$= -745(0.84 \times 0.25 + 0.7 \times 0.2) = -26 \, \text{N/m}^2$

Zone E: $p = -q_s[C_{a(ext)}(0.9) + C_{a(int)}(0.2)]$

$= -745(0.84 \times 0.9 + 0.7 \times 0.2) = -668 \, \text{N/m}^2$

Zone G: $p = -q_s[C_{a(ext)}(0.45) + C_{a(int)}(0.2)]$

$= -745(0.84 \times 0.45 + 0.7 \times 0.2) = -386 \, \text{N/m}^2$

Wind load case 2:

Zone A: $p = +q_s[C_{a(ext)}(0.5) + C_{a(int)}(0.3)]$

$= +745(0.84 \times 0.5 + 0.7 \times 0.3) = +469 \, \text{N/m}^2$

Zone C: $p = +q_s[C_{a(ext)}(0.3) + C_{a(int)}(0.3)]$

$= +745(0.84 \times 0.3 + 0.7 \times 0.3) = +344 \, \text{N/m}^2$

Zone E: $p = +q_s[C_{a(ext)}(-0.9) + C_{a(int)}(0.3)]$

$= +745(-0.84 \times 0.9 + 0.7 \times 0.3) = -407 \, \text{N/m}^2$

Zone G: $p = +q_s[C_{a(ext)}(-0.45) + C_{a(int)}(0.3)]$

$= +745(-0.84 \times 0.45 + 0.7 \times 0.3) = -125 \, \text{N/m}^2$

Determination of nodal wind loads
Spacing of truss $= 4 \, \text{m}$
Spacing of nodes on rafter $= 2.02 \, \text{m}$
Wind load at node $2 = 4 \times 2.02 \times \dfrac{(855 + 261)}{2} = 4509 \, \text{N}$

All other nodal loads are calculated in the same way, as shown in SK 3/2.

Dead Load
Assume own weight $= 0.2 \, \text{kN/m}^2$
Sheeting and purlins $= 0.15 \, \text{kN/m}^2$
Insulation $= 0.025 \, \text{kN/m}^2$
Fixings and fittings $= 0.025 \, \text{kN/m}^2$
Services $= 0.100 \, \text{kN/m}^2$

Total dead load $= 0.500 \, \text{kN/m}^2$
Nodal dead load $= 0.5 \times 4 \, \text{m} \times 15 \, \text{m} \div 8 \, \text{nodes} = 3.75 \, \text{kN}$

Analysis of Structures 89

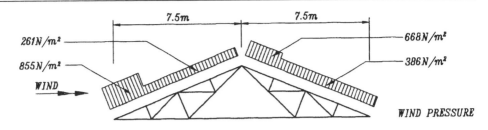

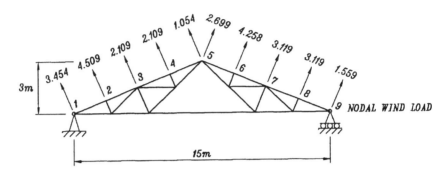

SK 3/2 Nodal wind loads due to wind load case 1.

Imposed load
No access has to be provided to the roof. As per Clause 4.3 of BS 6399: Part 3 the loading should be:

(1) the uniformly distributed snow load or
(2) a uniformly distributed load of $0.6\,\text{kN/m}^2$ on plan area.

Uniformly distributed snow load:

s_b = basic snow load = $0.5\,\text{kN/m}^2$ from Figure 1 of BS 6399: Part 3

Altitude of site is 100 m above mean sea level.

∴ s_0 = site snow load = s_b = $0.5\,\text{kN/m}^2$ as per BS 6399: Part 3
μ_1 = snow load shape coefficient = 0.8 from Figure 3 of BS 6399: Part 3
s_d = snow load on roof = $\mu_1 s_b$ = $0.4\,\text{kN/m}^2$

Design for a uniformly distributed imposed load of $0.6\,\text{kN/m}^2$.

Nodal imposed load = $\dfrac{0.6 \times 15\,\text{m} \times 4\,\text{m}}{8}$ = $4.5\,\text{kN}$

Step 4 Analysis of basic load cases
Basic load case 1 = dead load
Basic load case 2 = imposed load
Basic load case 3 = wind load case 1
Basic load case 4 = wind load case 2

3.1.3 Analysis by method of section

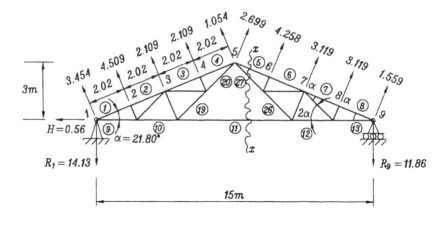

① = MEMBER NUMBER 1
1 = JOINT NUMBER 1

SK 3/3 Analysis of truss by method of section.

Consider the basic wind load case 1. At the support at joint 9, as shown in SK 3/3, the truss is restrained vertically but is allowed to slide horizontally. Therefore the horizontal reaction at joint 9 is zero.

Taking moments about joint 1 of all the external forces on the truss we equate to zero the following expression for equilibrium:

$(R_9 \times 15) - (4.509 \times 2.02) - [2.109 \times (4.04 + 6.06)] - (1.054 \times 8.08)$

$\quad - (1.559 \times 15 \cos \alpha) - [3.119 \times (15 \cos \alpha - 2.02)]$

$\quad - [3.119 \times (15 \cos \alpha - 4.04)] - [4.258 \times (15 \cos \alpha - 6.06)]$

$\quad - [2.699 \times (15 \cos \alpha - 8.08)] = 0$

This gives $R_9 = 11.86\,\text{kN}$

Net vertical component of wind load =

$\cos \alpha \times (3.454 + 4.509 + 2.109 + 2.109 + 1.054 + 2.699 + 4.258$

$\quad + 3.119 + 3.119 + 1.559) = 25.99\,\text{kN}$

$\therefore \quad R_1 = 25.99 - 11.86 = 14.13\,\text{kN}$

Net horizontal component of wind load =

$\sin \alpha \times (-3.454 - 4.509 - 2.109 - 2.109 - 1.054 + 2.699 + 4.258$

$\quad + 3.119 + 3.119 + 1.559) = +0.56\,\text{kN}$

$\therefore \quad$ Horizontal reaction at joint $1 = H_1 = -0.56\,\text{kN}$

The member numbers and the joint numbers are shown in SK 3/3. The internal force in any member is designated as S_i, where i is the member number of interest. When the internal force in a member acts towards a joint, the member is in compression; when it acts away from the joint, the member is in tension.

Consider the equilibrium of joint 9. Resolving and equating to zero all the internal and external forces about the vertical and the horizontal axes for equilibrium we get:

$$-S_8 \cos \alpha + S_{13} + 1.559 \sin \alpha = 0$$

$$S_8 \sin \alpha + 1.559 \cos \alpha - 11.86 = 0$$

From these two equations we get:

$$S_8 = 28.04 \, \text{kN (tensile)} \text{ and } S_{13} = 25.45 \, \text{kN (compressive)}$$

Similarly, considering the equilibrium of joint 8 we can conclude:

$$S_{21} = 3.12 \, \text{kN (tensile)} \text{ and } S_7 = 28.04 \, \text{kN (tensile)}$$

Consider the equilibrium of joint 15. Resolving vertically and horizontally we get:

$$-S_{22} \sin 2\alpha + S_{21} \cos \alpha = 0$$

$$\therefore \quad S_{22} = \frac{3.12 \cos \alpha}{\sin 2\alpha} = 4.2 \, \text{kN (compressive)}$$

$$S_{12} - S_{13} + S_{21} \sin \alpha + S_{22} \cos 2\alpha = 0$$

$$\therefore \quad S_{12} = 25.45 - 3.12 \sin \alpha - 4.2 \cos 2\alpha = 21.25 \, \text{kN (compressive)}$$

We cannot progress beyond this joint using the method of joints. Take a section x–x as shown in SK 3/3. The right-hand side of section x–x must be in equilibrium with the applied forces at joints 6, 7, 8, and 9 and the member internal forces of members 5, 27 and 11 designated by S_5, S_{27} and S_{11}.

Taking moments about joint 5 of all forces on the right of section x–x and equating to zero for equilibrium:

$$(-S_{11} \times 3) - (4.258 \times 2.02) - (3.119 \times 4.04) - (3.119 \times 6.06)$$

$$- (1.559 \times 8.08) + (11.86 \times 7.5) = 0$$

$$\therefore \quad S_{11} = 12.09 \, \text{kN (compressive)}$$

The rest of the joints can now be solved for by the method of equilibrium of joints.

3.1.4 Method of section using Maxwell diagram

Consider another basic load case, namely imposed load. The polygon of forces, funicular polygon and Maxwell diagram are shown in SK 3/4. We will now use a step-by-step method of drawing the Maxwell diagram, the force polygon and the funicular polygon by the method of section.

92 Structural Steelwork

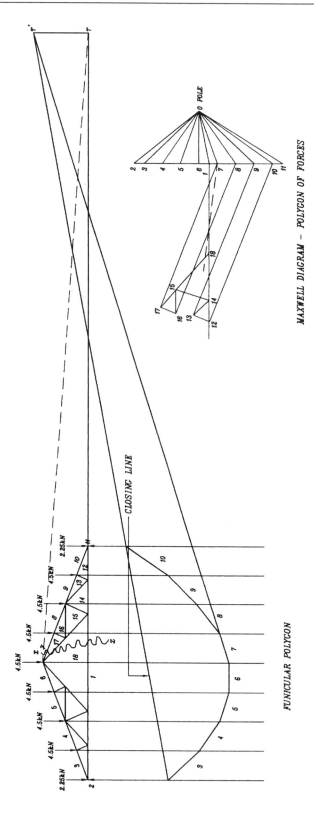

SK 3/4 Maxwell diagram using method of section.

Step 1 Draw the truss to scale. Choose any scale large enough to get an adequate directional parameter for all members.

Step 2 Draw on the nodes of the truss the known directions of all applied forces, including the reactions. If the direction of any reaction is not known to start with, it will become known at the end of the analysis. The direction of the reaction at the support with rollers is to be taken as vertical.

Step 3 Number or annotate alphabetically all the spaces between the loads, including the reactions, as shown in SK 3/4. The loads are designated vectorially by adjacent spaces. For example, the load on the apex of the truss shown in SK 3/4 is vertical and can be represented by a vertical vector $\overline{6-7}$ in the force polygon corresponding to the numbers of the two adjacent spaces on either side of the load.

Step 4 Number or annotate alphabetically each triangular space inside the truss as shown in SK 3/4. The internal forces in the members of the truss are vectorially represented by the two adjacent spaces on either side of the member. For example, the internal force in the central tie member of the truss in SK 3/4 is represented by a vector $\overline{1-18}$ in the Maxwell diagram.

Step 5 Draw the polygon of forces. Choose any scale large enough for the forces to be properly represented in magnitude and direction, and for unknown forces to be accurately scaled off from the Maxwell diagram. Start the polygon of forces with vector $\overline{2-3}$ in a direction vertically downwards from point 2 to point 3 representing the first applied load on the truss from the left-hand end. The distance from 2 to 3 in the force polygon is a representation of the magnitude of the force, using a convenient scale of forces (e.g. $x\,kN = 1\,mm$).

Step 6 Choose a pole 0 anywhere close to the polygon of forces and draw the rays from the pole to the ends of the vectorial representation of individual applied forces as shown in SK 3/4.

Step 7 Draw the funicular polygon under the truss by extending the lines of action of the applied forces. Start at the support which is restrained horizontally and vertically. Draw lines between the lines of action of forces under the truss in a systematic order parallel to rays in the polygon of forces where rays must correspond to spaces between the loads. For example, the first line of the funicular polygon under the truss in SK 3/4 is parallel to ray 0–3 of the force polygon because it corresponds to the space 3 between the loads in the diagram of the truss above. The next line parallel to ray 0–4 starts from the point where the first line parallel to ray 0–3 meets the line of action of the load $\overline{3-4}$. Following this principle complete the funicular polygon up to the last space between load direction lines. Close the funicular polygon by joining the last two points on the load lines.

Step 8 In the polygon of forces draw a line vertically from the last point after vectorial representation of all applied forces, to represent the direction of the reaction at the support with rollers. Draw a line from the pole 0 parallel to the closing line of the funicular polygon. These two lines meet at a point which gives the magnitude and direction of the vertical reaction at the support with rollers. The reaction at the other support is represented in magnitude and direction by the joining line between this point and the starting point of the polygon of forces.

In SK 3/4, the ray 0–1 in the force polygon is parallel to the closing line of the funicular polygon under the truss. The reaction at the roller support is represented in magnitude and direction by the vector $\overline{11-1}$ in the polygon of forces. The vector $\overline{1-2}$ represents the other reaction at the fully restrained end of the truss.

Step 9 Take a section x–x through the truss as shown in SK 3/4. Remove the right-hand portion of truss beyond section x–x. Assume that a fictitious member replaces the two members 7–17 and 17–18 of the truss. This fictitious member must act at the apex of the remaining left-hand portion of the truss for equilibrium. The left-hand portion of the truss is maintained in equilibrium by the load from the fictitious member, the load in tie member 1–18 and the reaction force 11–1, for which the direction and magnitude are known but the point of action is unknown. These three forces must meet at a point for equilibrium, which means that the line of action of the vertical reaction will be shifted from the position of the right-hand support. The magnitude and direction of the reaction will remain unchanged because the loading on the truss has not changed. Therefore, ray 0–1 in the polygon of forces will remain unchanged and hence the direction of the closing line in the funicular polygon under the truss must remain unchanged. Since there are no other forces in the system beyond space 7 in the truss diagram, the ray parallel to 0–7 must be projected to meet the original closing line of the funicular polygon to meet at a point r'. This point must lie on the line of action of the shifted vertical reaction of the right-hand support. Draw a vertical line from this point r' to meet the line of action of member 1–18 at r. For equilibrium, the fictitious member of the truss must have its line of action passing through this point. The line from r to the apex of the truss gives the direction of the fictitious member.

Step 10 Draw the Maxwell diagram by drawing a line parallel to the fictitious member from point 7 on the polygon of forces and another line parallel to member 1–18, starting from point 1. These two lines meet at point $\overline{18}$, defining vectorially the internal force in member 1–18 by vector $\overline{1-18}$. Similarly draw a line parallel to member 1–12 from point 1 and another line parallel to member 10–12 from point 10. These two lines meet at 12, giving vectorially the internal force in member 1–12. Solve the whole truss graphically by solving one space at a time.

Table 3.2 Table of internal forces in the truss.

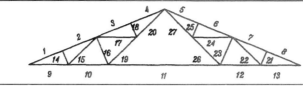

Member designation	Dead load (kN)	Imposed load (kN)	Wind load case 1 (kN)	Wind load case 2 (kN)
1	35.34[c]	42.41[c]	29.40[t]	11.52[c]
2	33.95[c]	40.74[c]	29.40[t]	11.52[c]
3	32.44[c]	39.07[c]	26.40[t]	10.89[c]
4	31.16[c]	37.39[c]	26.40[t]	10.89[c]
5	31.16[c]	37.39[c]	29.46[t]	1.89[t]
6	32.55[c]	39.07[c]	29.46[t]	1.89[t]
7	33.95[c]	40.74[c]	28.04[t]	0.47[t]
8	35.34[c]	42.41[c]	28.04[t]	0.47[t]
9	32.81[t]	39.38[t]	25.45[c]	16.85[t]
10	28.13[t]	33.75[t]	19.38[c]	12.42[t]
11	18.75[t]	22.50[t]	12.09[c]	4.60[t]
12	28.13[t]	33.75[t]	21.25[c]	1.11[t]
13	32.81[t]	39.38[t]	25.45[c]	0.25[c]
14	3.48[c]	4.18[c]	4.51[t]	3.29[c]
15	4.69[t]	5.63[t]	6.07[c]	4.42[t]
16	6.96[c]	8.36[c]	5.42[t]	5.81[c]
17	4.69[t]	5.63[t]	2.84[c]	3.74[t]
18	3.48[c]	4.18[c]	2.11[t]	2.78[c]
19	9.38[t]	11.25[t]	7.29[c]	7.82[t]
20	14.06[t]	16.88[t]	10.13[c]	11.57[t]
21	3.48[c]	4.18[c]	3.12[t]	1.01[t]
22	4.69[t]	5.63[t]	4.20[c]	1.36[c]
23	6.96[c]	8.36[c]	6.81[t]	2.59[t]
24	4.69[t]	5.63[t]	5.73[c]	2.89[c]
25	3.48[c]	4.18[c]	4.26[t]	2.15[t]
26	9.38[t]	11.25[t]	9.16[c]	3.49[c]
27	14.06[t]	16.88[t]	14.90[c]	6.38[c]

[c] = compression; [t] = tension in the member.

3.2 EXAMPLE 3.2: CONTINUOUS BEAM

Use of the building is as residential units. Residential occupancy class is Type 2, as per Table 5 of BS 6399: Part 1.

Step 1 Determine structural concept

There are two basic structural solutions possible for the roof of the building shown in SK 3/5. First, the columns may be capped off at the underside of the main beams and the main beams may be made continuous over the columns. This will mean several splices in the beam capable of moment and shear transfer. From the point of view of ease of fabrication, splicing and buildability it will also be

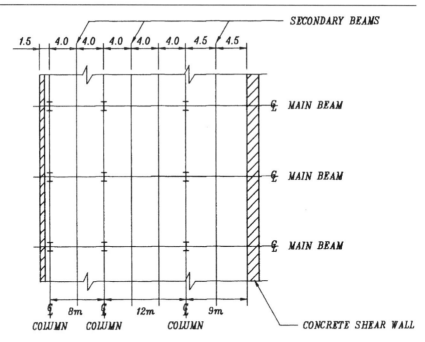

SK 3/5 Roof plan of a high-rise building.

necessary to keep the beam size unchanged over the whole length of the continuous beam. In this structural arrangement it may not be possible to take full advantage of the possible saving of material weight due to continuity of the beam. Second, the columns may be allowed to go up to the underside of the reinforced concrete roof slab, and the beams may be connected to the columns as simple supports. The sizes of the beams may follow the actual strength requirement depending on span and loading. Although this arrangement may be more economical and buildable, the continuous beam arrangement is adopted here to show the analytical procedure.

The main continuous beam is assumed built-in at the reinforced concrete shear wall, and the beam is simply supported on the columns. The columns carry vertical loads only and the lateral stability of the building is dependent on a system of shear walls. The vertical deformation of the columns is ignored. The secondary beams are simply supported on the main continuous beams. The reinforced concrete (RC) slab is supported by the secondary beams.

Step 2 **Determine loading**

Dead load (*DL*)

	200 mm thick RC slab	4.80 kN/m^2
	Average screed to falls (50 mm)	1.20 kN/m^2
	Insulation and waterproofing	0.02 kN/m^2
	Paving slabs	1.00 kN/m^2
	Services and false ceiling	0.18 kN/m^2
	Total	7.20 kN/m^2

Imposed load (*IL*)   Minimum on roof with access   1.50 kN/m^2

Concentrated dead load from secondary beams at 4.0 m centres:

= Self-weight + 4 m × 8 m × 7.2 kN/m² = 250 kN

Concentrated imposed load from secondary beams at 4.0 m centres:

= 4.0 m × 8.0 m × 1.5 kN/m² = 48 kN

Ultimate concentrated load = 1.4DL + 1.6IL = 430 kN
Similarly, for a secondary beam spaced at 4.5 m centres:

Ultimate load = 480 kN

Height of blockwork parapet is assumed to be 1.2 m
Weight of parapet, rendering and fascia panels = 2 kN/m
Dead load from the secondary beam at the tip of the cantilever:

= 0.75 m × 8.0 m × 7.2 kN/m² + 8.0 m × 2 kN/m = 60 kN including self-weight

Imposed load from secondary beam at the tip of the cantilever:

= 0.75 m × 8.0 m × 1.5 kN/m² = 9 kN

Ultimate load from secondary beam at the tip of the cantilever:

= 1.4 × 60 + 1.6 × 9 = 100 kN

Step 3 **Choose size of beam**
Maximum free span bending moment in the central 12.0 m span section of the continuous beam:

= 4.0 m × 430 kN = 1720 kNm (ultimate)

Approximate plastic modulus required for Grade 43 mild steel:

$$= \frac{1720 \times 10^6}{275} = 6.25 \times 10^6 \text{ mm}^3$$

The maximum bending moment in a continuous beam will be about 70% of this maximum free span bending moment.
Choose UB 610 × 305 × 149 kg/m with plastic modulus = S = 4.572 × 10⁶ mm³.

Step 4 **Draw the beam line diagram with loading**

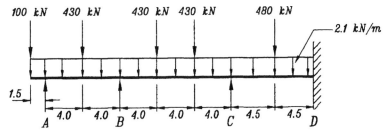

SK 3/6 Loading diagram on the continuous beam.

3.2.1 Analysis of a continuous beam by the three moment theorem

Step 5 *Determine free span (simply supported) bending moments*

Sign convention: +ve moments are sagging and −ve moments are hogging

Span AB:
Applied moment at $A = -[(100 \times 1.5) + (2.1 \times 1.5^2 \div 2)] = -152\,\text{kNm}$
Midspan moment due to concentrated load in span $= 430 \times 8 \div 4 = 860\,\text{kNm}$
Midspan moment due to distributed load in span $= 2.1 \times 8^2 \div 8 = 16.8\,\text{kNm}$

Span BC:
Midspan moment due to concentrated load in span $= 1720\,\text{kNm}$
Midspan moment due to distributed load in span $= 37.8\,\text{kNm}$

Span CD:
Midspan moment due to concentrated load in span $= 1080\,\text{kNm}$
Midspan moment due to distributed load in span $= 21.3\,\text{kNm}$

Step 6 *Find areas of free bending moment diagrams and centres of gravity*

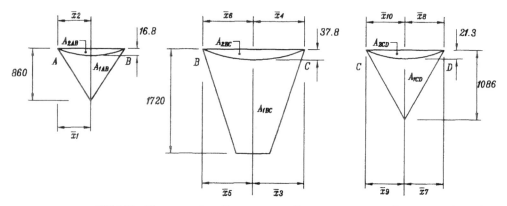

SK 3/7 **Free span bending moment diagrams: areas and centres of gravity.**

Areas of free span bending moment diagrams:

$A_{1AB} = 0.5 \times 860 \times 8 = 3440$ $\bar{x}_1 = 4.0\,\text{m}$

$A_{2AB} = (2 \div 3) \times 16.8 \times 8 = 89.6$ $\bar{x}_2 = 4.0\,\text{m}$

$A_{1BC} = 0.5 \times (12 + 4) \times 1720 = 13\,760$ $\bar{x}_3 = 6.0\,\text{m}$ and $\bar{x}_5 = 6.0\,\text{m}$

$A_{2BC} = (2 \div 3) \times 37.8 \times 12 = 302.4$ $\bar{x}_4 = 6.0\,\text{m}$ and $\bar{x}_6 = 6.0\,\text{m}$

$A_{1CD} = 0.5 \times 1080 \times 9 = 4860$ $\bar{x}_7 = 4.5\,\text{m}$ and $\bar{x}_9 = 4.5\,\text{m}$

$A_{2CD} = (2 \div 3) \times 21.3 \times 9 = 127.8$ $\bar{x}_8 = 4.5\,\text{m}$ and $\bar{x}_{10} = 4.5\,\text{m}$

Analysis of Structures

Step 7 *Establish the three moment equations and solve*

Consider spans AB and BC:

$$M_A \frac{l_{AB}}{I_{AB}} + 2M_B \left(\frac{l_{AB}}{I_{AB}} + \frac{l_{BC}}{I_{BC}}\right) + M_C \frac{l_{BC}}{I_{BC}}$$

$$= -6\left(\frac{A_{1AB}\bar{x}_1}{I_{AB}l_{AB}} + \frac{A_{2AB}\bar{x}_2}{I_{AB}l_{AB}} + \frac{A_{1BC}\bar{x}_3}{I_{BC}l_{BC}} + \frac{A_{2BC}\bar{x}_4}{I_{BC}l_{BC}}\right)$$

Moment of inertia I is constant for the beam.

$$\therefore \quad 8M_A + 40M_B + 12M_C$$

$$= -6\left(\frac{3440 \times 4}{8} + \frac{89.6 \times 4}{8} + \frac{13\,760 \times 6}{12} + \frac{302.4 \times 6}{12}\right)$$

$$= -52\,776$$

$$M_A = -152 \text{ kNm}$$

$$\therefore \quad M_B + 0.3M_C = -1289 \tag{3.1}$$

Consider spans BC and CD:

$$M_B l_{BC} + 2M_C(l_{BC} + l_{CD}) + M_D l_{CD}$$

$$= -6\left(\frac{A_{1BC}\bar{x}_5}{l_{BC}} + \frac{A_{2BC}\bar{x}_6}{l_{BC}} + \frac{A_{1CD}\bar{x}_7}{l_{CD}} + \frac{A_{2CD}\bar{x}_8}{l_{CD}}\right)$$

$$12M_B + 42M_C + 9M_D = -57\,150.6$$

$$M_B + 3.5M_C + 0.75M_D = -4762.6 \tag{3.2}$$

The rotation of the beam at support D is zero.

$$\theta_{DC} = \frac{M_{DC}l_{CD}}{3EI_{CD}} - \frac{M_{CD}l_{CD}}{6EI_{CD}} - \frac{A_{1CD}\bar{x}_9}{EI_{CD}l_{CD}} - \frac{A_{2CD}\bar{x}_{10}}{EI_{CD}l_{CD}} = 0$$

or $\quad -\dfrac{M_D l_{CD}}{3} - \dfrac{M_C l_{CD}}{6} - \dfrac{A_{1CD}\bar{x}_9}{l_{CD}} - \dfrac{A_{2CD}\bar{x}_{10}}{l_{CD}} = 0$

or $\quad 2M_D + M_C = -\dfrac{6}{l_{CD}^2}(A_{1CD}\bar{x}_9 + A_{2CD}\bar{x}_{10}) = -1662.6 \tag{3.3}$

Solving the equations we get the three unknowns M_B, M_C and M_D as follows:

$$M_B = -986.3 \text{ kNm}$$

$$M_C = -1009 \text{ kNm}$$

$$M_D = -326.8 \text{ kNm}$$

Step 8 **Draw the bending moment diagram of the beam**

SK 3/8 Bending moment diagram of the continuous beam.

For span AB, the bending moment at the centre of span is given by:

$$M_{AB} = 16.8 + 860 - \left(\frac{152 + 986.3}{2}\right) = 307.7 \text{ kNm}$$

Similarly, find span moments for all other spans.

Step 9 **Find shear forces and draw the shear force diagram**
Shear force at left of support $A = 100 + (2.1 \times 1.5) = 103.2 \text{ kN}$
Consider span AB:

Free span reaction at $A = (0.5 \times 430) + (2.1 \times 4) = 223.4 \text{ kN}$

Hyperstatic reaction at $A = -\left(\dfrac{986.3 - 152}{8}\right) = -104.3 \text{ kN}$

Shear force at right of support $A = 223.4 - 104.3 = 119.1 \text{ kN}$

Hyperstatic reaction is due to the difference of moment at the two ends of a member in a structure and this reaction balances the difference of moment by an equal and opposite couple acting at a lever arm equal to the length of the member. Always draw the bending moment diagram on the tension side of the structure. The direction of the hyperstatic reaction at the two ends of a member follows the direction of slope of the line joining the bending moments at the two ends of the member. If this line is sloping downwards in a beam at one end, the hyperstatic reaction at that end of the beam is acting downwards on the beam. Similarly, if this line is sloping upwards, then the hyperstatic reaction is acting upwards at that end of the beam. This is only true if the bending moment diagram is drawn on the tension side.

Shear force at left of support $B = 223.4 + 104.3 = 327.7 \text{ kN}$

Shear force at right of support B

$$= 430 + (2.1 \times 6) - \left(\frac{1009 - 986.3}{12}\right) = 440.7 \, \text{kN}$$

Reaction at support $B = 327.7 + 440.7 = 768.4 \, \text{kN}$

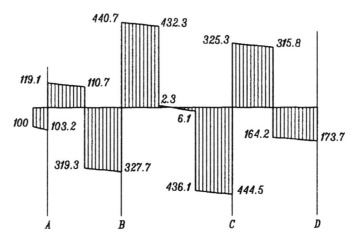

SK 3/9 Shear force diagram.

3.2.2 Analysis of a continuous beam by the method of moment distribution

Consider the same beam as analysed by the three moment theorem, and use the same loading for comparison. Steps 1 to 5 will remain unchanged.

Step 6 *Find fixed-end moments for the spans*

Consider span AB as a propped cantilever because the beam is free to rotate at joint A. The stiffness of span AB is also to be taken as three-quarters of I/l.

$$FEM_{BA} = +\frac{2.1 \times 8^2}{8} + \frac{3 \times 430 \times 8}{16} = 661.8 \, \text{kNm}$$

$$FEM_{BC} = -\frac{2.1 \times 12^2}{12} - \frac{430 \times 4 \times (12-4)}{12} = -1171.9 \, \text{kNm}$$

$$FEM_{CD} = -\frac{2.1 \times 9^2}{12} - \frac{480 \times 9}{8} = -554.2 \, \text{kNm}$$

Step 7 *Find distribution coefficients*

Distribution coefficients at joint B:

$$k_{AB} = \frac{3}{4} \frac{I_{AB}}{l_{AB}} = \frac{3}{32} I; \quad k_{BC} = \frac{I_{BC}}{l_{BC}} = \frac{1}{12} I$$

$$\frac{k_{AB}}{k_{AB} + k_{BC}} = \frac{\frac{3}{32}}{\frac{3}{32} + \frac{1}{12}} = 0.53$$

Distribution coefficient for span AB at joint B is 0.53 and for span BC at joint B is 0.47.

Distribution coefficients at joint C:

$$k_{CB} = \frac{I}{12}; \quad k_{CD} = \frac{I}{9}$$

$$\frac{k_{CB}}{k_{CB} + k_{CD}} = \frac{\frac{1}{12}}{\frac{1}{12} + \frac{1}{9}} = 0.43$$

The distribution coefficient for span BC at joint C is 0.43 and for span CD at joint C is 0.57.

Step 8 *Find joint moments by the moment distribution method*

Table 3.3 Moment distribution by Hardy-Cross method.

	A		B		C		D
Distribution factors			0.53	0.47	0.43	0.57	
Fixed-end moments	−152	0.0	+661.8	−1171.9	+1171.9	−554.2	+554.2
Distribution			+270.4	+239.7	−265.6	−352.1	0
Carry-over			−76	−132.8	+119.9	0	−176.1
Distribution			+110.7	+98.1	−51.6	−68.3	0
Carry-over			0	−25.8	+49.1	0	−34.2
Distribution			+13.7	+12.1	−21.1	−28	0
Carry-over			0	−10.6	+6.1	0	−14
Distribution			+5.6	+5.0	−2.6	−3.5	0
Carry-over			0	−1.3	+2.5	0	−1.8
Distribution			+0.7	+0.6	−1.1	−1.4	0
Totals	−152		+986.9	−986.9	+1007.5	−1007.5	−328.1

The initial fixed-end moments at joint B are +661.8 and −1171.9. The unbalanced moment at the joint is $(-1171.9 + 661.8) = -510.1$ kNm. To balance the joint a total positive moment equal to 510.1 kNm is distributed at joint B in proportion to the relative stiffness determined by the distribution factors. 510.1 kNm multiplied by 0.53 gives +270.4. No moment gets carried over to the free end, and nothing gets carried over from the fixed end. Half of the applied moment at A gets carried over to joint B as −76 kNm. Similarly, half of +239.7 distributed at B gets carried over to C as +119.9. These carried over moments create a moment imbalance at the joint again, requiring further distribution. This process of carry-over and distribution is performed a few times till the distribution moments become small. When all the moments at either side of a joint are added up, it gives the joint moment. Compared to the three moment method, this method is approximate and hence there is a small difference in the final value of the moments obtained.

3.2.3 Analysis of a continuous beam by the matrix method

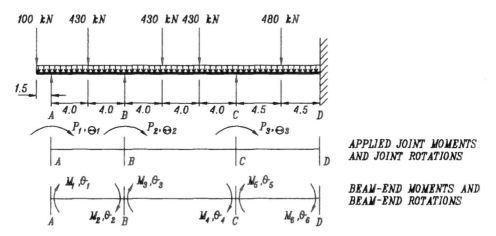

SK 3/10 Analysis of a continuous beam by the matrix method.

Find the external load matrix, which is the unbalanced joint moment at each joint. Denote these external loads as P.

$$P_1 = -152 + 441.2 = +289.2 \text{ kNm}$$
$$P_2 = +1171.9 - 441.2 = +730.7 \text{ kNm}$$
$$P_3 = -1171.9 + 554.2 = -617.7 \text{ kNm}$$

$$[P] = \begin{vmatrix} +289.2 \\ +730.7 \\ -617.7 \end{vmatrix}$$

The fixed-end moments used above have been determined before in the method demonstrating the moment distribution. The sign convention must be followed very carefully. The unbalanced fixed-end moments are applied to the joints by the members, and the clockwise moments are taken as positive.

Find the statics matrix by equilibrium of moments at the joints:

$$P_1 = M_1; \quad P_2 = M_2 + M_3; \quad P_3 = M_4 + M_5$$

$$[P] = [A][M]$$

$$[A] = \begin{vmatrix} 1 & 0 & 0 & 0 & 0 & 0 \\ 0 & 1 & 1 & 0 & 0 & 0 \\ 0 & 0 & 0 & 1 & 1 & 0 \end{vmatrix}$$

Similarly, find the deformation matrix by equating the joint rotations with beam-end rotations:

$\theta_1 = \Theta_1; \quad \theta_2 = \theta_3 = \Theta_2; \quad \theta_4 = \theta_5 = \Theta_3; \quad \theta_6 = 0$

$[\theta] = [B][\Theta]$

$$[B] = [A^T] = \begin{vmatrix} 1 & 0 & 0 \\ 0 & 1 & 0 \\ 0 & 1 & 0 \\ 0 & 0 & 1 \\ 0 & 0 & 1 \\ 0 & 0 & 0 \end{vmatrix}$$

Find stiffness matrix $[K]$

Span AB:
$\dfrac{4EI}{l} = \dfrac{4}{8} EI = 0.5 EI$

$\dfrac{2EI}{l} = \dfrac{2}{8} EI = 0.25 EI$

Span BC:
$\dfrac{4EI}{l} = \dfrac{4}{12} EI = 0.333 EI$

$\dfrac{2EI}{l} = \dfrac{2}{12} EI = 0.167 EI$

Span CD:
$\dfrac{4EI}{l} = \dfrac{4}{9} EI = 0.444 EI$

$\dfrac{2EI}{l} = \dfrac{2}{9} EI = 0.222 EI$

$$[K] = \begin{vmatrix} 0.5 & 0.25 & 0 & 0 & 0 & 0 \\ 0.25 & 0.5 & 0 & 0 & 0 & 0 \\ 0 & 0 & 0.333 & 0.167 & 0 & 0 \\ 0 & 0 & 0.167 & 0.333 & 0 & 0 \\ 0 & 0 & 0 & 0 & 0.444 & 0.222 \\ 0 & 0 & 0 & 0 & 0.222 & 0.444 \end{vmatrix} \quad \text{ignoring } EI$$

$[S] = [K][B]$

$$[S] = \begin{vmatrix} 0.5 & 0.25 & 0 \\ 0.25 & 0.5 & 0 \\ 0 & 0.333 & 0.167 \\ 0 & 0.167 & 0.333 \\ 0 & 0 & 0.444 \\ 0 & 0 & 0.222 \end{vmatrix}$$

$[D] = [A][S]$

$$[D] = \begin{vmatrix} 0.5 & 0.25 & 0 \\ 0.25 & 0.833 & 0.167 \\ 0 & 0.167 & 0.773 \end{vmatrix}$$

$$[F] = [D]^{-1}$$

$$[F] = \begin{vmatrix} 2.372068 & -0.74414 & 0.160764 \\ -0.74414 & 1.48827 & -0.32153 \\ 0.160764 & -0.32153 & 1.363124 \end{vmatrix}$$

$$[\Theta] = \frac{1}{EI}[F][P]$$

$$[\Theta] = \frac{1}{EI} \begin{vmatrix} 43.0 \\ 1070.9 \\ -1030.4 \end{vmatrix} \qquad \therefore \ [M] = \begin{vmatrix} +289.2 \\ +546.181 \\ +184.519 \\ -164.302 \\ -453.398 \\ -226.699 \end{vmatrix}$$

$$[M] = [S][\Theta]EI$$

Moment at joint $A = +289.2 - 441.2 = -152\,\text{kNm}$
Moment at joint B of span $AB = +546.181 + 441.2 = +987.4\,\text{kNm}$
Moment at joint C of span $BC = -164.302 + 1171.9 = +1007.6\,\text{kNm}$
Moment at joint D of span $CD = -226.7 + 554.2 = +327.5\,\text{kNm}$

These results are very similar to those obtained by other methods discussed earlier.

3.3 EXAMPLE 3.3: FRAME STRUCTURE

3.3.1 Analysis of a rigid frame by the method of moment distribution

A single span two-storey portal frame structure is chosen to demonstrate the method of moment distribution. SK 3/11 shows the

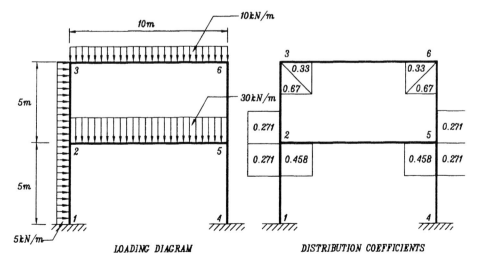

SK 3/11 Loading diagram of a two-storey portal frame.

106 Structural Steelwork

structure and the assumed loading. The loading is at serviceability limit state. The dead, imposed and wind loading are combined in one load case which is analysed.

Assume UB $610 \times 305 \times 149$ kg/m for the first floor beam, and for all other members assume UB $457 \times 191 \times 82$ kg/m.

Step 1 **Find distribution coefficients**
At first floor level:

$$k_{\text{beam}} = 124\,700 \div 1000 = 124.7; \quad k_{\text{columns}} = 37\,090 \div 500 = 74.2$$

This gives the beam distribution factor equal to 0.458 and the column distribution factor equal to 0.271 top and bottom. At the roof level the beam distribution factor is 0.33 and the column distribution factor is 0.67.

Step 2 **Find fixed-end moments**

$$FEM_{\text{floor}} = \frac{30 \times 10^2}{12} = 250 \,\text{kNm}$$

$$FEM_{\text{roof}} = \frac{10 \times 10^2}{12} = 83.3 \,\text{kNm}$$

$$FEM_{\text{column}} = \frac{5 \times 5^2}{12} = 10.42 \,\text{kNm}$$

Step 3 **Carry out moment distribution with props at joints 5 and 6 to prevent sway**

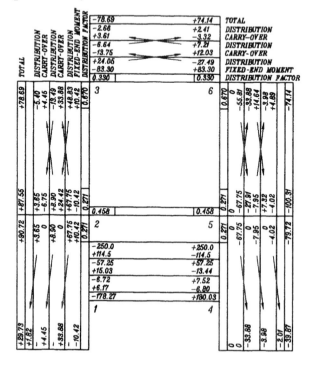

SK 3/12   Moment distribution with props at joints 5 and 6.

Step 4 Find the reactions at the props

The reactions at the props are the sum of the static reactions from applied horizontal loads assuming simple supports and the hyperstatic reactions found from the difference in end moments of the columns. The prop reactions P_5 at joint 5 and P_6 at joint 6 are determined as follows:

$$P_5 = -\left[5\,\text{m} \times 5\,\text{kN/m} + \left(\frac{-78.69 - 87.55 + 90.72 + 29.73}{5}\right.\right.$$
$$\left.\left. + \frac{+74.14 + 100.31 - 79.72 - 39.87}{5}\right)\right]$$

$= -26.814\,\text{kN}$ acting from right to left

$$P_6 = -\left[\frac{5\,\text{m} \times 5\,\text{kN/m}}{2} + \frac{(78.69 + 87.55 - 74.14 - 100.31)}{5}\right]$$

$= -10.858\,\text{kN}$ acting from right to left

The props will be removed one at a time. If the prop at joint 6 is removed, the frame will sway to the right between the first floor and roof level. This will give rise to fixed-end moments in columns 2–3 and 5–6 due to the relative displacement at the ends of the members. Similarly, if the prop at joint 5 is removed, keeping the prop at joint 6 in position, the frame will sway to the right at floor level only. This will give rise to fixed-end moments in columns 1–2, 2–3, 4–5 and 5–6. The amount of relative displacement required between floors is unknown but the forces required to cause these relative displacements must balance the propping reactions at joints 5 and 6.

Choose any small sway deflection Δ. The choice of this small deflection is entirely arbitrary. The value of the fixed-end moments due to the assumed value of Δ should be large enough to carry out the moment distribution successfully. Assume a relative sway between floors equal to 3.5 mm.

Fixed-end moment in the columns when $\Delta = 3.5\,\text{mm}$ is given by:

$$FEM_{\text{columns}} = \frac{6\Delta EI}{l^2} = \frac{6 \times \frac{3.5}{1000} \times 210 \times 10^6 \times 29\,410 \times 10^{-8}}{5^2}$$

$= 51.84\,\text{kNm}$

Carry out the moment distribution separately for the two cases: first when the prop is removed at joint 5 and then when the prop is removed at joint 6 (see SK 3/13 and SK 3/14).

When the prop is removed at joint 6, a horizontal force P_{1R} is applied at the roof level to produce the side sway. The value of this load P_{1R}, obtained after the moment distribution with a fictitious

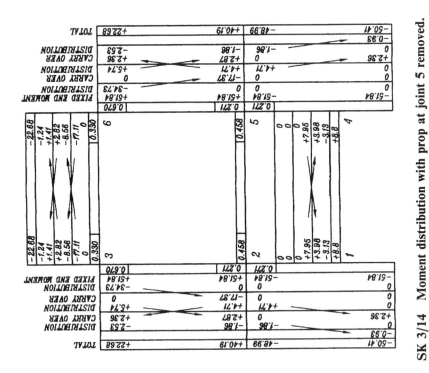

SK 3/14 Moment distribution with prop at joint 5 removed.

SK 3/13 Moment distribution with prop at joint 6 removed.

displacement, is given by:

$$P_{1R} = \frac{2 \times (20.11 + 30.34)}{5}$$

$= +20.18 \text{ kN}$ based on the column-end moments

This applied force at roof level also gives rise to a reaction P_{1F} at the prop at the floor level which is not removed. P_{1F} is given by:

$$P_{1F} = -\frac{2 \times (20.11 + 30.34 + 9.36 + 4.68)}{5}$$

$= -25.80 \text{ kN}$ based on the end moments

When the prop is removed at joint 5, a horizontal force P_{2F} is applied at the floor level to produce the side sway. The value of this load is obtained after the moment distribution with the prop at joint 5 removed, and is given by:

$$P_{2F} = \frac{2 \times (22.68 + 40.19 + 48.99 + 50.41)}{5}$$

$= +64.91 \text{ kN}$ based on the end moments

This applied force gives rise to a reaction P_{2R} at the prop at roof level which is given by:

$$P_{2R} = -\frac{2 \times (22.68 + 40.19)}{5} = -25.15 \text{ kN}$$

Assume that k_1 is a constant factor by which the force P_{1R} has to be multiplied to get an equal and opposite force to the reaction P_6. Similarly, k_2 is the factor by which the force P_{2F} has to be multiplied to get an equal and opposite force to the reaction P_5. These loads also cause reactions at the unremoved props. The combined action of these loads in the proportion k_1 and k_2 at the roof and floor level respectively must equal the propping reactions.

$$\therefore \quad P_{1R}k_1 + P_{2R}k_2 = 10.858$$

$$P_{2F}k_2 + P_{1F}k_1 = 26.814$$

or $\quad 20.18k_1 - 25.15k_2 = 10.858$

$\quad -25.80k_1 + 64.91k_2 = 26.814$

This gives $k_1 = 2.086$ and $k_2 = 1.242$.

Step 5 *Find bending moments at all joints*

The bending moment at any joint is given by:

(moment with props at 5 and 6) $+ k_1$

$\times$ (moment with prop at 6 removed) $+ k_2$

$\times$ (moment with prop at 5 removed)

Beam-end bending moments:

$$M_2 = -178.27 + (20.98 \times k_1) + (8.8 \times k_2) = -123.58 \, \text{kNm}$$

$$M_5 = +180.03 + (20.98 \times k_1) + (8.8 \times k_2) = +234.72 \, \text{kNm}$$

$$M_3 = -78.69 + (20.11 \times k_1) - (22.68 \times k_2) = -64.91 \, \text{kNm}$$

$$M_6 = +74.14 + (20.11 \times k_1) - (22.68 \times k_2) = +87.92 \, \text{kNm}$$

Similarly the column-end moments are found as follows:

Column 1–2, joint 1: $-23.12 \, \text{kNm}$; joint 2: $+49.40 \, \text{kNm}$
Column 2–3, joint 2: $+74.18 \, \text{kNm}$; joint 3: $+64.91 \, \text{kNm}$
Column 4–5, joint 4: $-92.72 \, \text{kNm}$; joint 5: $-121.04 \, \text{kNm}$
Column 5–6, joint 5: $-113.68 \, \text{kNm}$; joint 6: $-87.92 \, \text{kNm}$

3.3.2 Analysis of a rigid frame by the matrix method

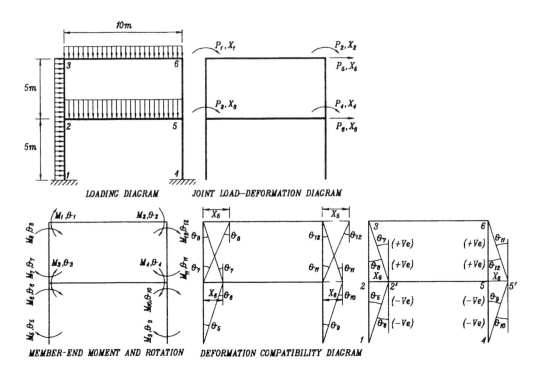

SK 3/15 The frame loading, deformation and compatibility diagrams.

For any indeterminate structure the degrees of freedom at the joints must be established before commencement of the analysis. In the frame illustrated in SK 3/15 the joints 2, 3, 5 and 6 can rotate and there are two degrees of freedom of side sway at the roof and at the floor level. The external force matrix should have these six components.

The statics matrix is composed as follows: P_1, P_2, P_3 and P_4 are the unbalanced fixed-end moments at joints 3, 6, 2 and 5, respectively, and P_5 and P_6 are the horizontal applied loads at the roof and floor level, respectively, resulting from the static reaction of loads on columns 1–2 and 2–3. In the absence of loads on these columns the loads P_5 and P_6 would be zero in the external force matrix.

From the calculation of fixed-end moments performed earlier for the moment distribution we can write the external force matrix as follows:

$$[P] = \begin{vmatrix} +72.88 \\ -83.30 \\ +250.0 \\ -250.0 \\ +12.5 \\ +25.0 \end{vmatrix}$$

These external loads are acting on the joints and clockwise moments are positive. Equating external and internal moments we get:

$P_1 = M_1 + M_8$

$P_2 = M_2 + M_{12}$

$P_3 = M_3 + M_6 + M_7$

$P_4 = M_4 + M_{10} + M_{11}$

$P_5 = -\frac{1}{5}(M_7 + M_8 + M_{11} + M_{12})$

$P_6 = \frac{1}{5}(M_7 + M_8 + M_{11} + M_{12} - M_5 - M_6 - M_9 - M_{10})$

$[P] = [A][M]$

$$[A] = \begin{vmatrix} 1 & 0 & 0 & 0 & 0 & 0 & 0 & 1 & 0 & 0 & 0 & 0 \\ 0 & 1 & 0 & 0 & 0 & 0 & 0 & 0 & 0 & 0 & 0 & 1 \\ 0 & 0 & 1 & 0 & 0 & 1 & 1 & 0 & 0 & 0 & 0 & 0 \\ 0 & 0 & 0 & 1 & 0 & 0 & 0 & 0 & 0 & 1 & 1 & 0 \\ 0 & 0 & 0 & 0 & 0 & 0 & -0.2 & -0.2 & 0 & 0 & -0.2 & -0.2 \\ 0 & 0 & 0 & 0 & -0.2 & -0.2 & 0.2 & 0.2 & -0.2 & -0.2 & 0.2 & 0.2 \end{vmatrix}$$

By equating the deformations at all the joints we get the compatibility of member deformations θ and joint deformations X. The joint deformations are either rotational or translational, depending on the degrees of freedom. The sign convention of joint deformations θ may be explained as follows. Joints 2 and 5 are connected by a horizontal member of high axial stiffness and as such it is assumed that the horizontal translation 2 to 2' is equal to 5 to 5', which is equal to X_6. Joints 1, 3, 4 and 6 are restrained horizontally. Owing to the rigid horizontal translation of the joints, the member ends should not rotate. To achieve this condition, member 1–2 at joint 1 must be rotated anticlockwise by θ_5 to coincide with the original state. Similarly, at joint 2′ member 2–1 must be rotated anticlockwise by θ_6.

$\theta_1 = X_1$

$\theta_2 = X_2$

$\theta_3 = X_3$

$\theta_4 = X_4$

$\theta_5 = -\dfrac{X_6}{L}$

$\theta_6 = X_3 - \dfrac{X_6}{L}$

$\theta_7 = X_3 - \dfrac{X_5}{L} + \dfrac{X_6}{L}$

$\theta_8 = X_1 - \dfrac{X_5}{L} + \dfrac{X_6}{L}$

$\theta_9 = -\dfrac{X_6}{L}$

$\theta_{10} = X_4 - \dfrac{X_6}{L}$

$\theta_{11} = X_4 - \dfrac{X_5}{L} + \dfrac{X_6}{L}$

$\theta_{12} = X_2 - \dfrac{X_5}{L} + \dfrac{X_6}{L}$

$[\theta] = [B][X]$ where $[B] = [A^T]$

Find the stiffness matrix $[K]$.

Floor beam:
$$\dfrac{4EI}{l} = \dfrac{4 \times 200 \times 10^6 \times 98\,500 \times 10^{-8}}{10} = 78\,800$$

$$\dfrac{2EI}{l} = 39\,400$$

Roof beam:
$$\dfrac{4EI}{l} = \dfrac{4 \times 200 \times 10^6 \times 29\,410 \times 10^{-8}}{10} = 23\,528$$

$$\dfrac{2EI}{l} = 11\,764$$

Column:
$$\dfrac{4EI}{l} = \dfrac{4 \times 200 \times 10^6 \times 29\,410 \times 10^{-8}}{5} = 47\,056$$

$$\dfrac{2EI}{l} = 23\,528$$

Analysis of Structures 113

The full stiffness matrix is given below:

23.528	11.764	0	0	0	0	0	0	0	0	0	0
11.764	23.528	0	0	0	0	0	0	0	0	0	0
0	0	78.800	39.400	0	0	0	0	0	0	0	0
0	0	39.400	78.800	0	0	0	0	0	0	0	0
0	0	0	0	47.056	23.528	0	0	0	0	0	0
0	0	0	0	23.528	47.056	0	0	0	0	0	0
0	0	0	0	0	0	47.056	23.528	0	0	0	0
0	0	0	0	0	0	23.528	47.056	0	0	0	0
0	0	0	0	0	0	0	0	47.056	23.528	0	0
0	0	0	0	0	0	0	0	23.528	47.056	0	0
0	0	0	0	0	0	0	0	0	0	47.056	23.528
0	0	0	0	0	0	0	0	0	0	23.528	47.056

We know that: $[M] = [K][\theta]$

Substituting we get: $[M] = [KB][X]$

We also know: $[P] = [A][M] = [AKB][X]$

From this we get: $[X] = [AKB]^{-1}[P]$

$[KB]$ is given below after matrix multiplication:

23.528	11.764	0	0	0	0
11.764	23.528	0	0	0	0
0	0	78.8	39.4	0	0
0	0	39.4	78.8	0	0
0	0	23.528	0	0	−14.1168
0	0	47.056	0	0	−14.1168
23.528	0	47.056	0	−14.1168	−14.1168
47.056	0	23.528	0	−14.1168	−14.1168
0	0	0	23.528	0	−14.1168
0	0	0	47.056	0	−14.1168
0	23.528	0	47.056	−14.1168	−14.1168
0	47.056	0	23.528	−14.1168	−14.1168

$[AKB]$ is given below after matrix multiplication:

70.584	11.764	23.528	0	−14.1168	14.1168
11.764	70.584	0	23.528	−14.1168	14.1168
23.528	0	172.912	39.4	−14.1168	0
0	23.528	39.4	172.912	−14.1168	0
−14.1168	−14.1168	−14.1168	−14.1168	11.29344	−11.29344
14.1168	14.1168	8.88E−16	8.88E−16	−11.29344	22.58688

$[AKB]^{-1}$ is given below after matrix inversion:

0.019996	0.001706	−0.00071	0.002512	0.03163	0.002252
0.001706	0.019996	0.002512	−0.00071	0.03163	0.002252
−0.00071	0.002512	0.007632	−0.00043	0.020265	0.009007
0.002512	−0.00071	−0.00043	0.007632	0.020265	0.009007
0.03163	0.03163	0.020265	0.020265	0.357494	0.139209
0.002252	0.002252	0.009007	0.009007	0.139209	0.111063

$[X]=[AKB]^{-1}[P]$ and $[M]=[KB][X]$ are given below after matrix multiplication:

$$[X] = 10^{-3} \begin{vmatrix} 0.961097 \\ -0.28389 \\ 2.231884 \\ -1.2937 \\ 7.619317 \\ 4.49324 \end{vmatrix}$$

$[M]$	+ FEM	= Results	
+19.27	−83.3	−64.03	(M_1)
+4.63	+83.3	+87.93	(M_2)
+124.90	−250.0	−125.10	(M_3)
−14.01	+250.0	+235.99	(M_4)
−10.92	−10.42	−21.34	(M_5)
+41.59	+10.42	+52.01	(M_6)
+83.51	−10.42	+73.09	(M_7)
+53.61	+10.42	+64.03	(M_8)
−93.87	0	−93.87	(M_9)
−124.31	0	−124.31	(M_{10})
−111.69	0	−111.69	(M_{11})
−87.93	0	−87.93	(M_{12})

These results are very similar to those obtained by the method of moment distribution.

3.4 EXAMPLE 3.4: ANALYSIS OF A HINGELESS ARCH

For the parabolic arch in SK 3/16 assume the following geometric parameters:

A_0 = area of the arch cross-section at the crown of the arch = 24.71×10^{-3} m^2

I_0 = moment of inertia of the arch cross-section at the crown = 750×10^{-6} m^4

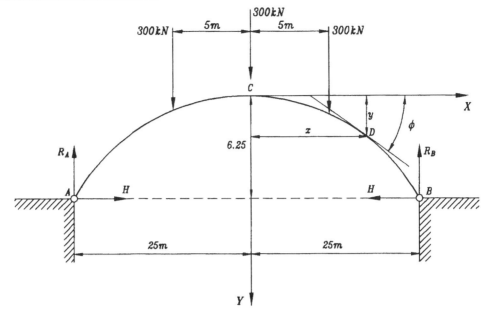

SK 3/16 Symmetrical parabolic hingeless arch with variable area and inertia.

A = area of the arch cross-section at any point on the arch rib where the inclination of the tangent to the horizontal is ϕ, and I is the moment of inertia of the arch cross-section at the same point

$$A = \frac{A_0}{\cos \phi}; \quad I = \frac{I_0}{\cos^3 \phi}$$

The origin of the arch is assumed to be at the crown, where x and y are zero. The equation of the centre-line of the parabolic arch is given by:

$$y = \frac{4fx^2}{l^2}$$

where f is the rise of the arch from the springing level and l is the span of the arch between supports. The span of the arch is 50 m and the rise is 6.25 m, giving a rise to span ratio of 1:8. The numerical integration method is used to find the internal forces in the arch rib subjected to a symmetrical loading of 300 kN at the centre of the arch and another loading case of a pair of loads at 5 m from the centre of the arch. As discussed in Chapter 2 in the section on arches, only half of the arch needs to be analysed because the shear vanishes at the crown due to symmetry. To use this analytical method, it should be remembered that only a pair of symmetrical loads can be analysed at a time. The central loading of 300 kN is split into two loads of 150 kN applied side by side.

The arch is divided into 10 segments. The centres of these segments are assumed to lie 2.5 m apart on the X-axis starting with the first segment at $x = 1.25$ m. The following expressions have been used to calculate the numbers given in Tables 3.4a, b and c:

$$\tan\phi = \frac{dy}{dx} = \frac{8fx}{l^2}; \quad \cos\phi = \frac{1}{\sqrt{1+\tan^2\phi}}$$

$$\Delta_s = \frac{dx}{\cos\phi}$$

$$d = \frac{\sum \Delta_s y/I}{\sum \Delta_s/I}$$

$$y_1 = y - d$$

$$H_0 = N_c = -\frac{\sum (M' y_1 \Delta_s/I) + \sum (N' \cos\phi \Delta_s/A)}{\sum (y_1^2 \Delta_s/I) + \sum (\cos^2\phi \Delta_s/A)}$$

$$M_0 = -\frac{\sum (M' \Delta_s/I)}{\sum (\Delta_s/I)}$$

$$M = M_c + N_c y + M'$$

$$M_c = M_0 - H_0 d$$

For the central load of 300 kN $M' = -150 \times x$ and $N' = 150 \times \sin\phi$. M' and N' are the bending moment and the direct thrust due to the applied load only.

This method of numerical integration can be used for any arch structure provided the energy-based formulae are derived for that particular type of arch. For the hingeless arch considered here, the results of the numerical integration are shown in Tables 3.4a, b and c.

Table 3.4a Determination of the elastic centre d.

x	y	$\cos\phi$	$\sin\phi$	Δ_s	A	I	Δ_s/I	$\Delta_s y/I$	y_1
1.3	0.0	1.0	0.0	2.5	0.0	0.0	3335.7	52.1	−1.9
3.8	0.1	1.0	0.1	2.5	0.0	0.0	3319.1	466.8	−1.8
6.3	0.4	1.0	0.1	2.5	0.0	0.0	3286.4	1283.8	−1.6
8.8	0.8	1.0	0.2	2.5	0.0	0.0	3238.6	2479.6	−1.1
11.3	1.3	1.0	0.2	2.6	0.0	0.0	3177.0	4020.8	−0.7
13.8	1.9	1.0	0.3	2.6	0.0	0.0	3103.1	5866.8	−0.1
16.3	2.6	1.0	0.3	2.6	0.0	0.0	3018.9	7971.8	0.7
18.8	3.5	0.9	0.4	2.7	0.0	0.0	2926.3	10 287.7	1.6
21.3	4.5	0.9	0.4	2.7	0.0	0.0	2827.1	12 766.3	2.6
23.8	5.6	0.9	0.4	2.8	0.0	0.0	2723.3	15 361.3	3.7
							30 955.6	60 556.9	
						$d =$		1.96	

Analysis of Structures 117

Table 3.4b Determination of moments and reactions for a central load of 300 kN

M'	N'	$M'y_1\Delta_s/I$	$N'\cos\phi\Delta_s/A$	$y_1^2\Delta_s/I$	$\cos\phi^2\Delta_s/A$	$M'\Delta_s/I$	M
−187.5	3.7	1 213 753	379.3	12 562.4	101.2	−625 443.5	520.6
−562.5	11.2	3 389 780	1132.3	10 941.5	100.6	−1 867 001	216.2
−937.5	18.6	4 823 749	1868.6	8055.7	99.7	−3 081 031	−17.4
−1312.5	25.9	5 060 960	2577.9	4591.0	98.2	−4 250 665	−180.3
−1687.5	32.9	3 702 529	3251.4	1515.3	96.3	−5 361 104	−272.6
−2062.5	39.8	420 030.1	3881.6	13.4	94.1	−6 400 166	−294.2
−2437.5	46.4	−5 036 017	4462.8	1414.0	91.5	−7 358 596	−245.1
−2812.5	52.7	−12 833 868	4991.4	7115.7	88.7	−8 230 152	−125.4
−3187.5	58.7	−23 063 745	5465.3	18 518.8	85.7	−9 011 486	65.1
−3562.5	64.4	−35 745 298	5884.0	36 968.1	82.6	−9 701 870	326.2
		−58 068 127	33 894.5	101 695.8	938.7	−55 887 515	

$H_0 = +565.4$ $M_0 = +1805.4$
$N_c = +565.4$ $M_s = +483.3$ $M_c = +699.3$

Table 3.4c Determination of moments and reactions for a pair of symmetrically placed loads.

M'	N'	$M'y_1\Delta_s/I$	$N'\cos\phi\Delta_s/A$	$y_1^2\Delta_s/I$	$\cos\phi^2\Delta_s/A$	$M'\Delta_s/I$	M
0	0.0	0	0	12 562.4	101.2	0	258.0
0	0.0	0	0	10 941.5	100.7	0	387.7
−375	37.2	1 929 499	3737.2	8055.7	99.7	−1 232 412	272.2
−1125	51.7	4 337 966	5155.9	4591.0	98.2	−3 643 427	−88.6
−1875	65.9	4 113 921	6502.8	1515.3	96.3	−5 956 782	−319.6
−2625	79.5	534 583.8	7763.1	13.4	94.1	−8 145 666	−420.9
−3375	92.7	−6 972 947	8925.7	1414.0	91.6	−10 188 825	−392.4
−4125	105.3	−18 823 006	9982.8	7115.7	88.7	−12 070 889	−234.2
−4875	117.3	−35 273 963	10 930.5	18 518.8	85.7	−13 782 273	53.7
−5625	128.7	−56 439 945	11 767.9	36 968.1	82.6	−15 318 742	471.4
		−1.07E+08	64 765.8	101 695.8	938.7	−70 339 017	

$H_0 = +1038.0$ $M_0 = +2272.3$
$N_c = +1038.0$ $M_s = +729.0$ $M_c = +241.8$

The results of the analyses may be summarised as follows:

- For the central load of 300 kN on the arch the internal forces are available from Table 3.4b.
- The horizontal thrust at the crown $= N_c = -(-565.45)\,\text{kN} = 565.4\,\text{kN}$.
- The bending moment at the crown $= M_c = -(-699.26)\,\text{kNm} = 699.26\,\text{kNm}$.
- The bending moment at any point on the arch is given in Table 3.4b in the column headed M ($M = M_c + N_c y + M'$).
- (M_s is the bending moment at the fixed support.)

Similarly, for the pair of 300 kN loads symmetrically spaced at 5.0 m about the centre of the arch, the results are given in Table 3.4c.

For a uniform rise of temperature of 1°C the internal forces in the arch will be determined as follows:

Coefficient of thermal expansion of steel $= \alpha = 12 \times 10^{-6}/°C$

$$H_0 = N_c = \frac{\alpha t l E/2}{\sum (y_1^2 \Delta_s/I) + \sum (\Delta_s \cos^2 \phi/A)}$$

$$= \frac{12 \times 10^{-6} \times 1 \times 50 \times 200 \times 10^6}{2 \times (101\,695.8 + 938.69)} = 0.60\,\text{kN/°C}$$

Because M' and N' are zero, M_0 is also 0.

$$M_c = M_0 - H_0 d = -0.60 \times 1.96 = -1.176\,\text{kNm/°C}$$

3.5 EXAMPLE 3.5: YIELD-LINE ANALYSIS OF A RECTANGULAR PLATE

Consider a rectangular steel plate rigidly supported on three sides and free on one long side, as shown in SK 3/17.

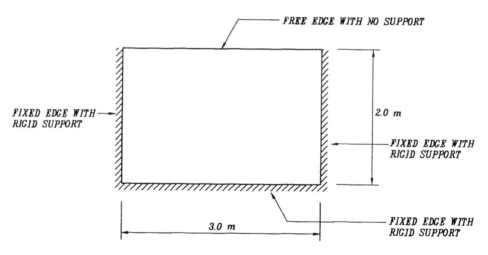

SK 3/17   Two-way spanning rectangular steel plate.

Step 1   Determine loading and select trial thickness
Assume that the plate is a part of a decking system in a dockside structure. The imposed loading is 20 kN/m². Assume 12.5 mm plate as a first trial.
Self-weight of the plate $= 1$ kN/m²
Ultimate loading on the plate $= (1.4 \times 1) + (1.6 \times 20) = 33.4$ kN/m²
Service loading on the plate $= 21$ kN/m²
Deflection should be limited to $l/200$ at service load. Stresses at service load must not go beyond yield.

Step 2 Assume yield-line location and draw free body diagrams

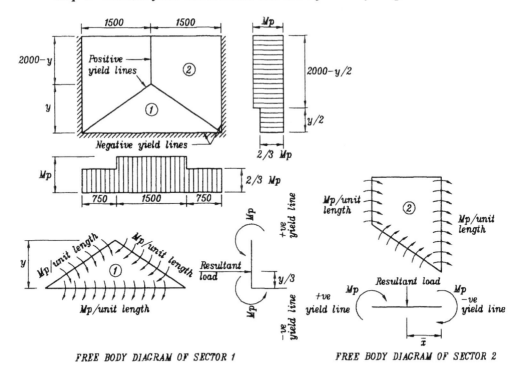

SK 3/18 Assumed yield-line location and free body diagrams.

Step 3 Determine plastic moment of resistance
Thickness of plate $= t = 12.5\,\text{mm}$

$$\text{Plastic moment of resistance of plate} = M_\text{P} = \frac{t^2}{4} f_\text{y} = \frac{12.5^2}{4} \times 275$$

$$= 10\,742\,\text{Nmm/mm}$$

Step 4 Determine ultimate unit resistance of each sector
Sectors 1 and 2 are as shown in SK 3/18. The ultimate unit resistance $= r_\text{u}$. The summation of moments in each sector is equated to zero for equilibrium. From the free body diagrams of each sector, the sum of all positive moments and negative moments on the yield-lines must balance the moment of the applied ultimate load about the line of rotation of the sector.

Consider sector 1

$$\sum M = (1500 \times M_\text{P}) + \left(2 \times \frac{2}{3} \times 750 \times M_\text{P}\right) + (1500 \times M_\text{P})$$

$$+ \left(2 \times \frac{2}{3} \times 750 \times M_\text{P}\right) = 5000 M_\text{P}$$

120 Structural Steelwork

$$\text{Resisting moment} = r_u A \bar{x} = r_u \times \frac{y}{2} \times 3000 \times \frac{y}{3} = 500 r_u y^2$$

$$\therefore \quad 5000 M_P = 500 r_u y^2; \quad \text{or} \quad r_u = \frac{10 M_P}{y^2}$$

Consider sector 2

$$\sum M = \left(2000 - \frac{y}{2}\right) M_P + \frac{2}{3}\frac{y}{2} M_P + \left(2000 - \frac{y}{2}\right) M_P + \frac{2}{3}\frac{y}{2} M_P$$

$$= 4000 M_P - \frac{y}{6} M_P$$

$$\text{Resisting moment} = r_u A \bar{x} = r_u \times \left\{[(2000 - y) \times 1500 \times 750]\right.$$

$$\left. + \left[\frac{y}{2} \times 1500 \times \frac{1500}{3}\right]\right\}$$

$$= r_u \times [(22.5 \times 10^8) - (75 \times 10^4 y)]$$

$$\therefore \quad 4000 M_P - \frac{y}{6} M_P = r_u[(22.5 \times 10^8) - (75 \times 10^4 \times y)]$$

$$\text{or} \quad r_u = \frac{\left(4000 - \frac{y}{6}\right) M_P}{(22.5 \times 10^8) - (75 \times 10^4 y)}$$

By equating r_u from both sectors and simplifying:

$$y^3 - 24\,000 y^2 - (45 \times 10^6 y) + (13.5 \times 10^{10}) = 0$$

A root of this equation is $y = 1649.2 \, \text{mm}$.

Step 5 *Find ultimate unit resistance*

$$r_u = \frac{10 M_P}{y^2} = \frac{10 \times 10\,742}{1649.2^2} = 0.0395 \, \text{N/mm}^2$$

$$= 39.5 \, \text{kN/m}^2 > 33.4 \, \text{kN/m}^2$$

The ultimate resistance is greater than the ultimate load.

3.6 EXAMPLE 3.6: SEISMIC ANALYSIS OF A TALL CANTILEVER STRUCTURE

The structure chosen for this exercise is a tall chimney made of steel. The geometry of the structure including its deflected profile is shown in SK 3/19.

The mass per unit length and the moment of inertia of the cross-section of the chimney will be assumed to be constant over its whole length. The principles described here may equally be applied to

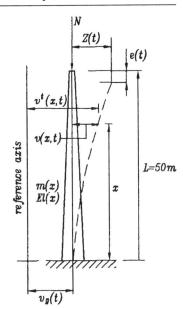

SK 3/19 The tall chimney treated as a single degree of freedom system.

cantilever structures with variable mass per unit length and variable moment of inertia. These two variables are usually functions of x, which is the vertical axis of the co-ordinate system.

The chimney structure will be approximated to a single degree of freedom (SDOF) system by this method of analysis. Assume that the chimney is subjected to a ground motion $v_g(t)$ and that there is a constant load N at the top of the structure. N is introduced in the analysis because the principles established may then be used for other structures (e.g. a water tank). The deflected shape has to be assumed as a function of x.

Assume that the shape function is $\psi(x)$. The amplitude of the motion is $v(x, t)$, which is a function of both x and t, where t is the time in seconds. Assuming a generalised co-ordinate system $Z(t)$, it can be said:

$$v(x, t) = \psi(x) Z(t)$$

where the shape function $\psi(x)$ is a dimensionless ratio of the local displacement to a reference displacement.

$$\psi(x) = \frac{v(x, t)}{Z(t)}$$

The energy principle, known as Hamilton's principle, is used to solve the problem of vibration. The kinetic energy is termed T and the potential energy is termed V. The kinetic energy of the chimney may be written as:

$$T = \frac{1}{2} \int_0^l m(x) [\dot{v}^t(x, t)]^2 \, dx$$

The strain energy of the chimney due to flexure may be expressed as:

$$V_f = \frac{1}{2} \int_0^l EI(x)[v''(x,t)]^2 \, dx$$

The derivatives with respect to x are marked by primes and those with respect to time are marked by dots. For example:

$$v'' = \frac{d^2 v}{dx^2} \quad \text{and} \quad \ddot{v} = \frac{d^2 v}{dt^2}$$

The vertical load N on top of the tower undergoes a vertical deflection equal to $e(t)$, which is a function of time. The parameter $e(t)$ may be determined as follows.

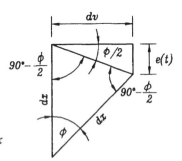

SK 3/20 The relationship between dv, dx and $e(t)$.

Consider the diagram in SK 3/20:

$$e(t) = dv \tan\left(\frac{\phi}{2}\right) = \frac{1}{2} dv \tan\phi \quad \text{for small values of } \phi$$

$$= \frac{1}{2} dv \frac{dv}{dx} = \frac{1}{2} \left(\frac{dv}{dx}\right)^2 dx$$

The potential energy of the load N is therefore given by:

$$V_N = -N \int_0^l e(t) = -\frac{N}{2} \int_0^l [v'(x,t)]^2 \, dx$$

The potential energy of the load N is reduced by the deformation $e(t)$ of the chimney, and therefore the negative sign is introduced.

There are no directly applied dynamic loads on the chimney. An earthquake produces a base motion. If damping is neglected in the analysis, then Hamilton's principle will take the form:

$$\int_{t_1}^{t_2} \partial(T - V) \, dt = 0$$

$$\therefore \int_{t_1}^{t_2} \left[\int_0^l m(x) \ddot{v}^t(x,t) \, \partial \dot{v}^t \, dx - \int_0^l EI(x) v''(x,t) \, \partial v'' \, dx \right.$$

$$\left. + N \int_0^l v'(x,t) \, \partial v' \, dx \right] dt = 0$$

where:

$$\dot{v}^t = \dot{v}_g + \dot{v}; \quad v'' = \psi'' Z; \quad v' = \psi' Z; \quad \dot{v} = \psi \dot{Z}$$

$$\partial \dot{v}^t = \partial \dot{v}; \quad \partial v'' = \psi'' \partial Z; \quad \partial v' = \psi' \partial Z; \quad \partial \dot{v} = \psi \partial \dot{Z}$$

Substituting in Hamilton's equation:

$$\int_{t_1}^{t_2} \left[\dot{Z} \partial \dot{Z} \int_0^l m(x)\psi^2 \, dx + \partial \dot{Z} \dot{v}_g(t) \int_0^l m(x)\psi \, dx \right.$$
$$\left. - Z \partial Z \int_0^l EI(x)(\psi'')^2 \, dx + NZ \partial Z \int_0^l (\psi')^2 \, dx \right] dt = 0$$

Integrating by parts this expression becomes:

$$\int_{t_1}^{t_2} \left[m^* \ddot{Z} + k^* Z - k_G^* Z - p_{\text{eff}}^*(t) \right] \partial Z \, dt = 0$$

where:

$$m^* = \int_0^l m(x)\psi^2 \, dx = \text{the generalised mass}$$

$$k^* = \int_0^l EI(x)(\psi'')^2 \, dx = \text{the generalised stiffness}$$

$$k_G^* = N \int_0^l (\psi')^2 \, dx = \text{the generalised geometric stiffness}$$

$$p_{\text{eff}}^*(t) = -\ddot{v}_g \int_0^l m(x)\psi \, dx = \text{the generalised effective load}$$

Since ∂Z is arbitrary, the expression in brackets must be equal to zero.

$$\therefore \quad m^* \ddot{Z} + \bar{k}^* Z - p_{\text{eff}}^*(t) = 0 \quad \text{where} \quad \bar{k}^* = k^* - k_G^*$$

This is the familiar equation of motion of an SDOF system.

For the chimney assume a shape function of vibration given by:

$$\psi(x) = 1 - \cos \frac{\pi x}{2l}$$

Also, mass per unit length is constant and equal to m. Therefore:

$$m^* = \int_0^l m(\psi)^2 \, dx = m \int_0^l \left(1 - \cos \frac{\pi x}{2l} \right)^2 dx = 0.228 ml$$

$$k^* = \int_0^l EI(\psi'')^2 \, dx = EI \int_0^l \left(\frac{\pi^2}{4l^2} \cos \frac{\pi x}{2l} \right)^2 dx = \frac{\pi^4}{32} \frac{EI}{l^3}$$

$$p_{\text{eff}}^*(t) = \ddot{v}_g(t) \int_0^l m\psi \, dx = m\ddot{v}_g(t) \int_0^l \left(1 - \cos\frac{\pi x}{2l}\right) dx$$

$$= 0.364 m l \ddot{v}_g(t) - \Lambda \ddot{v}_g(t)$$

$$k_G^* = N \int_0^l (\psi')^2 \, dx = N \int_0^l \left(\frac{\pi}{2l} \sin\frac{\pi x}{2l}\right)^2 dx = \frac{N\pi^2}{8l}$$

$$\bar{k}^* = k^* - k_G^* = \frac{\pi^4 EI}{32 l^3} - \frac{N\pi^2}{8l}$$

At the critical buckling load N_{cr} this stiffness will become zero.

$$\therefore \quad \frac{N_{\text{cr}} \pi^2}{8l} = \frac{\pi^4 EI}{32 l^3} \quad \text{which gives} \quad N_{\text{cr}} = \frac{\pi^2 EI}{4 l^2}$$

$$k_G^* = \frac{\pi^2 EI}{32 l^3} \frac{N}{N_{\text{cr}}} \quad \text{and} \quad \bar{k}^* = \frac{\pi^4 EI}{32 l^3}\left(1 - \frac{N}{N_{\text{cr}}}\right)$$

The equation of motion can be written as:

$$0.228 m l \ddot{Z}(t) + \frac{\pi^4 EI}{32 l^3}\left(1 - \frac{N}{N_{\text{cr}}}\right) Z(t) = 0.364 m l \ddot{v}_g(t)$$

For the chimney in SK 3/18 assume:

- Direct load $N = 0$ kN
- Constant mass per unit length $= m = 2.0 \dfrac{\text{kN s}^2}{\text{m}}$ (per metre length)
- Constant moment of inertia of the chimney cross-section $= I = 628 \times 10^{-4}$ m^4
- Young's modulus for steel $= E = 200$ kN/mm^2

Therefore:

$$m^* = 0.228 m l = 0.228 \times 2.0 \times 50 = 22.8 \frac{\text{kN s}^2}{\text{m}}$$

$$\bar{k}^* = \frac{\pi^4 EI}{32 l^3} = 305.9 \text{ kN/m}$$

$$\Lambda = 0.364 m l = 36.4 \frac{\text{kN s}^2}{\text{m}}$$

The circular frequency of the chimney $= \omega = \sqrt{\dfrac{\bar{k}^*}{m^*}}$

$$= \sqrt{\frac{305.9}{22.8}} = 3.66 \text{ rad/s}$$

The time period of oscillation $= T = \dfrac{2\pi}{\omega} = 1.72$ s

The cyclic frequency $= f = \dfrac{1}{T} = 0.58$ Hz

DAMPING VALUES (S): 0.5%, 2%, 5%, 7%, 10%

SK 3/21 Response spectra normalised to 1g.

Assume a ZPA (zero period acceleration) value of 0.25g for the seismic excitation.
Assume 2% damping for the steel chimney.
For a cyclic frequency of 0.58 Hz, S_v (pseudo-velocity) is obtained from the response spectra as $S_v = 0.25 \times 2.2 = 0.55$ m/s.

The maximum generalised co-ordinate displacement is given by:

$$Z_{max} = \frac{\Lambda}{m^*\omega} S_v = \frac{36.4 \times 0.55}{22.8 \times 3.66} = 0.240 \text{ m}$$

The maximum displacement of the chimney is therefore given by the expression:

$$v_{max} = 0.240 \times \left(1 - \cos\frac{\pi x}{2l}\right) \text{ m}$$

The maximum base shear is:
$$V_{max} = \frac{\Lambda^2}{m^*}\omega S_v = \frac{36.4^2 \times 3.66 \times 0.55}{22.8} = 117\,\text{kN}$$

The maximum distributed load on the chimney is given by:
$$w_{s,max} = \frac{m\psi(x)}{\Lambda} V_{max} = \frac{2 \times 117}{36.4}\left(1 - \cos\frac{\pi x}{2l}\right)$$
$$= 6.43 \times \left(1 - \cos\frac{\pi x}{2l}\right)\,\text{kN/m}$$

Assume there is a mass on top of the structure of weight N kN. In the equation of motion the stiffness to be used is then given by:
$$\bar{k}^* = \frac{\pi^4 EI}{32 l^3}\left(1 - \frac{N}{N_{cr}}\right)$$

where $N_{cr} = \pi^2 EI/4l^2$ and the total mass $M = m^* + N/g$.

Circular frequency $\omega = \sqrt{\dfrac{\bar{k}^*}{M}}$

Maximum base shear $= V_{max} = \left(\dfrac{\Lambda^2}{m^*} + \dfrac{N}{g}\right)\omega S_v$

where S_v is obtained from the response spectrum for $T = 2\pi/\omega$. The bending moment at the base of the chimney is given by:
$$M_{0,max} = \frac{\Lambda}{m^*} m\omega S_v \int_0^l \psi(x) x\,dx + \frac{N}{g}\omega S_v l$$
$$= \frac{\Lambda}{m^*} ml^2 \omega S_v \left(\frac{1}{2} + \frac{4}{\pi^2} - \frac{2}{\pi}\right) + \frac{N}{g}\omega S_v l$$
$$= 0.429 ml^2 \omega S_v + \frac{N}{g}\omega S_v l$$

At a height h above the base the bending moment and shear force are given by:
$$M_{h,max} = \frac{\Lambda}{m^*} m\omega S_v \int_h^l \psi(x) x\,dx + \frac{N}{g}\omega S_v(l - h)$$
$$V_{h,max} = \frac{\Lambda}{m^*} m\omega S_v \int_h^l \psi(x)\,dx + \frac{N}{g}\omega S_v$$

Carry out numerical integration to find m^*, k^*, Λ and ω if the mass per unit length and the moment of inertia vary with x. Divide the chimney or any tall cantilever structure into discrete segments, and use the summation principle to solve for ω.

3.7 EXAMPLE 3.7: PLASTIC ANALYSIS OF A PITCHED PORTAL FRAME

The frame to be analysed is shown in SK 3/22. These frames are to be spaced at 6.0 m centres.

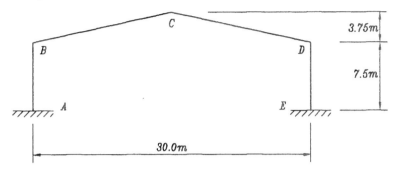

SK 3/22 Plastic analysis of a pitched portal.

Step 1 **Determine combined factored loads**

Dead load, including self-weight, sheeting, purlins, insulation, fixings and services $= 0.5\,\text{kN/m}^2$

Imposed load on inaccessible roof $= 0.6\,\text{kN/m}^2$

Assume uniformly distributed load on the rafter:

Dead load $= 0.5\,\text{kN/m}^2 \times 6\,\text{m} \times 1.4$ (load factor) $= 4.2\,\text{kN/m}$
Imposed load $= 0.6\,\text{kN/m}^2 \times 6\,\text{m} \times 1.6$ (load factor) $= 5.76\,\text{kN/m}$
Load combination LC1 $= 4.2 + 5.76 = 10\,\text{kN/m}$

Step 2 **Carry out elastic analysis**

Assume a uniform section of the rafter and the columns using any section property and find the bending moments with the help of a computer program.

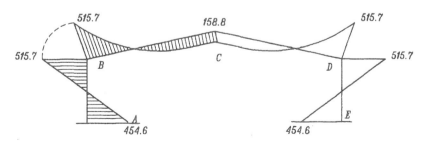

SK 3/23 Bending moment diagram with arbitrary uniform section properties.

Column member AB: Maximum bending moment $= 515.7\,\text{kNm}$
Coacting axial load $= 154.6\,\text{kN}$ (compression)
Coacting shear force $= 129.4\,\text{kN}$

128 Structural Steelwork

Rafter member *BC*: Maximum bending moment = 515.7 kNm
Coacting axial load = 163 kN
Coacting shear force = 118.6 kN

Step 3 **Choose trial sections for rafter and columns**
Do not allow the section to go beyond yield at any point at serviceability limit state.

Service loads = 3.0 kN/m² (dead) + 3.6 kN/m² (imposed) = 6.6 kN/m²

$\therefore$ Service bending moment in the rafter $= \dfrac{515.7}{10} \times 6.6 = 340$ kNm

Coacting shear force = 78.3 kN
Coacting axial load = 107.58 kN

Assuming that the shear force and the axial load will be carried by the web, the flanges only will be required to carry the total bending moment. Design strength of Grade 43 mild steel for thickness over 16 mm $= p_y = 265$ N/mm² as per BS 5950: Part 1.

Choose a depth of section of the rafter equal to span/30 $\approx 15/30 \approx 0.5$ m
Select a depth of section equal to 460 mm

$\therefore$ Minimum flange area required $= \dfrac{340 \times 10^6}{265 \times 460} = 2789$ mm²

Choose UB 457 × 152 × 82 kg/m with flange area equal to 153.5 × 18.9 = 2901.1 mm² for the rafter.

Plastic hinges should not be allowed to form simultaneously on the columns reducing the structure to a mechanism. Columns should be made stronger than the rafter.
Choose UB 457 × 191 × 98 kg/m for the columns. These sections will be used as a first trial in the analysis.
As per Clause 5.3.4 of BS 5950: Part 1 the sections must be classified as plastic according to Table 7.

Member properties (For section classification see Table 7 of BS 5950: Part 1)
Rafter: UB 457 × 152 × 82 kg/m; $B = 153.5$ mm; $b = 76.75$ mm; $T = 18.9$ mm; $t = 10.7$ mm; $D = 465.1$ mm; $S_x = 1802$ cm³; $Z = 1559$ cm³; $b/T = 4.06 < 8.5\varepsilon$; $d/t = 38.0 < 79\varepsilon/(0.4 + 0.6\alpha)$
Section classification – plastic; $A = 105$ cm²; $I = 36\,250$ cm⁴.
Column: UB 457 × 191 × 98 kg/m; $B = 192.8$ mm; $b = 96.4$ mm; $T = 19.6$ mm; $t = 12.8$ mm; $D = 467.4$ mm; $S_x = 2234$ cm³; $Z = 1956$ cm³; $b/T = 4.91 < 8.5\varepsilon$; $d/t = 37.2 < 39\varepsilon$
Section classification – plastic; $A = 125$ cm²; $I = 45\,770$ cm⁴.

Step 4 **Carry out analysis using trial sections and unit distributed loading**
Two separate analyses will be carried out with symmetrical uniformly distributed unit loading on the portal frame. The first analysis will

assume that the structure is fully elastic and that there are no plastic hinges. The second analysis will be carried out with two plastic hinges in the rafter at the connection with the columns. It is assumed that the plastic hinges in the rafter will form simultaneously in the rafter due to the symmetry. The next set of plastic hinges will turn the structure into a mechanism, thereby causing failure.

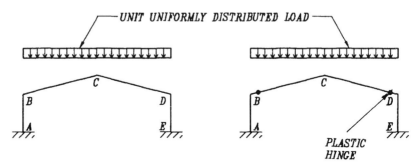

SK 3/24   Sketch of portal showing criteria for the two unit load analyses.

Table 3.5   Results of analyses.

Location: Column/ Rafter	Analysis 1: No plastic hinges				Analysis 2: Two plastic hinges			
	M (kNm)	N (kN)	V (kN)	Deflection (mm)	M (kNm)	N (kN)	V (kN)	Deflection (mm)
A: Column	$+48.14^{[s]}$	$15.46^{[c]}$	13.31	–	$+113.9^{[s]}$	$15.46^{[c]}$	15.19	–
B: Column	$-51.67^{[h]}$	$15.46^{[c]}$	13.31	–	0	$15.46^{[c]}$	15.19	–
B: Rafter	$-51.67^{[h]}$	$16.66^{[c]}$	11.77	–	0	$18.49^{[c]}$	11.32	–
Rafter +ve	$+19.72^{[s]}$	$12.91^{[c]}$	0	17.0	$66.0^{[s]}$	$14.74^{[c]}$	0	84.7
C: Rafter	$+14.38^{[s]}$	$12.91^{[c]}$	3.23	18.2	$59.0^{[s]}$	$14.74^{[c]}$	3.68	91.1

[s] = sagging; [h] = hogging; [c] = compression.

Step 5 **Calculate reduction of plastic moment of resistance of rafter section**
At the plastic hinges the simultaneous action of shear force and the direct compression will have some effect on the plastic moment of resistance of the section. From Table 3.5 it can be seen that, in response to 10 kN/m loading, the direct load on the rafter at location *B*, where the plastic hinge is postulated, will be approximately 166.6 kN. Similarly, at the same location, the shear force is likely to be 117.7 kN.

$$N = 166.6 \text{ kN}; \quad V = 117.7 \text{ kN}$$

$$f_v = \frac{V}{td_w} = \frac{117.7 \times 10^3}{10.7 \times [465.1 - (2 \times 18.9)]} = 25.7 \text{ N/mm}^2$$

d_w should be the depth of the web only in this analysis, assuming that the shear is taken by the web alone.

$$f_m = \sqrt{f_y^2 - 3f_v^2} = \sqrt{275^2 - (3 \times 25.7^2)} = 271 \, \text{N/mm}^2$$

Since the thickness of the web is less than 16 mm, the design strength is assumed to be 275 N/mm², in accordance with Table 6 of BS 5950: Part 1.

$$2a = \frac{N}{f_m t} = \frac{166.6 \times 10^3}{271 \times 10.7} = 57.4 \, \text{mm}; \quad \text{or} \quad a = 28.7 \, \text{mm}$$

The plastic moment of resistance of the web only is:

$$M_{pw} = \frac{t d_w^2}{4} f_y = \frac{10.7 \times [465.1 - (2 \times 18.9)]^2 \times 275 \times 10^{-6}}{4}$$

$$= 134.3 \, \text{kNm}$$

The design strength is taken to be equal to f_y because the thickness of web is less than 16 mm. The revised plastic moment of resistance of the rafter section is given by:

$$M' = M_P - \left(\frac{f_y - f_m}{f_y}\right) M_{pw} - t a^2 f_m$$

$$= (1802 \times 10^3 \times 265 \times 10^{-6}) - \left(\frac{275 - 271}{271}\right) \times 134.3$$

$$- (10.7 \times 28.7^2 \times 271 \times 10^{-6}) = 473.2 \, \text{kNm}$$

This plastic moment of resistance for the rafter section will be used assuming that there will not be any reduction of the capacity of the member as a whole due to lateral torsional buckling. The rafter will be adequately braced.

Step 6 *Calculate reduction of plastic moment of resistance of column section*
The second set of plastic hinges will form at the base of the columns at A. The axial load and shear in the column at the full ultimate loading of 10 kN/m can be estimated from Table 3.5. Approximately, the values are $N = 154.6$ kN and $V = 133.1$ kN. Following on from this:

$$f_v = \frac{133.1 \times 10^3}{12.8 \times [467.4 - (2 \times 19.6)]} = 24.3 \, \text{N/mm}^2$$

$$f_m = \sqrt{f_y^2 - 3f_v^2} = \sqrt{275^2 - (3 \times 24.3^2)} = 271.8 \, \text{N/mm}^2$$

$$2a = \frac{N}{f_m t} = \frac{154.6 \times 10^3}{271.8 \times 12.8} = 44.4\,\text{mm}; \quad a = 22.2\,\text{mm}$$

$$M_{pw} = \frac{t d_w^2}{4} f_y = \frac{12.8 \times (467.4 - 19.6)^2}{4} \times 275 \times 10^{-6} = 161.4\,\text{kNm}$$

The revised reduced plastic moment of resistance of the column section is given by:

$$M' = M_P - \left(\frac{f_y - f_m}{f_y}\right) M_{pw} - t a^2 f_m$$

$$= (2234 \times 10^3 \times 265 \times 10^{-6}) - \left(\frac{275 - 271.8}{275}\right) \times 161.4$$

$$- (12.8 \times 22.2^2 \times 271.8 \times 10^{-6})$$

$$= 588.4\,\text{kNm}$$

Step 7 *Determine final internal forces in the members*

The plastic hinges will form in the rafter at B when the loading is $473.2 \div 51.67 = 9.1\,\text{kN/m}$ on the rafter because unit uniformly distributed loading produces a bending moment equal to 51.67 kNm at B (see Table 3.5). Ultimate loading is equal to 10 kN/m. Therefore, the load remaining to be put on the rafter is equal to $10 - 9.1 = 0.9\,\text{kN/m}$.

The effect of the remaining load of 0.9 kN/m will be according to the results of Analysis 2 from Table 3.5. The final internal forces in the members are the results of Analysis 1 multiplied by 9.1 plus the results of Analysis 2 multiplied by 0.9. The serviceability limit state results are obtained by multiplying the results of Analysis 1 by 6.6.

Table 3.6 Table of final results.

Location: Column/Rafter	Serviceability limit state				Ultimate limit state			
	M (kNm)	N (kN)	V (kN)	Deflection (mm)	M (kNm)	N (kN)	V (kN)	Deflection (mm)
A: Column	+317.7[s]	102.0[c]	87.8	0	+540.6[s]	154.6[c]	134.8	0
B: Column	−341.0[h]	102.0[c]	87.8	30.2[x]	−473.2[h]	154.6[c]	134.8	62.6[x]
B: Rafter	−341.0[h]	110.0[c]	77.7	30.2[x]	−473.2[h]	168.2[c]	117.3	62.6[x]
Rafter +ve	+130.2[s]	85.2[c]	0	112.8[y]	+238.9[s]	130.7[c]	0	231.7[y]
C: Rafter	+94.9[s]	85.2[c]	21.3	120.2[y]	+184.0[s]	130.7[c]	32.7	247.7[y]

[s] = sagging; [h] = hogging; [c] = compression; [x] = X-direction; [y] = Y-direction.

132 Structural Steelwork

Step 8 Check stresses at serviceability limit state

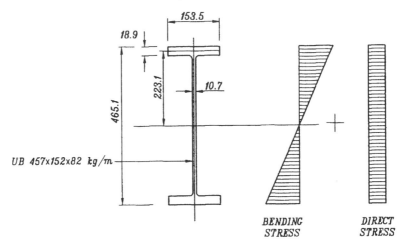

SK 3/25 Principal stress check at serviceability limit state.

Check stress in rafter section at B

Maximum direct stress in flange

$$= \frac{M}{Z} + \frac{N}{A} = \frac{341 \times 10^6}{1559 \times 10^3} + \frac{110 \times 10^3}{105 \times 10^2} = 229 \, \text{N/mm}^2$$

The maximum stress is less than $p_y = 265 \, \text{N/mm}^2$

Maximum principal stress at the junction of flange and web is given by:

$$\text{Direct stress} = \sigma_x = \frac{My}{I} + \frac{N}{A} = \frac{341 \times 10^6 \times 213.7}{36\,250 \times 10^4} + \frac{110 \times 10^3}{105 \times 10^2}$$

$$= 211.5 \, \text{N/mm}^2$$

$$\text{Shear stress} = \tau_x = \frac{V}{(D-2T)t} = \frac{77.7 \times 10^3}{[465.1 - (2 \times 18.9)] \times 10.7}$$

$$= 17 \, \text{N/mm}^2$$

$$\text{Principal stress} = \frac{\sigma_x}{2} + \sqrt{\left(\frac{\sigma_x}{2}\right)^2 + \tau_x^2}$$

$$= 212.9 \, \text{N/mm}^2 < 275 \, \text{N/mm}^2 = p_y$$

Maximum deflection at service limit state $= 120.2 \, \text{mm} = \dfrac{l}{250}$

All stresses are below the design strength at serviceability limit state.

Step 9 Calculate strength reserve

Plastic hinges will form in the column at A when the bending moment reaches 588.4 kNm, as determined in Step 6. The ultimate bending moment at A is 540.6 kNm from Table 3.6. The increase in bending moment for each kN/m of loading on the rafter at this stage is 113.9 kNm, as per Table 3.5 Analysis 2. Additional loading on the rafter beyond the ultimate loading required to cause failure is given by:

$$\frac{588.4 - 540.6}{113.9} = 0.42 \text{ kN/m}$$

At a loading of 10.42 kN/m, further plastic hinges may form in the columns at A and failure may take place. There is only 4.2% reserve strength in the structure above the ultimate load. There is no need to carry out any more refinement in the design to achieve economy.

Note: This portal frame is designed without haunches. Haunches may still be required at the connections of the rafters to the columns because it might not be possible to design the rafter connection to withstand the full plastic moment capacity, even with extended end plates.

3.7.1 Haunched pitched portal frame

The pitched portal frame shown in SK 3/22 is reanalysed with haunches at the connections of the rafters to the columns. The haunch geometry is shown in SK 3/26, with a reduced section of the rafter as a first trial. The plastic analysis is carried out assuming that the plastic hinges will form where the haunch ends on the rafter. It is assumed that the haunch is identical to the rafter in flange width and web thickness. The section classification of the haunch at its deepest part at the connection with the column should be at least compact as per Clause 5.3.4 of BS 5950: Part 1.

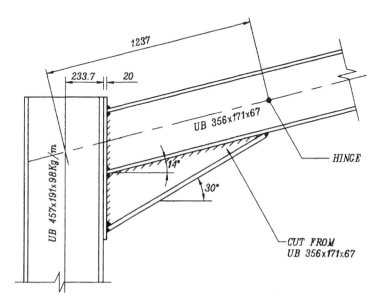

SK 3/26 The geometry of the haunch at the connection of rafter to column.

Step 3 Choose trial section for rafter and columns

Choose a reduced section of rafter UB 356 × 171 × 67 kg/m and use the same section for the column as before, i.e. UB 457 × 191 × 98 kg/m.

The first hinge will form in the rafter 1237 mm from the intersection of the centre-lines of the column and the rafter as shown in SK 3/26. The average increase of depth of the rafter in the haunch is 155 mm. In the analysis an allowance is made for this increased stiffness at the connection by modelling the haunched part of the rafter as a separate member with increased area and inertia.

Member properties
Rafter: UB 356 × 171 × 67 kg/m; $D = 364$ mm; $B = 173.2$ mm; $t = 9.1$ mm; $T = 15.7$ mm; $d = 312.2$ mm; $b/T = 5.52$; $d/t = 34.3$; $I = 19\,540$ cm^4; $A = 85.5$ cm^2; $Z = 1073$ cm^3; $S_x = 1213$ cm^3
Section classification is plastic.
Haunch: $A = 85.5 + (17.32 \times 1.57) + (15.5 \times 0.91) = 126.8$ cm^2

$$I \approx 19\,540 \times \left(\frac{519}{364}\right)^2 = 39\,724 \text{ cm}^4$$

Step 4 Carry out analysis with revised section properties

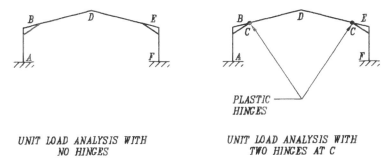

UNIT LOAD ANALYSIS WITH NO HINGES

UNIT LOAD ANALYSIS WITH TWO HINGES AT C

SK 3/27 Sketch of portal showing the two unit load analysis criteria.

Table 3.7 Results of analyses.

Location: Column/Rafter	Analysis 1: No plastic hinges				Analysis 2: Two plastic hinges at C			
	M (kNm)	N (kN)	V (kN)	Deflection (mm)	M (kNm)	N (kN)	V (kN)	Deflection (mm)
A: Column	+55.59[s]	15.46[c]	14.41	0	+122.65[s]	15.46[c]	18.01	0
B: Column	−52.47[h]	15.46[c]	14.41	6.01[x]	−12.41[h]	15.46[c]	18.01	23.9[x]
B: Rafter	−52.47[h]	17.73[c]	11.51	6.01[x]	−12.41[h]	21.22[c]	10.63	23.9[x]
C: Rafter	−38.98[h]	17.43[c]	10.31	5.95[y]	0	20.92[c]	9.43	25.2[y]
D: Rafter	+9.46[s]	13.98[c]	3.49	24.67[y]	+36.03	17.47[c]	4.37	96.2[y]

[s] = sagging; [h] = hogging; [c] = compression; [x] = X-direction; [y] = Y-direction.

Analysis of Structures   135

Step 5 *Calculate reduction of plastic moment of resistance of rafter section*
Assume maximum direct load and shear force acting simultaneously with the bending moment are as follows: $N = 17.43 \times 10 = 174.3$ kN and $V = 10.31 \times 10 = 103.1$ kN for the ultimate load of 10 kN/m. Therefore:

$$f_v = \frac{V}{td_w} = \frac{103.1 \times 10^3}{9.1 \times [364 - (2 \times 15.7)]} = 34.0 \, \text{N/mm}^2$$

$$f_y = p_y = 275 \, \text{N/mm}^2 \text{ from Table 6 of BS 5950: Part 1}$$

$$f_m = \sqrt{f_y^2 - 3f_v^2} = 268.6 \, \text{N/mm}^2$$

$$2a = \frac{N}{f_m t} = \frac{174.3 \times 10^3}{268.6 \times 9.1} = 71.3 \, \text{mm}; \quad a = 35.7 \, \text{mm}$$

$$M_{pw} = \frac{td_w^2}{4} f_y = \frac{9.1 \times [364 - (2 \times 15.7)]^2}{4} \times 275 \times 10^{-6} = 69.2 \, \text{kNm}$$

$$M' = M_P - \left(\frac{f_y - f_m}{f_y}\right) M_{pw} - ta^2 f_m$$

$$= (1213 \times 10^3 \times 275 \times 10^{-6}) - \left(\frac{275 - 268.6}{275}\right)$$

$$\times 69.2 - (9.1 \times 35.7^2 \times 268.6 \times 10^{-6}) = 328.8 \, \text{kNm}$$

This plastic moment of resistance of the rafter section will be used in the analysis.

Step 6 *Calculate reduction of plastic moment of resistance of the column section*
This step remains unchanged from the previous analysis of the portal without haunches. Plastic moment of resistance of the column section is 588.4 kNm.

Step 7 *Find final internal forces in the structure*
The plastic hinges will form in the rafter at location C when the loading has reached $328.8 \div 38.98 = 8.44$ kN/m, because the unit uniformly distributed load produces a 38.98 kNm bending moment at location C on the rafter.

Load remaining to be put on the rafter $= 10 - 8.44 = 1.56$ kN/m.

The effect of this additional load will be as shown by the results of Analysis 2 in Table 3.7.

The final internal forces in the members are the results of Analysis 1 multiplied by 8.44 plus the results of Analysis 2 multiplied by 1.56. The serviceability limit state results are obtained by multiplying the results of Analysis 1 by 6.6.

Check: The bending moment at A in the column at the ultimate limit state is equal to $(55.59 \times 8.44) + (122.65 \times 1.56) = 660.5$ kNm $>$ 588.4 kNm. This is in excess of the plastic moment of resistance of the column, hence failure may take place before the ultimate load is reached. There are three possible solutions:

(1) Increase the depth of the haunch, which also increases the length of the haunch.
(2) Increase the size of the rafter.
(3) Increase the size of the column.

The most cost effective alternative could be to increase the length and depth of the haunch. The depth may be increased to a maximum value of 400 mm if the rafter section has to be used to form the haunch. Assume a haunch depth of 400 mm and a length of 1500 mm. Repeat the calculations of Steps 3 to 7. The structure should be able to carry the ultimate loading with this revised geometry.

Step 8 *Carry out stress checks at serviceability limit state*

Step 9 *Calculate reserve of strength*

Step 10 *Design restraint system*
Haunched portals designed by the method of plastic analysis must have the restraints as specified in Section 5.5 of BS 5950: Part 1. The following list shows the necessary steps.

- Provide torsional restraint at the end of the haunch as per Clause 5.3.5.
- Find minimum distance of the adjacent restraints from the hinge restraint as per Clause 5.3.5.
- Provide note on the drawing regarding the fabrication restriction as per Clause 5.3.7.
- Check if stiffeners are required at the hinge location as per Clause 5.3.6.
- Check sway stability with notional horizontal force applied at the top of the column, when the deflection must not exceed $h/1000$, h being the height of the column.
- Check haunch restraints as per Figure 10 of BS 5950: Part 1.
- Check column stability as per Clause 5.5.3.4.
- Check rafter stability as per Clause 5.5.3.5.

Chapter 4
Design of Structures

4.1 PRINCIPAL ISSUES

The criteria which govern the design of structures may be summarised as follows:

Fitness for purpose is generally covered by the overall geometry of the structure and its components. It should be possible to have unrestricted and unhindered use of the structure for its intended purpose.

Safety and reliability are assured by following the Codes of Practice for loading, materials, design, manufacture, erection, corrosion protection and fire resistance.

Durability is satisfied by the choice of the right materials for the purpose, including protection systems compatible with the environmental exposure.

Value for money is a very important criterion. The designer should take into account the whole life cost of the structure, including the maintenance requirements and the cost of demolition/decommissioning.

External appearance of any structure changes over time. Designers should make appropriate allowances in their choice of materials and finishes to avoid rapid degradation of appearance.

User comfort is influenced by vibration of the structure as a result of wind, road/rail traffic and vibrating machinery. Large deflections cause alarm to users. Designers should take appropriate measures to minimise this discomfort.

Robustness comes with the chosen structural form. It is a measure of the additional inherent strength of the structure to withstand accidental overloads. Failure of a component part of the structure must not initiate global collapse.

Buildability is perhaps the most important issue – it includes the manufacture, handling, transport and erection of components. Good design would most necessarily mean good buildability. It has an enormous impact on overall cost of the structure and to understand the principles of buildability it is necessary to know the manufacturing process, the handling problems and the transport restrictions. It should be borne in mind that the lowest material weight does not necessarily mean the lowest cost. The fundamental design considerations which influence buildability are discussed below.

4.2 MATERIAL GRADE SELECTION AND SECTION TYPE SELECTION

The use of small quantities of different grades and section sizes should be avoided because it becomes more expensive to purchase from stockholders. Large quantities of sections within a serial size can be purchased directly from the steel mills in lengths up to 15 m. Buying from a stockholder could also increase wastage because only standard stock lengths of material are generally available.

The problem with purchasing directly from the steel mills lies in the fact that the current rolling programme of a particular section size may already have been fully subscribed and the next programme of rolling of the same size may not suit the contract requirements. Popular section sizes in large quantities, if selected by the designers, may tend to increase the total tonnage of material but could improve the buildability in terms of overall cost and time. Small quantities of high grade material have stiffer price penalties.

The cost penalty for Design Grade 50 is small compared to its strength advantage over Grade 43. Grade 50 should be specified unless the design is governed by deflection only. Angles are quite difficult to obtain in Grade 50. Structural hollow sections are very efficient for axial compression and torsional capacity. These sections are 60–80% more expensive than open sections and involve connection by welding. They frequently require additional stiffening at connections because the wall thickness may not be adequate. Also, such sections are only available in standard mill lengths, which means extra wastage and/or additional splicing by butt welding. Seamless tubes with thicker walls are much more expensive. The materials for all the fittings need not be related to the material of the members. They should be chosen by the fabricator on the basis of the design requirements for strength and performance.

4.3 MANUFACTURING PROCESS

The modern manufacturing process is described very briefly here to give an appreciation of the steps involved. It starts with blast cleaning of material from stock and the application of a prefabrication primer. (Some fabricators choose to carry out the blast cleaning after cutting the sections to size and after carrying out the holing operations.) After blast cleaning and priming, the next stage is cutting, which can be done in many ways. The most popular technique is to use a circular saw in conjunction with a conveyor system. This can be done fairly accurately by utilising a measuring device.

Band sawing, motor-operated hacksawing, guillotining, flame cutting and plasma cutting are other techniques practised in modern fabrication shops. Plates are generally flame cut or plasma cut. Flats and angles are generally cropped using specialised handling and cutting machines.

The next operation is holing. In modern fabrication shops CNC (computer numerically controlled) beam-drilling lines are used. There are usually three drill heads, which can carry out simultaneous drilling operations in the top and bottom flanges and the web of a beam. This system operates in conjunction with a cutting and conveyor network operated by CNC software. All fittings and short members are drilled manually. Fabricators generally prefer punching to drilling holes. This is normally permitted for untensioned bolts up to a material thickness equal to the hole diameter.

Flange thinning, stripping and coping are other operations carried out under CNC instructions or by marking manually and flame cutting. Vertical surface milling is sometimes used with deep sections to achieve the desired accuracy of length. Horizontal surface milling of bearing surfaces is also carried out to achieve specified flatness criteria.

Profile preparation of circular hollow-section (CHS) ends for welding is generally done by plasma cutting machines under CNC instruction. Manually, this end preparation is an expensive and time-consuming operation involving flame cutting and grinding.

Joining of components in the shop is done by welding or bolting. Jigs and templates are used for positioning the components to be joined. Semi-automatic metal inert gas (MIG) welding techniques are used with wire-fed welding equipment.

4.4 CONNECTION DESIGN

The raw material cost of the finished product is only 35–50%. The rest of the cost is in fabrication and handling in the shop. The connections

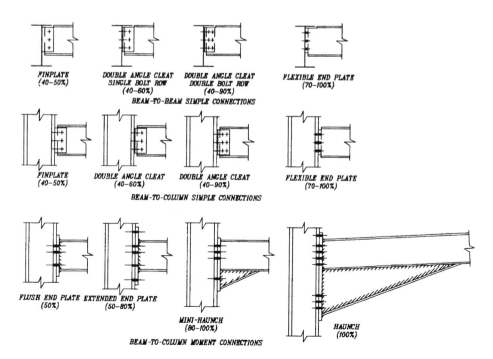

SK 4/1 Example connections showing degrees of complexity and percentage shear carrying capacity of the connected member.

can influence up to 65% of the cost of the production. Member sizes should be chosen on the basis of connection requirements rather than the actual design requirement (i.e. if an increase in the size of the member could simplify the connections, one should think seriously about increasing the size of member). Design drawings should show the philosophy of connection assumed by the designer such that the complexity is fully understood at tender stage and the best price can be obtained. This also avoids future claims and delays in the event of complications during fabrication.

4.4.1 Simple connections

These are the cheapest to fabricate but, depending on the shear load they transfer, their complexity may go up and increase the fabrication costs accordingly. In SK 4/1 the generally accepted types of simple connections are shown with a reference to their shear carrying capacity and complexity. The carrying capacity in percentage terms relates to the shear capacity of the member being connected. In this type of low-cost connection philosophy one would find primary beams with end cleats or part-depth end plates for connections to columns. Secondary beams will have notched ends and be drilled in the web for connection to fin plates. The columns will have simple base plates and splices with flanges drilled for primary beams and webs, with fin plates attached for the connection of secondary beams. This philosophy has the lowest fabrication cost.

4.4.2 Moment connections

These connections can be moderately or highly complex, depending on the moment and shear they transfer. The fabrication costs go up quite significantly in line with the degree of complexity. In SK 4/1 the standard types of moment connections are shown with reference to their degree of complexity and their moment carrying capacity. The carrying capacity in percentage terms relates to the ultimate plastic moment capacity of the member being connected.

In a moderately complex moment connection one will find primary beams with end plates butt welded to the top flange or extended end plates with fillet welds. The primary beam will have fin plates or holes for the connection of secondary beams. The secondary beams will be notched and will have end plates or a double row of angle cleats. The columns may have base plates without stiffeners and tensile stiffening of the flanges may be required at the primary beam connection. Secondary beam-to-web of column connection will be by means of double angle cleats.

In a highly complex moment connection one will find primary beams with haunched end plates and will have holes for the connection of secondary beams by double angle cleats or end plates. The secondary beams will have flexible end plates or double angle cleats and they will be notched. Columns will have base plates with stiffeners, and at the connection with primary beams they will have tensile as well as compressive stiffeners. The secondary beams with end plate connections to the webs of the columns may have welded T-stubs if fin plates are not adequate.

Note: The cost of fabrication of a high-cost moment connection could be six to eight times more than that of a simple connection.

4.4.3 Trusses and open-web girders

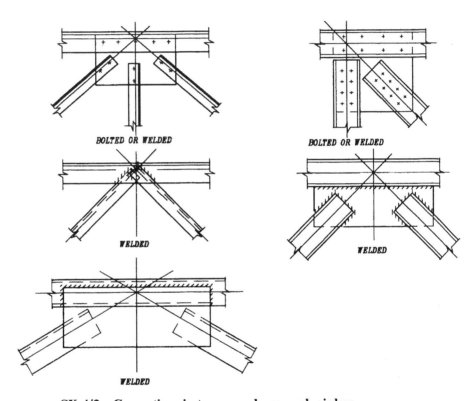

SK 4/2 Connections in trusses and open-web girders.

Low-cost connections
All monoplanar trusses with two continuous booms and internal bracing where the joints are assumed 'pinned' may have low-cost connection provided the booms are made of UB, UC, RSJ, ST, RSC, RSA or RHS with gusset plates bolted or welded at the intersections with the internal bracing. The internal bracing is generally made of RSA, RSC, RSJ, UC or UB, using bolts for connections to the gusset plates. When RSJ, UC or UB are used, flange stripping may be necessary for connection to gusset plates, requiring extra fabrication work. The cost of fabrication goes up with welded internals. The boom faces for connections of the internals should be wider than the internals to allow fillet welds all round. The welded connections require a lot more handling in the shop.

Medium-cost connections
If monoplanar trusses require stiffening of the booms at the connections because the forces are high, then the fabrication costs are increased considerably. Any moment connection at the internal bracing, as in a Vierendeel truss, will increase the complexity of the connection. The types of sections for the booms and the internals remain the same as in a low-cost connection option. Welded connections at the intersections requiring stiffening add even more to the cost of fabrication.

High-cost connections
All monoplanar trusses made of circular hollow sections with welded intersections require full profile preparation for the internals. Some connections may require the thickening of walls (by 'canning' or by the use of 'saddles') to prevent high local stress concentrations. Through-fitted and welded gussets are also used. These connections require a great deal of time, effort and handling in the shops.

If the trusses are in more than one plane then the complexity of the connection is more severe. Jigging in three dimensions adds to the cost of fabrication. Problems associated with multi-positional welding and numerous handling and transportation processes should be carefully considered by the designers at the conceptual design stage.

In SK 4/2 the different types of connections of the truss are illustrated. It should be emphasised again that a low raw material weight is not necessarily the lowest cost option.

A check list of actions and design considerations for designers is included here to show how to achieve a design which satisfies the good buildability guidelines. The check list is split into four parts, each one covering the considerations relevant to the overall activity during that stage. Designers will benefit from this check list because at each design stage it will enable them to take into account the problems which could be encountered later in the manufacturing, transportation and erection stages.

4.5 CHECK LIST OF ACTIONS AND DESIGN CONSIDERATIONS

(1) Conceptual design stage
- Use as few different section sizes as possible in any one structure.
- Use same grade of steel in individual projects.
- Use Grade 50 steel whenever possible unless deflection governs the design.
- Structural hollow sections are 60–80% more expensive and they should be used with care.
- High strength friction grip (HSFG) bolts should be used only where it is absolutely necessary from the design point of view.
- Do not mix welding and bolting in the same piece to be fabricated.
- Consider weldability of materials before specifying welding.
- Use rolled steel beams in place of fabricated latticed girders wherever possible.
- Avoid using multi-planar trusses.

(2) Detailed design stage
- Do not specify small quantities of high grade steel.
- Increase member size to avoid stiffening at connections.
- Select section sizes for members of a truss such that local stiffening at intersections is avoided.
- Avoid haunched beam-end connections.
- Avoid tension and compression stiffening in columns – use heavier sections instead.
- Avoid using 'canning' or 'saddles' in CHS truss fabrication – use thicker walled sections instead.
- Avoid using RSJ, UB and UC as truss internals because flange stripping may be necessary at the intersections.
- Use bolted internals in a truss in preference to welded internals.
- Do not specify intermittent fillet welds where there is a possibility of moisture ingress.
- Remember, a continuous smaller size fillet weld is cheaper than an intermittent fillet weld.
- Do not over-specify weld size.
- Drawings must show the design philosophy of the connections.
- Do not use different grades of bolt of the same diameter.
- Standardise the bolts to be used – stick to M20 Grade 8.8 for shear connection and M24 Grade 8.8 for moment connections wherever possible.
- Use fillet welds of up to 12 mm leg length in preference to butt welds of equivalent strength.
- Use site bolting in preference to site welding.

(3) Manufacturing stage
- Design single end cuts square to the member length.
- Use one hole diameter on any one piece.
- Align holes on axis square to the member length.
- Do not use staggered holes.
- Use adequate side clearance for holes in webs.
- Rationalise the range of fitting sizes.
- Do not specify the grade of material for the fittings – leave it to the fabricator.
- Provide adequate access for welding.
- Consider the effects of distortion due to welding.
- Weld inspection and acceptance criteria must be clearly specified.
- Specify single coat of protection system during fabrication.
- Allow the fabricator to choose the method of corrosion protection. Specify only the performance criteria.
- Always check the limits of transportation before designing components and parts. See Tables 4.4 and 4.5 at the end of the chapter.
- Avoid using structural hollow sections for long members which are beyond the standard length available from stockholders.
- Limit the size of frames to be transported to less than 5 m wide and 27.4 m long.

- Be aware that dimensional limits apply not only to transport — handling in a fabricator's workshop, off-site painting yards and galvanising facilities will also impose size restrictions.
- Stability of part frames used for manufacture and transport should be considered at the design stages.

(4) *Erection stage*
- Permit the use of fully threaded bolts wherever possible.
- Always specify washers to protect the surface finish.
- Specify that bolts, nuts and washers should be pre-coated with the designed corrosion protection system before delivery to the site.
- Do not specify a corrosion protection system unless it is required by the environmental exposure conditions.
- Make sure that the corrosion protection system is compatible with the fire protection scheme.
- Do not use composite construction with welded through-deck stud shear connectors to the top flanges of beams in environments where corrosion protection is required. It is impossible to paint the top flanges after the metal decking has been installed and shear connectors have been fitted.
- Do not specify shop-applied decorative finishes because they will be damaged in handling during transport and erection.
- Specify the procedure for removal of temporary lifting points in the structural components.
- Consider adding temporary support to an incomplete structure during manufacture, handling, transport and erection.

4.6 ULTIMATE LIMIT STATE DESIGN

There are three basic types of design. They are described below.

(1) *Simple design*
Beam-to-column connections are assumed simple and do not develop bending moments. Structural members intersecting at a point are assumed pin jointed. Lateral stability of the structure is maintained by vertical bracing or shear wall.

(2) *Rigid design*
Beam-to-column connections are assumed rigid, and the connection geometry is not altered by the application of load on the structure. The connections are designed to ensure full continuity of members.

(3) *Semi-rigid design*
It is assumed that some degree of stiffness is offered by the joints. The beam-to-column connections are designed to withstand up to 10% of the maximum free span bending moment in the beam. This 10% restraint moment at the end connections is taken into account to find the design bending moment in the beams. Columns are designed to carry this additional restraint moment from the connection. The structure should be braced in both directions for lateral stability.

BS 5950 permits semi-rigid designs where a limited amount of moment transfer can be provided by the connection. In order to limit the high cost of complicated connection details, it is possible to limit the connection stiffness based on a simple connection geometry as in an extended end plate connection or flange cleat connection. Elastic analysis results with factored loads will predict much higher connection forces, but by limiting these connection forces to the pre-designed connection capability, one could achieve a degree of economy.

The advantages are:
- Reduction in the depth of the beam.
- Reduction in the cost of the frame.
- Reduction in fabrication cost by avoiding complexity.
- Improved exploitation of composite construction where the mid-span plastic moment capacity in a beam is very high.

The disadvantages are:
- Greater deflection at service loads.
- With a high proportion of live load, unless the designer is very careful, joints may go plastic at service load and ratchetting will occur, resulting in large unacceptable deflections over a period of time.
- Full service load stress checks are necessary to make sure all components at the joints are below yield.
- The design is unacceptable for seismic loading because the ductility of the structure depends on the formation of plastic hinges which must *not* form at the connections because the connections are brittle compared with the member, where the hinges should be allowed to form. In seismic design the connections should be made significantly stronger than the member capacity.
- At least 70% of the elastically analysed moments in a rigid unbraced frame must be catered for by the connections, otherwise at service loads the components of the joint may go beyond yield.
- Much less available reserve strength against accidental overload – sacrifice of robustness.

4.6.1 Step-by-step method of semi-rigid design

Step 1 Carry out elastic analysis with factored loads and find the design bending moments at the beam–column junctions.

Step 2 Reduce design bending moments at the connection to 70% of the elastically analysed value.

Step 3 Design a simple moment connection with either end plates or flange angle cleats, whichever is functionally acceptable. Check that no column stiffening is required.

Step 4 Determine plastic moment capacity of the designed connections.

Step 5 Determine plastic moment capacities of the sections in the frame at critical points.

Step 6 Carry out unit load analyses of the frame assuming progressive formation of hinges. Several analyses of the frame will be required, depending on the degree of redundancy. The last hinge should make the structure into a mechanism. An example of the unit load analyses is shown in the plastic frame analyses in Chapter 3.

Step 7 Firstly load the complete undamaged structure till the first set of hinges is formed. This first set of hinges must never form at or below the service load. No part of the structure should go beyond yield at service load.

Step 8 After the first set of hinges has formed, use the results of the analysis with unit loading of the frame in Step 6 with these hinges and apply further loading till the next set of hinges forms. Carry on using the next revised frame with more hinges from the previous one till such time as you have reached the ultimate loading or the structure has become a mechanism.

Step 9 Check that at ultimate loading the structure is not a mechanism. Find the reserve strength or robustness by applying further load beyond the ultimate load till failure occurs.

Note: The unit load frame analysis with a progressively increasing number of hinges is a very useful tool because the rest of the analyses can be done simply by hand.

4.6.2 Load combinations at the ultimate limit state as per BS 5950: Part 1

Basic load cases:
DL = Dead load
IL = Imposed load
CLV = Crane vertical load
CLH = Crane horizontal load
TLR = Thermal load due to rise in temperature
TLF = Thermal load due to fall in temperature
WL = Wind load
NHF = Notional horizontal force (Clause 2.4.2.3 of BS 5950: Part 1)

The combinations of basic load cases with appropriate load factors at the ultimate limit state are given in Table 4.1. Load cases 11, 12, 13, 14, 37, 38, 39 and 40 are required to check uplift and overturning of the structure. Load cases 41 and 42 are required to check the lateral stability of the structure with a notional horizontal force. The notional horizontal force is applied at a floor level. This is taken as equal to the greater of either 1% of the factored dead load at the floor level or 0.5% of the combined factored dead and imposed loads at the floor level. The load factors to be used for this computation are 1.4 for the dead load and 1.6 for the imposed load.

Design of Structures   147

Table 4.1 Load combinations at ultimate limit state.

Load case (LC)	DL	IL	CLV	CLH	TLR	TLF	WL	NHF
1	1.4	1.6			1.2			
2	1.4	1.6				1.2		
3	1.2	1.2			1.2		1.2	
4	1.2	1.2			1.2		−1.2	
5	1.2	1.2				1.2	1.2	
6	1.2	1.2				1.2	−1.2	
7	1.4				1.2		1.4	
8	1.4				1.2		−1.4	
9	1.4					1.2	1.4	
10	1.4					1.2	−1.4	
11	1.0				1.2		1.4	
12	1.0				1.2		−1.4	
13	1.0					1.2	1.4	
14	1.0					1.2	−1.4	
15	1.2		1.2	1.2	1.2		1.2	
16	1.2		1.2	−1.2	1.2		1.2	
17	1.2		1.2	1.2	1.2		−1.2	
18	1.2		1.2	−1.2	1.2		−1.2	
19	1.2		1.2	1.2		1.2	1.2	
20	1.2		1.2	−1.2		1.2	1.2	
21	1.2		1.2	1.2		1.2	−1.2	
22	1.2		1.2	−1.2		1.2	−1.2	
23	1.2	1.2	1.2	1.2	1.2			
24	1.2	1.2	1.2	−1.2	1.2			
25	1.2	1.2	1.2	1.2		1.2		
26	1.2	1.2	1.2	−1.2		1.2		
27	1.4		1.4	1.4	1.2			
28	1.4		1.4	−1.4	1.2			
29	1.4		1.4	1.4		1.2		
30	1.4		1.4	−1.4		1.2		
31	1.4		1.6		1.2			
32	1.4		1.6			1.2		
33	1.4			1.6	1.2			
34	1.4			−1.6	1.2			
35	1.4			1.6		1.2		
36	1.4			−1.6		1.2		
37	1.0			1.6	1.2			
38	1.0			−1.6	1.2			
39	1.0			1.6		1.2		
40	1.0			−1.6		1.2		
41	1.4	1.6						1.0
42	1.4	1.6						−1.0

4.6.3 Patterned loading

With significant live load on the span a patterned loading sequence could produce higher design bending moments. The distribution of live loads on spans to produce the maximum effect may be carried out following the simple guidelines in Table 4.2.

Table 4.2 Guidelines for patterned loading on beams.

Type of structure	Location of bending moment	Distribution of live loads to produce maximum bending moments
Continuous beam	Support: −ve hogging	Adjacent spans loaded with one span unloaded on either side
	Ultimate span: +ve sagging	Ultimate span loaded and no load on penultimate span
	Any other span: +ve sagging	Span loaded but adjacent spans unloaded
Single/multistorey frame structure	Beam bending moments	Same as for continuous beam
	End column moments	Adjacent span loaded and penultimate span unloaded
	Central column moments	Larger adjacent span loaded and the other adjacent span unloaded

4.6.4 Structural stability against lateral loads

At the concept stage of the design of a structure the type of global framework for lateral stability should be determined. The requirements of lateral stability provided by braced-bay frames, rigid frames and shear walls have to be decided at the concept stage. Some guidance about the requirements of any chosen global structural form is given in Table 4.3.

4.6.5 Stability of multistorey rigid frames

Elastic design of sway frames
(1) If the effective lengths of the columns are obtained by the limited frame method as described earlier, then the stability of the frame is satisfied.
(2) If the effective lengths of the columns are taken as $1.0L$, then the moments in the columns due to the horizontal loads should be multiplied by a correction factor to include the effects of vertical loads present. This factor may be taken as:

$$\frac{\lambda_{cr}}{\lambda_{cr} - 1}$$

where λ_{cr} = elastic critical load factor. To determine λ_{cr}, the sway index ϕ_s has to be found for each storey height. By elastic analysis of the rigid frame against horizontal loads find deflections at every storey level. The horizontal load should be taken as 0.5% of all factored vertical loads at the level under consideration. The sway index is given by:

$$\phi_s = \frac{\delta_u - \delta_l}{h}$$

Design of Structures 149

Table 4.3 Bracing systems.

Ties	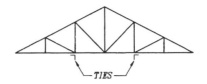
Rafter bracing	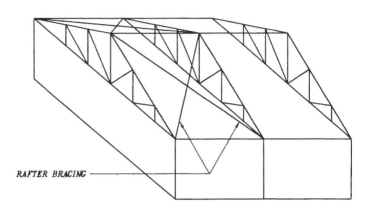
Lateral bracing of roof truss	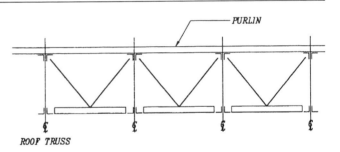
Wind girders: gable and longitudinal plan bracing	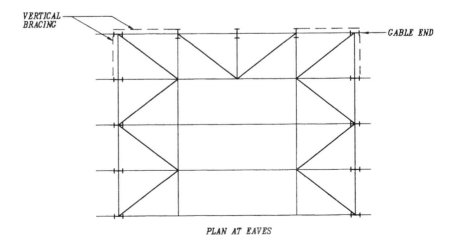

150 Structural Steelwork

Table 4.3 (contd)

Vertical bracing: longitudinal and transverse (plan)

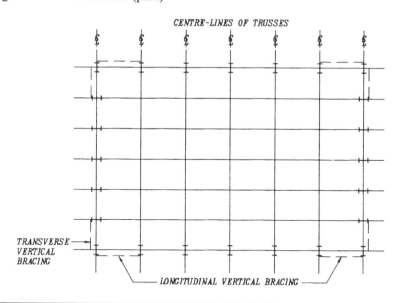

Vertical bracing elevations of Z- and X-bracing

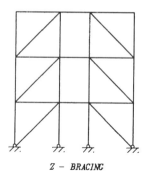 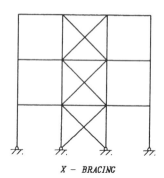

Vertical bracing elevations of V- and K-bracing

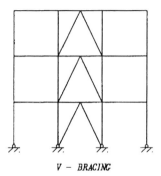 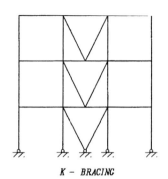

Table 4.3 (contd)

Eccentric vertical bracing for ductile seismic response

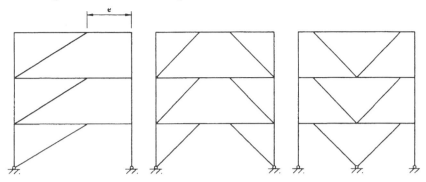

Von Mises criteria should be satisfied due to presence of high shear loads

$$\left(\frac{N}{N_u}\right)^2 + \left(\frac{V}{V_u}\right)^2 \leq 1.0; \quad N_u = Af_y; \quad V_u = 0.55 dt_w f_y$$

Anti-sag systems

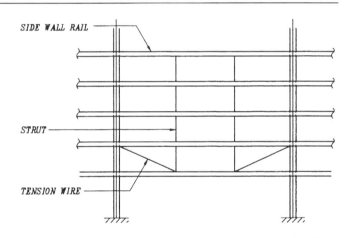

Torsional restraints and tension and compression flange restraints

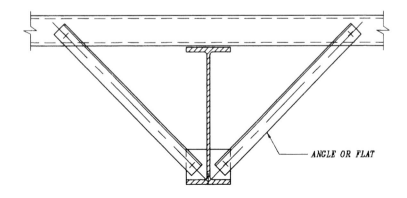

Table 4.3 (contd)

Types of framing	Bracing/ties/restraints required
Single-storey single bay or multi-storey multi-bay portal frames	Rafter bracing; wind girders at gable end; longitudinal vertical bracing; anti-sag systems; tension/compression flange restraints; torsional restraints at plastic hinge
Open-web latticed girders and roof trusses pin-jointed to column	Ties; rafter bracing; wind girders at gable end; longitudinal wind girders; longitudinal vertical bracing; transverse vertical bracing; lateral bracing of roof truss; anti-sag systems
Multistorey building frames	Longitudinal vertical bracing; transverse vertical bracing; plan bracing at each floor level
Multistorey building frames with shear walls	Plan bracing at each floor level connecting to shear wall; vertical bracing also required if shear wall is not central to structure footprint

where δ_u = horizontal deflection at the top of the column
δ_l = horizontal deflection at the bottom of the column
h = height of the storey or the column

Find the maximum value of the storey index called $\phi_{s,max}$ and λ_{cr} is given by:

$$\lambda_{cr} = \frac{1}{200\phi_{s,max}}$$

Plastic design of sway frames
(1) Carry out full elastoplastic sway analysis using different patterned loading scenarios. Use the limited frame method to determine the effective lengths of columns, in which case the graphs relating to the condition where the frame is braced against side sway may be used. The stability of the frame will be deemed to be satisfied.
(2) If the following conditions are satisfied, then the stability will also be deemed to be satisfied:

Condition 1: $\lambda_{cr} \geq 4.6$

Condition 2: $\lambda_p \geq \dfrac{0.9\lambda_{cr}}{(\lambda_{cr} - 1)}$ when $4.6 \leq \lambda_{cr} < 10$

Condition 3: $\lambda_p \geq 1$ when $\lambda_{cr} \geq 10$

λ_p is the factor by which each of the factored loads have to be multiplied to cause ultimate plastic collapse of the frame.

Design of Structures 153

Table 4.4 Vehicles complying with The Motor Vehicles (Construction and Use) Regulations 1986 SI 1078/36.

Notification and/or permission to move	Authorities and notice period			Weight (tonnes)		Width (m)	Length (m)	Comments
	Police	Road and bridge	Dept of Transport	Gross vehicle weight	Effective payload (max.)			
Unrestricted running	—	—	—	<38	Approx. 26	<2.9	<15.5 o/a	Maximum size of ordinary articulated vehicle including the tractor unit
Unrestricted running	—	—	—	<38	Approx. 26	0.305 lateral projection	<18.3 o/a or 1.83 front or 3.05 rear projection	If the load projects more than 1.83 m front or 3.05 m rear an attendant is required
Notice to authorities required	2 days clear	—	—	<38	Approx. 26	>2.9 <4.3	>18.3 <27.4 >3.05 front or rear projection	If the load is wider than 3.5 m, or longer than 18.3 m, or projects more than 1.83 m front or 3.05 m rear an attendant is required
Notice to authorities required	2 days clear	—	—	<38	Approx. 26	>4.3 <5.0	<27.4	Attendant is required. Speed limit of 20 mph (excluding motorways)
Notice to authorities required	2 days clear	—	6 weeks min.	<38	Approx. 26	>5.0 <6.1	<27.4	Attendant is required. Speed limit of 20 mph (excluding motorways). The maximum width of 6.1 m and length of 27.4 m can be exceeded if authorised by Special Order issued by the Secretary of State

Table 4.5 Vehicles operating within Motor Vehicles (Authorisation of Special Types) General Order 1979.

Notification and/or permission to move	Authorities and notice period			Weight (tonnes)		Width (m)	Length (m)	Comments
	Police	Road and bridge	Dept of Transport	Gross vehicle weight	Effective payload (max.)			
Unrestricted running	—	—	—	<38	Approx. 26	<2.9	<18.3 o/a	If the load projects more than 1.83 m front or 3.05 m rear an attendant is required
Notice to authorities required	2 days clear	—	—	<38	Approx. 26	>2.9 <3.5	>18.3 o/a <27.4 >3.05 front or rear projection	If the loads projects more than 1.83 m front or 3.05 m rear an attendant is required
Notice to authorities required	2 days clear	—	—	<38	Approx. 26	>3.5 <4.3	<27.4	An attendant is required
Notice to authorities required	2 days clear	—	—	<38	Approx. 26	>4.3 <5.0	<27.4	Attendant is required. Speed limit of 20 mph (excluding motorways)
Notice to authorities required	2 days clear	—	6 weeks min.	<38	Approx. 26	>5.0 <6.1	<27.4	Attendant is required. Speed limit of 20 mph (excluding motorways)
Notice to authorities required	2 days clear	2 days clear	—	>38 <80	Approx. 65	<6.1	<27.4	Speed limits of: 40 mph motorways 35 mph dual carriageways 30 mph other roads
Notice to authorities required	2 days clear	5 days	—	>80 <150	>65 <120 approx.	<6.1	<27.4	Speed limits of: 30 mph motorways 25 mph dual carriageways 20 mph other roads
Notice to authorities required	2 days clear	5 days	6 weeks min.	>150	126 approx.	>6.1	>27.4	The maximum weight of 150 tonnes, width of 6.1 m and length of 27.4 m can be exceeded if authorised by Special Order issued by the Secretary of State

Chapter 5
Design of Struts

5.1 AXIAL CAPACITY OF A COLUMN OR A STRUT

The capacity of a column or a strut to carry axial compression depends on the following factors generally:

(1) Its end conditions – how it is attached to other parts of the structure or foundation.
(2) Its material strength.
(3) Its cross-sectional shape.
(4) Its overall length.
(5) Its manufacturing process.
(6) The structural concept of the global structure of which it is only a part (i.e. braced, unbraced, sway frames cantilever etc.).
(7) The presence of local transverse loading or eccentric axial loading.
(8) The presence of simultaneously acting bending moment and shear, as in multistorey rigid frames.

5.2 TYPES OF FAILURE OF A COLUMN OR STRUT

Purely axial loading on a column may cause failure for the following reasons:

(1) *Squashing* – as in a very small stocky column where the yield stress p_y may be applied uniformly over the whole cross-sectional area.
(2) *Flexural buckling* – as in a slender column where the column deflects laterally about the weaker principal axis and the failure occurs before the entire cross-section can be loaded to the yield stress p_y.
(3) *Torsional buckling* – as in an unsymmetrical shape made of thin plate thickness where the member twists about its longitudinal axis and failure occurs before the section can attain the yield stress p_y. This type of failure is very uncommon and is normally encountered in cold-formed thin sections.
(4) *Local buckling* – as in a section where the proportion of individual elements is such that the whole member may not behave as a slender column (as in case 2 above) but local elements may buckle before the yield stress p_y is attained. In these cases, the local elements prone to buckling, because of their geometric proportions, must be designed with reduced design strength p_y in compression.

(5) *Local failure* – as in a compound section where the compound column as a whole is stronger than its main components. In a compound column the main components are tied together by lacing or battens. The main components may buckle about their weak minor axis between the points of restraint by the lacing or batten system.

5.3 DESIGN BASIS OF COLUMNS AND STRUTS

5.3.1 Compressive strength (p_c)

Basically there is only one consideration in the design of a column or strut, namely its slenderness. The slenderness affects the flexural buckling, the local buckling and the local failure of a compound strut.

The Euler theory of a pin-jointed column predicts a critical axial load under ideal conditions. If other important factors are taken into consideration, this critical load for failure is revised to give the current considerations in BS 5950: Part 1. The factors which influence the theoretical elastic critical load are:

- lack of straightness
- eccentricities of loading which are not accountable in design
- material variability
- material imperfections
- geometric imperfections
- residual stresses due to the manufacturing process

These effects cannot be fully addressed by numerical methods. Experimental observations have given rise to the adoption of the design curves in BS 5950: Part 1. These factors change from shape to shape and also according to the selection of manufacturing process. BS 5950: Part 1 therefore recommends the use of four different curves of compressive strength p_c for four types of struts. These curves lie below the theoretical Euler curve for compressive strength of a pin-jointed column.

5.3.2 Boundary conditions

The effective length of a column, which influences the slenderness, is dependent on the boundary conditions. The degrees of freedom at the boundary are translational and rotational. Depending on the available freedom at the ends, an approximate method has been devised whereby a factor is applied to the length of the strut to convert it to the theoretical pin-jointed elastic critical case, as in the Euler theory. The consideration of the buckling mode is important when the effective length is determined. A single curvature buckling is expected in a pin-jointed column whereas a double curvature is expected when there is an intermediate restraint. When the ends are rotationally restrained the column will buckle in such a manner that the distance between the points of contraflexure is half the total length. Similarly, for a cantilever column the effective length is twice its normal length. The combination of rotational and translational freedoms at ends gives rise to many methods of finding the effective lengths of columns or struts.

5.3.3 Combined axial compression and bending moment

Local capacity check
The design basis for the local capacity check is exactly the same as in a member subjected to combined axial tension and bending. To understand this philosophy see Section 6.3.

Overall buckling check
The principal issues are:

(1) Axial capacity of the member in compression.
(2) Secondary moments due to applied bending moment and applied axial compression.
(3) Overall buckling about the minor axis due to applied moments.

The complex nature of the problem makes it extremely difficult to analyse numerically. The axial compression by itself causes elastic instability in a slender column. The axial capacity of the member depends on the slenderness ratio and the section classification. Similarly, the buckling resistance moment also depends on the effective length of compression flange and section classification. A mathematical explanation of this complex process may be given as follows:

Assume that a constant moment M is applied to the member and causes a deflection δ at the centre of the column. This deflection gives rise to an additional moment $F\delta$ where F is the axial compression.

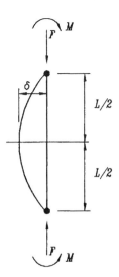

SK 5/1 Column with bending and axial compression.

The total maximum moment in the section can be stated as $M + F\delta$. This maximum moment can be approximated to the following:

$$M_{max} = \frac{M}{\left(1 - \dfrac{F}{P_{cr}}\right)}$$

where P_{cr} = elastic critical load = $\pi^2 EI/L^2$
M = constant applied moment on column section
F = axial compression in column
δ = deflection at centre of column
M_{max} = maximum moment in column section due to moment and direct load
L = effective length of column between restraints

The elastic critical load decreases rapidly with the increase in the effective length of the member. This results in a rapid increase in the secondary moment, as can be seen from the expressions given above. The failure due to axial load alone on the member is controlled by the slenderness about the minor axis or a flexural buckling failure at a load equal to P_{cy}. Similarly, the failure due to bending moment about the major axis is governed by the lateral torsional buckling of the compression flange about its minor axis and this failure moment is termed M_b, as explained in Chapter 7. If these two failure conditions are imposed in the expressions derived before, we get an interaction equation given by:

$$\frac{F}{P_{cy}} + \frac{M}{\left(1 - \dfrac{F}{P_{crx}}\right) M_b} \leq 1.0$$

where P_{cy} = flexural buckling failure axial load
M_b = lateral torsional buckling failure moment
P_{crx} = elastic critical load corresponding to effective length L_x
P_{cry} = elastic critical load corresponding to effective length L_y

Note: The additional moment is dictated by the Euler critical buckling load about the major axis.

If we extend the argument to both axes at the same time we will get the following expression which is the basis of the code formulation:

$$\frac{F}{P_{cy}} + \frac{M_x}{\left(1 - \dfrac{F}{P_{crx}}\right) M_b} + \frac{M_y}{\left(1 - \dfrac{F}{P_{cry}}\right) M_{cy}} \leq 1.0$$

where M_{cy} = plastic moment capacity of the section about the minor axis.

Conservatively the code also allows the following unity check for a member subjected to axial compression and biaxial bending.

$$\frac{F}{A_g p_c} + \frac{m M_x}{M_b} + \frac{m M_y}{p_y Z_y} \leq 1.0$$

where A_g = gross sectional area
       p_c = compressive strength of the column as a strut
       p_y = design strength
       Z_y = elastic section modulus for moments about the minor axis
       m = uniform moment factor

The theoretical derivation of the interaction formula is based on uniform moments M_x and M_y acting on the whole member about the major and minor axes respectively. The factor m is introduced to account for the variation of moment along the length of the member. This factor is called the uniform moment factor and can be obtained from Table 18 of BS 5950: Part 1.

5.4 STEP-BY-STEP DESIGN PROCEDURE OF COLUMNS/STRUTS

Step 1 **Select type of strut and trial section**
At the early concept stage of the design the major issues governing the choice of a type of strut should be considered. These issues are addressed in Table 5.1. The star ratings provide general guidance only and may differ from site to site, so feedback from the manufacturers of the components is most important. It should be borne in mind that Grade 50 steel offers substantial economy over Grade 43, provided the serviceability limit state of deflection does not govern the design.

Step 2 **Determine combined ultimate axial loading**
Follow the load combination procedures after structural analyses with basic loads as described in Chapters 3 and 4. Determine the ultimate axial compression, choosing load combinations and factors from Table 4.1.

Step 3 **Determine ultimate bending moments and shear forces**
At the concept stage of the design, the type of construction will have been decided – continuous or simple. Calculate the bending moments in a compression member according to the construction method chosen:

(1) Continuous construction (as in a rigid frame structure): carry out structural analyses to determine bending moments and shear forces.
(2) Simple construction (as in a braced frame construction): determine the nominal bending moments following the rules below:
    (a) For a beam supported on a column cap plate assume that the beam-end reaction is applied on the face of the column or the edge of the packing plate, if used. The face of the column towards the span of the beam should be taken to determine the eccentricity of the applied load on the column.
    (b) For a roof truss supported on a column cap plate, the eccentricity of the vertical load from the truss may be ignored if the connection between the truss and the column cap plate is simple and does not transfer any moments.

Table 5.1 Selection of a type of truss.

Application	Type of strut	Positive aesthetic impact	Cost economy of materials	Cost economy of fabrication	Cost economy of connections	Ease of availability	Ease of corrosion protection	Ease of handling in shop	Ease of transport
Light trusses	Angle	*	***	***	***	Grade 50 difficult	***	**	**
Light bracings	Compound angle	*	***	**	***	Grade 50 difficult	*	**	**
Light columns	Structural tees	*	***	***	***	***	***	**	**
	Channels	**	**	**	**	***	**	***	***
	Circular hollow section	***	*	**	**	**	***	***	***
Large trusses	Circular hollow section	***	*	**	**	**	***	***	***
Heavy bracings	Rectangular hollow section	***	*	*	*	**	***	***	***
Medium columns	Compound angles	*	**	**	**	Grade 50 difficult	*	**	***
	Compound channels	**	**	**	**	***	*	**	***
	Universal columns	**	***	***	***	***	**	***	***
Industrial buildings	Compound UC	*	**	*	*	***	*	*	***
Heavy columns	Stiffened box	***	***	*	*	**	***	*	***
	Fabricated section	**	***	*	**	***	***	*	***
Multistorey frames	UC	Generally hidden	***	***	***	***	***	***	***
Braced	UB	Generally hidden	**	***	***	***	***	***	***
Unbraced	Fabricated section	Generally hidden	***	*	**	***	***	*	***

* = low; ** = medium; *** = high.

(c) For a beam connected to the side of a column, the eccentricity of the vertical load should be assumed to be 100 mm, or the distance from the centre of the column to the centre of a stiff bearing, whichever is greater.

(d) Ignore the effect of eccentric end connections for struts made of angles, single channels and single T-sections, provided the slenderness ratios are limited to values given in Table 5.3 in Step 6.

(e) For laced struts, battened struts, batten-starred angle struts, and battened parallel angle struts, allow for eccentricity of end connections.

(f) Columns in multistorey buildings using simple construction may not be designed for patterned loading as described in Table 4.2. All beams should be fully loaded to determine maximum column loads. The nominal moments due to beam eccentricity should be added to moments from partial fixity (semi-rigid design approach), and this total moment should be divided between column lengths above and below the joint in proportion to the column stiffness I/l. These nominal moments at any level may be divided equally between the columns above and below if the ratio of stiffness of the columns does not exceed 1.5. These nominal moments have no effect below or above the level considered.

Step 4 **Determine local bending moments in latticed girders and trusses**
Where the exact positions of purlins or other concentrated point loads on the rafter relative to the intersections of the web members are not known, the local bending may be taken as:

$$M_{local} = \frac{WL}{6}$$

where W = ultimate point load from purlin or any other source
  L = length of the rafter between the points of intersection with web members
  M_{local} = local bending moment in rafter due to concentrated load of purlin

Step 5 **Determine effective length of strut**
The most important element in the design of a strut is its effective length in both orthogonal principal axes. Table 5.2 may be used to find the effective lengths.

Limited frame method
Base stiffness
Stiffness of the foundation at the base of a column may be taken as:

(1) Equal to the column stiffness if it is rigidly connected in the direction being considered.
(2) Equal to 10% of column stiffness if it is nominally connected in the direction being considered.
(3) Equal to zero if a proper pinned or rocker connection is used.

162 Structural Steelwork

Table 5.2 Effective lengths of struts.

Description of structure		Effective length
Braced, pinned at top, pinned at base. Column effectively held in position at both ends but not restrained in direction.	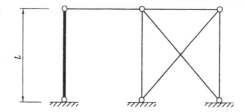	$L_E = 1.0L$
Braced, pinned at top, fixed at base. Effectively held in position at both ends and in direction at one end.	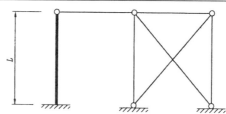	$L_E = 0.85L$
Unbraced, fixed at base, pinned at the top. Effectively held in position and direction at one end but not restrained in position or direction at the other end.	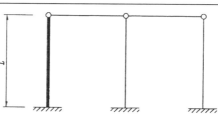	$L_E = 2.0L$
Braced, partially restrained by beams at both ends. Effectively held in position at both ends and partially restrained in direction at both ends.	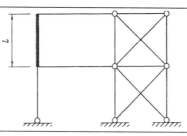	$L_E = 0.85L$
Braced, fixed at base and rigidly restrained by a very stiff beam at the top. Effectively held in position and direction at both ends.	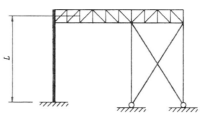	$L_E = 0.7L$
Unbraced, fixed at base and rigidly restrained by a very stiff beam at the top. Effectively held in position and direction at one end and held in direction at the other end.	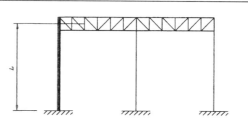	$L_E = 1.2L$

Design of Struts

Table 5.2 (contd)

Description of structure		Effective length
Unbraced, fixed at base and partially restrained by a beam at the top. Effectively held in position and direction at one end and partially restrained in direction at the other end.	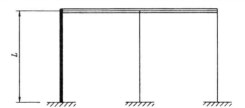	$L_E = 1.5L$
Stanchions for single-storey buildings as per BS 5950: Part 1. Effectively held in position and direction about both axes at the base, not held in position at the top along the Y–Y axis, or direction about the X–X axis, but effectively held in position along the X–X axis by braced bays, and also held in position along the X–X axis at the crane girder level by braced bays. Partial frame method may be used if the column is rigidly connected to a stiff roof truss or latticed girder at the top. If at the base of the column the directional restraint about the Y–Y axis is not available then effective length factors will be increased from 0.85 to 1.0.	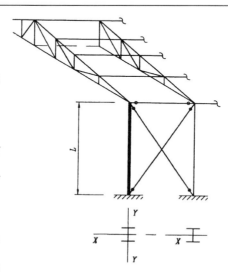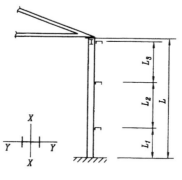	$L_{EX} = 1.50L$ $L_{EY} = 0.85L$ $L_{EX} = 1.50L$ $L_{EY} = 0.85L_1$, $1.0L_2$ or $1.0L_3$, whichever is greatest
	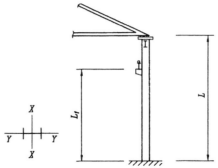	$L_{EX} = 1.50L$ $L_{EY} = 0.85L_1$

Table 5.2 (contd)

Description of structure		Effective length
Stanchions for single-storey buildings as per BS 5950: Part 1 (contd).	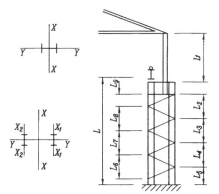	Upper roof leg: $L_{EX} = 1.5L_1$; $L_{EY} = 1.0L_1$ Lower roof leg: $L_{EY} = 0.85L$ $L_{EX1} = L_2$, L_3, L_4 or L_5 Lower crane leg: $L_{EY} = 0.85L$ $L_{EX2} = L_6$, L_7, L_8 or L_9 Combined column: $L_{EX} = 1.5L$; $L_{EY} = 0.85L$
Bracing systems generally: vertical and plan bracing Single diagonal bracing		$L_E = 1.0L$
Cross bracing, connected at intersection		$L_E = 0.7L$
Lacing systems generally: columns and struts Single intersection lacing	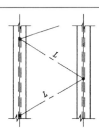	$L_E = 1.0L$
Double intersection lacing, connected at intersection	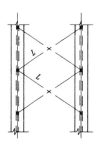	$L_E = 0.7L$

Design of Struts   165

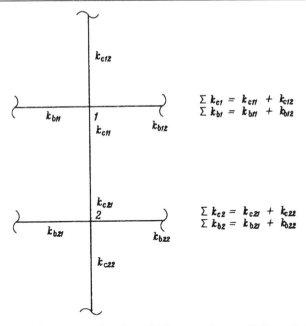

SK 5/2   Determination of joint estraint coefficients (column between joints 1 and 2 in a multistorey rigid frame).

Beam stiffness
Stiffness of a beam connected rigidly to a column may be taken as:

(1) Equal to the beam stiffness I/l if a concrete floor is supported by the beam.
(2) Equal to $0.5I/l$ if the frame is braced and there is no concrete floor.
(3) Equal to $1.5I/l$ if the frame is free to sway and there is no concrete floor.
(4) Equal to zero if the beam is carrying 90% or more of its reduced plastic moment capacity (reduction due to combined moment and axial load).

I = moment of inertia of section of beam and l = span of beam.

Column stiffness
The column stiffness may be taken as equal to I/l, where l = storey height and I = moment of inertia of section of column.

Joint restraint coefficient (see SK 5/2)

(1) $k_i = \dfrac{\text{Total column stiffness at joint } i}{\text{Total stiffness of all members at the joint}}$

$$k_1 = \frac{\sum k_{c1}}{\sum k_{c1} + \sum k_{b1}}; \quad k_2 = \frac{\sum k_{c2}}{\sum k_{c2} + \sum k_{b2}}$$

where k_{ci} = column stiffness at joint i
k_{bi} = beam stiffness at joint i
k_i = joint restraint coefficient of joint i

(2) $k_i = 1$ if the column considered for effective length is carrying more than 90% of its reduced plastic moment capacity at that joint i.

Relative stiffness

The relative stiffness k_3 of the effective bracing in any storey may be determined as follows:

(1) Rigid-jointed frames braced against side sway $k_3 = \infty$.
(2) Rigid-jointed frames with unrestricted side sway $k_3 = 0$.
(3) Rigid-jointed frames with no bracing but with wall panel built in the plane of the frame and extending the full height of the storey:

$$k_3 = \frac{h^2 \sum S_P}{80E \sum K_c} \leq 2$$

where h = storey height
E = modulus of elasticity of steel
$\sum K_c$ = sum of all the column stiffness at that level of the frame
$\sum S_P$ = sum of spring stiffness of the wall panels

$$= \sum \frac{0.6h/b}{\{1 + (h/b)^2\}^2} tE_P$$

b = width of each panel
t = thickness of each panel
E_P = modulus of elasticity of the material of panels

Effective length from SK 5/3

(1) When $k_3 = \infty$ use joint restraint coefficients k_1 and k_2 to find L_E/L from SK 5/3a.
(2) When $k_3 = 0$ use joint restraint coefficients k_1 and k_2 to find L_E/L from SK 5/3b.
(3) When $k_3 = 1$ use joint restraint coefficients k_1 and k_2 to find L_E/L from SK 5/3c.
(4) When $k_3 = 2$ use joint restraint coefficients k_1 and k_2 to find L_E/L from SK 5/3d.
(5) Use interpolation to find L_E/L for intermediate values of k_3 between 0, 1 and 2.

Note: A restraint is considered effective if it lies at an angle of not more than 45° to the plane of buckling considered.

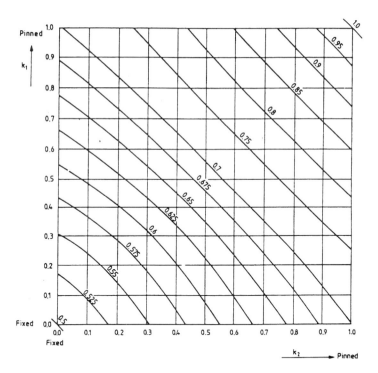

SK 5/3a Effective length ratio L_E/L for a column in a rigid-jointed frame braced against sidesway for $k_3 = \infty$.

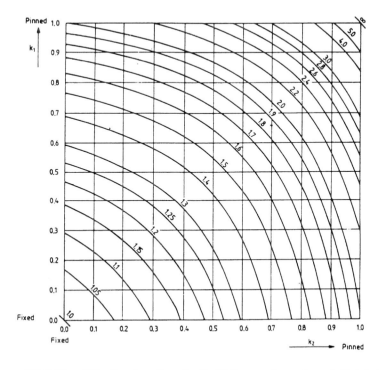

SK 5/3b Effective length ratio L_E/L for a column in a rigid-jointed frame with unrestricted sidesway for $k_3 = 0$.

Reproduced by kind permission of British Standards Institution from BS 5950: Part 1.

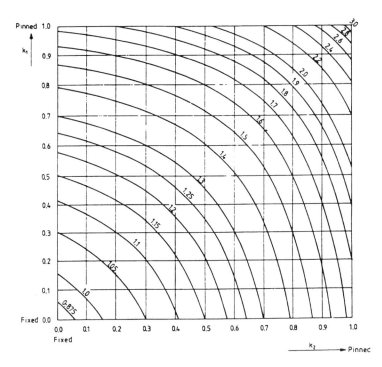

SK 5/3c Effective length ratio L_E/L for a column in a rigid-jointed frame with partial sway bracing of relative stiffness $k_3 = 1.0$.

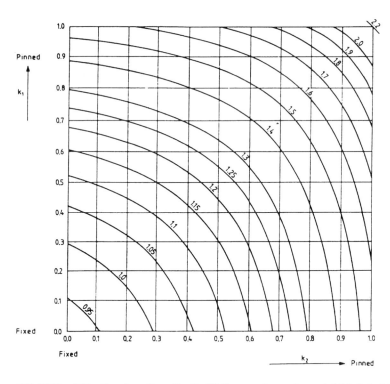

SK 5/3d Effective length ratio L_E/L for a column in a rigid-jointed frame with partial sway bracing of relative stiffness $k_3 = 2.0$.

Reproduced by kind permission of British Standards Institution from BS 5950: Part 1.

Table 5.3 Slenderness ratios for angles, channels and T-sections.

Type of section	End connection	Slenderness ratio
Single angle	At least two bolts in line or welded	$\lambda = 0.85 L_{vv}/r_{vv}$ $= 0.7 L_{vv}/r_{vv} + 15$ $= 1.0 L_{aa}/r_{aa}$ $= 0.7 L_{aa}/r_{aa} + 30$ $= 0.85 L_{bb}/r_{bb}$ $= 0.7 L_{bb}/r_{bb} + 30$ (the greatest of)
Single angle *See note on page 173	Single bolt	$\lambda = 1.0 L_{vv}/r_{vv}$ $= 0.7 L_{vv}/r_{vv} + 15$ $= 1.0 L_{aa}/r_{aa}$ $= 0.7 L_{aa}/r_{aa} + 30$ $= 1.0 L_{bb}/r_{bb}$ $= 0.7 L_{bb}/r_{bb} + 30$ (the greatest of)
Double angles parallel or back to back	Bolted or welded	For each main component: $$\lambda_c = \frac{L_c}{r_{min}} \leq 50$$ r_{min} = minimum slenderness ratio of one main component

Table 5.3 (contd)

Type of section	End connection	Slenderness ratio
Double angle back to back separated		
Parallel angle battened	At least two bolts in line connected to one leg or equivalent weld	$\lambda = 1.0 L_{xx}/r_{xx}$ $= 0.7 L_{xx}/r_{xx} + 30$ $= \left[\left(0.85 \dfrac{L_{yy}}{r_{yy}}\right)^2 + \lambda_c^2\right]^{\frac{1}{2}}$ $= 1.4 \lambda_c$ (the greatest of) See page 169 for the definition of λ_c
Double angle back to back in contact		

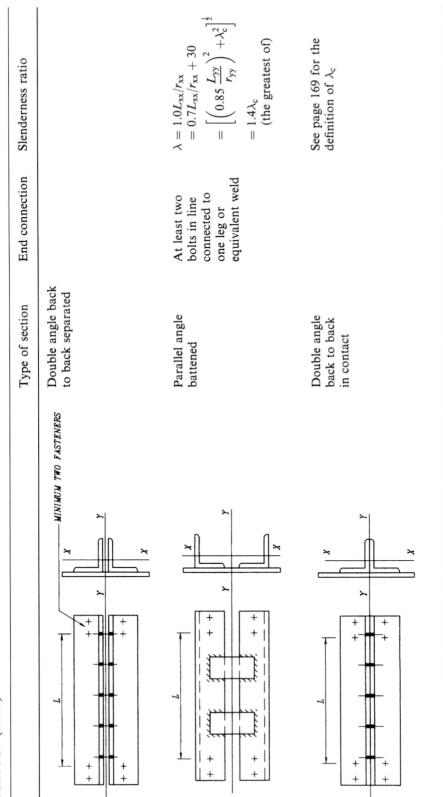

Table 5.3 (contd)

Type of section	End connection	Slenderness ratio
Double angle back to back separated		$\lambda = 1.0 L_{xx}/r_{xx}$ $= 0.7 L_{xx}/r_{xx} + 30$ $= \left[\left(1.0 \dfrac{L_{yy}}{r_{yy}} \right)^2 + \lambda_c^2 \right]^{\frac{1}{2}}$ $= 1.4 \lambda_c$ (the greatest of)
Parallel angle battened	One bolt connected to one leg of one angle	See page 169 for the definition of λ_c
Double angle back to back in contact		
*See note on page 173		

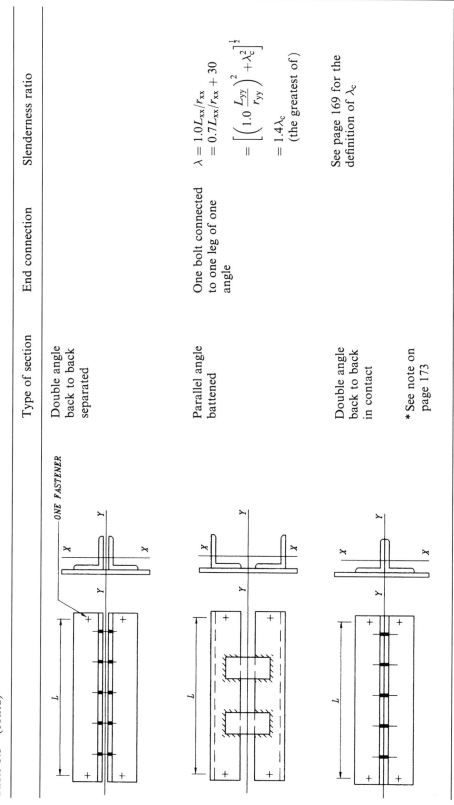

Table 5.3 (contd)

Type of section	End connection	Slenderness ratio
Double angle back to back separated	One leg of both angles connected by minimum two bolts in line or equivalent weld	$\lambda = 0.85 L_{xx}/r_{xx}$ $= 0.7 L_{xx}/r_{xx} + 30$ $= \left[\left(1.0 \dfrac{L_{yy}}{r_{yy}}\right)^2 + \lambda_c^2\right]^{\frac{1}{2}}$ $= 1.4\lambda_c$ (the greatest of) See page 169 for the definition of λ_c
Double angle back to back separated *See note on page 173	Both angles connected by one bolt to end gusset	$\lambda = 1.0 L_{xx}/r_{xx}$ $= 0.7 L_{xx}/r_{xx} + 30$ $= \left[\left(1.0 \dfrac{L_{yy}}{r_{yy}}\right)^2 + \lambda_c^2\right]^{\frac{1}{2}}$ $= 1.4\lambda_c$ (the greatest of) See page 169 for the definition of λ_c
Single channel	Connected to a gusset or another member by two or more rows of minimum two bolts in each row or equivalent weld	$\lambda = 0.85 L_{xx}/r_{xx}$ $= 1.0 L_{yy}/r_{yy}$ $= 0.7 L_{yy}/r_{yy} + 30$ (the greatest of)

Design of Struts 173

Table 5.3 (contd)

Type of section	End connection	Slenderness ratio
Single channel	Connected to a gusset or another member by one row of minimum two bolts	$\lambda = 1.0 L_{xx}/r_{xx}$ $= 1.0 L_{yy}/r_{yy}$ $= 0.7 L_{yy}/r_{yy} + 30$ (the greatest of)
Single T-section	Connected to gusset or another member by two or more rows of minimum two bolts in each row	$\lambda = 1.0 L_{xx}/r_{xx}$ $= 0.7 L_{xx}/r_{xx} + 30$ $= 0.85 L_{yy}/r_{yy}$ (the greatest of)
Single T-section	Connected to a gusset or another member by one row of minimum two bolts	$\lambda = 1.0 L_{xx}/r_{xx}$ $= 0.7 L_{xx}/r_{xx} + 30$ $= 1.0 L_{yy}/r_{yy}$ (the greatest of)

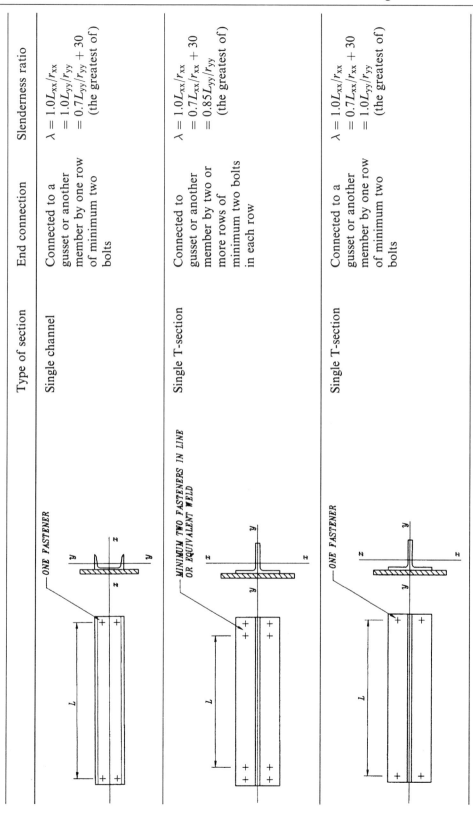

Note: If angles are connected by one bolt only to one leg of the angle, the capacity P_{LC} found in Step 11 should be multiplied by 0.8.

Step 6
Determine the slenderness ratio λ
The slenderness ratio λ is defined as the effective length with respect to an axis of buckling divided by the radius of gyration of the column section about the same axis.

$$\lambda_x = \frac{L_{EX}}{r_x} \quad \text{and} \quad \lambda_y = \frac{L_{EY}}{r_y}$$

L_{EX} and L_{EY} are the effective lengths about the two orthogonal axes. The rules for finding λ for single angles, double angles, single channels and single T-sections are given in Table 5.3. These values of slenderness ratio may be used if the eccentricity of end connections is ignored.

Step 7 Check maximum slenderness ratio
The slenderness ratio λ should not exceed the following values:

(1) Members resisting dead imposed and crane loads: 180.
(2) Members resisting self-weight and wind only: 250.
(3) Member is normally a tie but wind causes a reversal of stress: 350.
    (Deflection due to self-weight should be limited to $l/1000$ – otherwise, design for axial load plus bending due to deformation.)

Step 8 Check classification of section
This should be checked with reference to Clause 3.5.2 and Table 7 of BS 5950: Part 1.

Step 9 Determine strength reduction factors for slender elements
This should be determined with reference to Clauses 3.6.3 and 3.6.4 and Table 8 of BS 5950: Part 1. See Worked Example 5.2 for an explanation of this principle.

Step 10 Determine compressive strength

$$\text{Find} \quad \lambda_0 = 0.2\sqrt{\left(\frac{\pi^2 E}{p_y}\right)}$$

where p_y = design strength of steel
E = modulus of elasticity of steel = $205\,\text{kN/mm}^2$

Table 5.4 Design strength p_y as per BS 5950: Part 1.

Thickness of any part of the compressive strut less than or equal to:	Grades of steel		
	Grade 43	Grade 50	Grade 55
16 mm	275 N/mm²	355 N/mm²	450 N/mm²
40 mm	265 N/mm²	345 N/mm²	–
63 mm	255 N/mm²	335 N/mm²	400 N/mm²
80 mm	245 N/mm²	325 N/mm²	–
100 mm	235 N/mm²	315 N/mm²	–

Find the Robertson constant *a* from Table 5.5.

Table 5.5 Robertson constant (*a*).

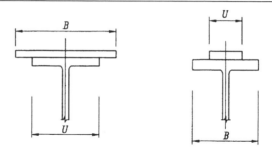

Type of section	Buckling about:	
	X–X axis	Y–Y axis
Structural hollow section	2	2
Rolled I-section	2	3.5
Rolled I-section with welded cover plate $0.25 < U/B < 0.8$	2	3.5
Rolled I- or H-section with welded cover plate $U/B \geq 0.8$ up to 40 mm thickness	3.5	2
Rolled I- or H-section with welded cover plate $U/B \geq 0.8$ over 40 mm thickness	5.5	3.5
Rolled I- or H-section with welded cover plate $U/B \leq 0.25$ up to 40 mm thickness	3.5	5.5
Rolled I- or H-section with welded cover plate $U/B \leq 0.25$ over 40 mm thickness	3.5	8.0
Rolled H-section up to 40 mm thickness with or without cover plate $0.25 < U/B < 0.8$	3.5	5.5
Rolled H-section over 40 mm thickness with or without cover plate $0.25 < U/B < 0.8$	5.5	8.0
Welded box section up to 40 mm thickness	3.5	3.5
Welded box section over 40 mm thickness	5.5	5.5
Round, square or flat bar up to 40 mm thickness	3.5	3.5
Round, square or flat bar over 40 mm thickness	5.5	5.5
Rolled angle, channel, T-section, battened or laced compound sections	5.5	5.5

Find the Perry coefficient, which is given by:

$$\eta = 0.001a(\lambda - \lambda_0)$$

Find the Euler strength, which is given by:

$$p_E = \frac{\pi^2 E}{\lambda^2}$$

Find the compressive strength p_c, which is given by:

$$p_c = \frac{p_E p_y}{\phi + (\phi^2 - p_E p_y)^{1/2}} \quad \text{where } \phi = \frac{p_y + (\eta + 1)p_E}{2}$$

For sections fabricated from plates by welding, the design strength p_y should be reduced by 20 N/mm^2. For slender elements of a section, p_c should be found by using a reduced value of p_y as found in Step 9.

Alternatively, use Tables 25, 26 and 27 of BS 5950: Part 1 to find the compressive strength p_c. Use a reduced value of p_y as appropriate.

Step 11 *Determine compressive resistance*

For plastic, compact and semi-compact sections the compressive resistance P_c is given by:

$$P_c = A_g p_c \quad \text{where } A_g = \text{gross cross-sectional area}$$

For slender section, each element of the section may have a different p_c depending on the classification of the element and its reduced p_y. The compressive resistance will be the sum of the compressive resistance of each of the component elements.

Step 12 *Unity check of columns with nominal moments*

Columns with nominal moments in simple construction may be checked using the unity equation of Clause 4.7.7 of BS 5950: Part 1.

Find L, which is the distance between points on a strut or a column where the column is restrained about both axes.

$$\lambda_{LT} = \text{equivalent slenderness ratio of the column} = 0.5\left(\frac{L}{r_y}\right)$$

r_y = radius of gyration about the minor axis of the column or strut

$$\lambda_{LO} = \text{limiting equivalent slenderness ratio} = 0.4\left(\frac{\pi^2 E}{p_y}\right)^{\frac{1}{2}}$$

η_{LT} = Perry coefficient = $0.007 (\lambda_{LT} - \lambda_{LO}) \geq 0$

(for a rolled section)

$= 2 \times 0.007 \times \lambda_{LO}$ (for a welded section)

$\leq 2 \times 0.007(\lambda_{LT} - \lambda_{LO}) \geq 0$

$\geq 0.007 (\lambda_{LT} - \lambda_{LO}) \geq 0$

M_P = plastic moment capacity = $S_x p_y$ where S_x = plastic modulus of section

$$M_E = \text{elastic critical moment} = \frac{M_P \pi^2 E}{\lambda_{LT}^2 p_y}$$

$$M_b = \text{buckling resistance moment} = \frac{M_E M_P}{\phi_B + (\phi_B^2 - M_E M_P)^{\frac{1}{2}}}$$

$$\text{where } \phi_B = \frac{M_P + (\eta_{LT} + 1)M_E}{2}$$

Alternatively, the buckling resistance moment of the column or the strut can be found by reference to Clauses 4.3.7.3 and 4.3.7.4 and Tables 11 and 12 of BS 5950: Part 1.

The following relationship should be satisfied:

$$\frac{F_c}{A_g p_c} + \frac{M_x}{M_{bs}} + \frac{M_y}{p_y Z_y} \leq 1$$

where F_c = ultimate compressive load
M_x = ultimate nominal bending moment about the X–X axis (major axis) acting simultaneously with F_c
M_y = ultimate nominal bending moment about the Y–Y axis (minor axis) acting simultaneously with F_c
A_g = gross cross-sectional area
p_c = compressive strength as found in Step 10 (see also Step 11)
p_y = design strength as found in Step 10
Z_y = elastic section modulus about the Y–Y axis (minor axis)
$M_{bs} = M_b$ = buckling moment of resistance of the simple column.

For circular hollow sections, rectangular hollow sections and box sections with uniform wall thicknesses $M_{bs} = p_y S_x = M_P$, provided the slenderness ratio λ does not exceed the values given in Table 38 of Appendix B of BS 5950: Part 1.

Step 13 *Unity checks for columns/struts with moments*
This unity check has to be done in two stages. Firstly the local section capacity check has to be carried out, and then an overall member buckling check has to be satisfied.

Local section capacity check
Check shear capacity P_v where:

$$P_v = 0.6 p_y A_v$$

A_v = effective shear area
   = tD for rolled sections
   = td for built-up sections
   = $0.9A$ for solid bars and plates
   = $\left(\dfrac{D}{D+B}\right)A$ for rectangular hollow sections
   = $0.6A$ for circular hollow sections

where t = thickness of the web
D = overall depth of section
d = depth of web only
A = area of section

Find the applied shear force F_v from Step 3.

For $F_v \leq 0.6 P_v$
M_c for plastic and compact sections is given by:

$$M_{cx} = p_y S_x \leq \mu_f p_y Z_x \quad \text{and} \quad M_{cy} = p_y S_y \leq \mu_f p_y Z_y$$

where, μ_f = average load factor for ultimate limit state.

For $F_{vx} > 0.6P_{vx}$ and $F_{vy} > 0.6P_{vy}$

$$M_{cx} = p_y(S - S_{vx}\rho_{1x}) \leq \mu_f p_y Z_x$$

and

$$M_{cy} = p_y(S - S_{vy}\rho_{1y}) \leq \mu_f p_y Z_y$$

$$\rho_{1x} = \frac{2.5F_{vx}}{P_{vx}} - 1.5 \quad \text{and} \quad \rho_{1y} = \frac{2.5F_{vy}}{P_{vy}} - 1.5$$

F_{vx} and F_{vy} are applied shears about the X–X and Y–Y axes respectively. P_{vx} and P_{vy} are section shear capacities about the X–X and Y–Y axes. S_{vx} and S_{vy} are the plastic moduli of the shear area only about X–X and Y–Y axes respectively.

For all values of $F_v < P_v$
M_c for semi-compact and plastic sections is given by:

$$M_{cx} = p_y Z_x \quad \text{and} \quad M_{cy} = p_y Z_y$$

For slender sections use reduced value of p_y as found in Step 9.

Note: M_c for plastic and compact sections is reduced in the presence of high shear only in the orthogonal direction in which the shear exceeds the limit of $0.6P_v$.

Local capacity check

$$\frac{F}{A_g p_y} + \frac{M_x}{M_{cx}} + \frac{M_y}{M_{cy}} \leq 1$$

where F = ultimate axial compressive load on a section in the member
  M_x = ultimate bending moment about X–X axis (major axis) at the same section in the member acting simultaneously
  M_y = ultimate bending moment about Y–Y axis (minor axis) at the same section in the member acting simultaneously
  A_g = the gross cross-sectional area of the section of the member
  p_y = the design strength
  M_{cx} = plastic moment capacity of the section in the absence of axial compressive load about the X–X axis (major axis)
  M_{cy} = plastic moment capacity of the section in the absence of axial compressive load about the Y–Y axis (minor axis)

Overall buckling check
Determine buckling resistance moment M_b about the major axis.

λ = slenderness ratio found in Step 6

$\lambda_{LT} = nuv\lambda$

Design of Struts 179

where $n = 1.0$ conservatively
   $u =$ buckling parameter from published steel section tables
   or $= 0.9$ conservatively for rolled sections
   or $= 1.0$ conservatively for any other section
   $v =$ slenderness factor from Table 14 of BS 5950: Part 1

To find v determine λ/x and N

where $x =$ torsional index from steel section tables
   or $= D/T$ conservatively along with u determined conservatively
   $D =$ overall depth of section
   $T =$ thickness of flange

$$N = \frac{I_{cf}}{I_{cf} + I_{tf}}$$

$I_{cf} =$ second moment of inertia of the compression flange about the minor axis
$I_{tf} =$ second moment of inertia of the tension flange about the minor axis

$N = 0.5$ for sections with equal flanges.

Having found λ_{LT} follow the method in Step 12 to determine M_b about the major axis. The buckling resistance moment for a single angle should be taken as:

$M_b = 0.8 p_y Z$ for $L/r_{vv} \leq 100$
$\quad = 0.7 p_y Z$ for $L/r_{vv} > 100 \leq 180$
$\quad = 0.6 p_y Z$ for $L/r_{vv} > 180 \leq 300$

For uniform sections with equal flanges
Find β, which is the ratio of the smaller end moment to the larger end moment on a span equal to the unrestrained length.

$$\beta = \frac{M_1}{M_2}$$

where M_1 is the smaller moment at end 1 and M_2 is the larger moment at end 2.

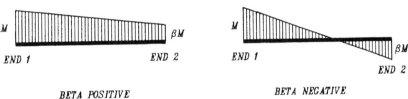

MEMBER END MOMENTS WHEN $\beta \leq |1.0|$

SK 5/4 The sign convention for β.

β is positive when both end moments are of the same sign (i.e. hogging or sagging). β is negative if the end moments are of opposite sign.

Find equivalent uniform moment factor m given by:

$$m = 0.57 + 0.33\beta + 0.10\beta^2 \geq 0.43 \text{ for uniform sections with equal flanges}$$

$m = 1.0$ for all other sections

Overall buckling check

$$\frac{F}{A_g p_c} + \frac{mM_x}{M_b} + \frac{mM_y}{p_y Z_y} \leq 1$$

where $p_c =$ the compressive strength as found in Step 10
$F =$ ultimate axial compressive load at a section in the member
$M_x =$ ultimate bending moment about $X-X$ axis (major axis) at the same section in the member acting simultaneously
$M_y =$ ultimate bending moment about $Y-Y$ axis (minor axis) at the same section in the member acting simultaneously
$A_g =$ the gross cross-sectional area at the section of the member
$p_y =$ the design strength
$M_b =$ buckling resistance moment

Alternatively

$$\frac{mM_x}{M_{ax}} + \frac{mM_y}{M_{ay}} \leq 1; \quad M_{ax} = M_{cx} \frac{\left(1 - \frac{F}{P_{cx}}\right)}{\left(1 + \frac{0.5F}{P_{cx}}\right)}$$

$$\text{or } = M_b\left(1 - \frac{F}{P_{cy}}\right) \text{ whichever is smaller}$$

$$\text{and } M_{ay} = M_{cy} \frac{\left(1 - \frac{F}{P_{cy}}\right)}{\left(1 + \frac{0.5F}{P_{cy}}\right)}$$

where $P_{cx} = A_g p_{cx}$ and $P_{cy} = A_g p_{cy}$

To find p_{cx} and p_{cy} as shown in Step 10, use λ_x and λ_y corresponding to L_{EX}/r_{xx} and L_{EY}/r_{yy} respectively. Use reduced value of p_y where appropriate.

Note: For small values of M_x and M_y do not use this alternative method.

Note: Box sections of uniform wall thickness need not be checked for lateral torsional buckling if the slenderness ratio λ does not exceed the values given in Table 5.6.

Design of Struts 181

Table 5.6 Limiting values of slenderness ratio for rectangular box sections.

D/B	1	2	3	4
λ	∞	$\dfrac{350 \times 275}{p_y}$	$\dfrac{225 \times 275}{p_y}$	$\dfrac{170 \times 275}{p_y}$

D = overall depth of box section; B = overall breadth of box section.

Step 14 *Design of lacing systems*
General rules:

(1) Lacing should not vary in section or inclination throughout the length of the member.
(2) Inclination of the lacing should be between 40° and 70° to the axis of the member.
(3) Lacing can be double intersection (cross) or single intersection. Single intersection lacing should not be opposed in direction on opposite faces of the main members. This would give rise to torsion in the member.
(4) Tie panels in the form of battens should be provided at the ends of the lacing system.
(5) Slenderness ratio λ_c should not be greater than 50 for individual main members connected by the lacing, where L_c (effective length) is the length between the points of intersections of the lacing with the main members. Radius of gyration of the main members about the minor axis is used to compute $\lambda_c = L_c/r_{yy}$.

Select a type and section of lacing to satisfy the general rules. Find effective lengths of the lacing bars from the sketches in Table 5.2 in Step 5. Check that the slenderness ratio of the lacing bars do not exceed 180.

Find shear force V across a plane perpendicular to the axis of the member, which is given by:

$$V = \left(\frac{2.5}{100}\right)F + V_s$$

where F = ultimate maximum axial compression in the member
V_s = transverse shear in the member at any section acting with F

The lacing bar should be checked for an axial tension or compression equal to N_L, which is given by:

$$N_L = \frac{V}{\sin\phi}\frac{1}{J}$$

where J = number of lacing bars cut by a plane perpendicular to the axis of the member
ϕ = angle of inclination of the lacing bars to the axis of the member
P_{LC} = capacity of lacing bar in compression = $A_g p_c$
A_g = the gross cross-sectional area of the lacing bar

The compressive strength p_c may be obtained from Table 27(c) of BS 5950: Part 1 corresponding to the slenderness ratio of the lacing bar and its design strength p_y.

Note: If the lacing system is connected to the main member by one fastener only, then the capacity P_{LC} should be multiplied by 0.8.

The capacity of the lacing bar in tension $P_{LT} = A_e p_y$, where A_e is the effective area of the section, which equals the gross area less deduction for holes, if any. Check that P_{LC} and P_{LT} are both greater than N_L.

The load in the tie panels at the end of lacing is $N_L \sin \phi$. The tie panel should be designed as a batten, as in Step 15.

Step 15 *Design of battens*
General rules:

(1) The joints between the main components and the battens must be rigid. Use welded connections or connection by at least two bolts along the axis of the main member on each main member.
(2) Battens should be provided opposite each other in each plane at the ends of the members and at the point of lateral restraints.
(3) Intermediate battens on opposite faces of main members should be uniformly spaced throughout the length of the member.
(4) The maximum slenderness ratio λ_c of the main member between end welds or end fasteners of adjacent battens should not exceed 50. ($\lambda_c = L_c/r_{yy}$ – see point (5) of Step 14.)
(5) The slenderness ratio of the compound column about an axis perpendicular to the plane of the battens λ_b is given by:

$$\lambda_b = \sqrt{(\lambda_m^2 + \lambda_c^2)} \geq 1.4\lambda_c$$

where $\lambda_m = L_E/r$, L_E = effective length of the compound column about an axis perpendicular to the plane of the battens, and r = radius of gyration of the compound column about the same axis.
(6) The thickness of the plate used as batten should not be less than 1/50 of the minimum distance between welds or fasteners on the batten connecting the main members, i.e. the free length of batten.
(7) The slenderness ratio of the batten should not be less than 180 where the length of the batten is taken as the minimum distance between welds or fasteners on the batten connecting the main members, i.e. the free length of batten.
(8) Width of the end battens should not be less than the distance between the centroids of the main members of the compound column.
(9) Width of the intermediate battens should not be less than half the distance between the centroids of the main members of the compound column.
(10) Width of any batten should not be less than twice the width of the narrower connected element of the main members of the compound column.

Design of Struts   183

Select a type and section of end batten and intermediate batten. Decide on a spacing of the intermediate battens. Satisfy maximum slenderness ratio λ_c of the main members.

Check thickness, slenderness and width limitations of the battens.

Find transverse shear force V across a plane perpendicular to the axis of the compound column, which is given by:

$$V = 0.025F + V_s$$

where F = ultimate maximum axial compressive force in the compound column
V_s = transverse shear in the compound column at a batten position acting with F

The batten should be checked for axial load tension or compression equal to $V/2$ for two parallel planes of battens.

P_{BC} = capacity of the batten in compression = $A_g p_c$

A_g is the gross cross-sectional area and p_c is the compressive strength, which can be obtained from Table 27(c) of BS 5950: Part 1 corresponding to the slenderness ratio and design strength p_y of the batten.

The capacity of the batten in tension P_{BT} is given by:

$$P_T = A_e p_y$$

where A_e is the effective area of the batten after deduction of holes, if any. Check that P_{BC} and P_{BT} are both greater than $V/2$.

5.5 WORKED EXAMPLES

5.5.1 Example 5.1a: Design the internal compression member of a roof truss

Choose the roof truss in Example 3.1 in Chapter 3. Select member 16.

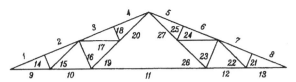

SK 5/5 Truss showing the member selected for design.

Step 1 **Select type of strut and a trial section**

By reference to Table 5.1, the most cost-effective option should be an angle in Grade 43 steel. From inspection of the basic load table, it is evident that maximum compression is of the order of 25 kN. Assuming a compressive strength p_c equal to 125 N/mm^2, the minimum area required for the strut will be 200 mm^2. The design will be based on the minimum slenderness ratio.

Choose RSA $50 \times 50 \times 6 \times 4.47$ kg/m as a trial section.

184 Structural Steelwork

Step 2 *Determine combined ultimate axial loading*
Basic load cases:

Dead load (DL) 6.96 kN (compression)
Imposed load (IL) 8.36 kN (compression)
Wind load (WL) 5.81 kN (compression) or 5.42 kN (tension)

Combination 1: $1.4DL + 1.6IL = 23.12$ kN (compression)
Combination 2: $1.2DL + 1.2IL + 1.2WL = 25.36$ kN (compression)
Combination 3: $1.4DL + 1.4WL = 17.88$ kN (compression)

Step 3 *Determine ultimate bending moments and shear forces*
Ignore eccentric end connection and use the slenderness ratio given in Table 5.3.

Step 4 *Determine local bending moments*
Not required.

Step 5 *Determine effective length*
$L_E = 1.0L = 1550$ mm to centre of last row of fasteners.

Step 6 *Determine the slenderness ratio λ*
Refer to Table 5.3.
Single angle connected by a single fastener.

$$L_{aa} = L_{bb} = L_{vv} = 1550 \text{ mm}; \quad r_{aa} = r_{bb} = 15 \text{ mm}; \quad r_{vv} = 9.7 \text{ mm}$$

$$\lambda = 1.0 L_{vv}/r_{vv} = 1550/9.7 = 160$$

$$= 0.7 L_{vv}/r_{vv} + 15 = 127$$

$$= 1.0 L_{aa}/r_{aa} = 1550/15 = 103$$

$$= 0.7 L_{aa}/r_{aa} + 30 = 102$$

$$\therefore \quad \lambda = 160, \quad \text{which is the greatest}$$

Step 7 *Check minimum slenderness ratio*
Check $\lambda = 160 < 180$. Satisfied.

Step 8 *Check classification of section*
$b/T = d/T = 50/6 = 8.3 < 8.5\varepsilon$ where $\varepsilon = \sqrt{(275/p_y)} = 1$
The section classification is plastic.

Step 9 *Determine strength reduction factor*
Not required.

Step 10 *Determine compressive strength*
Use Table 27(c) of BS 5950: Part 1.

$$p_c = 61 \text{ N/mm}^2 \text{ for } \lambda = 160 \text{ and } p_y = 275 \text{ N/mm}^2$$

Step 11 Determine compression resistance

$$P_c = A_g p_c \times 0.8 \text{ (see Table 5.3 and note on page 173)}$$
$$= 569 \times 61 \times 0.8 = 27.8 \text{ kN} > 25.36 \text{ kN (see Step 2)}$$

Section check is satisfied.

Steps 12 to 15 Not required.

Chosen section (RSA 50 × 50 × 6 connected by one fastener at each end) satisfies all conditions of loading.

5.5.2 Example 5.1b: Same member as in Example 5.1a but use CHS Grade 50

Choose circular hollow section (CHS) 42.4 × 2.6 × 2.6 kg/m as first trial.

Area = 325 mm²; radius of gyration = 14.1 mm; $p_y = 355 \text{ N/mm}^2$; effective length = 1550 mm; $\lambda = L/r = 110 < 180$
Refer to Table 7 of BS 5950: Part 1 for section classification.

$$D/t = 42.4/2.6 = 16.31 < 40\varepsilon^2 = 31 \quad \text{where } \varepsilon = \sqrt{(275/p_y)} = 0.88$$

The section classification is plastic.

Compression strength from Table 27(a) of BS 5950: Part 1

$$p_c = 145 \text{ N/mm}^2 \text{ for } \lambda = 110 \text{ and } p_y = 355 \text{ N/mm}^2.$$

Compression resistance $P_c = 325 \times 145 = 47\,100 \text{ N} > 25.36 \text{ kN}$.

CHS 42.4 × 2.6 × 2.6 kg/m satisfies all loading conditions. By using CHS there is a potential saving in weight of material of up to 40%, but the CHS may cost 60% more in price of material and fabrication.

5.5.3 Example 5.2a: Design the rafter of a roof truss

Choose the roof truss in Example 3.1 in Chapter 3. Select member number 1 (see SK 5/5).

Step 1 Select type of strut and a trial section

By reference to Table 5.1, the most cost-effective option should be a T-section in Grade 50 steel. From inspection of the basic load table, it is evident that the maximum compression in the member is of the order of 120 kN. Assuming $p_c = 120 \text{ N/mm}^2$, the minimum area of section required is 1000 mm².

Choose 102 × 127 × 13 kg/m structural T-section in Grade 50 steel.

$B = 101.9$ mm; $d = 128.5$ mm; $t = 6.1$ mm; $T = 8.4$ mm; $Z_f = 75.3 \times 10^3$ mm³; $Z_t = 26.6 \times 10^3$ mm³; $r_{xx} = 39.6$ mm; $r_{yy} = 21.4$ mm; $A = 1610$ mm²

Step 2 Determine combined ultimate axial loading
Basic load cases:

Dead load (*DL*)   35.34 kN (compression)
Imposed load (*IL*) 42.41 kN (compression)
Wind load (*WL*)   11.52 kN (compression)   and 29.40 kN (tension)

Combination 1: $1.4DL + 1.6IL = 117$ kN (compression)
Combination 2: $1.2DL + 1.2IL + 1.2WL = 107$ kN (compression)
Combination 3: $1.4DL + 1.4WL = 66$ kN (compression)

Step 3 Determine ultimate bending moment and shear forces
Ignore eccentric end connections.

Step 4 Determine local bending moments
Assume a purlin is located half way between nodes. The ultimate end reaction of this purlin on the rafter is given by: spacing of truss = 4.0 m

Ultimate dead load and imposed load of sheeting and purlins
$$= (1.4 \times 0.15) + (1.6 \times 0.6) = 1.17 \text{ kN/m}^2$$

End reaction of purlin $= W = 1.17 \times 4\text{ m} \times 2.02\text{ m}$ (spacing of purlins)
$$= 10 \text{ kN}$$

Design local bending moment $= WL/6 = (10 \times 2.02)/6 = 3.4$ kNm, where L is taken as the spacing of the nodes of the truss. This local bending moment in the rafter may be taken to be hogging at the intersection points with the internal members and sagging midway between the nodes.

Step 5 Determine effective length
Distance between nodes on the rafter = 2.02 m
Effective length $L = 2020$ mm

Step 6 Determine slenderness ratio λ
$L_{xx} = L_{yy} = 2020$ mm; $r_{xx} = 39.6$ mm; $r_{yy} = 21.4$ mm

$$\lambda = \frac{2020}{21.4} = 94$$

Step 7 Check maximum slenderness ratio
$\lambda = 94 < 180$ for a strut. Satisfied.

Step 8 Check classification of section
Refer to Table 7 of BS 5950: Part 1 for section classification.

$$\varepsilon = \left(\frac{275}{p_y}\right)^{\frac{1}{2}} = 0.88$$

$$\frac{b}{T} = \frac{101.9}{2 \times 8.4} = 6.1 < 8.5\varepsilon = 7.48$$

$$\frac{d}{t} = \frac{128.5}{6.1} = 21.1 > 19\varepsilon = 16.72$$

The flanges of the T-section are classified as plastic but the stem is classified as slender.

Step 9 Determine strength reduction factors for slender elements
Compression reduction factor
As per Table 8 of BS 5950: Part 1 the strength reduction factor for the stem of a T-section is given by:

$$\frac{14}{(d/t\varepsilon) - 5} = \frac{14}{(21.1/0.88) - 5} = 0.738$$

which gives $p_y = 0.738 \times 355 = 262 \text{ N/mm}^2$.

Web moment and axial load combined reduction factor
The factor ε has to be increased such that the section classification becomes at least semi-compact. After finding this increased ε, the design strength p_y corresponding to this increased ε has to found. In all strength calculations for this element using design strength p_y this revised value will be used.

$$\varepsilon_{\text{revised}} = \frac{21.1}{16.72} \times 0.88 = 1.11$$

which gives

$$p_y = \frac{275}{\varepsilon_{\text{revised}}^2} = 223.2 \text{ N/mm}^2$$

Use $p_y = 223.2 \text{ N/mm}^2$ for all computations involving compression in the stem of the T-section.

Step 10 Determine compressive strength p_c
For a flange classified as plastic use $p_y = 355 \text{ N/mm}^2$

$$\lambda_0 = 0.2 \left(\frac{\pi^2 E}{p_y} \right)^{\frac{1}{2}} = 0.2 \times \left(\frac{\pi^2 \times 205 \times 10^3}{355} \right)^{\frac{1}{2}} = 15$$

The Robertson constant a is found from Table 5.5 to be 5.5 for both axes of buckling.

Perry coefficient $\eta = 0.001 a (\lambda - \lambda_0)$

$$= 0.001 \times 5.5 \times (94 - 15) = 0.4345$$

Euler strength $p_E = \dfrac{\pi^2 E}{\lambda^2} = \dfrac{\pi^2 \times 205 \times 10^3}{94^2} = 229 \text{ N/mm}^2$

$$\phi = \frac{p_y + (\eta + 1) p_E}{2} = \frac{355 + 1.4345 \times 229}{2}$$

$$= 341.8 \text{ N/mm}^2$$

Design strength $p_c = \dfrac{p_E p_y}{\phi + (\phi^2 - p_E p_y)^{\frac{1}{2}}}$

$$= \frac{229 \times 355}{341.8 + [341.8^2 - (229 \times 355)]^{\frac{1}{2}}} = 153 \text{ N/mm}^2$$

For stem of T-section classified as slender using $p_y = 223.2\,N/mm^2$
Calculating in the same manner as for the flanges we get:

$\lambda_0 = 19$; $a = 5.5$; $\eta = 0.4125$; $p_E = 229\,N/mm^2$; $\phi = 273.2\,N/mm^2$;
$p_c = 119.7\,N/mm^2$

Step 11 **Determine compression resistance**
$P_c = A_{gf}p_{cf}$ (for area of flange) $+ A_{gw}p_{cw}$ (for area of stem)
$A_{gw} = d \times t = 128.5 \times 6.1 = 784\,mm^2$; $A_{gf} = A - A_{gw} = 1610 - 784 = 826\,mm^2$

$\therefore\quad P_c = [(826 \times 153) + (784 \times 119.7)] \times 10^{-3} = 220.2\,kN$

Step 12 **Unity check for columns with nominal moments**
Not required.

Step 13 **Unity check for columns or struts with moments**
$p_y = 223\,N/mm^2$ for the stem of the T-section.

Shear capacity of the stem $= P_v = 0.6\,p_y A_v = 0.6 \times 223 \times 784 \times 10^{-3} = 104.9\,kN$

$F_v = W/2 = 5\,kN$ (see Step 4) $< 0.6 P_v$

For a slender section: $M_{cx} = p_y Z_x = 223 \times 26.6 \times 10^3 \times 10^{-6} = 5.93\,kNm$

Local capacity check

$$\frac{F}{A_g p_y} + \frac{M_x}{M_{cx}} = \frac{117}{[(784 \times 223) + (826 \times 355)] \times 10^{-3}} + \frac{3.4}{5.93}$$

$= 0.823 < 1$ (Satisfied)

See Steps 2 and 4 for F and M_x.

Overall buckling check
Determine buckling resistance M_b about the major axis: $\lambda = 94$; $p_y = 223.2\,N/mm^2$

Equivalent slenderness ratio $\lambda_{LT} = nuv\lambda$

$n = 1.0$; $u = 0.641$ (from published tables); $x = 15.4$ (from published tables); $\lambda/x = 6.1$
$N = 0$ (for the stem of a T-section in compression)
$v = 1.21$ from Table 14 of BS 5950: Part 1

$\therefore\quad \lambda_{LT} = 1 \times 0.641 \times 1.21 \times 94 = 72.9$

Limiting equivalent slenderness ratio $\lambda_{LO} = 0.4\sqrt{(\pi^2 E/p_y)} = 38.1$

Perry coefficient $\eta_{LT} = 0.007(\lambda_{LT} - \lambda_{LO}) \geq 0$
$\qquad\qquad\qquad\qquad = 0.007 \times (72.9 - 38.1) = 0.24$

$M_P = S_x p_y = 47.6 \times 10^3 \times 223.2 \times 10^{-6} = 10.62\,kNm$

Design of Struts 189

Elastic critical moment M_E is given by:

$$M_E = \frac{M_P \pi^2 E}{\lambda_{LT}^2 p_y} = \frac{10.62 \times \pi^2 \times 205 \times 10^3}{72.92 \times 223.2} = 18.12\,\text{kNm}$$

$$\phi_B = \frac{M_P + (\eta_{LT} + 1)M_E}{2} = \frac{10.62 + (1.24 \times 18.12)}{2} = 16.5\,\text{kNm}$$

Buckling resistance moment M_b is given by:

$$M_b = \frac{M_E M_P}{\phi_B + (\phi_B^2 - M_E M_P)^{\frac{1}{2}}} = \frac{18.12 \times 10.62}{16.5 + [16.5^2 - (18.12 \times 10.62)]^{\frac{1}{2}}}$$

$$= 7.57\,\text{kNm}$$

$$\frac{F}{P_c} + \frac{M_x}{M_b} = \frac{117}{220.2} + \frac{3.4}{7.57} = 0.98 < 1$$

∴ Overall buckling check is satisfied.

Steps 14 and 15 Not required.

The T-section chosen ($102 \times 127 \times 13\,\text{kg/m}$) is adequate for all conditions of loading.

5.5.4 Example 5.2b: Design the same rafter as in Example 5.2a using a rectangular hollow section

Choose $100 \times 50 \times 3.2 \times 7.18\,\text{kg/m}$ RHS.
Area of section $= 914\,\text{mm}^2$; $r_{xx} = 35.8\,\text{mm}$; $r_{yy} = 20.70\,\text{mm}$; $Z_{xx} = 23.5\,\text{cm}^3$; $S_x = 29.2\,\text{cm}^3$.

Steps 1 to 5 as before.

Step 6 *Slenderness ratio λ*

$$\lambda = \frac{2020}{20.7} = 98 < 180$$

Step 7 *Check maximum slenderness ratio*
Slenderness ratio $\lambda = 98 < 180$ (satisfied)

Step 8 *Section classification*

$$b = 50 - 3t = 50 - (3 \times 3.2) = 40.4\,\text{mm}$$

$$d = D - 3t = 100 - (3 \times 3.2) = 90.4\,\text{mm}$$

$$b/T = 40.4/3.2 = 12.6 < 28\varepsilon = 28 \times 0.88 = 24.6$$

$$d/t = 90.4/3.2 = 28.25 < 39\varepsilon = 39 \times 0.88 = 34.3$$

The section is classified as plastic.

Step 9 *Determine strength reduction factor*
Not required.

Step 10 Determine compressive strength
$p_y = 355 \text{ N/mm}^2$; $\lambda = 98$;

$\therefore p_c = 176 \text{ N/mm}^2$ from Table 27(c) of BS 5950: Part 1.

Step 11 Determine compression resistance
$P_c = 914 \times 176 \times 10^{-3} = 160.9 \text{ kN} > F = 117 \text{ kN}$

Step 13 Unity check
Shear is negligible.

$$M_{cx} = p_y S_x = 355 \times 29.2 \times 10^3 \times 10^{-6} = 10.37 \text{ kNm} < \mu_f p_y Z_x$$
$$= 12.51 \text{ kNm}$$

$$\mu_f = \frac{\text{ultimate loading}}{\text{service loading}} = \frac{117}{77.75} = 1.5$$

Local capacity check

$$\frac{F}{A_g p_y} + \frac{M}{M_{cx}} = \frac{117}{914 \times 355 \times 10^{-3}} + \frac{3.4}{10.37} = 0.69 < 1 \text{ (Satisfied)}$$

Overall buckling check
Box sections of uniform wall thickness need not be checked for lateral torsional buckling effects provided the slenderness ratio λ is not greater than the limiting values given in Table 5.6.

$$\frac{D}{B} = \frac{100}{50} = 2$$

Limiting value of $\lambda = \dfrac{350 \times 275}{p_y} = 271$, which is not exceeded.

Note: There is a potential saving in weight of material of up to 45% if RHS is used in place of the structural T-section, but the price of raw material and the additional fabrication costs could be 60% higher.

5.5.5 Example 5.3: Design the vertical leg of a portal frame

Consider the pitched portal frame in Example 3.7. Design the frame using the elastic analysis results factored to ultimate limit state.

Step 1 Select type of strut and trial section
In the elastic analysis in Example 3.7 UB $457 \times 191 \times 98$ kg/m has been chosen for the column. Select Grade 50 steel.

$A = 125 \text{ cm}^2$; $D = 467.4 \text{ mm}$; $B = 192.8 \text{ mm}$; $t = 11.4 \text{ mm}$; $T = 19.6 \text{ mm}$; $d = 407.9 \text{ mm}$; $r_{xx} = 19.1 \text{ cm}$; $r_{yy} = 4.33 \text{ cm}$; $Z_x = 1960 \text{ cm}^3$; $Z_y = 243 \text{ cm}^3$; $S_x = 2230 \text{ cm}^3$; $S_y = 378 \text{ cm}^3$; $u = 0.88$; $x = 25.8$.

Design of Struts 191

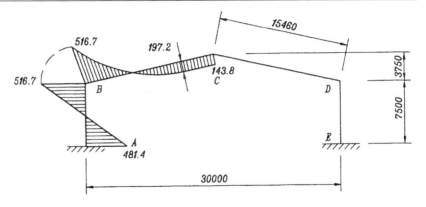

SK 5/6 Sketch of portal showing bending moment diagram.

Step 2 **Determine combined ultimate axial load**
From Example 3.7: $F = 154.6$ kN.

Step 3 **Determine ultimate bending moment and shear force**
See Step 4 of Example 3.7 (Table 3.5):

At base of column	$+481.40$ kNm
At top of column	-516.70 kNm
Shear force	133.10 kN

Step 4 Not required.

Step 5 **Determine effective length**
(See Table 5.2.) One could use an effective length $L_E = 1.5L$ when the frame is unbraced, fixed at the base and partially restrained in direction at the top.

By limited frame method

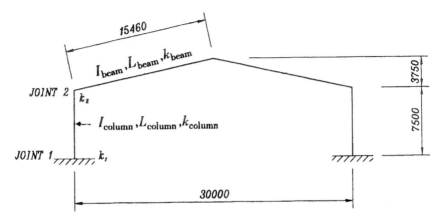

SK 5/7 Sketch of portal frame with stiffness for limited frame method.

Assume the stiffness of foundation is equal to the column stiffness.

$$k_1 = \frac{k_C}{k_C + k_{base}} = 0.5$$

The sway frame critical buckling mode of the beam is in double curvature. (See page 165.)

$$I_{beam} = 36\,250\,\text{cm}^4; \quad L_{beam} = 15\,460\,\text{mm}$$

$$1.5\,\frac{I_{beam}}{L_{beam}} = \frac{1.5 \times 36\,250 \times 10^4}{15\,460} = 35\,171\,\text{mm}^3$$

$$\frac{I_{column}}{L_{column}} = \frac{45\,770 \times 10^4}{7500} = 61\,027\,\text{mm}^3$$

$$\therefore k_2 = \frac{61\,027}{61\,027 + 35\,171} = 0.63$$

$k_3 = 0$ for unrestricted side sway. Using SK 5/3b or Figure 24 of BS 5950: Part 1, it is found:

$$\frac{L_{Ex}}{L} = 1.55 \quad \therefore L_{Ex} = 1.55 \times 7500 = 11\,625\,\text{mm}$$

Assume two longitudinal restraints connecting all the portal frames in the building at mid-height and at the top of the column, which connect to a vertical bracing system. Assume that the column is rigidly fixed to the foundation about both orthogonal axes. The top section of the column above the restraint is held in position along the X–X axis at the top and held in position and direction at the bottom, where it is continuous.

About the minor axis $k_1 = 0.5$ at the bottom and $k_2 = 1.0$ at the top for the top half of the column.

The frame is fully braced about the minor axis and using SK 5/3a or Figure 23 of BS 5950: Part 1:

$$\frac{L_{Ey}}{L} \approx 0.85 \quad \text{which gives} \quad L_{Ey} = 0.85 \times \left(\frac{7500}{2}\right) = 3188\,\text{mm}$$

Step 6 Slenderness ratio λ

$$\lambda_x = \frac{L_{Ex}}{r_{xx}} = \frac{11\,625}{191} = 61; \quad \lambda_y = \frac{L_{Ey}}{r_{yy}} = \frac{3188}{43.3} = 74$$

Step 7 Check maximum slenderness ratio
$\lambda_{max} = 74 < 180$ (Satisfied)

Step 8 Classification of section
Assume $p_y = 355 \text{ N/mm}^2$ for web < 16 mm and $p_y = 345 \text{ N/mm}^2$ for flanges > 16 mm.

$$\varepsilon_{\text{web}} = \sqrt{\frac{275}{355}} = 0.88 \quad \text{and} \quad \varepsilon_{\text{flange}} = \sqrt{\frac{275}{345}} = 0.89$$

$$\frac{b}{T} = 4.92 < 7.5\varepsilon_f = 6.7 \text{ (Flange is plastic)}$$

$$\frac{d}{t} = 35.8 > 39\varepsilon_w = 34.3 \text{ (Web is slender)}$$

At or near the point of contra-flexure the whole section of the web is in compression and therefore, by definition, the web of the rolled section is classed as slender.

Step 9 Determine strength reduction factor
The web should be designed with a reduced value of design strength p_y.

$$\varepsilon_{\text{revised}} = \frac{0.88 \times 35.8}{34.3} = 0.92$$

$$\therefore \quad p_{y,\text{revised}} = \frac{275}{(0.92)^2} = 325 \text{ N/mm}^2$$

Step 10 Determine compressive strength
Axis of buckling is about the Y–Y axis. Use Table 27(b) of BS 5950: Part 1.

$\lambda = 74$; $p_y = 345 \text{ N/mm}^2$; $p_c = 226 \text{ N/mm}^2$ for flange.
$\lambda = 74$; $p_y = 325 \text{ N/mm}^2$; $p_c = 217 \text{ N/mm}^2$ for web.

Step 11 Determine compression resistance
Area of web = $A_{gw} = [467.4 - (2 \times 19.6)] \times 11.4 = 4881 \text{ mm}^2$
Area of flange = A_{gf} = total area $- 4881 = 12\,500 - 4881 = 7619 \text{ mm}^2$
Compression resistance = $P_c = [(4881 \times 217) + (7619 \times 226)] \times 10^{-3} = 2781 \text{ kN}$

Step 12 Not required.

Step 13 Unity check with axial load and moment
Local capacity check

$$P_v = 0.6 p_y A_v = 0.6 \times 217 \times 11.4 \times 467.4 \times 10^{-3} = 694 \text{ kN}$$

$$F_v = 133.1 < 0.6 \times 694 = 416.4 \text{ kN}$$

Shear is low. The section is classified as slender because the web is slender.

$$M_{cx} = p_y Z_x = 345 \times 1960 \times 10^3 \times 10^{-6} = 676.2 \, \text{kNm} < M_x$$

$$\frac{F}{A_g p_y} + \frac{M_x}{M_{cx}} = \frac{154.6 \times 10^3}{(4881 \times 325) + (7619 \times 345)} + \frac{516.7}{676.2}$$

$$= 0.80 < 1 \, \text{(Satisfied)}$$

Overall buckling check

$$\lambda_{LT} = nuv\lambda; \quad \lambda = 74; \quad x = 25.8; \quad n = 1.0; \quad u = 0.88;$$

$$\lambda/x = 2.87; \quad N = 0.5$$

∴ From Table 14 of BS 5950: Part 1 $v = 0.91$.

$$\lambda_{LT} = 1 \times 0.88 \times 0.91 \times 74 = 59.3$$

From Table 11 of BS 5950: Part 1 $p_b = 241 \, \text{N/mm}^2$ for $\lambda_{LT} = 59.3$ and $p_y = 325 \, \text{N/mm}^2$.

$$M_b = S_x p_b = 2230 \times 10^3 \times 241 \times 10^{-6} \, \text{kNm} = 537.4 \, \text{kNm}$$

Determine the value of β:

$$M_1 = +481.4 \, \text{kNm}; \quad M_2 = -516.7 \, \text{kNm}$$

$$\beta = \frac{M_1}{M_2} = -\frac{481.4}{516.7} = -0.932$$

∴ $m = 0.57 + 0.33\beta + 0.10\beta^2 \geq 0.43$

$$= 0.35 \geq 0.43$$

or $m = 0.43$

$$\frac{F}{A_g p_c} + \frac{mM_x}{M_b} = \frac{154.6}{2781} + \frac{0.43 \times 516.7}{537.4} = 0.47 < 1 \, \text{(Satisfied)}$$

Alternative method of overall buckling check
Find p_{cx} and p_{cy} about X–X and Y–Y axes respectively.
From Step 6: $\lambda_x = 61$ and $\lambda_y = 74$.
Take $p_y = 345 \, \text{N/mm}^2$ for flanges and $325 \, \text{N/mm}^2$ for the web.
From Table 27(a) of BS 5950: Part 1 find for $\lambda_x = 61$:

p_{cx} for flange $= 289 \, \text{N/mm}^2$ and p_{cx} for web $= 275 \, \text{N/mm}^2$.

From Table 27(b) of BS 5950: Part 1 find for $\lambda_y = 74$:

p_{cy} for flange $= 226 \, \text{N/mm}^2$ and p_{cy} for web $= 217 \, \text{N/mm}^2$.

See step 11:

$P_{cx} = [(4881 \times 275) + (7619 \times 289)] \times 10^{-3} = 3544\,\text{kN}$

$P_{cy} = 2781\,\text{kN}$

$$M_{ax} = M_{cx} \frac{\left(1 - \dfrac{F}{P_{cx}}\right)}{\left(1 + \dfrac{0.5F}{P_{cx}}\right)} = 676.2 \times \frac{\left(1 - \dfrac{154.6}{3544}\right)}{\left(1 + \dfrac{0.5 \times 154.5}{3544}\right)}$$

$= 632.9\,\text{kNm}$

or $= M_b\left(1 - \dfrac{F}{P_{cy}}\right) = 537.4 \times \left(1 - \dfrac{154.6}{2781}\right) = 507.5\,\text{kNm}$

Alternative overall buckling check:

$$\frac{mM_x}{M_{ax}} = \frac{0.43 \times 516.7}{507.5} = 0.44 < 1\ \text{(Satisfied)}$$

Conclusion: the section chosen ($457 \times 191 \times 98\,\text{kg/m}$ UB) is adequate for the purpose.

5.5.6 Example 5.4: Design the corner column of a multistorey building

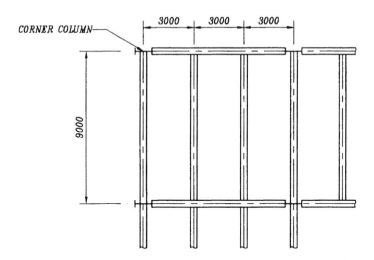

SK 5/8 Floor plan showing typical column and beam layout.

Design a corner column at the ground floor level in a simple beam–column construction of a multistorey building, where side sway is prevented by shear walls and plan bracing. The beams are carrying concrete floors. The main beams sit on very stiff brackets connected to the flanges of the columns. The secondary beams are connected to the web of the columns by web cleats. The storey height is 5 metres.

Step 1 **Select type of strut and a trial size**
Table 5.1 shows that a cost-effective option could be a UC in Grade 50 steel.

From basic load cases it is seen that maximum axial compression is approximately 4000 kN. The minimum sectional area required, assuming $p_c = 200 \text{ N/mm}^2$, is given by:

$$A_{\min} = \frac{4000 \times 10^3}{200} = 20\,000 \text{ mm}^2$$

Select UC $305 \times 305 \times 158$ kg/m as a trial size in Grade 50 steel.

Step 2 **Determine combined ultimate axial loading**
For simple construction it is not necessary to carry out analysis with patterned loading on the beams to find the maximum load on the column. Assume all beams are fully loaded.

Basic loads – end reactions on the corner column
Main beam:
   At roof level   Dead load = 150 kN Imposed load = 75 kN
   At levels 3, 4 and 5 Dead load = 200 kN Imposed load = 75 kN
   At levels 2 and 1   Dead load = 200 kN Imposed load = 150 kN
Secondary beam:
   At roof level   Dead load = 75 kN   Imposed load = 37.5 kN
   At levels 3, 4 and 5 Dead load = 100 kN Imposed load = 37.5 kN
   At levels 2 and 1   Dead load = 100 kN Imposed load = 75.0 kN

Maximum ultimate axial load on the column at ground floor level F_c is given by:

$$\begin{aligned}F_c = {} & 1.4 \times [150 + (3 \times 200) + (2 \times 200) + 75 + (3 \times 100) \\ & + (2 \times 100)] + 1.6 \times [75 + (3 \times 75) + (2 \times 150) + 37.5 \\ & + (3 \times 37.5) + (2 \times 75)] = 3855 \text{ kN}\end{aligned}$$

Step 3 **Determine ultimate bending moments and shear forces**
Assume that the columns are continuous, with full strength splices where required. For the column at ground floor, only eccentricity of beam reactions at the first floor level has to be considered for simple construction. Assume 200 mm wide stiff bearing for the main beams.
   Eccentricity of the main beam reaction to the centre of column = 100 mm + $D/2$ = 263.6 mm about the major axis of the column. (D = overall depth of column.)
   Eccentricity of the secondary beam reaction = 100 mm (minimum to be allowed in the design).
   The beams are not designed with partial fixity at the columns.
   Assume that the column size is $305 \times 305 \times 137$ kg/m at the first floor level. The column stiffnesses at the ground floor and first floor level are given by:

$$\frac{I_1}{L_1} = \frac{32\,800}{500} = 65.6 \quad \text{at first floor;}$$

$$\frac{I_G}{L_G} = \frac{38\,700}{500} = 77.4 \quad \text{at ground floor}$$

The ratio of stiffness is not more than 1.5, therefore the moment due to beam eccentricities at first floor level may be equally divided between the columns immediately above and below that level. These moments will have no effect on columns at any other level.

Ultimate main beam reaction at the first floor level $= (1.4 \times 200) + (1.6 \times 150) = 520$ kN.

Ultimate secondary beam reaction at the first floor level $= (1.4 \times 100) + (1.6 \times 75) = 260$ kN.

The bending moments at the beam–column joint at the first floor level are given by:

$$M_{xj} = 520 \times 0.2636 = 137.1 \text{ kNm}$$

$$M_{yj} = 260 \times 0.100 = 26 \text{ kNm}$$

Column bending moments are half of these moments:

$$M_x = 137.1/2 = 68.6 \text{ kNm}$$

$$M_y = 26/2 = 13 \text{ kNm}$$

Step 4 Not required.

Step 5 *Determine effective length*
It is assumed that the column is pinned at the base, braced against side sway, continuous at first floor level, restrained in position and direction at the top, and restrained in position only at the bottom. By reference to the sketches in Table 5.2 it can be concluded that $L_E = 0.85L$.

By limited frame method:
  Column braced against side sway, hence $k_3 = \infty$
  Column pinned at the base, hence $k_2 = 1$
  Column continuous at the top, hence $k_1 = 0.5$
  ∴ Effective length $L_E = 0.825L$ from SK 5/3a.
  Conservatively, take $L_E = 0.85L = 0.85 \times 5000 = 4250$ mm

Step 6 *Determine slenderness ratio*
$L_{Ex} = L_{Ey} = 4250$ mm; $r_{xx} = 139$ mm; $r_{yy} = 78.9$ mm.

$$\therefore \lambda_x = \frac{4250}{139} = 30.6; \quad \lambda_y = \frac{4250}{78.9} = 53.9$$

Step 7 *Check slenderness limit*
$\lambda_{max} = 53.9 < 180$ (Satisfied)

Step 8 Check classification of section
$\varepsilon = 0.88$ for $p_y = 355 \text{ N/mm}^2$ and $\varepsilon = 0.89$ for $p_y = 345 \text{ N/mm}^2$

$\dfrac{b}{T} = 6.21 < 8.5\varepsilon = 7.57$

(The section is classified as plastic)

$\dfrac{d}{t} = 15.7 < 39\varepsilon = 34.3$

Step 9 Not required.

Step 10 Determine compressive strength
Flange thickness $T = 25.0 \text{ mm}$.

Assume $p_y = 345 \text{ N/mm}^2$ governed by the flange thickness.

$$\lambda_0 = 0.2\left(\dfrac{\pi^2 E}{p_y}\right)^{\frac{1}{2}} = 15.3$$

Robertson constant $a_x = 3.5$ for buckling about X–X axis
$\phantom{\text{Robertson constant }}a_y = 5.5$ for buckling about Y–Y axis

Perry factor $\eta_x = 0.001 a_x (\lambda_x - \lambda_0) = 0.0536$
$\phantom{\text{Perry factor }}\eta_y = 0.001 a_y (\lambda_y - \lambda_0) = 0.2123$

Euler strength $p_{Ex} = \dfrac{\pi^2 E}{\lambda_x^2} = 2160.8 \text{ N/mm}^2$

$$p_{Ey} = \dfrac{\pi^2 E}{\lambda_y^2} = 696.4 \text{ N/mm}^2$$

$$\phi_x = \dfrac{p_y + (\eta_x + 1)p_{Ex}}{2} = 1310.8 \text{ N/mm}^2$$

$$\phi_y = \dfrac{p_y + (\eta_y + 1)p_{Ey}}{2} = 594.6 \text{ N/mm}^2$$

$$p_{cx} = \dfrac{p_{Ex} p_y}{\phi_x + (\phi_x^2 - p_{Ex} p_y)^{\frac{1}{2}}} = 324.5 \text{ N/mm}^2$$

$$p_{cy} = \dfrac{p_{Ey} p_y}{\phi_y + (\phi_y^2 - p_{Ey} p_y)^{\frac{1}{2}}} = 258.0 \text{ N/mm}^2$$

Step 11 Determine compression resistance
$P_{cx} = A_g p_{cx} = 20\,100 \times 324.5 \times 10^{-3} = 6522 \text{ kN}$
$P_{cy} = A_g p_{cy} = 20\,100 \times 258.0 \times 10^{-3} = 5185.8 \text{ kN}$

Step 12 Unity check for columns with nominal moments
L = distance between points where the column is restrained about both axes = 5.0 m.

Equivalent slenderness ratio λ_{LT} is given by:

$$\lambda_{LT} = 0.5\left(\frac{L}{r_y}\right) = \frac{0.5 \times 5000}{78.9} = 31.7$$

$$\lambda_{LO} = 0.4\left(\frac{\pi^2 E}{p_y}\right)^{\frac{1}{2}} = 30.6$$

Perry coefficient $\eta_{LT} = 0.007\,(\lambda_{LT} - \lambda_{LO}) \geq 0$
$\phantom{Perry coefficient \eta_{LT}} = 0.0077$

Plastic moment capacity of section about the major axis
$= M_P = S_x p_y = 2680 \times 345 \times 10^{-3} = 924.6\,\text{kNm}$

Elastic critical moment M_E is given by:

$$M_E = \frac{M_P \pi^2 E}{\lambda_{LT}^2 p_y} = 5396\,\text{kNm}$$

$$\phi_B = \frac{M_p + (\eta_{LT} + 1)M_E}{2} = 3181\,\text{kNm}$$

Buckling moment of resistance M_{bs} is given by:

$$M_{bs} = \frac{M_E M_P}{\phi_B + (\phi_B^2 - M_E M_P)^{\frac{1}{2}}} = 916.1\,\text{kNm}$$

Buckling unity check for simple column:

$$\frac{F_c}{A_g p_c} + \frac{M_x}{M_{bs}} + \frac{M_y}{p_y Z_y} = \frac{3855}{5185.8} + \frac{68.6}{916.1} + \frac{13}{345 \times 806 \times 10^{-3}}$$

$$= 0.865 < 1\,\text{(Satisfied)}$$

Conclusion: the chosen section (UC $305 \times 305 \times 158$ kg/m) is adequate for the purpose.

5.5.7 Example 5.5: Design the compound column of an industrial building with a heavy-duty crane

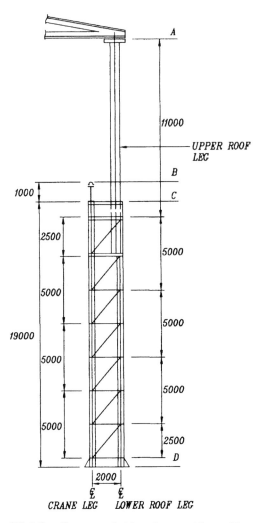

SK 5/9 Compound side column with roof leg and crane leg in a steel plant.

After analysis, the basic load table is as follows:

Location	Dead load (kN)	Imposed load (kN)	Wind load vertical (kN)	Wind load horizontal (kN)	Crane load vertical (kN)	Crane load horizontal (kN)
A (Truss)	240	180	60	48	–	–
B (Crane)	60	–	–	–	1840	72
AC (Cladding)	20	–	–	90	–	–
CD (Cladding)	30	–	–	150	–	–

Design of Struts 201

Assume that the upper roof stanchion is made from UB 914 × 419 × 388 kg/m. To get adequate end clearance for the crane, the minimum centre-to-centre distance between the lower roof stanchion and the crane stanchion is 2000 mm.

The compound column will be checked against only one load combination to show the procedure. There are several possible load combinations, with the basic loads as given in Table 4.1. Most of these conditions have to be properly investigated.

Step 1 **Select type of strut and trial sections**
For a very heavy industrial building column select fabricated section in Grade 50 steel. It is assumed that the lacing will be spaced at about 2500 mm along the longitudinal axis of the compound stanchion.

$\lambda_c = L_c/r_{yy}$ should be kept below 50 for an individual component of the compound column.

$$\therefore \quad r_{yy,\text{minimum}} = \frac{2500}{50} = 50 \text{ mm}$$

The maximum anticipated load on the crane leg by inspection of the basic load table is of the order of 6000 kN. Assuming a design strength p_c of 200 N/mm², the minimum area required for this leg will be of the order of 30 000 mm².

Minimum I_{yy} required is $(r_{\min})^2 \times \text{area} = 50^2 \times 30\,000 \text{ mm}^4 = 75 \times 10^6 \text{ mm}^4$.

Assume a depth of section $d = 800$ mm and the trial section is shown in SK 5/10.

Area of crane leg $= (2 \times 300 \times 25) + (750 \times 25)$
$= 33\,750 \text{ mm}^2 \equiv 2.65 \text{ kN/m}$
Area of lower roof leg $= (2 \times 300 \times 20) + (760 \times 25)$
$= 31\,000 \text{ mm}^2 \equiv 2.43 \text{ kN/m}$

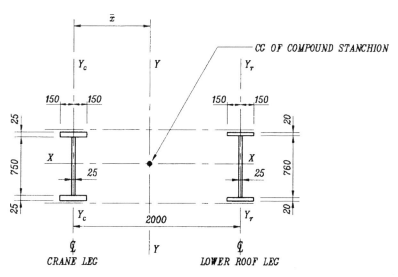

SK 5/10 Section through the compound column showing trial size.

Crane leg
Moment of inertia about X–X axis $= I_{xx}$

$$= 2 \times \left[\frac{1}{12} \times 300 \times 25^3 + 300 \times 25 \times (400 - 12.5)^2\right]$$

$$+ \frac{1}{12} \times 25 \times 750^3 = 3132 \times 10^6 \text{ mm}^4$$

Moment of inertia about Y_c–Y_c axis $= I_{yy}$

$$= 2 \times \left[\frac{1}{12} \times 25 \times 300^3\right] + \frac{1}{12} \times 750 \times 25^3 = 113.48 \times 10^6 \text{ mm}^4$$

Lower roof leg
Moment of inertia about X–X axis $= I_{xx} = 2740 \times 10^6 \text{ mm}^4$

Moment of inertia about Y_r–Y_r axis $= I_{yy} = 90.99 \times 10^6 \text{ mm}^4$

Centre of gravity of the compound stanchion is given by:

$$\bar{x} = \frac{31\,000 \times 2000}{33\,750 + 31\,000} = 958 \text{ mm from the centre-line of the crane leg}$$

Compound column
Moment of inertia about X–X axis $= (3132 \times 10^6) + (2740 \times 10^6) = 5872 \times 10^6 \text{ mm}^4$

Moment of inertia about Y–Y axis $= (113.48 \times 10^6) + (33\,750 \times 948^2) + (90.99 \times 10^6) + 31\,000 \times 1052^2) = 64\,843 \times 10^6 \text{ mm}^4$

Radius of gyration $\quad r = \sqrt{\frac{I}{A}}$

Crane leg
$r_{xx} = 304.6$ mm; $r_{yy} = 58.0$ mm

Lower roof leg
$r_{xx} = 297.3$ mm; $r_{yy} = 54.2$ mm

Compound column
$r_{xx} = 301.1$ mm; $r_{yy} = 1000$ mm

Step 2 **Determine combined ultimate axial loading**
For the purpose of this example, consider only one load combination.

Combination 1: 1.2 DL + 1.2 CLV + 1.2 CLH + 1.2 WLV + 1.2 WLH
(1) Vertical load from the upper roof leg $= 1.2 \times (240 + 60 + 20) = 384$ kN acting 467.5 mm from the centre-line of the lower roof leg.
F_{c1} = load on crane leg $= (384 \times 467.5)/2000 = 90$ kN
F_{r1} = load on lower roof leg $= 384 - 90 = 294$ kN

(2) Bending moment at the base of combined column due to wind
$= 1.2 \times [(48 \times 30) + (90 \times 5.5) + (90 \times 19) + (150 \times 9.5)]$
$= 6084 \, \text{kNm}$
$F_{c2} = $ load on crane leg $= 6084/2 = \pm 3042 \, \text{kN}$
$F_{r2} = $ load on lower roof leg $= \pm 3042 \, \text{kN}$

(3) Bending moment at the base of combined column due to crane surge
$= 1.2 \times 72 \times 20 = 1728 \, \text{kNm}$
$F_{c3} = $ load on crane leg $= 1728/2.0 = \pm 864 \, \text{kN}$
$F_{r3} = $ load on lower roof leg $= \pm 864 \, \text{kN}$

(4) Self-weight and sheeting on lower roof leg
$F_{r4} = 1.2 \times [30 + (2.43 \times 19)] = 70 \, \text{kN}$

(5) Crane vertical load and self-weight of crane leg
$F_{c4} = 1.2 \times [1840 + (2.65 \times 19)] = 2260 \, \text{kN}$

Maximum compressive load on the crane leg $= F_c = F_{c1} + F_{c2} + F_{c3} + F_{c4} = 6256 \, \text{kN}$

Maximum compressive load on the roof leg $= F_r = F_{r1} + F_{r2} + F_{r3} + F_{r4} = 4270 \, \text{kN}$

Step 3 *Determine ultimate bending moments and shear forces*
There is no bending moment in each component of the compound column. The maximum bending moment about the Y–Y axis of the compound column at the base of the column is:

$$-384 \times (1.042 - 0.4675) + 6084 + 1728 - (70 \times 1.042)$$
$$+ (2260 \times 0.958) = 9683.5 \, \text{kNm}$$

Maximum vertical load on the compound column $= 384 + 70 + 2260 = 2714 \, \text{kN}$

Maximum transverse shear $= 1.2 \times (48 + 90 + 150) = 345.6 \, \text{kN}$

Step 4 Not required.

Step 5 *Determine effective lengths*
See Table 5.2.

Lower roof leg:
Axis X–X: $L_E = 0.85L = 0.85 \times 19\,000 = 16\,150 \, \text{mm}$
Axis Y–Y: $L_E = 1.0L_c = 2500 \, \text{mm}$

Crane leg:
Axis X–X: $L_E = 0.85L = 16\,150 \, \text{mm}$
Axis Y–Y: $1.0L_c = 2500 \, \text{mm}$

Compound column:
Axis X–X: $L_E = 0.85L = 16\,150 \, \text{mm}$
Axis Y–Y: $L_E = 1.5L = 1.5 \times 19\,000 = 28\,500 \, \text{mm}$

Step 6 Determine slenderness ratio

Lower roof leg:
Axis X–X: $\lambda_{xx} = 16\,150/297.3 = 54.3$
Axis Y–Y: $\lambda_{yy} = 2500/54.2 = 46.1$

Crane leg:
Axis X–X: $\lambda_{xx} = 16\,150/304.6 = 53.0$
Axis Y–Y: $\lambda_{yy} = 2500/58 = 43.1$

Compound column:
Axis X–X: $\lambda_{xx} = 16\,150/301.1 = 53.6$
Axis Y–Y: $\lambda_{yy} = 28\,500/1000 = 28.5$

Step 7 Check maximum slenderness ratio
Roof leg: $\lambda_{c,max} = 46.1 < 50$ (Satisfied)
Crane leg: $\lambda_{c,max} = 43.1 < 50$ (Satisfied)
Compound stanchion $\lambda_{max} = 53.6 < 1.4 \times \lambda_c = 1.4 \times 46.1 = 64.5$
Assume slenderness ratio of the compound stanchion is equal to 64.5.

Step 8 Check classification of section

and

Step 9 Determine strength reduction factors
For sections which are built up by welding and are in compression, the design strength p_y should be reduced by $20\,\text{N/mm}^2$. All elements are above 16 mm but less than 40 mm in thickness. For Grade 50 steel the design strength $p_y = 345 - 20 = 325\,\text{N/mm}^2$.

$$\varepsilon = \sqrt{\frac{275}{325}} = 0.92$$

Crane leg
$b = 137.5$ mm; $T = 25$ mm; $t = 25$ mm; $d = 750$ mm

Flanges: $b/T = 5.5 < 7.5\varepsilon = 6.9$ (Element is class 1 plastic)
Web: $d/t = 30 > 28\varepsilon = 25.76$ (Element is slender)

Slenderness reduction of design strength p_y:

$$\varepsilon_{revised} = 0.92 \times \frac{30}{25.76} = 1.0714$$

$$p_{y,revised} = \frac{275}{1.0714^2} = 240\,\text{N/mm}^2$$

$$\therefore \quad p_{y,flange} = 325\,\text{N/mm}^2$$

$$p_{y,web} = 240\,\text{N/mm}^2$$

Roof leg
$b = 137.5$ mm; $T = 20$ mm; $t = 25$ mm; $d = 760$ mm

Flanges: $b/T = 6.875 < 7.5\varepsilon = 6.9$ (Element is class 1 plastic)
Web: $d/t = 30.4 > 28\varepsilon = 25.8$ (Element is slender)

Design of Struts 205

Slenderness reduction of design strength p_y:

$$\varepsilon_{\text{revised}} = 0.92 \times \frac{30.4}{25.76} = 1.0857$$

$$p_{y,\text{revised}} = \frac{275}{1.0875^2} = 233 \, \text{N/mm}^2$$

$$\therefore \quad p_{y,\text{flange}} = 325 \, \text{N/mm}^2$$

$$p_{y,\text{web}} = 233 \, \text{N/mm}^2$$

Step 10 *Determine compressive strength p_c*

Leg	Flange						Web					
	X–X			Y–Y			X–X			Y–Y		
	λ	p_y	p_c	λ	p_y	p_c	λ	p_y	p_c	λ	p_y	p_c
Crane leg	53.0	325	269	43.1	325	267	53.0	240	205	43.1	240	209
Roof leg	54.3	325	267	46.1	325	264	54.3	233	198	46.1	233	196
Comb. leg	53.6	325	267	64.5	325	219	53.6	233	198	64.5	233	168

Use Table 27(b) for X–X axis and Table 27(c) for Y–Y axis from BS 5950: Part 1.

Step 11 *Determine compression resistance*

Crane leg
Flange: $p_{c,\text{min}} = 267 \, \text{N/mm}^2$; area $= 2 \times 300 \times 25 = 15\,000 \, \text{mm}^2$
  Compression resistance $= P_{cf} = 15\,000 \times 267 = 4005 \times 10^3 \, \text{N}$
Web: $p_{c,\text{min}} = 205 \, \text{N/mm}^2$; area $= 750 \times 25 = 18\,750 \, \text{mm}^2$
  Compression resistance $P_{cw} = 18\,750 \times 205 = 3844 \times 10^3 \, \text{N}$
Total resistance $= P_c = P_{cf} + P_{cw} = 7849 \, \text{kN} > 6256 \, \text{kN}$ (see Step 2)

Roof leg
Flange: $p_{c,\text{min}} = 264 \, \text{N/mm}^2$; area $= 2 \times 300 \times 20 = 12\,000 \, \text{mm}^2$
  Compression resistance $= P_{cf} = 12\,000 \times 264 = 3168 \times 10^3 \, \text{N}$
Web: $p_{c,\text{min}} = 196 \, \text{N/mm}^2$; area $= 760 \times 25 = 19\,000 \, \text{mm}^2$
  Compression resistance $P_{cw} = 19\,000 \times 196 = 3724 \times 10^3 \, \text{N}$
Total resistance $= P_c = P_{cf} + P_{cw} = 6892 \, \text{kN} > 4270 \, \text{kN}$ (see Step 2)

Compound column
Flange: $p_{c,\text{min}} = 219 \, \text{N/mm}^2$; area $= 15\,000 + 12\,000 = 27\,000 \, \text{mm}^2$
  Compression resistance $= P_{cf} = 27\,000 \times 219 = 5913 \times 10^3 \, \text{N}$
Web: $p_{c,\text{min}} = 168 \, \text{N/mm}^2$; area $= 18\,750 + 19\,000 = 37\,750 \, \text{mm}^2$
  Compression resistance $P_{cw} = 37\,750 \times 168 = 6342 \times 10^3 \, \text{N}$
Total resistance $= P_c = P_{cf} + P_{cw} = 5913 + 6342 = 12\,255 \, \text{kN}$

Step 12 **Unity check of column in simple construction**

The unity check is carried out for the compound column as one unit. The crane leg is the compression flange of this compound section. The web of this compound section is the lacing system. The web of the crane leg may be considered as an internal element of the compression flange. The section classification of the web may be carried out as follows:

$b = 750 \text{ mm}; \quad b/T = 30 > 28\varepsilon = 24.92$

∴ Strength reduction factor for the flange

$$= \frac{21}{\dfrac{b}{T\varepsilon} - 7}$$

$= 0.786$ as per Table 8 of BS 5950: Part 1

Design strength for the web acting as a flange
$= 345 \times 0.786 = 271.2 \text{ N/mm}^2$

Equivalent slenderness of the compound column =

$\lambda_{LT} = 0.5\lambda_{max} = 64.5 \times 0.5 = 32.2$ (see Step 7)

$\lambda_{LO} = 0.4\sqrt{\dfrac{\pi^2 E}{p_y}} = 0.4 \times \sqrt{\dfrac{\pi^2 \times 205 \times 10^3}{271}} = 34.6$

$\lambda_{LO} > \lambda_{LT}$

∴ $p_c = p_y = 345 \text{ N/mm}^2$ and 271.2 N/mm^2 for the flanges and the web respectively

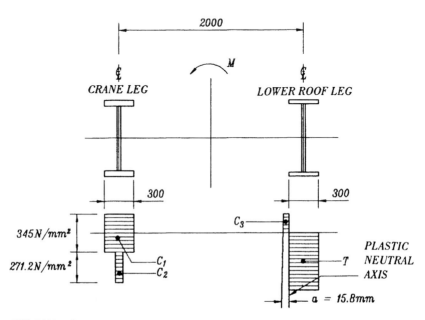

SK 5/11 Stress diagram at plastic state as a compound column.

Design of Struts 207

The crane leg in compression can provide a resistance in bending of the compound column given by:

(Area of flanges × 345) + (Area of web × 271.2)

$= (15\,000 \times 345) + (18\,750 \times 271.2) = 10\,260\,\text{kN}$

The roof leg in tension can provide a resistance in bending of the compound column given by:

Area of roof leg × 345 = 31 000 × 345 = 10 695 kN

This means that the plastic neutral axis must lie in the roof leg because 10 695 kN is higher than 10 260 kN. Assume that the neutral axis is a mm from the tip of the flange of the roof leg in tension, as shown in SK 5/11.

C_1 = Area of flanges of crane leg × 345 = 5175 kN

C_2 = Area of web of crane leg × 271.2 × 5085 kN

$C_3 = 2 \times 20 \times a \times 345 \times 10^{-3} = 13.8 \times a\,\text{kN}$

T = Total area of roof leg $- (2 \times 20 \times a) \times 345 \times 10^{-3}$

$= 10\,695 - (13.8 \times a)\,\text{kN}$

$C_1 + C_2 + C_3 = T$

or $10\,260 + (13.8 \times a) = 10\,695 - (13.8 \times a)$

or $a = 15.8\,\text{mm}$

The plastic moment of resistance of the compound column is obtained by taking moments of these internal forces about the centre-line of the crane leg:

$$M_\text{P} = \left[2 \times 20 \times (300 - 15.8) \times 345 \times \left(2000 + \frac{15.8}{2}\right)\right]$$

$$+ [760 \times 25 \times 345 \times 2000]$$

$$- [2 \times 15.8 \times 345 \times 20 \times (2000 - 150 + 7.9)]$$

$$= 20\,580 \times 10^6\,\text{Nmm}$$

$$\eta_\text{LT} = 2 \times 0.007 \times \lambda_\text{LO} = 0.4844 \leq 2 \times 0.007 \times (\lambda_\text{LT} - \lambda_\text{LO})$$

$$= -ve = 0$$

$\therefore\ M_\text{b} = M_\text{P}$

Unity check: $\dfrac{F_\text{c}}{P_\text{c}} + \dfrac{M_\text{x}}{M_\text{b}} = \dfrac{2714}{12\,255} + \dfrac{9683.5}{20\,580} = 0.69 < 1$ (Satisfied)

Step 13   Not required.

208 Structural Steelwork

Step 14 *Design of lacing system*
Inclination of the lacing system is given by:

$$\tan \alpha = \frac{2.5}{2.3} = 1.08695; \quad \alpha = 47.4°$$

The inclination of the lacing system to the axis of the column ϕ is therefore:

$$\phi = 90° - 47.4° = 42.6° > 40° \text{ (Satisfied)}$$

Length of the diagonal lacing bar = 3400 mm. $\lambda = l/r$ should not exceed 180.

$$r_{\text{minimum}} = \frac{3400}{180} = 19 \text{ mm}$$

$$\text{Transverse shear} = \frac{2.5}{100} \times F + V_s$$

$$= \frac{2.5 \times 2714}{100} + 345.6 = 413.4 \text{ kN (see Step 3)}$$

$$\text{Load on the diagonal lacing bar} = \frac{413.4}{2} \times \frac{1}{\sin \phi} = 305.4 \text{ kN}$$

Assuming $p_c = 100 \text{ N/mm}^2$ for the lacing bar, $A_{\text{minimum}} = 3054 \text{ mm}^2$

Select a channel section 203 × 89 × 29.78 kg/m
Area = 3790 mm²; $r_{\text{min}} = 26.4$ mm.

$$\lambda = \frac{l}{r_{\text{min}}} = \frac{3400}{26.4} = 128.8$$

From Table 27(c) of BS 5950: Part 1 for $\lambda = 128.8$ and $p_y = 355 \text{ N/mm}^2$, $p_c = 94 \text{ N/mm}^2$.

∴ Compression resistance
$$= P_{\text{LC}} = A_g p_c = 3790 \times 94$$
$$= 356.3 \times 10^3 \text{ N} > 305.4 \text{ kN (Satisfied)}$$

Step 15 Not required.

Step 16 *Check deflection at service load (wind or crane surge)*
Check that the horizontal deflection due to service crane surge and wind load combined (not strictly necessary) at the top of the cantilever column is less than the height divided by 300. Use the section properties of the compound column found in Step 1.

Upper roof leg
$I_{xx} = 7193 \times 10^{-6} \text{ m}^4$; length = 11 m from compound stanchion.
Horizontal load at top of this column = 48 kN and uniformly distributed load on the side of this column = 90 kN.

Deflection at the top of this roof leg only is given by:

$$\delta_1 = \frac{PL^3}{3EI} + \frac{WL^3}{8EI} = \frac{48 \times 11^3}{3 \times 205 \times 7193} + \frac{90 \times 11^3}{8 \times 205 \times 7193} = 0.02459\,\text{m}$$

Lower compound column
$I_{yy} = 64\,873 \times 10^{-6}\,\text{m}^4$; length $= 19\,\text{m}$ from foundation to crane gantry level.
Horizontal shear at top of this column $= 72 + 90 + 48 = 210\,\text{kN}$.
Distributed load on the side of this column $= 150\,\text{kN}$.
Deflection at top of this compound stanchion is given by:

$$\delta_2 = \frac{210 \times 19^3}{3 \times 205 \times 64\,873} + \frac{150 \times 19^3}{8 \times 205 \times 64\,873} = 0.04577\,\text{m}$$

Slope at the top of this compound column is given by:

$$\theta = \frac{PL^2}{2EI} + \frac{WL^2}{6EI} = \frac{210 \times 19^2}{2 \times 205 \times 64\,873} + \frac{150 \times 19^2}{6 \times 205 \times 64\,873}$$

$$= 0.003529\,\text{radians}$$

$\delta_3 = 0.003529 \times 11 = 0.0388\,\text{m}$

$\delta = \delta_1 + \delta_2 + \delta_3 = 0.109\,\text{m} > H/300 = 30/300$

$= 0.10\,\text{m}$ (May be allowed)

Chapter 6
Design of Ties

6.1 PRINCIPAL ISSUES

(1) The effective area of the section in tension.
(2) The design strength of the material p_y.
(3) The bending moments due to eccentric connections.
(4) The reduced contribution in tension of unconnected legs of simple tension members like angles, channels and T-sections.
(5) The effect of stress concentration around holes or discontinuities.
(6) Compound tension members made up by lacing or batten.

6.2 DESIGN BASIS

The effective area is determined by finding the net area of an element of a section (i.e. deducting the area of holes for connections) and multiplying by a correction factor to allow for some stress concentration effects. At or near the holes, the tensile stress in a purely tensile member may be allowed to go up to the ultimate stress. On this basis, the correction factor is higher in the case of Grade 43 steel because the ratio of tensile stress at the ultimate stress level to that at yield is higher in Grade 43 steel. This factor becomes 1 for Grade 55 steel. Such an increase in stress around holes is not allowed in simple tension members, where the effect of eccentric connection is neglected. The basis of finding the effective area in a simple tension member connected by one element of its section is by using the net area of its connected element and adding a proportion of the area of the unconnected element. For other types of tension members, full account of any eccentric connection must be taken. The capacity in tension of a member is simply its effective area multiplied by its design strength.

6.3 COMBINED AXIAL TENSION AND BENDING MOMENT

6.3.1 Principal issues

(1) Axial capacity of the member in tension.
(2) Plastic moment capacity of the section about both axes.
(3) Lateral torsional buckling of member due to bending moment.

The axial capacity can be found, as usual, by multiplying the effective area by the design strength. The plastic moment of resistance depends on the section classification and the design strength. The lateral torsional buckling effects are much less pronounced when the member is in tension, but, depending on the level of axial tension compared to its axial capacity, this effect may be severe if applied tension loads are small. The steel design code does not require checking of the effect of lateral torsional buckling when a member is under axial tension, but designers should use their own judgement when dealing with slender members with low axial tension and high bending moments. The residual compressive stress in the compression flange after allowing for axial tension may give rise to elastic instability owing to buckling.

6.3.2 Design basis

Using a fully elastic approach the maximum tensile stress at a corner of the member subjected to biaxial bending and tension should not exceed the design strength p_y. This expressed in stress notation gives:

$$\sigma_c + \sigma_x + \sigma_y \leq p_y$$

where σ_c = direct stress in tension
σ_x = stress due to applied moment M_x
σ_y = stress due to applied moment M_y
p_y = design strength

Dividing both sides by p_y and also expressing the stresses in terms of the applied direct tension (F), applied moments (M_x and M_y), effective area (A_e) and section moduli (Z_x and Z_y), the equation can be rewritten in the form given below:

$$\frac{F}{A_e p_y} + \frac{M_x}{p_y Z_x} + \frac{M_y}{p_y Z_y} \leq 1$$

where F = applied direct tension
M_x = applied moment about the X–X axis
M_y = applied moment about the Y–Y axis
A_e = effective area in tension
Z_x = section modulus about the X–X axis
Z_y = section modulus about the Y–Y axis

The plastic moment capacity of the section may be used instead of the elastic moment of resistance. Depending on the section classification and the design strength, M_{cx} and M_{cy} should replace $p_y Z_x$ and $p_y Z_y$ respectively. Therefore:

$$\frac{F}{A_e p_y} + \frac{M_x}{M_{cx}} + \frac{M_y}{M_{cy}} \leq 1$$

where M_{cx} = plastic moment of resistance of section about X–X axis
M_{cy} = plastic moment of resistance of section about Y–Y axis

This modified equation is in the code BS 5950: Part 1, but this is only a local section capacity check and not a full member capacity check. The lateral torsional buckling effects of members with axial tension should be checked with moments alone, ignoring the axial tension which has a beneficial effect. The equivalent uniform moment factor m should be used to represent the effect of the variation of bending moment over the whole length of the member. The unity equation of the member with moments only is given by:

$$\frac{mM_x}{M_b} + \frac{mM_y}{p_y Z_y} \leq 1$$

where m = equivalent uniform moment factor (see Step 13 in Chapter 5)
M_b = buckling resistance moment of the member (see Step 13 in Chapter 5)

Alternatively, for greater economy of plastic and compact sections the local capacity may be checked by reference to published tables where values of M_{rx} and M_{ry} are given corresponding to the level of axial load present. The following relationship should be satisfied:

$$\left(\frac{M_x}{M_{rx}}\right)^{Z_1} + \left(\frac{M_y}{M_{ry}}\right)^{Z_2} \leq 1$$

Z_1 is a constant taken as:

2.0 for I- and H-sections
2.0 for solid and hollow circular sections
$\frac{5}{3}$ for solid and hollow rectangular sections
1.0 for all other cases

Z_2 is a constant taken as:

1.0 for I- and H-sections
2.0 for solid and hollow circular sections
$\frac{5}{3}$ for solid and hollow rectangular sections
1.0 for all other cases

M_{rx} = reduced local section moment capacity about the X–X axis in the presence of axial load
M_{ry} = reduced local section moment capacity about the Y–Y axis in the presence of axial load

The values of M_{rx} and M_{ry} are derived by interaction formulation based on the principles explained in Chapter 4. The constants Z_1 and Z_2 are based on experimental results. It should be borne in mind that this is only a local section capacity check and the overall member buckling is ignored.

6.4 STEP-BY-STEP DESIGN OF MEMBERS IN TENSION

Step 1 Select type of tie and trial section

The guidance given for struts in Table 5.1 is also valid for ties. However, the star ratings are for general guidance and may vary from site to site. Use Grade 50 steel whenever possible; angles are difficult to obtain in higher grades of steel. Structural hollow sections used as ties cost substantially more than other rolled sections.

The trial section is best chosen by dividing the axial load by the design strength p_y to get a minimum area required, and then, by making an allowance for the holes for connection, an approximate area of section can be determined. Select from the published section tables a section which has an area closest to this area. When a combined moment and axial tension load is present, a further allowance has to be made in the area required for the trial section. Check that the member slenderness ratio does not fall below 350 where reversal of stress can occur due to wind.

Step 2 **Determine ultimate axial tension**
From analysis of basic load cases, find the combination which causes the maximum axial tension.

Step 3 **Determine ultimate bending moments and shear forces**
From analysis of basic load cases, find the various combinations of axial tensions with bending moments and shear forces. Use the load combination principle given in Table 4.1.

Step 4 **Find moments due to eccentric connection**
Except for angles, channels and T-sections designed as simple tension members, all other members should be designed with an allowance for eccentric connection.

Step 5 **Find net area of an element of a tension member with fasteners**

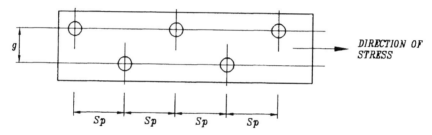

SK 6/1 Determination of net area from staggered holes in elements.

Net area A_n is gross area of the element A_g less the deduction for bolt holes.

(1) For holes in a row perpendicular to the line of action of the tensile force on the element the net area A_n is given by:

$$A_n = A_g - \sum t \times d$$

where t = thickness of element
d = diameter of hole in a row perpendicular to the direction of tensile force

(2) For staggered holes on the element, as shown in SK 6/1, the net area A_n is the smaller of A_{n1} or A_{n2} given by:

$$A_{n1} = A_g - \sum t \times d$$

where all holes are considered in one row perpendicular to the line of action of the tensile load, and:

$$A_{n2} = A_g - \sum t \times d + \frac{S_p^2 t}{4g} \leq A_g$$

Design of Ties 215

where all holes are considered in a line inclined to the direction of load. (The correction is applied to allow for the increased strength due to stagger.)

Here S_p = spacing of holes along the direction of load
g = the distance measured between the holes at right angles to the direction of the tensile load. For an angle with holes on both legs, the gauge length g may be taken as the sum of the back marks less the leg thickness.

Step 6 **Determine effective area of an element of a section with fasteners**
The code allows the stress across the net area to go beyond yield to approach the ultimate tensile strength. Therefore, to find the effective area from the net area, an increase of the net area may be allowed by multiplying it by a factor. The ratio of ultimate strength to yield strength is higher for lower grades of steel and therefore this factor is higher for lower grades of steel as given in Table 6.1.

A_e = effective area = $K_e \times A_n \leq A_g$
U_s = specified minimum ultimate tensile strength
Y_s = specified minimum yield strength

Table 6.1 Values of K_e.

Grade 43	Grade 50	Grade 55	Any Grade
1.2	1.1	1.0	$0.75 \times (U_s/Y_s) \leq 1.2$

Step 7 **Determine effective area of simple tension members**
The members in Table 6.2 are connected eccentrically to their centroidal axes. To account for some bending moments due to the

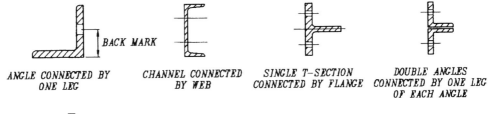

ANGLE CONNECTED BY ONE LEG   CHANNEL CONNECTED BY WEB   SINGLE T-SECTION CONNECTED BY FLANGE   DOUBLE ANGLES CONNECTED BY ONE LEG OF EACH ANGLE

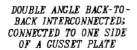

DOUBLE ANGLE BACK-TO-BACK INTERCONNECTED; CONNECTED TO ONE SIDE OF A GUSSET PLATE   DOUBLE ANGLE BACK-TO-BACK INTERCONNECTED; CONNECTED TO BOTH SIDES OF A CENTRAL GUSSET PLATE

SK 6/2 Connection diagrams of simple tension members.

Table 6.2 Effective area of simple tension members.

Section	Connection	Effective area A_e	Remarks
Single angle	Connected by one leg	$A_{nc} + A_{go}\left(\dfrac{3A_{nc}}{3A_{nc} + A_{go}}\right)$	A_{nc} = net area of connected leg A_{go} = gross area of unconnected leg
Single channel	Connected by web	$A_{nw} + 2A_{gf}\left(\dfrac{3A_{nw}}{3A_{nw} + 2A_{gf}}\right)$	A_{nw} = net area of web A_{gf} = gross area of flange
Single T-section	Connected by flange	$A_{nf} + A_{gs}\left(\dfrac{3A_{nf}}{3A_{nf} + A_{gs}}\right)$	A_{nf} = net area of flange A_{gs} = gross area of stem
Double angle ties without interconnection	Connected by one leg of each angle	$2\left[A_{nc} + A_{go}\left(\dfrac{3A_{nc}}{3A_{nc} + A_{go}}\right)\right]$	A_{nc} = net area of connected leg A_{go} = gross area of unconnected leg
Double angle back-to-back interconnected*	Connected to one side of a gusset or a section	$A_{nw} + A_{go}\left(\dfrac{5A_{nc}}{5A_{nc} + A_{go}}\right)$	A_{nw} = net area of connected legs A_{go} = gross area of unconnected legs
Single angle	Connected by both legs	Use net area A_n of both legs	See Step 5
Single channel	Connected by both flanges	Use net area A_n of both flanges plus area of web	See Step 5
Single T-section	Connected by stem or both flanges and stem	Use sum of net area A_n of all elements	See Step 5
Double angle ties back-to-back interconnected*	Connected to both sides of a central gusset	Use sum of net area A_n of all elements	See Step 5

*Interconnection of angles is required at regular intervals of length L_c through solid packing or in contact such that the slenderness ratio $\lambda_c = L_c/r_y \leq 80$ for each angle component.

eccentricity, correction factors are introduced. The net area of the elements after deduction of holes is used to find the effective area for simple members. Do not apply the factor K_e to these elements.

Note: Double angles back to back should be separated by a distance not exceeding the aggregate thickness of the legs. Solid packing pieces should be used when they are designed as interconnected.

Step 8 *Determine tension capacity*

Determine design strength p_y (see Step 10, Table 5.4 in Chapter 5). The tension capacity P_t is then:

$$P_t = A_e p_y$$

where A_e = effective area as found in Step 6 or Step 7 as appropriate.

Step 9 *Check classification of section*

This should be checked with reference to Clause 3.5.2 and Table 7 of BS 5950: Part 1.

Step 10 *Check lateral torsional buckling for moment only*

Follow Step 12 of Chapter 7.

Step 11 *Unity check of tension members with moments*

The condition is:

$$\frac{F}{A_e p_y} + \frac{M_x}{M_{cx}} + \frac{M_y}{M_{cy}} \leq 1$$

where F = the applied ultimate tension load
M_x = the applied ultimate moment about the major axis acting simultaneously with F
M_y = the applied ultimate moment about the minor axis acting simultaneously with F
M_{cx} = the plastic moment capacity of the section about the major axis
M_{cy} = the plastic moment capacity of the section about the minor axis

The shear capacity P_v is given by:

$$P_v = 0.6 p_y A_v$$

where A_v = effective shear area = tD for rolled sections
$\qquad\qquad\qquad\quad$ = td for built-up sections
$\qquad\qquad\qquad\quad$ = $0.9A$ for solid bars and plates
$\qquad\qquad\qquad\quad$ = $\left(\dfrac{D}{D+B}\right)A$ for rectangular hollow sections
$\qquad\qquad\qquad\quad$ = $0.6A$ for circular hollow sections

t = thickness of web
D = overall depth of section
d = depth of web only
A = area of section

The applied shear force F_v comes from Step 3.

$F_v \leqslant 0.6P_v$: plastic and compact sections

The plastic section moment capacity for plastic and compact sections is given by:

$$M_{cx} = p_y S_x \leq \mu_f p_y Z_x; \quad M_{cy} = p_y S_y \leq \mu_f p_y Z_y$$

where μ_f = average load factor for ultimate limit state

$F_{vx} > 0.6P_{vx}$ and $F_{vy} > 0.6P_{vy}$: plastic and compact sections

$$M_{cx} = p_y(S - S_{vx}\rho_{1x}) \leq \mu_f p_y Z_x; \quad M_{cy} = p_y(S - S_{vy}\rho_{1y}) \leq \mu_f p_y Z_y$$

$$\rho_{1x} = \frac{2.5 F_{vx}}{P_{vx}} - 1.5; \quad \rho_{1y} = \frac{2.5 F_{vy}}{P_{vy}} - 1.5$$

where F_{vx} = applied shear about the X–X axis
F_{vy} = applied shear about the Y–Y axis
P_{vx} = section shear capacity about the X–X axis
P_{vy} = section shear capacity about the Y–Y axis
S_{vx} = plastic modulus of shear area only about X–X axis
S_{vy} = plastic modulus of shear area only about Y–Y axis

For all values of $F_v < P_v$: semi-compact and slender sections

The plastic section moment capacity M_c for semi-compact and slender sections is given by:

$$M_{cx} = p_y Z_x; \quad M_{cy} = p_y Z_y$$

For slender elements in compression, use a reduced value of p_y as per Clause 3.6 and Table 8 of BS 5950: Part 1.

Note: The plastic moment capacity of section M_c for plastic and compact sections is reduced in the presence of high shear only in the orthogonal direction in which the shear exceeds the limit of $0.6P_v$.

Step 12 *Design of lacing or batten system*
General rules: lacing

(1) Lacing should not vary in section or inclination throughout the length of the member.
(2) Inclination of the lacing should be between 40° and 70° to the axis of the member.
(3) Lacing can be double intersection (cross) or single intersection. Single intersection lacing should not be opposed in direction on opposite faces of the main members. This will give rise to torsion in the member.
(4) Tie panels in the form of battens should be provided at the ends of the lacing system.
(5) Slenderness ratio $\lambda_c = L_c/r_y$ should not be greater than 80 for individual main members connected by the lacing, where L_c (effective length) will be the length between the points of intersection of the lacing with the main members and r_y is the radius of gyration of the main members about the minor axis.

Design of Ties

General rules: batten

(1) The joints between the main components and the battens must be rigid. Use welded connection or connection by at least two bolts along the axis of the main member on each main member.
(2) Battens should be provided opposite each other in each plane at the ends of the members and at the points of lateral restraint.
(3) Intermediate battens on opposite faces of main members should be uniformly spaced throughout the length of the member.
(4) The maximum slenderness ratio λ_c of the main member between end welds or the end fastener of adjacent battens should not exceed 80.
(5) The thickness of the plate used as battens should not be less than $\frac{1}{50}$ of the minimum distance between welds or fasteners on the batten connecting the main members.
(6) The slenderness ratio of the batten should not be less than 180, where the length of the batten is taken as the minimum distance between welds or fasteners on the batten connecting the main members.
(7) Width of the end battens should not be less than the distance between the centroids of the main components.
(8) Width of the intermediate battens should not be less than half the distance between the centroids of the main components.
(9) Width of any batten should not be less than twice the width of the narrower connected element of the main members.

The transverse shear force acting parallel to the planes of lacing or batten, V, is the greater of V_1 or V_2, as given by:

$$V_1 = 1\% \ F; \quad V_2 = V_s$$

where F = the ultimate maximum axial tensile force in the compound member
V_s = the maximum ultimate transverse shear in a direction parallel to the planes of the lacing or battens

The lacing bar should be checked for an axial tension or compression equal to N_L given by:

$$N_L = \frac{V}{\sin\phi} \frac{1}{J}$$

where J = number of lacing bars cut by a plane perpendicular to the axis of the member
ϕ = angle of inclination of the lacing bars to the axis of the member

The capacity of lacing bar in compression P_{LC} is given by:

$$P_{LC} = A_g p_c$$

where A_g = gross cross-sectional area of lacing bar
p_c = compressive strength, which may be obtained from Table 27(c) of BS 5950: Part 1 corresponding to the slenderness ratio of the lacing bar and its design strength p_y

Note: If the lacing system is connected to the main member by one fastener only, then the capacity P_{LC} should be multiplied by 0.8.

The capacity of lacing bar in tension P_{LT} is given by:

$$P_{LT} = A_e p_y$$

where A_e = effective area of the section
= gross area less deduction for holes, if any

The batten should be checked for axial load tension or compression equal to $V/2$ for two parallel planes of battens.

The capacity of batten in compression P_{BC} is given by:

$$P_{BC} = A_g p_c$$

where A_g = gross cross-sectional area
p_c = the compressive strength to be obtained from Table 27(c) of BS 5950: Part 1 corresponding to the slenderness ratio and design strength p_y of batten

The capacity of batten in tension P_{BT} is given by:

$$P_{BT} = A_e p_y$$

where A_e = effective area of batten after deduction of holes, if any

Check that P_{BC} and P_{BT} are both greater than $V/2$.

6.5 WORKED EXAMPLES

6.5.1 Example 6.1: Design the main tie of the roof truss in Example 3.1

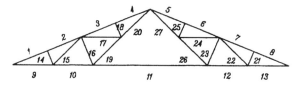

SK 6/3 Roof truss of Example 3.1.

Basic load cases for member 9:

Dead load = 32.81 kN (tension)
Imposed load = 39.38 kN (tension)
Wind load = 25.45 kN (tension)

Step 1 **Select type of tie and trial section**

Refer to Table 5.1, where some guidance is given about the choice of type of section. An angle section in Grade 43 steel could be the most cost-effective solution. The maximum tension in the member is of the order of 117 kN, so the minimum area of section required = $(117 \times 10^3)/275 = 425\,\text{mm}^2$.

Design of Ties

There is no reversal of stress in any part of the tie in the truss and therefore there is no limitation of the slenderness ratio of the member.

Try RSA $75 \times 50 \times 6 \times 5.65\,\text{kg/m}$ in Grade 43 steel with area of section equal to $719\,\text{mm}^2$.

Step 2 **Determine ultimate axial tension**
Combination 1:   1.4 (dead load) + 1.6 (imposed load)
$$= (1.4 \times 32.81) + (1.6 \times 39.38) = 108.9\,\text{kN}$$
Combination 2:   1.2 (dead load) + 1.2 (imposed load)
+ 1.2 (wind load)
$$= (1.2 \times 32.81) + (1.2 \times 39.38) + (1.2 \times 25.45)$$
$$= 117.2\,\text{kN}$$
Combination 3:   1.4 (dead load) + 1.4 (wind load)
$$= (1.4 \times 32.81) + (1.4 \times 25.45) = 81.60\,\text{kN}$$

Step 3 **Determine ultimate bending moment and shear**
Not required.

Step 4 **Determine moments due to eccentric connection**
Not required for angles, channels and T-section used as simple tension members.

Step 5 **Determine net area of an element of a section with fasteners**
Assume one bolt hole of diameter 22 mm for an M20 Grade 4.6 bolt of capacity 39.2 kN in the 6 mm thick leg of the angle. This capacity is enough for all internal members connecting to the tie.

Net area $= A_n = A_g - \sum t \times d$

A_g = gross area of element
= area of the longer leg of the angle tie
= (length of the long leg $- \frac{1}{2}$ thickness of leg) × thickness
$= (75 - 3) \times 6 = 432\,\text{mm}^2$

Given t = thickness of leg = 6 mm and d = diameter of hole = 22 mm:

$$\therefore \quad A_n = 432 - (6 \times 22) = 300\,\text{mm}^2$$

Step 6 **Determine effective area of an element**
Not required for simple tension members.

Step 7 **Determine effective area of simple tension members**
See Table 6.2. For a single angle, the effective area is given by:

$$A_e = A_{nc} + A_{go}\left(\frac{3A_{nc}}{3A_{nc} + A_{go}}\right)$$

Here, A_{go} = area of outstanding leg $= (50 - 3) \times 6 = 282\,\text{mm}^2$.

$$\therefore \quad A_e = 300 + 282 \times \left(\frac{3 \times 300}{(3 \times 300) + 282}\right) = 514.7\,\text{mm}^2$$

Step 8 Determine tension capacity
Design strength $p_y = 275 \text{ N/mm}^2$. Therefore:

$$P_t = A_e p_y = 514.7 \times 275 \times 10^{-3} = 141.5 \text{ kN} < 117.2 \text{ kN (Satisfied)}$$

Steps 9, 10, 11 and 12 are not required because there is no bending in the member.

6.5.2 Example 6.2: Design the tie of a 75 m span latticed girder for the roof of an aircraft hangar

The spacing of the nodes in the girder is 5 m and the girder is 4.5 m deep. A runway beam is connected to the tie half-way between the nodes. The capacity of the hoist is 2 tonnes.

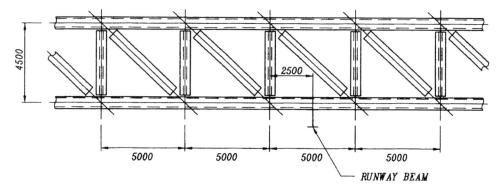

SK 6/4 Part elevation of latticed girder with runway beam between nodes.

Basic loads: Dead load = 612 kN (tension)
Imposed load = 940 kN (tension)
Wind load = 586 kN (tension) and
513 kN (compression)

Under the worst adverse loading condition the tie may suffer reversal of stress. The tie is assumed continuous.

Bending moment due to runway beam and hoist:
= 29 kNm (sagging) at midpoint between nodes
= 13.4 kNm (hogging) at the nodes

Step 1 Select type of tie and a trial section
Refer to Table 5.1. Compound channels in Grade 50 steel should be the most cost-effective solution. From inspection of basic loads, the maximum load in tension is of the order of 2500 kN. Assuming $p_y = 355 \text{ N/mm}^2$, the area of section required is given by:

$$A_{min} = \frac{2500 \times 10^3}{355} = 7050 \text{ mm}^2$$

Design of Ties 223

Select two channels 254 × 89 × 35.74 kg/m battened. The area of each channel is 4550 mm². The plastic moment capacity of each channel is given by:

$$S_x p_y = 414 \times 355 \times 10^{-3} = 147\,\text{kNm}$$

This is well in excess of the requirement.

Because there is a possible reversal of stress due to wind, the maximum slenderness ratio must be below 350. Given $L = 5000\,\text{mm}$ and $r_{xx} = 98.8\,\text{mm}$, which is the minimum for the combined section.

$$\therefore \lambda_{\max} = \frac{L}{r_{xx}} = 50.6 < 350 \quad (\text{Satisfied})$$

The channels are placed face to face as shown in SK 6/5. The back-to-back distance is kept the same as the depth of the vertical compression member in the latticed girder, which is 203 mm.

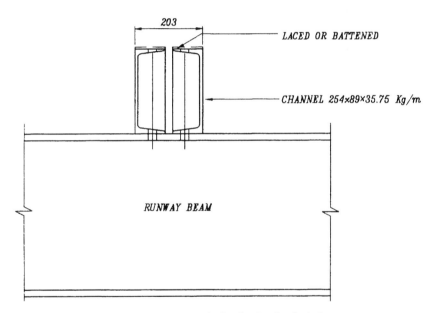

SK 6/5 The trial section of the tie in the latticed girder.

Step 2 *Determine ultimate axial tension*
Combination 1: $1.4DL + 1.6IL = (1.4 \times 612) + (1.6 \times 940)$
$\qquad\qquad\qquad = 2360.8\,\text{kN}$
Combination 2: $1.2DL + 1.2IL + 1.2WL = 1.2 \times (612 + 940 + 586)$
$\qquad\qquad\qquad = 2566\,\text{kN}$
Combination 3: $1.4DL + 1.4WL = 1.4 \times (612 + 586) = 1677.2\,\text{kN}$

Step 3 *Determine ultimate bending moments and shear force*
Ultimate bending moment due to runway beam hoist in Combination 2:
$\qquad = 1.2 \times 29 = 34.8\,\text{kNm (sagging)}$
$\qquad = 1.2 \times 13.4 = 16.1\,\text{kNm (hogging)}$

224 Structural Steelwork

Maximum ultimate bending moment due to runway beam hoist:
= 1.6 × 29 = 46.4 kNm (sagging)
= 1.6 × 13.4 = 21.4 kNm (hogging)

Shear force may be regarded as negligible.

Step 4 **Determine moment due to eccentric connection**
Not required.

Step 5 **Determine net area of an element of a section**
Assume that the runway beam is connected to the bottom flange of the channels using 4 M12 Grade 8.8 bolts. The runway beam is assumed spliced at the girder and each section is connected by two bolts to the flange of a channel.
Deduct the equivalent of one 14 mm diameter hole from each flange.

$$A_g = \text{gross area of flange} = \left(B - \frac{t}{2}\right)T$$

$$= \left(88.9 - \frac{9.1}{2}\right) \times 13.6 = 1147 \, \text{mm}^2$$

A_n = net area of flange = $1147 - (13.6 \times 14) = 957 \, \text{mm}^2$

Step 6 **Determine effective area of an element**
$A_e = K_e A_n$ and $K_e = 1.1$ from Table 6.1.

$$\therefore \quad A_e = 1.1 \times 957 = 1053 \, \text{mm}^2 < 1147 \, \text{mm}^2$$

Loss of area due to holes in each channel = $1147 - 1053 = 94 \, \text{mm}^2$.
Therefore, the total effective area of the tension member is:

$$A_e = 2 \times (4550 - 94) = 8912 \, \text{mm}^2$$

Step 7 Not required.

Step 8 **Determine tension capacity**
$P_t = A_e p_y = 8912 \times 355 \times 10^{-3} = 3164 \, \text{kN} > 2566 \, \text{kN}$ (Satisfied)

Step 9 **Check classification of section**

$$\frac{b}{T} = 6.54 < 8.5\varepsilon = 7.48$$

$$\frac{d}{T} = 21.4 < \frac{79\varepsilon}{0.4 + 0.6\alpha}$$

∴ section classification is plastic.

Step 10 **Check lateral torsional buckling for maximum bending moment**
$M_{x,max} = 46.4 \, \text{kNm}$.

For individual channel sections: $I_{yy} = 302\,\text{cm}^4$; Area $= 45.5\,\text{cm}^2$; $C_y = 2.42\,\text{cm}$.
Moment of inertia about the Y–Y axis of the compound tie is given by:

$$I_{yy} = 2 \times \left[302 + 45.5 \times \left(\frac{20.3}{2} - 2.42\right)^2\right] = 6041.5\,\text{cm}^4$$

$r_{yy} = \sqrt{(I_{yy}/A)} = \sqrt{(6041.5 \div 45.5)} = 11.52\,\text{cm}$
$r_{xx} = 9.89\,\text{cm}$ from steel section table
$L_{xx} = L_{yy} = 5000\,\text{mm}$
$\therefore \lambda_{max} = 500 \div 9.89 = 50.6$

Assume $x = D/T = 254 \div 13.6 \approx 20$.
$\therefore v = 0.47$ for $N = 0.5$ from Table 14 of BS 5950: Part 1

Take $u = 0.9$.
$\gamma = M/M_o = -2$, approximately; $\beta = 1.0$
$\therefore$ from Table 15 of BS 5950: Part 1 $n = 0.81$

$\lambda_{LT} = nuv\lambda = 0.81 \times 0.9 \times 0.47 \times 50.6 = 17$
$p_b = 355\,\text{N/mm}^2$ for $\lambda_{LT} = 17$ from Table 11 of BS 5950: Part 1
$\therefore M_b = M_{cx} = S_x p_b = 2 \times 414 \times 355 \times 10^{-3} = 294\,\text{kNm} > 46.4\,\text{kNm}$
(Satisfied)
where $\bar{M} = mM_x = M_x$ because $m = 1$.

Step 11 *Unity check of tension members with moment*
For combination 2 in Step 2, the unity check is as below:

$$\frac{F}{A_e p_y} + \frac{M_x}{M_{cx}} = \frac{2566}{3164} + \frac{34.8}{294} = 0.93 < 1 \text{ (Satisfied)}$$

Step 12 *Design the batten system*
Shear force in the batten system $= V_1 = F/100 = 25.7\,\text{kN}$
Use flats of same grade of steel as the channels: $p_y = 355\,\text{N/mm}^2$
Area of flat required $= 25\,700 \div 355 = 72\,\text{mm}^2$
Limiting value of maximum slenderness of main members $= 80$
Minimum radius of gyration of each channel $= 2.58\,\text{cm}$
$\therefore$ *maximum spacing of the battens* $= 80 \times 2.58 = 200\,\text{cm}$
Use battens at 100 cm centres.
Minimum thickness of battens $=$ length of batten $\div 50 = 200 \div 50 = 4\,\text{mm}$
Width of batten should be twice the width of the narrower main component.
$\therefore$ *minimum width of batten* $= 2 \times 89 = 178\,\text{mm}$
Use 5 mm × 200 mm battens at 1000 mm centres.
Area of battens $= 5 \times 200 = 1000\,\text{mm}^2$

Minimum moment of inertia of battens $= \dfrac{1}{12} \times 200 \times 5^3 = 2083\,\text{mm}^4$

Minimum radius of gyration of batten $= \sqrt{\dfrac{2083}{1000}} = 1.44\,\text{mm}$

Minimum slenderness of batten $= 200 \div 1.44 = 139 < 180$

Step 13 ***Local capacity check***
At the attachment of the runway beam to the flange of the channel, a local capacity check should be carried out. The tensile stress in the bottom flange of the channel due to combined axial tension and bending is very close to the design strength p_y. If vertical stiffeners are provided in the channel at the point of attachment of the runway beam, then further checks are not necessary.

Chapter 7
Design of Beams

7.1 PRINCIPAL ISSUES

The principal issues governing the design of beams are as follows:

(1) The section capacity at the point of maximum moment is the first criterion to be investigated. This section capacity depends on the design strength of the material, the geometry of the section and the slenderness of the elements in the section. This section capacity also depends on the shear force that is acting simultaneously.

(2) Lateral torsional buckling of the compression flange as a whole is another very important consideration in the design of a beam. This relates to the degree of restraint of the compressive flange, the distribution of bending moment along the length of the member, the slenderness ratio of the member about the minor axis, the torsional index of the section as a whole, the buckling parameter of the section and the slenderness factor of the compression flange relative to the tension flange.

(3) Local buckling of the compression flange, where the flange could buckle locally due to the application of a concentrated load, is another major issue. This may cause failure in the event of a sudden change in the line of action of compressive force in the flange. This phenomenon is accompanied by web compression buckling. Web stiffeners are needed to overcome this problem.

(4) Similarly, local failure of the web due to buckling, compression, crushing or stress concentrations around openings may occur and web stiffeners are also required to overcome these problems.

(5) Beams have to be checked for excessive deflection. It is necessary to limit deflection for user comfort, aesthetics and protection of secondary attachments. Even if a small amount of compressive force is present, the secondary effect of this force can be amplified several times by large in-plane deflections, causing failure.

7.2 DESIGN BASIS

7.2.1 Local capacity

This check is carried out to find if the maximum bending moment at any section is within the plastic moment of resistance of the section. An important factor which governs the local capacity is the slenderness of the elements in the cross-section. The section is normally made of plate elements. The thickness of the plate in the elements determines if the failure will occur at or before the full plastic moment of resistance is reached. The full plastic moment of resistance is reached when the stress at every part of the section reaches the design strength p_y. The fundamental issues here are the overhangs of flanges from the web and the depth to thickness ratio of the web.

There are factors which influence the local buckling of the elements in a cross-section, namely the manufacturing process (i.e. rolled or welded), the depth of neutral axis, the material strength and the presence of high shear load at the section.

The classification of sections in BS 5950: Part 1 addresses most of the issues mentioned above. The sections are classified as plastic, compact, semi-compact and slender. The basis of their classification is by the slenderness of their constituent elements. The slenderness is measured as the ratio of the longest dimension over the shortest dimension. A factor relating to the material strength is also introduced. It is proportional to the inverse of the square root of the design strength, which means that the Grade 43 materials have a factor of unity. The higher the design strength, the higher the probability that the whole section will not become fully plastic and this is reflected in the limitation on the slenderness ratio. Because of the presence of residual stresses and, also, possible geometric irregularities, welded built-up sections suffer from a further reduction in the limiting ratios for compactness.

Class 1 plastic sections are defined as sections in which a plastic hinge can form with full hinge rotation capability. In a plastic analysis of a rigid redundant frame, plastic hinges are allowed to form progressively as incremental load is applied. The first hinge to form in the structure must have sufficient rotational ductility to allow subsequent hinges to form in the structure. So, for plastic analysis of a structure the section must be class 1. These sections have local capacity such that the full design strength p_y can develop at all points in the section, resulting in a plastic moment of resistance equal to $M_{cx} = S_x p_y$, where S_x is the plastic modulus of the section.

Class 2 compact sections are similar to class 1 sections in all respects except that the full rotational ductility may not be available. Plastic analysis may not be valid using a member with compact section. The design strength p_y may develop at all points in the cross-section, but, very soon after, a collapse mechanism may form with slightly more rotation resulting in a considerable degradation of rotational stiffness. The plastic moment of resistance for these sections may also be taken as $M_{cx} = S_x p_y$, as in class 1.

Class 3 semi-compact sections show only partial plastic behaviour, depending on the slenderness of the elements. The compression flange and/or the web can be slender when compared to the limiting values given in Table 7 of BS 5950: Part 1. The design strength may be achieved at the extreme fibres but the rest of the section may not see full p_y before failure occurs by local buckling of the slender element. The moment capacity of the section is taken as $M_{cx} = Z_x p_y$, where Z_x is the elastic modulus of the section.

Class 4 slender sections may be governed more by local buckling than the design strength p_y. The design strength has to be reduced to accommodate this deficiency. Clause 3.6 and Table 8 of BS 5950: Part 1 give rules to find the reduced design strength. The moment capacity is take as $M_{cx} = Z_x p_y$ where p_y is the reduced value of the design strength of the slender element.

For semi-compact and slender sections the stress distribution in the web is triangular, whereas for plastic and compact sections the stress distribution is rectangular at the ultimate limit state.

7.2.2 Lateral torsional buckling

In the case of a column, failure may take place suddenly at the critical axial compression due to elastic instability. Similarly, a beam loaded in the plane of its web may fail at the elastic critical load by deflecting sideways and by twisting about its longitudinal axis. It can be shown by theoretical analysis that the elastic critical load for failure of the beam depends on the boundary conditions at the supports, the unrestrained length of the compression flange, the torsional index of the section of the beam, the shape of the bending moment diagram and the type of loading. Some other factors which are equally important and cannot be properly accounted for in the theoretical analysis are the initial out-of-correctness of the geometry of the beam, the residual stresses in the beam as a result of the manufacturing process, the eccentricity of applied loading etc. The formulas in the BS 5950 code of practice for checking the lateral torsional buckling of a beam are based on theoretical analysis as well as laboratory testing, and take into account the effects of all other factors contributing to elastic instability.

The check for lateral torsional buckling of a beam includes finding the equivalent slenderness λ_{LT}, which is equal to the minor axis slenderness λ of the beam multiplied by the appropriate correction factors. These correction factors include u and v, which are functions of the geometry of the section of the beam, and n, a slenderness correction factor which depends on the nature of loading on the beam. Conservatively, this n may be taken as equal to 1.

M_E = elastic critical moment
λ_{LT} = equivalent slenderness ratio of beam
p_y = design strength of material of beam
M_P = plastic moment of resistance of the beam section = $p_y S_x$

The relationship between M_E and λ_{LT} is given by:

$$M_E = \frac{M_P \pi^2 E}{\lambda_{LT}^2 p_y}$$

$$\text{or} \quad \lambda_{LT} = \sqrt{\frac{\pi^2 E}{p_y}} \times \sqrt{\frac{M_P}{M_E}}$$

The bending strength p_b may be taken as equal to design strength p_y if the equivalent slenderness ratio of the beam corrected for section geometry and loading condition is less than or equal to λ_{L0}, which is given by:

$$\lambda_{L0} = \text{limiting equivalent slenderness ratio} = 0.4\sqrt{\frac{\pi^2 E}{p_y}}$$

This means that a beam with a value of $\sqrt{(M_P/M_E)} \leq 0.4$ may be designed without consideration of lateral torsional buckling effects. Also, if the lateral restraints of the compression flange of the beam are designed in such a way that λ_{LT} does not exceed λ_{L0}, then the lateral torsional buckling effects can again be ignored.

7.2.3 Buckling of web

The behaviour of the solid web in shear is similar to that of open-web latticed girders in which distinct lines of compression and tension diagonal members are used to carry the vertical shear force. Web buckling due to vertical shear occurs in thin webs along lines of compressive stress acting diagonally within the web. Assuming a is the spacing of transverse stiffener and d is the web depth, as the aspect ratio (a/d) increases, the length of the diagonal compressive strut increases. This results in a decrease of the elastic critical load along the diagonal in compression. The elastic critical shear stress derived from elastic theory is expressed as follows:

Case 1: $a/d \leq 1$ $\quad q_c = \left[0.75 + \dfrac{1}{(a/d)^2}\right]\left[\dfrac{1000}{(d/t)}\right]^2$

Case 2: $a/d > 1$ $\quad q_c = \left[1.0 + \dfrac{0.75}{(a/d)^2}\right]\left[\dfrac{1000}{(d/t)}\right]^2$

q_c = elastic critical shear stress
a = spacing of transverse stiffeners
d = depth of web

t = thickness of web
V_{cr} = shear buckling resistance of web = $q_{cr}dt$
q_{cr} = critical shear strength which depends on the web slenderness factor
λ_w = web slenderness factor = $\sqrt{[0.6(p_{yw}/q_c)]}$
p_{yw} = design strength of web

The relationship between λ_w and the critical shear strength q_{cr} for design purposes is given by:

(1) If $\lambda_w \leq 0.8$, then $q_{cr} = 0.6p_{yw}$ and no web buckling is expected where the maximum shear stress is kept below this level.
(2) If $0.8 < \lambda_w < 1.25$, then $q_{cr} = 0.6p_{yw}[1 - 0.8(\lambda_w - 0.8)]$
(3) $\lambda_w \geq 1.25$, then $q_{cr} = q_c$

7.2.4 Tension-field action in thin webs

It has been shown experimentally that the shear buckling resistance of thin webs can be considerably improved if properly designed transverse stiffeners are used and the flanges are capable of carrying an additional force due to tension-field action. The flanges must be anchored properly at the ends to allow the tension-field action to develop. The tension-field action in solid webs of beams may be used for all beams except gantry girders. The rules of detailing as specified in BS 5950: Part 1 should be adhered to.

V_b = shear buckling resistance of a stiffened panel of web = $q_b dt$

q_b = buckling shear strength

$$= q_{cr} + \frac{v_b}{2[(a/d) + \sqrt{1 + (a/d)^2}]}$$

v_b = basic tension field strength of web

$$= \sqrt{p_{yw}^2 - 3q_{cr}^2 + \phi_t^2} - \phi_t$$

$$\phi_t = \frac{1.5 q_{cr}}{\sqrt{1 + (a/d)^2}}$$

If there is residual capacity in the flanges, i.e. they are not fully stressed at the point of interest, then the shear resistance of the web may be further enhanced to:

$$V_b = (q_b + q_f \sqrt{K_f}) dt \leq 0.6 p_{yw} dt$$

where $K_f = \dfrac{M_{pf}}{4M_{pw}}\left(1 - \dfrac{f}{p_{yf}}\right)$

q_f = flange dependent shear strength

$= \left(4\sqrt{3}\sin\dfrac{\theta}{2}\sqrt{\dfrac{v_b}{p_{yw}}}\right)0.6p_{yw}$

$\theta = \tan^{-1}\dfrac{d}{a}$

M_{pf} = plastic moment capacity of the smaller flange about its centroidal axis parallel to the flange

M_{pw} = plastic moment capacity of web about its centroidal axis perpendicular to the web

f = mean longitudinal stress in the smaller flange due to applied loads

7.3 STEP-BY-STEP DESIGN OF BEAMS

Step 1 Determine design strength

Table 7.1 Design strength p_y.

Grade of steel	Thickness (mm)	Value of p_y (N/mm^2)
43	≤16	275
	≤40	265
	≤63	255
	≤100	245
50	≤16	355
	≤40	345
	≤63	340
	≤100	325
55	≤16	450
	≤25	430
	≤40	415
	≤63	400

Step 2 Carry out analysis

Calculate ultimate loading and load combinations. Assume size of members. Determine boundary conditions. Analyse using manual methods or a computer. Rules for trial size of plate girder:

$D = \dfrac{L}{12}$

$A_f = \dfrac{M}{Dp_y}$

$t = \dfrac{D}{150}$

where L = span of beam
 A_f = area of flange
 M = maximum ultimate bending moment

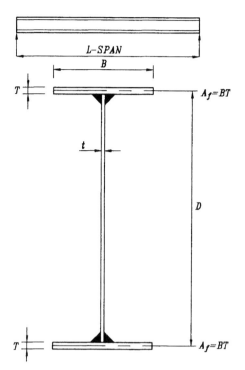

SK 7/1 Determination of trial size of plate girder.

Step 3 **Determine internal forces**
Determine bending moment for relevant load combinations. Determine shear force for relevant load combinations. Find maximum coacting direct loads with bending moments.

Step 4 **Select trial section and carry out section classification**
Rules for choice of section with equal flanges and normal loads:

(1) When the compression flange is fully restrained along its length, choose a section from the Steelwork Design Guide to BS 5950: Part 1 (pages 128 to 133 for Grade 43 steel; pages 256 to 261 for Grade 50) if using UB sections. Use a section with M_{cx} greater than the maximum applied moment M.
(2) When the compression flange is not fully restrained laterally and the beam is loaded between lateral restraints, use the equivalent slenderness method. Find n from Table 15 or 16 of BS 5950: Part 1. Choose a section from the Steelwork Design Guide to BS 5950: Part 1 using appropriate n, L_e and grade of steel such that the buckling resistance moment M_b is greater than the maximum applied moment M. In this case the equivalent uniform moment factor m is taken as equal to 1.

(3) When the compression flange is not fully restrained laterally and there is no load between lateral restraints on the beam, the equivalent moment method of design may be adopted. From the bending moment diagram between restraints find m from Table 18 of BS 5950: Part 1. Choose a section from the Steelwork Design Guide to BS 5950: Part 1 which has M_b greater than mM. In this case $n=1$ and M_b is obtained using the effective length of the beam $= L_e$.

M_b = buckling resistance moment of beam
m = equivalent uniform moment factor
L_e = effective length of beam

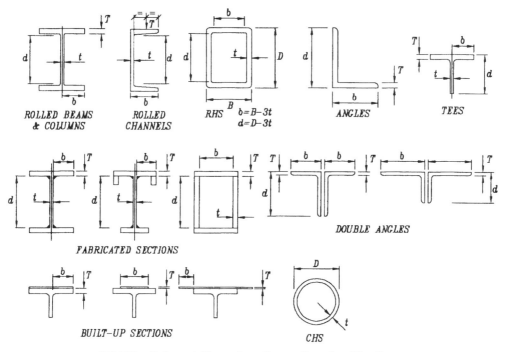

SK 7/2   **Relevant dimensions for section classification.**

For the chosen section, determine $\varepsilon = \sqrt{(275/p_y)}$.
Find d/t and b/T.
Find from Table 7 of BS 5950: Part 1 the section classification: plastic (1), compact (2), semi-compact (3) or slender (4).

Check web for shear buckling if $d/t > 63\varepsilon$.

For webs of semi-compact sections find:

$$R = \frac{N}{Ap_y}$$

where N = direct axial load (+ve, compression and −ve, tension)
            A = area of section

(1) When R is positive

$$\frac{d}{t} \leq \frac{120\varepsilon}{1+1.5R} \quad \text{and} \quad \leq \left(\frac{41}{R} - 13\right)\varepsilon$$

for sections built up by welding, and

$$\frac{d}{t} \leq \frac{120\varepsilon}{1+1.5R} \quad \text{and} \quad \leq \left(\frac{41}{R} - 2\right)\varepsilon \text{ for rolled sections}$$

(2) When R is negative

$$\frac{d}{t} \leq \frac{120\varepsilon}{(1+R)^2} \quad \text{and} \quad \leq 250\varepsilon \text{ for built up and rolled sections}$$

Step 5   Determine section properties

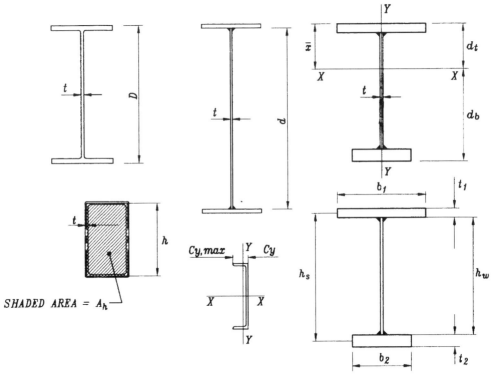

SK 7/3   Determination of sectional properties.

A = area of the section
A_v = shear area of the section
  = tD for rolled I-, H- and channel sections
  = td for built-up sections and boxes
  = $0.9A$ for solid bars and plates

$$= \left(\frac{D}{D+B}\right) A \text{ (for rectangular hollow sections)}$$

= $0.6A$ for circular hollow sections

$\bar{x}$ = location of the neutral axis from the major axis (X–X)
$\bar{y}$ = location of the neutral axis from the minor axis (Y–Y)
I_x = second moment of area about the neutral axis parallel to the major axis (X–X)
I_y = second moment of area about the neutral axis parallel to the minor axis (Y–Y)
r_x = radius of gyration about the major axis (X–X)

$$= \sqrt{I_x/A}$$

r_y = radius of gyration about the minor axis (Y–Y)

$$= \sqrt{I_y/A}$$

Z_{xb} = elastic modulus about the major axis to the bottom fibre (see SK 7/3)

$$= I_x/d_b$$

Z_{xt} = elastic modulus about the major axis to the top fibre (see SK 7/3)

$$= I_x/d_t$$

Z_y = elastic modulus about the minor axis

$$= 2I_y/B \text{ (for sections symmetrical about the } Y\text{–}Y \text{ axis)}$$

B = width of the larger flange or width of section

$$= I_y/C_y^* \text{ (for unsymmetrical sections – see SK 7/3)}$$

*Use $C_{y,\max}$ for minimum Z and maximum stress

S_x = plastic modulus about the major axis
S_y = plastic modulus about the minor axis
J = torsion constant
$$= \tfrac{1}{3}(t_1^3 b_1 + t_2^3 b_2 + t_w^3 h_w) \text{ (for flanged section – see SK 7/3)}$$

$$= \frac{t^3 h}{3} + \frac{4 A_h^2 t}{h} \text{ (for rectangular hollow sections – see SK 7/3)}$$

$$N = \frac{I_{cf}}{I_{cf} + I_{tf}}$$

I_{cf} = second moment of inertia of compression flange about the minor axis
I_{tf} = second moment of inertia of tension flange about the minor axis

$$H = \frac{h_s^2 t_1 t_2 b_1^3 b_2^3}{12(t_1 b_1^3 + t_2 b_2^3)}$$

u = buckling parameter
x = torsional index

For flanged sections symmetrical about the minor axis:

$$u = \left(\frac{4S_x^2 \gamma}{A^2 h_s^2}\right)^{\frac{1}{4}}$$

$$x = 0.566 h_s \left(\frac{A}{J}\right)^{\frac{1}{2}}$$

$$\gamma = \left(1 - \frac{I_y}{I_x}\right)$$

For flanged sections symmetrical about the major axis:

$$u = \left(\frac{I_y S_x^2 \gamma}{A^2 H}\right)^{\frac{1}{4}}$$

$$x = 1.132 \left(\frac{AH}{I_y J}\right)^{\frac{1}{2}}$$

For box sections:

$$\Phi_b = \left(\frac{S_x^2 \gamma'}{AJ}\right)^{\frac{1}{2}}$$

$$\gamma' = \left(1 - \frac{I_y}{I_x}\right)\left(1 - \frac{J}{2.6 I_x}\right)$$

Step 6 **Check shear** $d/t \leq 63\varepsilon$
P_v = shear capacity = $0.6 p_y A_v$
Check $P_v \geq F_{v,\max}$ (applied maximum shear)

Step 7 **Check minimum web thickness for built-up section**
Without intermediate stiffeners

Thickness of web = $t \geq \dfrac{d}{250}$ or $\dfrac{d}{250}\left(\dfrac{p_{yf}}{345}\right)$ (whichever is the greater)

p_{yf} = design strength of compression flange

With transverse stiffeners
a = spacing of stiffeners
d = depth of girder

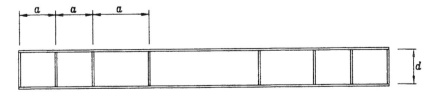

SK 7/4 Spacing of stiffeners in a stiffened girder.

When $a > d$, $t \geq \dfrac{d}{250}$ or $\dfrac{d}{250}\left(\dfrac{p_{yf}}{345}\right)$

When $a \leq d$, $t \geq \dfrac{d}{250}\left(\dfrac{a}{d}\right)^{\frac{1}{2}}$

When $a \leq 1.5d$, $t \geq \dfrac{d}{250}\left(\dfrac{p_{yf}}{455}\right)^{\frac{1}{2}}$

p_{yf} is the design strength of the compression flange.

Step 8 **Check shear buckling of thin webs d/t > 63ε**
V_{cr} = shear buckling resistance $\geq F_v$ (applied shear force)

Design with no web stiffeners
$V_{cr} = q_{cr}dt$
q_{cr}, the critical shear strength, may be obtained from Table 21(a) to (d) of BS 5950: Part 1 as appropriate, taking $a/d = \infty$.

Design with web stiffeners but not using tension-field action
$V_{cr} = q_{cr}dt$
q_{cr} may be obtained from Table 21(a) to (d) of BS 5950: Part 1 as appropriate, taking a suitable value of a/d.

Design with web stiffeners and using tension field action
V_b (shear buckling resistance) $\geq F_v$
$V_b = q_b dt$
q_b, the basic shear strength, may be obtained from Table 22(a) to (d) of BS 5950: Part 1 as appropriate, taking a suitable value of a/d.

Note: (1) When end panels are designed using tension-field action the bearing stiffeners and the end posts must be designed to resist additional moments and shears. See Step 18 for stiffener design.
(2) The shear buckling resistance of gantry girders will be calculated without using tension-field action (Clause 4.11.4 of BS 5950: Part1).

Step 9 **Section moment capacity with low shear and d/t ≤ 63ε**
Check $F_v \leq 0.6 P_v$

P_v is obtained from Step 6

Check both conditions:
    Maximum moment with coacting shear
    Maximum shear with coacting moment

For plastic (1) and compact (2) sections
The section moment capacity M_c is given by:

$M_c = p_y S \leq k p_y Z$

where $k = \dfrac{W_u}{W_s}$

W_u = moment due to factored ultimate load
W_s = moment due to unfactored service load
S = plastic section modulus about the relevant axis
Z = elastic section modulus about the relevant axis

For semi-compact (3) sections
$M_c = p_y Z$

For slender (4) sections
$M_c = R p_y Z_c$ or $p_y Z_t$ (whichever is the lower)
The reduction factor R may be obtained from Table 8 of BS 5950: Part 1.
Z_t = elastic section modulus of tension flange
Z_c = elastic section modulus of compression flange

Check applied moment $M \leq M_c$

Step 10 *Section moment capacity with high shear and $d/t \leq 63\varepsilon$*

Check $0.6 P_v < F_v < P_v$

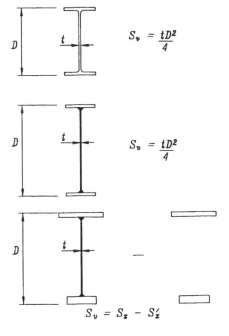

SK 7/5 Plastic modulus of shear area.

For plastic (1) and compact sections (2)
$$M_c = p_y(S - S_v\rho_1) \leq 1.2 p_y Z$$
$$\rho_1 = \frac{2.5 F_v}{P_v} - 1.5$$

S_v = plastic modulus of shear area A_v (for sections with equal flanges)
  = plastic modulus of the gross section less the plastic modulus of the section remaining after deducting the shear area (for sections with unequal flanges)

For semi-compact (3) sections
$M_c = p_y Z$

For slender (4) sections
$M_c = R p_y Z_c$ or $p_y Z_t$ (whichever is the lower)
The reduction factor R may be obtained from Table 8 of BS 5950: Part 1.
Z_t = elastic section modulus relative to tension flange
Z_c = elastic section modulus relative to compression flange

Check applied moment $M \leq M_c$

Step 11 **Section moment capacity with slender web $d/t > 63\varepsilon$**
Moment and axial load will be carried by flanges and shear will be carried by the web.

For plastic (1) and compact (2) sections
$M_c = p_y S_f \leq k p_y Z_f$
  S_f = plastic modulus of flanges only, ignoring web
  Z_f = elastic modulus of flanges only, ignoring web
$$k = \frac{\text{moment due to factored ultimate load}}{\text{moment due to unfactored service load}}$$

For semi-compact (3) sections
$M_c = p_y Z_f$

For slender (4) sections
$M_c = R p_y Z_{fc}$ or $p_y Z_{ft}$ (whichever is the lower)
The reduction factor R may be obtained from Table 8 of BS 5950: Part 1.
Z_{fc} = elastic section modulus of the compression flange, when flanges only are considered, ignoring web
Z_{ft} = elastic section modulus of the tension flange, when flanges only are considered, ignoring web

Step 12 **Member moment capacity (lateral torsional buckling)**
Destabilising load is one which is applied to the top flange of a beam at a point which is not laterally braced or restrained, and the load and the flange are both free to deflect laterally because the top flange is not laterally and rotationally restrained. Lateral restraint of the compression flange of a beam becomes effective only when the restraint is connected to an appropriate bracing system which transfers the load to the support of the structure.

Torsional restraint effectively holds the flanges of the beam in position relative to each other.

Determination of effective lengths
(1) For beams with no lateral restraint in span the effective length L_E will be as shown in Table 7.2.

Table 7.2 Effective length L_E of beam.

Condition at support		Loading condition	
		Normal	Destabilising
Case 1	Compression flange laterally restrained Full torsional restraint Both flanges restrained to rotate on plan	0.7L	0.85L
Case 2	Compression flange laterally restrained Full torsional restraint Both flanges partially restrained to rotate on plan	0.85L	1.0L
Case 3	Compression flange laterally restrained Full torsional restraint Both flanges free to rotate on plan	1.0L	1.2L
Case 4	Compression flange laterally unrestrained No torsional restraint Both flanges free to rotate on plan Bottom flange only fixed to support	1.0L + 2D	1.2L + 2D
Case 5	Compression flange laterally unrestrained No torsional restraint Both flanges free to rotate on plan Bottom flange simply resting on support	1.2L + 2D	1.4L + 2D

D is the depth of beam; L is as illustrated in SK 7/6.

(2) For beams with intermediate lateral restraint in the span, the effective length L_E will be $1.0L$ for normal loads and $1.2L$ for destabilising loads, where L is the distance between the lateral restraints.
(3) For a section of beam between a lateral restraint and a support, use mean value of L_E as obtained from point (1) for condition at support and from (2) for intermediate lateral restraint.
(4) For effective lengths of cantilever beams use SK 7/7.

Note: When torsional restraint is provided by torsional stiffener see Step 18 for the design of torsional stiffeners.

Determine λ = slenderness ratio = L_E/r_y
Determine λ/x and N (obtain x and N from Step 5)
Determine v = slenderness factor from Table 14 of BS 5950: Part 1 or, alternatively:

$$v = \left[\left(4N(1-N) + \frac{1}{20}\left(\frac{\lambda}{x}\right)^2 + \psi^2\right)^{\frac{1}{2}} + \psi\right]^{-\frac{1}{2}}$$

where $\psi = 0.8(2N-1)$ for $N > 0.5$ and $\psi = 1.0(2N-1)$ for $N < 0.5$

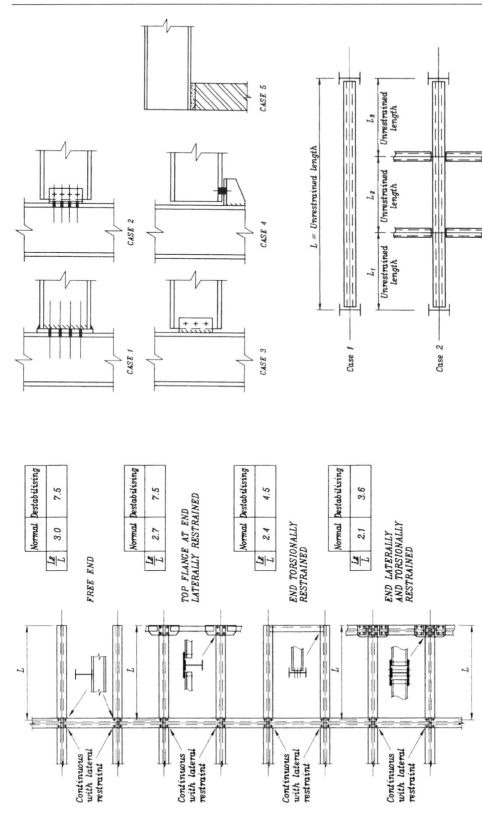

SK 7/6 Condition of beam at support and unrestrained length L.

SK 7/7 Effective lengths of cantilever beams based on conditions at supports and free ends.

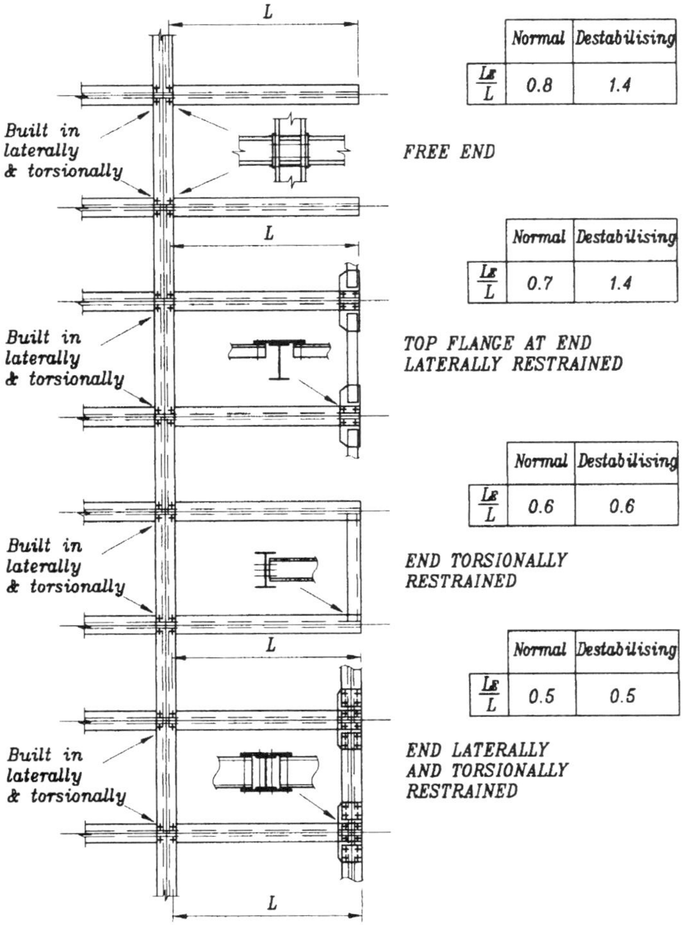

Design of Beams 243

$\frac{L_E}{L}$	Normal	Destabilising
	1.0	2.5

FREE END

$\frac{L_E}{L}$	Normal	Destabilising
	0.9	2.5

TOP FLANGE AT END LATERALLY RESTRAINED

$\frac{L_E}{L}$	Normal	Destabilising
	0.8	1.5

END TORSIONALLY RESTRAINED

$\frac{L_E}{L}$	Normal	Destabilising
	0.7	1.2

END LATERALLY AND TORSIONALLY RESTRAINED

Continuous with lateral restraint & torsional restraint

SK 7/7 (contd)

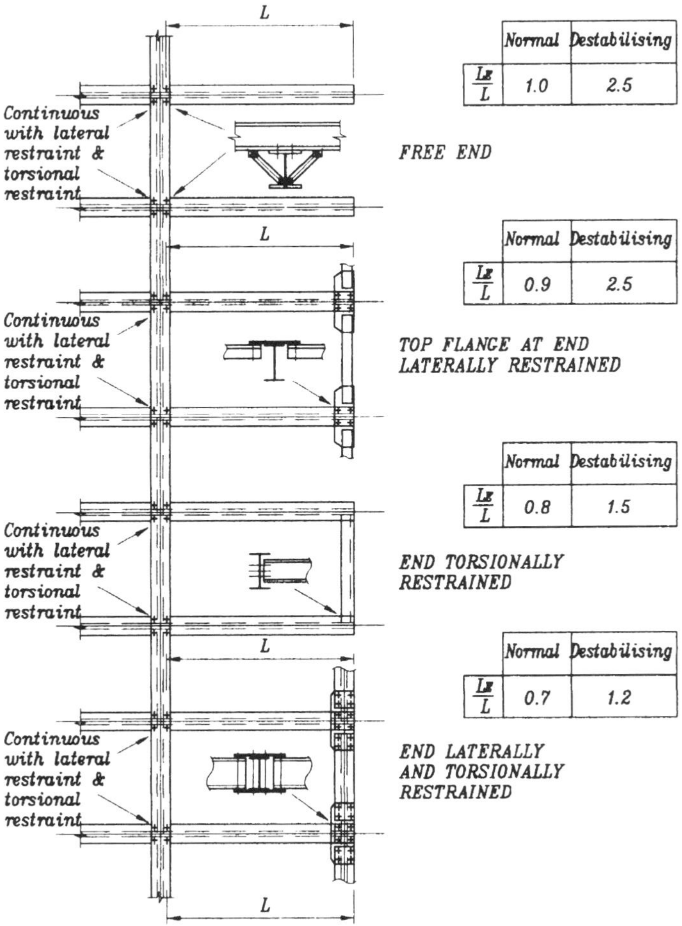

$\frac{L_E}{L}$	Normal	Destabilising
	0.8	1.4

FREE END

$\frac{L_E}{L}$	Normal	Destabilising
	0.7	1.4

TOP FLANGE AT END LATERALLY RESTRAINED

$\frac{L_E}{L}$	Normal	Destabilising
	0.6	0.6

END TORSIONALLY RESTRAINED

$\frac{L_E}{L}$	Normal	Destabilising
	0.5	0.5

END LATERALLY AND TORSIONALLY RESTRAINED

Built in laterally & torsionally

SK 7/7 (contd)

If there is no load in the span or in the unrestrained length of the beam, then the equivalent moment factor m obtained from Tables 13 and 18 of BS 5950: Part 1 converts the end moments of the length of the beam under consideration into a uniform equivalent moment.

Similarly, if there is applied load on the beam between restraints, then a slenderness factor n is used, which is obtained from Tables 15 and 16 of BS 5950: Part 1.

Find the equivalent uniform moment factor m from Tables 13 and 18 of BS 5950: Part 1 or, alternatively, use:

$m = 0.57 + 0.33\beta + 0.10\beta^2 \geq 0.43$ for beams of uniform section with equal flanges

$m = 1.0$ for all other sections

where β is the ratio of the smaller end moment to the larger end moment on a span equal to the unrestrained length.

When m is determined from Table 18 of BS 5950: Part 1, then n is taken as equal to 1.0. Similarly, when n is determined from Tables 15 and 16 of BS 5950: Part 1, then m is taken as equal to 1.0.

$M_0 =$ bending moment at mid-length on a simply supported span equal to the unrestrained length

$\gamma = M/M_0$

(hogging moments are positive and sagging moments are negative)

$M =$ numerically larger end moment of the beam in the unrestrained length

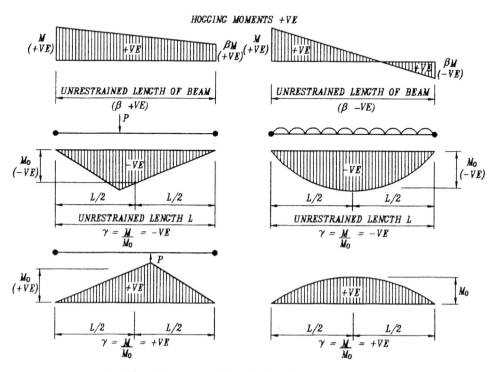

SK 7/8 Sign convention of β and γ.

- For members subject to destabilising forces m and n are always taken equal to 1.0
- For members with unequal flanges the factors m and n are taken equal to 1.0
- For cantilevers without intermediate lateral restraint m and n are taken equal to 1.0

Find slenderness correction factor n from Tables 13, 15 and 16 of BS 5950: Part 1.
Determine u from Step 5.
λ_{LT} = equivalent slenderness = $nuv\lambda$

Select appropriate p_y based on plate thickness
The limiting equivalent slenderness ratio λ_{LO} is given by:

$$\lambda_{LO} = 0.4\left(\frac{\pi^2 E}{p_y}\right)^{\frac{1}{2}}$$

The Perry coefficient η_{LT} is as follows:

For rolled section $\eta_{LT} = 0.007(\lambda_{LT} - \lambda_{LO}) \geq 0$

For welded section $\eta_{LT} = 2 \times 0.007 \times \lambda_{LO}$

$$\leq 2 \times 0.007(\lambda_{LT} - \lambda_{LO}) \geq 0$$

$$\geq 0.007(\lambda_{LT} - \lambda_{LO}) \geq 0$$

M_P = plastic moment capacity = $S_X p_y$
S_X = plastic modulus of section

M_E = elastic critical moment = $\dfrac{M_P \pi^2 E}{\lambda_{LT}^2 p_y}$

M_b = buckling resistance moment = $\dfrac{M_E M_P}{\phi_B + (\phi_B^2 - M_E M_P)^{\frac{1}{2}}}$

where $\phi_B = \dfrac{M_P + (\eta_{LT} + 1)M_E}{2}$

Alternatively find bending strength p_b from Table 11 of BS 5950: Part 1
$M_b = S_x p_b$

Check $\bar{M} = mM \leq M_b$

M = maximum ultimate moment in the member or portion of the member under consideration
p_b = bending strength

Note: For gantry girders, m and n should be taken as equal to 1.0.

246 Structural Steelwork

Alternative conservative method for rolled steel equal flange sections
Find p_b from Table 19 of BS 5950: Part 1 for the appropriate p_y using the following:

$\lambda = L_E/r_y$ and x from Step 5
$M_b = S_x p_b$

Check $M \leq M_b$

$M =$ applied maximum ultimate moment in the member or portion of member under consideration

Step 13 **Check web buckling: unstiffened web**
(a) Are the following conditions satisfied?
(1) Load is applied through flange, as in bearing on bottom flange or external loads on top flange.
(2) At the point of application of load, the flange is restrained from rotating relative to the web.
(3) At the point of application of load, the flange is restrained from movement relative to the other flange.

If so, slenderness of the web $= \lambda = 2.5d/t$

$d =$ depth of web
$t =$ thickness of web

Find p_c from Table 27(c) of BS 5950: Part 1 using λ and an appropriate p_y for the web thickness.

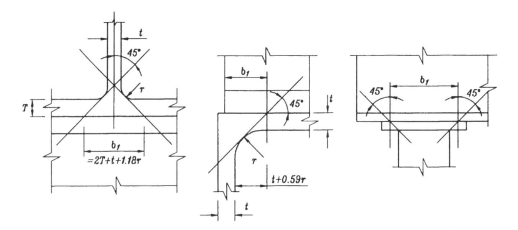

SK 7/9 Dimensions to check web buckling with unstiffened web.

Determine b_1 and n_1 following SK 7/9 and SK 7/10.

$b_1 =$ stiff bearing length which cannot deform appreciably in bending
$n_1 =$ the length obtained by dispersion at 45° through half the depth of the section

Determine $P_w =$ buckling resistance of unstiffened web $= (b_1 + n_1)tp_c$

Check $P_w \geq F_v =$ applied load through flange $=$ reaction at the bearing.

Use stiffeners if this condition is not satisfied.

Design of Beams 247

*(b) Are conditions (2) and/or (3) not satisfied?
Then slenderness of the web* $= \lambda = 3.5d/t$

Find p_c from Table 27(c) of BS 5950: Part 1 using λ and an appropriate p_y for the web thickness.

Determine b_1 and n_1 following SK 7/9.

Determine P_w = buckling resistance of unstiffened web $= (b_1 + n_1)tp_c$

Check $P_w \geq F_v$ = applied load through flange.

Use stiffeners if this condition is not satisfied.

Step 14 Check web bearing: unstiffened web

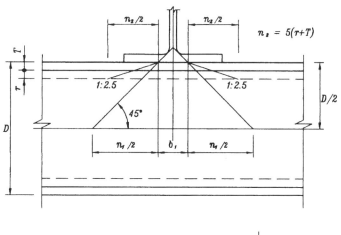

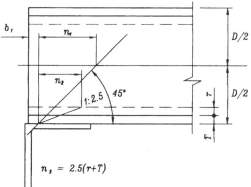

SK 7/10 Dimensions to check web bearing with unstiffened web.

P_{LW} = local capacity of web at bearing or under concentrated load
$= (b_1 + n_2)tp_{yw}$

where p_{yw} = design strength of web
n_2 = the length obtained by dispersion through the flange to the flange-to-web connection at a slope of 1:2.5 (see SK 7/10)

Check $P_{LW} \geq R_L$ = maximum ultimate reaction or maximum applied load on flange.

Use bearing stiffener if this condition is not satisfied.

Step 15 Web check between stiffeners

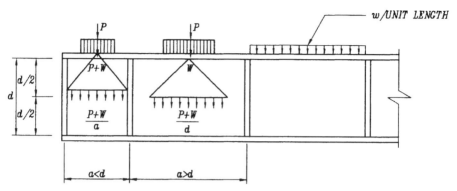

SK 7/11 Load configurations on stiffened girder for web check between stiffeners.

This is required only when loads are applied between web stiffeners. Consider a panel of the beam between two stiffeners loaded with P, W and w, as described below.

P = applied point load between stiffeners
W = total distributed load over a length shorter than panel dimension a
w = distributed load per unit length over the whole panel a
t = thickness of web
E = modulus of elasticity of steel
f_{ed} = compressive stress in web at the compression edge of web
p_{ed} = compressive strength due to edge loading

$$f_{ed} = \left[\frac{(P+W)}{a} + w\right]\frac{1}{t} \quad \text{or} \quad = \left[\frac{(P+W)}{d} + w\right]\frac{1}{t}$$

(whichever is the greater)

When the compression flange is restrained against rotation relative to web:

$$p_{ed} = \left[2.75 + \frac{2}{(a/d)^2}\right]\frac{E}{(d/t)^2}$$

When the compression flange is not restrained against rotation relative to web:

$$p_{ed} = \left[1 + \frac{2}{(a/d)^2}\right]\frac{E}{(d/t)^2}$$

Check $p_{ed} \geq f_{ed}$

Use closer spacing of stiffeners if this condition is not satisfied.

Step 16 Intermediate transverse web stiffener

Maximum outstand of stiffener = $19t_s\sqrt{(275/p_y)}$
Maximum value of outstand to be considered in the design calculations = $13t_s\sqrt{(275/p_y)}$

Check minimum stiffness

Transverse web stiffeners *not subject to external loads or moments* should have a minimum stiffness I_s given by:

$$I_s \geq 0.75 dt^3 \quad \text{for } a \geq \sqrt{2}d$$

$$I_s \geq \frac{1.5 d^3 t^3}{a^3} \quad \text{for } a < \sqrt{2}d$$

d = depth of web
t = minimum web thickness required for spacing of stiffeners equal to a using tension field action

$$= \frac{F_v}{q_b d}$$

F_v = applied shear force
q_b = basic shear strength obtained from Tables 22(a) to (d) of BS 5950: Part 1
I_s = minimum stiffness of transverse web stiffener

$$= \frac{t_s}{12}\left[(2b_s + t)^3 - t^3\right] \quad \text{(for the type of stiffener shown in SK 7/12)}$$

t_s = thickness of stiffener plate
t = thickness of web
b_s = width of stiffener plates

Calculate I_s for other types of stiffeners from first principles.

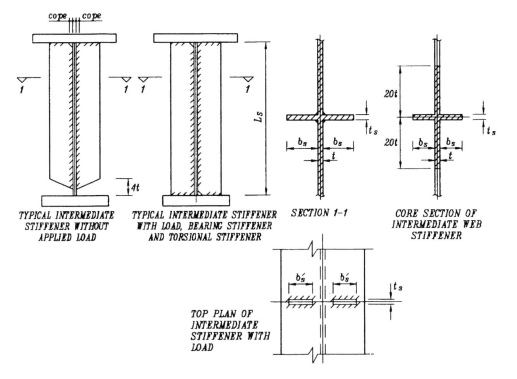

SK 7/12 Web stiffeners.

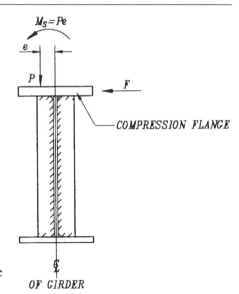

SK 7/13 Lateral and eccentric loads on stiffeners.

Transverse web stiffeners *subject to lateral loads or moments* due to eccentricity of transverse loads relative to the web should have a minimum stiffness I_s given by:

$$I_s \geq 0.75dt^3 + \frac{2FD^3}{Et} + \frac{M_sD^2}{Et} \quad \text{for } a \geq \sqrt{2}d$$

$$I_s \geq \frac{1.5d^3t^3}{a^2} + \frac{2FD^3}{Et} + \frac{M_sD^2}{Et} \quad \text{for } a < \sqrt{2}d$$

F = lateral load on the stiffener applied at the compression flange of beam
M_s = moment on the stiffener due to eccentricity of transverse load relative to web
E = modulus of elasticity of steel
D = overall depth of section
$I_s = \frac{t_s}{12}[(2b_s + t)^3 - t^3]$ (for the type of stiffener shown in SK 7/12)

Calculate I_s from first principles for other types of stiffeners.

Note: When transverse load is in line with the web, no increase in stiffness of transverse stiffener is required.

Check buckling of intermediate transverse web stiffeners
Transverse web stiffeners *not subject to external loads* should be checked for stiffener force F_q given by:

$$F_q = V - V_s \leq P_q$$

V = maximum shear force adjacent to the stiffener
V_s = shear buckling resistance of the web panel without tension-field action = V_{cr}
$V_{cr} = q_{cr}dt$ (see Step 8)
P_q = the buckling resistance of the intermediate web stiffener

Determine buckling resistance P_q of intermediate web stiffener

The core section of an intermediate plate stiffener is shown in SK 7/12 for which the equivalent area of the web stiffener A_{sc} is given by:

$$A_{sc} = 40t^2 + 2b_s t_s$$

b_s should be taken as equal to the actual b_s or $13t_s\sqrt{(275/p_y)}$ (whichever is the smaller). For the type of stiffener shown in SK 7/12:

$$I_{sc} = \frac{1}{12} t_s(2b_s + t)^3 + \frac{1}{6} t^3 \left(20t - \frac{t_s}{2}\right)$$

$$r_{sc} = \left(\frac{I_{sc}}{A_{sc}}\right)^{\frac{1}{2}}$$

$$\lambda_q = \frac{0.7 L_s}{r_{sc}}$$

L_s = length of the stiffener
r_{sc} = radius of gyration of the equivalent core section of the transverse web stiffener

Corresponding to λ_q find p_{cq} from Table 27(c) of BS 5950: Part 1.

$$P_q = p_{cq} A_{sc}$$

Note: *Reduce p_y by 20 N/mm² if the stiffener is welded to a welded section.*

Intermediate transverse web stiffeners *subject to loads and moment* should satisfy the following unity equation:

$$\frac{F_q - F_x}{P_q} + \frac{F_x}{P_x} + \frac{M_s}{M_{ys}} \leq 1$$

Use positive values of $F_q - F_x$.
F_q is the stiffener force $= V - V_s$ (use positive values).
P_q is the buckling resistance of an unloaded intermediate web stiffener based on the compressive strength p_{cq} of a strut using Table 27(c) of BS 5950: Part 1. The effective section is the core area of the stiffener with an effective length of web equal to 20 times the web thickness on either side of the centre-line of the stiffener.

F_x = external load on the stiffener
P_x = buckling resistance of a load carrying stiffener
M_s = moment on the stiffener due to eccentrically applied load
M_{ys} = moment capacity of the stiffener based on its elastic modulus
V = maximum shear force adjacent to the stiffener
V_s = shear buckling resistance of the web panel without tension field action = V_{cr}

$$\lambda_x = \frac{k L_s}{r_{sc}}$$

L_s = length of the stiffener
r_{sc} = radius of gyration of the equivalent section of the transverse stiffener

At the point of application of the load, if the flange is restrained against rotation in the plane of the stiffener by other structural elements, $k = 0.7$.

At the point of application of the load, if the flange is free to rotate, $k = 1.0$.

Corresponding to λ_x find p_{cx} from Table 27(c) of BS 5950: Part 1.

$P_x = p_{cx} A_{sc}$

A_{sc} = equivalent area of the web stiffener = $40t^2 + 2b_s t_s$

Note: Reduce p_y by 20 N/mm² if the stiffener is welded to a welded section.

For the plate stiffeners shown in SK 7/12, M_{ys} is given by:

$$M_{ys} = p_y \frac{I_{sc}}{\left(b_s + \dfrac{t}{2}\right)}$$

Calculate from first principles M_{ys} and I_{sc} for other types of stiffeners.

Note: $(F_q - F_x)$ to be taken equal to zero if F_x is greater than F_q.

Check bearing of intermediate transverse web stiffener
Assuming that 80% of the load on the flange is transferred by direct bearing on the contact surface of the stiffeners and the flange, the following condition should be satisfied:

$$A > \frac{0.8 F_x}{p_{ys}}$$

where A = area of stiffener in contact with the flange = $2b'_s t_s$
 F_x = external load
 p_{ys} = design strength of the stiffener
 b'_s = width of stiffener in contact with flange

Step 17 **Bearing stiffener without tension-field action**
R_L is the end reaction of the beam at the bearing. If R_L exceeds the local capacity of web, as calculated in Step 14, then a bearing stiffener is needed.
Maximum outstand of stiffener = $19 t_s \sqrt{(275/p_y)}$
Maximum value of outstand to be considered in the design
 = $13 t_s \sqrt{(275/p_y)}$
Assuming 80% of the end reaction is transferred to the contact surface, check $A \geq 0.8 R_L / p_y$.

 A = area of stiffener in contact with flange = $2b'_s t_s$
 M_s = applied moment due to transverse eccentricity of R_L
 F = lateral load at the bearing
 I_s = second moment of area of stiffeners only about the centre-line of the web
 E = elastic modulus of steel

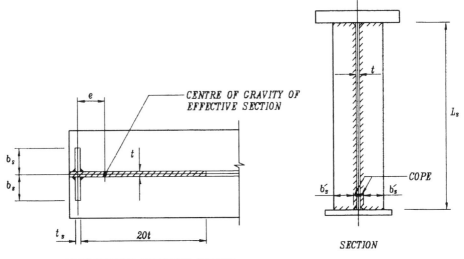

SK 7/14 Bearing stiffeners.

A_{sc} = equivalent area of stiffeners with or without including 20 times the thickness of web

I_{sc} = equivalent moment of inertia with or without including 20 times the thickness of web

p_{cx} = compressive strength of strut corresponding to λ_x

$$I_s = \frac{t_s}{12}[(2b_s + t)^3 - t^3] \text{ (for stiffeners shown in SK7/14)}$$

$$\geq 0.75dt^3 + \frac{2FD^3}{Et} + \frac{M_s D^2}{Et} \quad \text{for } a \geq \sqrt{2}d$$

$$\geq \frac{1.5d^3 t^3}{a^2} + \frac{2FD^3}{Et} + \frac{M_s D^2}{Et} \quad \text{for } a < \sqrt{2}d$$

Calculate I_s from first principles for other types of stiffeners.

Note: Conservatively the effective section of web ($20t$) may not be included to find the buckling resistance of bearing stiffeners.

For the stiffener shown in SK 7/14:

$A_{sc} = 20t^2 + 2b_s t_s$ if the web is included

$\quad\quad = 2b_s t_s$ if the web is not included

$$I_{sc} = \frac{1}{12} t_s (2b_s + t)^3 + \frac{1}{12} t^3 \left(20t - \frac{t_s}{2}\right) \text{ if the web is included}$$

$$= \frac{1}{12} t_s (2b_s + t)^3 \text{ if the web is not included}$$

b_s should be taken as equal to the actual b_s or $13t_s \sqrt{(275/p_y)}$ (whichever is the smaller).

The radius of gyration about the minor axis of the effective section is given by:

$$r_{sc} = \left(\frac{I_{sc}}{A_{sc}}\right)^{\frac{1}{2}}$$

Calculate from first principles for other types of stiffeners.

$$\lambda_x = \frac{kL_s}{r_{sc}}$$

where λ_x = slenderness ratio of the effective section of bearing stiffener
L_s = length of bearing stiffener

At the point of application of the load, if the flange is restrained against rotation in the plane of the stiffener by other structural elements, $k = 0.7$.

At the point of application of the load, if the flange is free to rotate, $k = 1.0$.

Corresponding to λ_x find P_{cx} from Table 27(c) of BS 5950: Part 1.

Check $P_x = A_{sc} p_{cx} \geq R_L$ when the web is not included in the effective section of the bearing stiffener.

Check $P_{LW} + A p_y \geq R_L$

See Step 14 for P_{LW}.

P_x = buckling resistance of a load carrying stiffener
P_{LW} = local capacity of web at bearing or under concentrated load

Note: *Reduce* p_y *by 20 N/mm² if the stiffener is welded to a welded section.*

A bearing stiffener is designed as a column. For a bearing stiffener it is important to note that the applied reaction R_L at the bearing is eccentric along the centre-line of the web to the equivalent effective section, including 20 times the thickness of the web. If it is necessary to include the effective section of the web in the calculation of the buckling strength of the bearing stiffener, then the centre of gravity of the equivalent effective section should be found and the eccentricity of the reaction R_L from the centre of gravity of the effective section along the centre-line of web should be treated as an applied moment on the section. The procedure of analysis is as follows:

- Design as a column in simple construction as per Clause 4.7.7 of BS 5950: Part 1.
- Determine the applied moment M on the effective section of the bearing stiffener = $R_L e$ (where e is the eccentricity of reaction R_L from the centre of gravity of effective section along the centre-line of web). See SK 7/14.

- Determine the plastic modulus S_x of the effective section about the equal area axis perpendicular to the web.
- Determine effective slenderness $\lambda_{LT} = 0.5(L_s/r_{sc})$ (where L_s is the length of the bearing stiffener and r_{sc} is the radius of gyration about the minor axis of the effective section of the stiffener).
- Determine the bending strength p_b from Table 12 of BS 5950: Part 1 using λ_{LT} and the design strength p_y.
- Determine $M_b = S_x p_b$
- Check the unity ratio: $\dfrac{R_L}{A_{sc} p_{cx}} + \dfrac{M}{M_b} \leq 1.0$

(r_{sc} and A_{sc} are determined with the web included)

Step 18 Bearing stiffeners using tension-field action

Use of transverse stiffeners results in an increase in shear buckling resistance V_{cr} of the web due to decreased values of the a/d ratio. The transverse web stiffeners also increase the shear strength of web due to the development of tension field action. If a panel of web between transverse stiffeners buckles in shear due to the diagonal compressive stress, then the tension field develops along the other diagonal to carry the additional shear force.

In SK 7/15 the solid web girder is shown to behave like a typical N-truss with tension diagonals. Prior to the buckling of the compression diagonal, equal compressive and tensile principal stresses develop in the plane of the web. The tension field appears after the buckling along the compression diagonal. The progressive shortening along the compression diagonal results in plastic hinge formation in the flanges.

The shear buckling resistance V_b is composed of the basic shear strength q_b, which combines the buckling strength q_{cr} and the post-buckling strength due to tension-field action. The elastic critical shear stress of the panel $a \times d$ is called q_{cr}. The shear buckling resistance also has a component based on the flange-dependent shear stress q_f. The flanges are required to carry the end anchor forces of the tension diagonal after development of tension-field action. The flanges must have some spare capacity to carry this anchoring force in addition to the stresses due to moment and axial load. Hence this flange-dependent shear strength is multiplied by a factor $(\sqrt{K_f})$ to allow for the state of stress in the flange due to the bending moment and axial load present at the point of interest. The smaller flange should be considered to determine the state of stress.

V_b = shear buckling resistance
q_b = basic shear strength
q_f = flange-dependent shear strength
q_{cr} = critical shear strength
a = spacing of transverse stiffeners
d = depth of web
t = thickness of web
p_{yw} = design strength of web
p_{yf} = design strength of flange

Conservatively, $V_b = q_b dt$, where the flange-dependent shear strength factor q_f is ignored. Otherwise, V_b is given by:

$$V_b = [q_b + q_f(K_f)^{\frac{1}{2}}]dt \leq 0.6 p_y dt$$

$$q_b = q_{cr} + \frac{y_b}{2\left[\frac{a}{d} + \left\{1 + \left(\frac{a}{d}\right)^2\right\}^{\frac{1}{2}}\right]}$$

y_b = basic tension field strength = $(p_{yw}^2 - 3q_{cr}^2 + \phi_t^2)^{\frac{1}{2}} - \phi_t$

$$\phi_t = \frac{1.5 q_{cr}}{\left[1 + \left(\frac{a}{d}\right)^2\right]^{\frac{1}{2}}}$$

$$q_f = \left[4\sqrt{3} \sin\left(\frac{\theta}{2}\right)\left(\frac{y_b}{p_{yw}}\right)^{\frac{1}{2}}\right] 0.6 p_{yw}$$

$$\theta = \tan^{-1}\left(\frac{d}{a}\right)$$

The factor K_f depends on the mean longitudinal stress f in the smaller flange due to the applied moment and axial load.

$$K_f = \frac{M_{pf}}{4 M_{pw}}\left(1 - \frac{f}{p_{yf}}\right)$$

M_{pf} = plastic moment capacity of the smaller flange about its own equal area axis parallel to the flange
$= 0.25 b T^2 p_{yf}$
b = width of the smaller flange
T = thickness of the smaller flange
p_{yf} = design strength of flange
M_{pw} = plastic moment capacity of the web about its own equal area axis perpendicular to the web $= 0.25 d^2 t p_{yw}$
d = depth of web
t = thickness of web
p_{yw} = design strength of web
f = mean longitudinal stress in the smaller flange due to moment and axial load on the beam at the point of interest

Note: The mean longitudinal stress f in the smaller flange may be taken as the highest stress state in the length of a panel bounded by two stiffeners at a spacing equal to a.

The tension-field forces require the anchorage of the elements forming the boundary of the panel of the web. The anchorage force H_q from the tension-field action is given by:

$$H_q = 0.75 dt p_y \left[1 - \frac{q_{cr}}{0.6 p_y}\right]^{\frac{1}{2}}$$

$$f_v = \frac{F_v}{dt}$$

If $f_v < q_b$ then the value of H_q may be multiplied by $(f_v - q_{cr})/(q_b - q_{cr})$.

F_v = applied maximum shear force in the panel under consideration
H_q = anchorage force parallel to the longitudinal horizontal axis of the beam

Case 1: $a/d \leq 1$ $\quad q_c = \left[0.75 + \dfrac{1}{(a/d)^2}\right]\left[\dfrac{1000}{(d/t)}\right]^2$

Case 2: $a/d > 1$ $\quad q_c = \left[1.0 + \dfrac{0.75}{(a/d)^2}\right]\left[\dfrac{1000}{(d/t)}\right]^2$

q_c = elastic critical shear stress
a = spacing of transverse stiffeners
d = depth of web
t = thickness of web
V_{cr} = shear buckling resistance of web = $q_{cr}dt$
q_{cr} = critical shear strength, which depends on the web slenderness factor
λ_w = web slenderness factor = $\sqrt{[0.6(p_{yw}/q_c)]}$
p_{yw} = design strength of web

The relationship between λ_w and the critical shear strength q_{cr} for design purposes is given by:

(1) If $\lambda_w \leq 0.8$, $q_{cr} = 0.6p_{yw}$ and no web buckling is expected where the maximum shear stress is kept below $0.6p_{yw}$.
(2) If $0.8 < \lambda_w < 1.25$, $q_{cr} = 0.6p_{yw}[1 - 0.8(\lambda_w - 0.8)]$.
(3) If $\lambda_w \geq 1.25$, $q_{cr} = q_c$.

Using a/d, d/t and p_y:
 Find q_{cr} from Table 21 of BS 5950: Part 1.
 Find q_b from Table 22 of BS 5950: Part 1.
 Find q_f from Table 23 of BS 5950: Part 1.

The longitudinal anchor force H_q is resisted by the end panel of the girder or the end stiffener as a vertical beam spanning between the two flanges of the beam. The longitudinal horizontal shear force and the bending moment in this vertical beam are given by:

$$R_{tf} = \dfrac{H_q}{2}$$

$$M_{tf} = \dfrac{H_q d}{10}$$

Note: To calculate R_{tf} and M_{tf}, use a value of a/d for the panel nearest to the bearing which has been designed using tension-field action.

There are three possible ways that girders may be designed using tension-field action.

Case 1

Intermediate panels (B and C) are designed by using tension-field action; the end panel (A) is not. The shear force F_v in the end panel is limited to the shear buckling resistance V_{cr} of the panel, which depends on the spacing of stiffeners in the end panel. The end panel is designed as a vertical beam between the two flanges resisting the shear R_{tf} and bending moment M_{tf}.

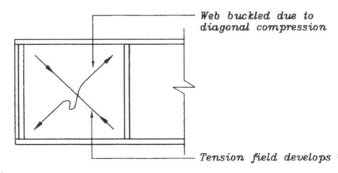

(a) PART ELEVATION OF GIRDER

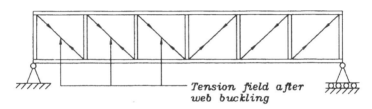

(b) EQUIVALENT N-GIRDER

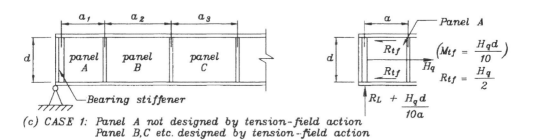

(c) CASE 1: Panel A not designed by tension-field action
Panel B,C etc. designed by tension-field action

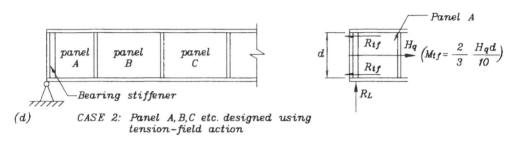

(d) CASE 2: Panel A,B,C etc. designed using tension-field action

SK 7/15 Web panels designed using tension-field action.

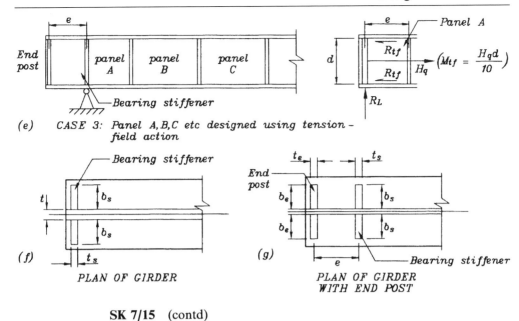

(e) CASE 3: Panel A,B,C etc designed using tension-field action

(f) PLAN OF GIRDER

(g) PLAN OF GIRDER WITH END POST

SK 7/15 (contd)

Select the stiffener spacing (a) for the end panel such that a/d for the end panel closest to the support gives q_{cr} (from Table 21 of BS 5950: Part 1) greater than or equal to the shear stress f_v.

Note: R_{tf} and M_{tf} are obtained for the panel which is designed using tension-field action. To calculate R_{tf} and M_{tf} use the values of a/d for that panel.

The flanges of the vertical beam spanning between the flanges of the girder are the two vertical stiffeners bounding panel A. The bending moment M_{tf} is resisted by these two stiffeners acting as flanges at a lever arm equal to a, the spacing of the stiffeners. Therefore, the additional vertical forces in the stiffeners = M_{tf}/a.

Follow Step 17 completely but instead of R_L use $(R_L + M_{tf}/a)$.

Check that end panel A has adequate capacity for horizontal shear R_{tf}.
To check the shear capacity of the vertical beam the following procedure should be used.

- Determine a/t ratio of the vertical beam where the depth of the vertical beam is the spacing a of the stiffener in end panel A.
- If this ratio is greater than 63ε, the web is classed as slender and Tables 21(a) to (d) of BS 5950: Part 1 can then be used to find q_{cr}.
- Determine $V_{cr} = q_{cr}at$ and check that $V_{cr} \geq R_{tf}$. Note that the depth of this beam is the spacing a of the stiffeners in end panel A.
- If the web of the vertical beam is not slender then check $0.6p_{yw}ta \geq R_{tf}$, where a is the spacing of the stiffeners in the end panel A.

Case 2
Intermediate panels (B and C) and the end panel (A) are designed using tension-field action and a single bearing stiffener.

The single end bearing stiffener in this case has to resist the tension-field force of end panel A. This stiffener, spanning vertically between the flanges, has to resist the vertical end reaction R_L of the girder as axial compression, a bending moment equal to $\frac{2}{3}M_{tf}$ about the vertical axis and a horizontal shear force equal to R_{tf}.

- Design as a column in simple construction as per Clause 4.7.7 of BS 5950: Part 1.
- Determine applied moment M on the effective section of bearing stiffener $= R_L e$ (where e is the eccentricity of reaction R_L from the centre of gravity of effective section along the longitudinal axis of the beam when the effective section of the bearing stiffener includes 20 times the thickness of the web).
- Determine plastic modulus S_x of the effective section about the equal area axis perpendicular to the web.
- Determine effective slenderness $\lambda_{LT} = 0.5(L_s/r_{sc})$ (where L_s is the length of the bearing stiffener and r_{sc} is the radius of gyration about the minor axis of the effective section of the stiffener).
- Determine the bending strength p_b from Table 12 of BS 5950: Part 1 using λ_{LT} and the design strength p_y.
- Determine $M_b = S_x p_b$.
- Check the unity ratio: $\dfrac{R_L}{P_x} + \dfrac{M + \frac{2}{3}M_{tf}}{M_b} \leq 1.0$.

Determine P_x according to method described in Step 17.

Conservatively the effective section of the bearing stiffener may not include the contribution from 20 times the thickness of web along the longitudinal axis of the beam. In that case the term M vanishes from the unity equation.

Check that the bearing stiffener has adequate shear capacity against horizontal shear R_{tf}: $0.6p_{yw}(2b_s t_s) \geq R_{tf}$.

Note: The width and the thickness of the bearing stiffener should not exceed those of the smaller and thinner flange of the beam. In practice, a single bearing stiffener designed using tension-field action becomes very thick and uneconomical.

Case 3
Intermediate panels (B and C) and the end panel (A) are designed using tension-field action and also using a pair of end stiffeners spaced e apart.

In this case a bearing stiffener is placed in line with the centre-line of the bearing and the web is allowed to project beyond the centre-line of the bearing by a distance e, where the end post is located. The end post and the bearing stiffener jointly form the vertical beam which carries the anchorage force H_q from tension-field action. This system requires the availability of space beyond the centre-line of the bearing for the web to project, and in most practical applications it becomes difficult to accommodate.

The vertical beam has to resist the horizontal shear R_{tf} and the moment M_{tf} resulting from tension-field action. The bearing reaction R_L is resisted by the bearing stiffener. The length of web equal to e projecting over the bearing resists the horizontal shear R_{tf}. To check the shear capacity of the vertical beam the following procedure should be used.

- Determine the e/t ratio of the vertical beam where the depth of the vertical beam is the spacing e of the bearing stiffener and the end post.
- If this ratio is greater than 63ε, the web is classed as slender and Tables 21(a) to (d) of BS 5950: Part 1 can then be used to find q_{cr}.
- Determine $V_{cr} = q_{cr} et$ and check that $V_{cr} \geq R_{tf}$. Note that the depth of this beam is the spacing e of the end posts.
- If the web of the vertical beam is not slender then check $0.6 p_{yw} te \geq R_{tf}$, where e is the spacing of bearing stiffener and the end post (see SK 7/15), p_{yw} is the design strength of the web and t is the thickness of the web.

The moment M_{tf} is resisted by the vertical beam, with flanges as the bearing stiffener and the end post spaced at a lever arm equal to e.

Design of end post
The width of the end post b_e should be taken as equal to the actual b_e or $13t_e \sqrt{(275/p_y)}$ (whichever is the smaller).

t_e = thickness of end post
b_s = width of bearing stiffener
t_s = thickness of bearing stiffener
t = thickness of web

The vertical end beam has width of flanges equal to $2b_e + t$ and $2b_s + t$ and the depth of web is e. The procedure for determining the buckling moment of resistance of this vertical beam is as follows.

- Determine the vertical equal area axis of this beam and determine the plastic modulus S_x of the section of the vertical beam about the equal area axis.
- Determine the section area A_s of this vertical beam and determine the moment of inertia I_s of the vertical beam section about the minor axis, which is horizontal through the centre-line of the web along the longitudinal axis of the beam.
- $A_s = 2(b_e + b_s) + et$ and $I_s = [(2b_e + t)^3 t_e + (2b_s + t)^3 t_s + et^3]/12$.
- Determine the radius of gyration $r_s = \sqrt{(I_s/A_s)}$.
- Determine the slenderness ratio of the vertical beam $= \lambda = L_s/r_s$, where L_s is the length of the stiffener.
- Determine $N = \dfrac{(2b_e + t)^3 t_e}{(2b_e + t)^3 t_e + (2b_s + t)^3 t_s}$.

- Determine $x = e/t_e$ or e/t_s (whichever is the smaller).
- Determine λ/x and, using the value of N, find the slenderness factor v from Table 14 of BS 5950: Part 1.
- Determine the equivalent slenderness $\lambda_{LT} = nuv\lambda$, where n and u may be taken equal to 1.0.
- Using λ_{LT} and the design strength p_y find the bending strength p_b from Table 12 of BS 5950: Part 1.
- Determine the buckling resistance moment $M_b = S_x p_b$.
- Check $M_{tf} \leq M_b$.

Design of bearing stiffener

The width of the bearing stiffener b_s should be taken as equal to the actual b_s or $13 t_s \sqrt{(275/p_y)}$ (whichever is the smaller).

Check $A \geq 0.8 R_L/p_y$, assuming that 80% of the load is transferred at the contact surface.

F = lateral load at the bearing
A = area of stiffener in contact with flange = $2 b'_s t_s$
t_s = thickness of the bearing stiffener

$$I_s = \frac{t_s}{12}[(2b_s + t)^3 - t^3]$$

$$\geq 0.75 d t^3 + \frac{2FD^3}{Et} \quad \text{for } a \geq \sqrt{2} d$$

$$\geq \frac{1.5 d^3 t^3}{a^2} + \frac{2FD^3}{Et} \quad \text{for } a < \sqrt{2} d$$

For $e \geq 20t$:

$$A_{sc} = 40 t^2 + 2 b_s t_s$$

$$I_{sc} = \frac{1}{12} t_s (2 b_s + t)^3 + \frac{1}{12} t^3 \left(40 t - \frac{t_s}{2}\right)$$

$$r_{sc} = \left(\frac{I_{sc}}{A_{sc}}\right)^{\frac{1}{2}}$$

For $e < 20t$:
Conservatively, ignore the contribution of 20 times the thickness of web on either side of the centre-line of the stiffener. Otherwise, the reaction R_L may be assumed to act eccentrically to the equivalent effective section and the section has to be analysed accordingly.

$$A_{sc} = (2 b_s + t) t_s$$

$$I_{sc} = \frac{1}{12} t_s (2 b_s + t)^3$$

$$r_{sc} = \left(\frac{I_{sc}}{A_{sc}}\right)^{\frac{1}{2}}$$

$$\lambda_x = \frac{k L_s}{r_{sc}}$$

Design of Beams 263

At the point of application of load, if the flange is restrained against rotation in the plane of the stiffener by other structural elements, $k = 0.7$.

At the point of application of load, if the flange is free to rotate, $k = 1.0$.

Corresponding to λ_x find p_{cx} from Table 27(c) of BS 5950: Part 1.

Check $P_x = A_{sc} p_{cx} \geq R_L$

Check $P_{LW} + A p_y \geq R_L$

See Step 14 for P_{LW}.

Note: *Deduct 20 N/mm² from p_y if the stiffener is welded to a welded section.*

Webs with openings
- If any dimension of the opening exceeds 10% of the minimum dimension of the panel where it occurs, then the panel shear capacity must not exceed V_{cr}.
- The tension-field action of a panel with an opening should be ignored.
- The panel adjacent to a panel with an opening should be designed as an end panel.

Step 19 **Torsion stiffener**
When stiffeners are required to provide torsional restraint at the supports of the beam, the following condition must be satisfied:

$$I_s \geq 0.34 \alpha_s D^3 T_c$$

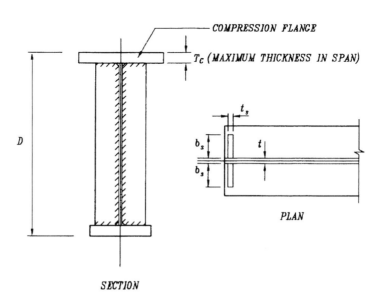

SK 7/16 Torsion stiffener.

where $I_s = \dfrac{t_s}{12}[(2b_s + t)^3 - t^3]$ (for stiffener shown in SK 7/16)

$\lambda = \dfrac{L_E}{r_y}$ (from Step 12)

$\alpha_s = 0.006$ for $\lambda \leq 50$

$\alpha_s = \dfrac{0.3}{\lambda}$ for $50 < \lambda \leq 100$

$\alpha_s = \dfrac{30}{\lambda^2}$ for $\lambda > 100$

$T_c =$ maximum thickness of the compression flange

Step 20 Connection of stiffeners to web

Connection of intermediate stiffeners

$w_s =$ shear per unit length between the web and stiffener to be resisted by weld

$F =$ external load on the stiffener in kN

$w = \dfrac{t^2}{5b_s} + \dfrac{Fb_s t_s}{(A_{sc})L_s}$ kN/mm per leg of stiffener

$A_{sc} =$ equivalent area of stiffener as found in Step 16

when t, b_s, t_s and L_s are in mm for the type of stiffener shown in SK 7/14.
Calculate from first principles for other types of stiffeners.

Connection of bearing stiffener

$R_L =$ reaction in kN

$w_s = R_L b_s t_s / A_{sc} L_s$ or $b_s t_s p_y / L_s$ kN/mm (whichever is the smaller for the stiffener shown in SK 7/14)

Step 21 Biaxial bending
Local capacity check

$$\dfrac{M_x}{M_{cx}} + \dfrac{M_y}{M_{cy}} \leq 1.0$$

See Step 9, Step 10 and Step 11 as appropriate for M_{cx} and M_{cy}

Overall buckling check

$$\dfrac{mM_x}{M_b} + \dfrac{mM_y}{p_y Z_y} \leq 1$$

See Step 12 for M_b

Step 22 Deflection due to imposed load
Load factor on imposed load $= 1.0$ for deflection checks at serviceability limit state.

Table 7.3 Limits of deflection of beams.

Type of beam	Deflection limit
Cantilever	$L/180$ vertical
Beams carrying plaster	$L/360$ vertical
All other beams	$L/200$ vertical
Gantry girders (for static wheel loads)	$L/600$ vertical
Gantry girders (due to crane surge)	$L/500$ horizontal

Step 23 **Local stresses**

Check local bearing stresses under wheel in gantry girders.

Allow 45° dispersion of wheel load through rail of depth H_R and flange thickness of the gantry girder T.

$$X_R = 2(H_R + T)$$

Assuming continuously welded rail, check bearing stress f_w given by:

$$f_w = \frac{W_v}{X_R t} \leq p_y$$

where W_v is the wheel load and t is the thickness of web.

Note: If there are joints in the rail, then load from the wheel cannot disperse at 45° either side of the joint.

Check also web buckling under wheel.

See Step 13. Use $\lambda = 3.5(d/t)$.

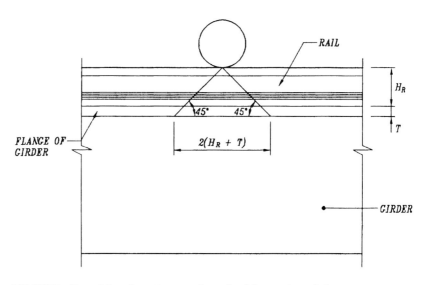

SK 7/17 Local bearing stress under wheel in gantry girder.

Step 24 **Connection of flange to web in welded plate girders**

A_{ft} = area of top flange (SK 7/18)
A_{fb} = area of bottom flange (SK 7/18)
I_x = moment of inertia about major axis
Q_{xt} = first moment of area of top flange about neutral axis
Q_{xb} = first moment of area of bottom flange about neutral axis
V = ultimate shear at the section under consideration
h_t = distance from neutral axis to centroid of top flange
h_b = distance from neutral axis to centroid of bottom flange
$Q_{xt} = A_{ft} h_t$
$Q_{xb} = A_{fb} h_b$
V_{xt} = horizontal shear per unit length at the connection between top flange and web
$$= \frac{V Q_{xt}}{I_x}$$
V_{xb} = horizontal shear per unit length at the connection between bottom flange and web
$$= \frac{V Q_{xb}}{I_x}$$

The capacity of two runs of fillet weld or butt weld should be greater than V_{xt} or V_{xb} as appropriate. The capacity of weld runs in kN/mm can be obtained from Table 11.3.

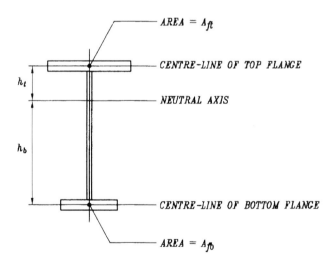

SK 7/18 Connection of flange to web of plate girder.

Note: For built-up gantry girders it may be assumed that the wheel loads of the crane are transferred from the top flange to the web by the weld between the top flange and the web.

W_v = wheel load in kN
$X_R = 2(H_R + T)$ (see Step 23)

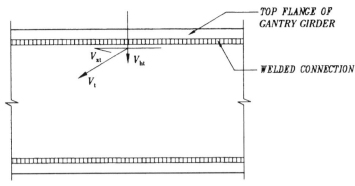

SK 7/19 Connection of top flange in gantry girder.

V_{ht} = vertical shear per unit length due to wheel load in gantry girder
$$= \frac{W_v}{X_R} \text{ kN/mm}$$
V_t = vector sum V_{xt} and V_{ht}
$$= \sqrt{V_{xt}^2 + V_{ht}^2} \text{ kN/mm}$$

Select capacity of two runs of fillet weld which is greater than V_t.

Step 25 *Check special requirements for gantry girders*
Find the following parameters (see SK 7/20):
    Crane live load including hook load = W_{cap}
    Weight of crab = W_{cb}
    Weight of crane bridge = W_c
    Span of crane bridge = L_c
    Wheel spacing in end carriage = a_w
    Minimum hook approach = a_h
    Total number of wheels = N

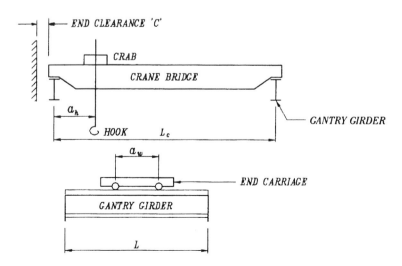

SK 7/20 Geometric layout of gantry girder.

Class of utilisation U_1 to U_9 (Table 1 of BS 2573: Part 1).
State of loading Q_1 to Q_4 (Table 2 of BS 2573: Part 1).
Nominal load spectrum factor K_p (Table 2 of BS 2573: Part 1).
Maximum number of operating cycles (Table 1 of BS 2573: Part 1).
Impact factor I (Table 4 of BS 2573: Part 1).

The maximum unfactored static wheel load W_{us} is given by:

$$W_{us} = \frac{2}{N}\left[\frac{W_c}{2} + (W_{cb} + W_{cap})\frac{(L_c - a_h)}{L_c}\right]$$

CLV = vertical unfactored load on the gantry girder per wheel of crane = IW_{us}

CLH_s = horizontal surge acting transversely at the level of the rail on the girder = $0.1(W_{cb} + W_{cap})/N$

CLH_c = crabbing action of the travel of the crane, acting horizontally in the transverse direction as two equal and opposite forces at the level of the rail = $L_c W_{us}/40 a_w \geq W_{us}/20$

CLH_t = longitudinal traction per wheel = $0.05 W_{us}$

$DL = W_g$ = self-weight of gantry girder + walkway etc.

IL = imposed load on the walkway

Refer to Chapter 4 for load combinations LC15 to LC40.

Note: Crabbing action need not be considered for gantry girders carrying loading classes Q_1 and Q_2.

Note: By inspection most of these load combinations may be ignored.

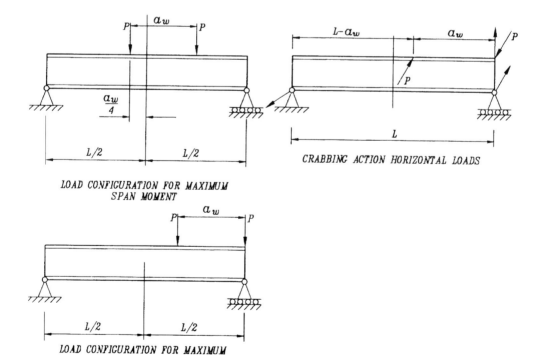

SK 7/21 Loading diagrams on gantry girders.

Design of Beams 269

Note: Fatigue checks may be necessary for crane gantry girders. See Example 7.4.5.

For a compound flange the following ratios should be maintained (see SK 7/22).

(1) $\dfrac{B/2}{T} < 8.5\varepsilon$

(2) $\dfrac{B}{T_p} < 25\varepsilon$

(3) $\dfrac{(B_p - B)/2}{T_p} < 8.5\varepsilon$

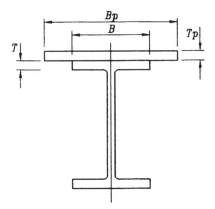

SK 7/22 Properties of compound section of gantry girders.

7.4 WORKED EXAMPLES

7.4.1 Example 7.1: Beam supporting the floor of a workshop

Span of beam = 6.0 m centre-to-centre of bearing on concrete walls
Centre-to-centre of beams on plan = 3.0 m
Assume 10 mm floor plates.
Imposed loading on the floor = 5 kN/m²
The beam is carrying two legs of a water tank with four legs and containing 18 m³ of water.

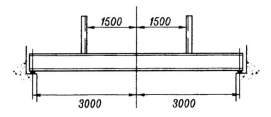

SK 7/23 Elevation of floor beam supporting two legs of a four-legged water tank.

Step 1 ***Determine design strength*** p_y
Choose design Grade 43. See Table 6 of BS 5950: Part 1.
$p_y = 275 \, \text{N/mm}^2$ up to 16 mm thickness of material
$p_y = 265 \, \text{N/mm}^2$ up to 40 mm thickness of material

Step 2 ***Carry out analysis***
Determine loading:
Self-weight of flooring $= 1.2 \, \text{kN/m}^2$
Self-weight of floor on beam $= 1.2 \times 3 \, \text{m} = 3.6 \, \text{kN/m}$
Self-weight of beam assumed $= 1 \, \text{kN/m}$
Self-weight of water tank $= 16 \, \text{kN}$
Concentrated dead load of water tank per leg of tank $= 4 \, \text{kN}$
Imposed load from floor on beam $= 5 \times 3 \, \text{m} = 15 \, \text{kN/m}$
Concentrated load of water in tank per leg of tank $= 45 \, \text{kN}$
Ultimate distributed load $= (1.4 \times 3.6) + (1.4 \times 1) + (1.6 \times 15)$
$\quad = 30.44 \, \text{kN/m}$
Ultimate concentrated load from each leg of tank $= 1.4 \times (4 + 45)$
$\quad = 68.6 \, \text{kN}$
Service distributed load $= 4.6 \, \text{kN/m}$
Service concentrated load per leg of tank $= 49 \, \text{kN}$

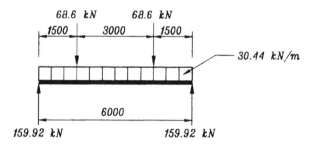

SK 7/24 Loading diagram for the beam.

Ultimate reaction at supports $= 68.6 + (30.44 \times 3) = 159.92 \, \text{kN}$
Ultimate bending moment at midspan:
$\quad W_u = (68.6 \times 1.5) + (30.44 \times 6^2 \div 8) = 239.88 \, \text{kNm}$
Service bending moment at midspan:
$\quad W_s = (19.6 \times 6^2 \div 8) + (49 \times 1.5) = 161.7 \, \text{kNm}$

A section with a plastic moment of resistance M_{cx} greater than 239.88 kNm is required. Select from the steel section handbook UB $457 \times 152 \times 52 \, \text{kg/m}$ with $M_{cx} = 300 \, \text{kNm}$.

Step 3 ***Draw bending moment and shear force diagrams***

Note: For a simply supported beam using rolled section and designed on the basis of maximum bending moment and shear force, these diagrams are not necessary.

Design of Beams 271

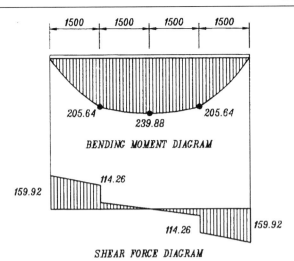

SK 7/25 Bending moment and shear force diagrams.

Step 4 *Carry out section classification*
$p_y = 275 \, \text{N/mm}^2$ for UB $457 \times 152 \times 52 \, \text{kg/m}$

$$\varepsilon = \sqrt{\frac{275}{p_y}} = 1$$

$\dfrac{b}{T} = 6.99 < 8.5\varepsilon$ and $\dfrac{d}{t} = 53.6 < 79\varepsilon$ (see Table 7 of BS 5950: Part 1)

Section classification is plastic (class 1).

Step 5 *Find section properties*
From the section handbook the following section properties are found:
$A = 66.5 \, \text{cm}^2$
$A_v = tD = 7.6 \times 449.8 = 3418 \, \text{mm}^2$
$I_x = 21\,300 \, \text{cm}^4$; $I_y = 645 \, \text{cm}^4$; $r_x = 17.9 \, \text{cm}$; $r_y = 3.11 \, \text{cm}$
$Z_x = 949 \, \text{cm}^3$; $Z_y = 84.6 \, \text{cm}^3$; $S_x = 1090 \, \text{cm}^3$; $S_y = 133 \, \text{cm}^3$
$u = 0.859$; $x = 43.9$

Step 6 *Check allowable shear force*
$P_v = 0.6 p_y A_v$
$= 0.6 \times 275 \times 3418 \times 10^{-3} \, \text{kN} = 564 \, \text{kN} > 159.92 \, \text{kN}$ (see Step 2)

Steps 7 and 8 Not required.

Step 9 *Section moment capacity with low shear*
$F_v = 159.92 \, \text{kN}$ (maximum)
$0.6 P_v = 0.6 \times 564 = 338.4 \, \text{kN} > F_v$

Design for moment capacity is based on low shear force.
The section classification is plastic (see Step 4).

$$k = \frac{W_u}{W_s} = \frac{239.88}{161.7} = 1.48$$

See Step 2 for W_u and W_s.
$M_{cx} = p_y S_x = 275 \times 1090 \times 10^3 \times 10^{-6} = 299.75\,\text{kNm}$
$kp_y Z_x = 1.48 \times 275 \times 949 \times 10^3 \times 10^{-6} = 386.24\,\text{kNm}$
$M_x = 239.88 < M_{cx} = 299.75 < kp_y Z_x = 386.24$ (Satisfied)

Steps 10 and 11 Not required.

Step 12 *Member moment capacity (lateral torsional buckling)*
The compression flange of this simply supported beam is laterally restrained by the floor plate. The lateral torsional buckling of this beam is not critical.

Step 13 *Web buckling of unstiffened web*
At the bearing, the bottom flange rests on concrete and is restrained from rotating by bolts fixed to the concrete. The top flange is restrained from movement relative to the bottom flange by the continuous floor plates and edge angles. Therefore, the slenderness ratio λ of the effective web in bearing is given by:

$$\lambda = 2.5\,\frac{d}{t} = 2.5 \times 53.6 = 134$$

From Table 27(c) of BS 5950: Part 1, using $\lambda = 134$ and $p_y = 275\,\text{N/mm}^2$, $p_c = 81\,\text{N/mm}^2$
From SK 7/26, $b_1 = 100\,\text{mm}$ and $n_1 = 225\,\text{mm}$.

$P_w = (b_1 + n_1) t p_c$

$\qquad = (100 + 225) \times 7.6 \times 81 \times 10^{-3} = 200\,\text{kN} > F_v = 159.92\,\text{kN}$

(Satisfied)

No bearing stiffeners are required.

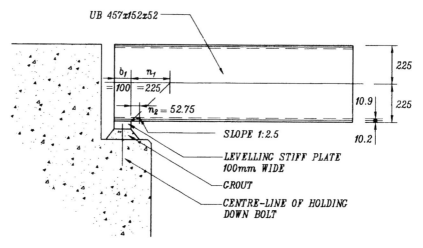

SK 7/26 Detail of beam at the bearing.

Step 14 Web bearing of unstiffened web

From SK 7/26, n_2 is derived as follows:

$$n_2 = 2.5 \times (10.9 + 10.2) = 52.75$$

$$P_{LW} = (b_1 + n_2)tp_{yw} = (100 + 52.75) \times 7.6 \times 275 \times 10^{-3}$$

$$= 319.2 \, \text{kN} > F_v$$

Web strength in bearing is satisfactory.

Steps 15 to 21 Not required.

Step 22 Deflection due to imposed loads

Distributed service imposed load $= 15 \, \text{kN/m} = w$
Concentrated service load from the weight of water $= 45 \, \text{kN}$ per leg of tank $= P$
Deflection at midspan δ is given by:

$$\delta = \frac{5}{384} \frac{wL^3}{EI} + \frac{PL^3}{6EI} \left[\frac{3a}{4L} - \left(\frac{a}{L} \right)^3 \right]$$

$$= \left[\frac{5}{384} \times \frac{15 \times 6^3 \times 10^9}{205 \times 21\,300 \times 10^4} \right]$$

$$+ \left(\frac{45 \times 6^3 \times 10^9}{6 \times 205 \times 21\,300 \times 10^4} \times \left[\frac{3 \times 1.5}{4 \times 6} - \left(\frac{1.5}{6} \right)^3 \right] \right)$$

$$= 7.34 \, \text{mm} < L/200 = 6000/200 = 30 \, \text{mm}$$

Deflection limit is satisfied.

7.4.2 Example 7.2: Main beam in a multistorey building using simple construction

Centre-to-centre of columns $= 12 \, \text{m}$.
Secondary beams are at 3 m centres.
Main beam is assumed as simply supported on the column and laterally restrained at the column.
The compression flange of the main beam is laterally restrained at the secondary beams.
Main beams are at 6 m centres.
Overall depth of concrete floor slab $= 200 \, \text{mm} \equiv 0.2 \times 25 = 5 \, \text{kN/m}^2$.
Imposed load on floor $= 5 \, \text{kN/m}^2$ during use after completion.
Construction live load $= 5 \, \text{kN/m}^2$.

Step 1 Determine design strength p_y

Select steel = Grade 50.

$p_y = 355 \, \text{N/mm}^2$ for thickness of material up to 16 mm
$p_y = 345 \, \text{N/mm}^2$ for thickness of material up to 40 mm.

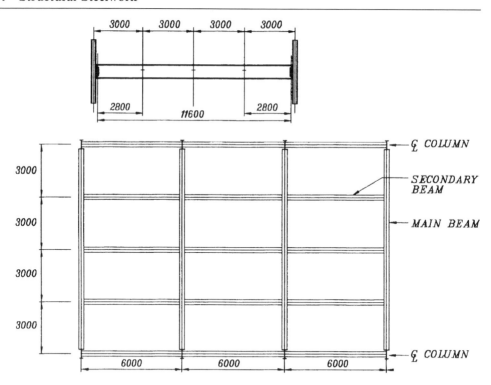

SK 7/27 General arrangement of beam.

Step 2 **Carry out analysis**
Ultimate loads on the beam are assumed to be the same at the construction stage as they are during use after completion.
Span of beam = centre of bearings = 11.6 m
Self-weight of beam assumed = 1.4 kN/m
Concentrated load from secondary beam:
  Dead load = 90 kN ($5\,\text{kN/m}^2 \times 3\,\text{m} \times 6\,\text{m} = 90\,\text{kN}$)
  Imposed load = 90 kN

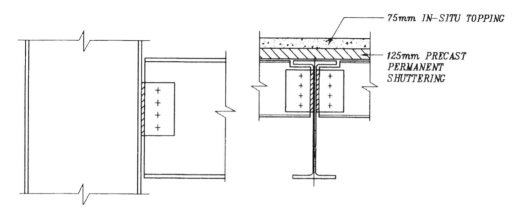

SK 7/28 Details of beam–column and beam–beam connection.

Design of Beams

Ultimate limit state load case1 = $1.4DL + 1.6IL$
Ultimate end reaction of beam at A and $E =$

$$\left(1.4 \times 1.4\,\text{kN/m} \times \frac{11.6}{2}\right) + [1.5 \times (1.4 \times 90 + 1.6 \times 90)]$$

$$= 416.4\,\text{kN}$$

Ultimate bending moment at B and $D =$

$$(416.4 \times 2.8) - \left(1.4 \times 1.4 \times \frac{2.8^2}{2}\right) = 1158.2\,\text{kNm}$$

$(1.4 \times 90) + (1.6 \times 90) = 270\,\text{kN}$

Ultimate bending moment at $C =$

$$\left(416.4 \times \frac{11.6}{2}\right) - (270 \times 3) - \left(1.4 \times 1.4 \times \frac{11.6^2}{8}\right) = 1572.2\,\text{kNm}$$

Serviceability limit state end reaction of beam at A and $E =$

$$\left(11.6 \times \frac{1.4}{2}\right) + \left(180 \times \frac{3}{2}\right) = 278.12\,\text{kN}$$

Serviceability limit state bending moment at $C =$

$$\left(278.12 \times \frac{11.6}{2}\right) - (180 \times 3) - \left(\frac{1.4 \times 116^2}{8}\right) = 1049.5\,\text{kNm}$$

Step 3 *Draw bending moment and shear force diagrams*

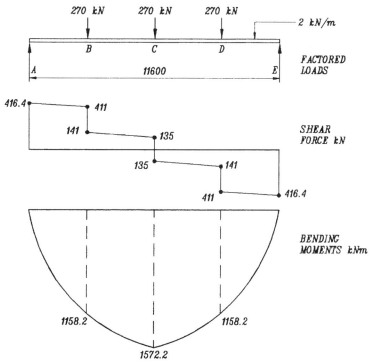

SK 7/29 Bending moment and shear force diagrams of the beam.

Step 4 *Choose trial section and carry out section classification*
Maximum bending moment in the beam $= 1572.2$ kNm
Required plastic modulus without consideration of any buckling $=$

$$S_x = \frac{M}{p_y} = \frac{1572.2 \times 10^6}{345} = 4557 \times 10^3 \text{ mm}^3$$

Select section UB 686 × 254 × 152 kg/m (with $S_x = 5000 \text{ cm}^3$)
Thickness of web $= t = 13.2$ mm
Thickness of flange $= T = 21.0$ mm
$b/T = 6.06$ and $d/t = 46.6$
$p_y = 345 \text{ N/mm}^2$ for flange and $= 355 \text{ N/mm}^2$ for web

For flange: $\varepsilon = \sqrt{(275/345)} = 0.89$

For web: $\varepsilon = \sqrt{(275/355)} = 0.88$

$b/T = 6.06 < 8.5\varepsilon$ and $d/t = 46.6 < 79\varepsilon$
Therefore, the section classification is plastic (class 1) as per Table 7 of BS 5950: Part 1.

Step 5 *Section properties*
Area of section $= A = 194 \text{ cm}^2$
Shear area of section $= tD = 13.2 \times 687.6 = 9076 \text{ mm}^2$
Moment of inertia of section: $I_x = 150\,000 \text{ cm}^4$ and $I_y = 5780 \text{ cm}^4$
Radius of gyration of section: $r_x = 27.8$ cm and $r_y = 5.46$ cm
Elastic section modulus: $Z_x = 4370 \text{ cm}^3$ and $Z_y = 454 \text{ cm}^3$
Plastic section modulus: $S_x = 5000 \text{ cm}^3$ and $S_y = 710 \text{ cm}^3$
$u = 0.871$; $x = 35.5$; $N = 0.5$

Step 6 *Check shear capacity*
$P_v = 0.6 p_y A_v$
$= 0.6 \times 355 \times 9076 = 1933 \times 10^3 \text{ N} > F_v = 416.4 \text{ kN}$ (Satisfied)

Steps 7 and 8 Not required.

Step 9 *Section moment capacity with low shear*
Ultimate maximum shear force $= F_v = 416.4 \text{ kN} < 0.6 P_v = 0.6 \times 1933 = 1159.8$ kN
Therefore, the design is based on low shear force.

$$k = \frac{W_u}{W_s} = \frac{1572.2}{1049.5} = 1.5 \quad \text{(see Step 2)}$$

$$M_{cx} = p_y S_x = 345 \times 5000 \times 10^3 \times 10^{-6} = 1725 \text{ kNm}$$

$$k p_y Z_x = 1.5 \times 345 \times 4370 \times 10^3 \times 10^{-6}$$
$$= 2261 \text{ kNm} > M_{cx} \quad \text{(Satisfied)}$$

Steps 10 Not required.
Step 11

Design of Beams 277

Step 12 ***Member moment capacity M_b (lateral torsional buckling)***
At the construction stage the floor slab is unable to offer lateral restraint to the beam. See SK 7/27. Self-weight of the beam is negligible. Consider the beam as unloaded between restraints.
It is assumed that: (1) there is no destabilizing load on the flange; (2) at the supports A and D, there is lateral restraint and torsional restraint; and (3) the flanges are partially restrained against rotation.

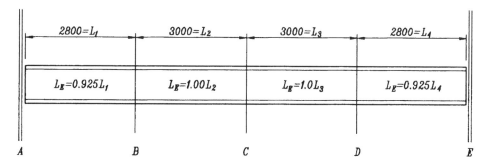

SK 7/30 Effective length of beam.

Effective length $L_E = 1.0L = 3000$ mm between B and C and C and D.
Effective length of the beam for sections AB and DE may be obtained as an average because of the end conditions at A and B.
Based on end condition at A the effective length $L_E = 0.85L$.
Based on end condition at B the effective length $L_E = 1.0L$.
Therefore effective length of sections AB and $DE =$

$$\frac{1.85L}{2} = 0.925L = 0.925 \times 2800 = 2590 \text{ mm}$$

Check sections BC and CD, where the bending moment is at the maximum and the effective length is also greatest.

$$\lambda = \frac{L_E}{r_y} = \frac{3000}{54.6} = 55 \text{ for sections } BC \text{ and } CD$$

$$\frac{\lambda}{x} = \frac{55}{35.5} = 1.55$$

$N = 0.5$ for symmetrical sections
$v = 0.97$ (from Table 14 of BS 5950: Part 1)
$u = 0.871$ (as found in Step 5)

$$\beta = \frac{M_1}{M_2} = \frac{1158.2}{1572.2} = 0.74$$

(which gives $m = 0.82$ from Table 18 of BS 5950: Part 1)
$n = 1.0$ (from Table 13 of BS 5950: Part 1)
$\lambda_{LT} = nuv\lambda = 1.0 \times 0.871 \times 0.97 \times 55 = 46.5$
$p_y = 345$ N/mm^2 for flange
$p_b = 300$ N/mm^2 (corresponding to $\lambda_{LT} = 46.5$ and $p_y = 345$ N/mm^2
  from Table 11 of BS 5950: Part 1)
$M_b = S_x p_b = 5000 \times 10^3 \times 300 \times 10^{-6} = 1500$ kNm

278 Structural Steelwork

Equivalent moment $= \bar{M} = mM = 0.82 \times 1572.2 = 1289.2\,\text{kNm} < M_\text{b}$

The section chosen is satisfactory.

By inspection, the other sections of the beam need not be checked for lateral torsional buckling.

Step 13 **Web buckling of unstiffened web**
This check is not required because the connection of the beam to the column is made by the use of web fin plates (see SK 7/28).

Step 14 **Web bearing of unstiffened web**
This check is not required.

Steps 15 to 21 Not required

Step 22 **Deflection due to applied imposed load**

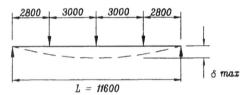

SK 7/31 **Deflection due to imposed loads.**

At the serviceability limit state, with a load factor equal to 1.0, the concentrated end reactions of the secondary beams on the main beam equal 90.0 kN from each secondary beam.

$$\delta_{\max} = \frac{PL^3}{48EI} + \frac{PL^3}{6EI}\left[\frac{3a}{4L} - \left(\frac{a}{L}\right)^3\right]$$

$$= \frac{90 \times 11.6^3 \times 10^9}{48 \times 205 \times 150\,000 \times 10^4}$$

$$+ \left(\frac{90 \times 11.6^3 \times 10^9}{6 \times 205 \times 150\,000 \times 10^4} \times \left[\frac{3 \times 2800}{4 \times 11\,600} - \left(\frac{2800}{11\,600}\right)^3\right]\right)$$

$$= 22.2\,\text{mm} < l/360 = 32.2\,\text{mm}$$

Deflection of the beam due to imposed load is within limits.

7.4.3 Example 7.3: Main beam with full end fixity to concrete wall

The clear span of beam between concrete walls is 12 m. Secondary beams are spaced at 3 m centres and are resting on the top flange of the main beam, but do not provide it with any lateral restraint. The loading from the secondary beams may be considered as destabilizing.

Step 1 **Determine design strength p_y**
Select steel = Grade 50.
$p_y = 355\,\text{N/mm}^2$ for thickness of material up to 16 mm
$p_y = 345\,\text{N/mm}^2$ for thickness of material up to 40 mm

Design of Beams 279

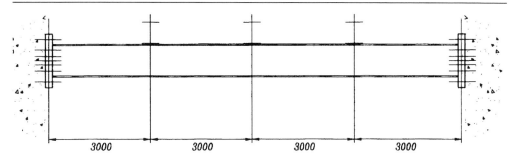

SK 7/32   Elevation of main beam with full end fixity to concrete walls.

Step 2 *Carry out analysis*
Assume a self-weight of beam equal to 2.5 kN/m. Unfactored concentrated loads from secondary beams are as follows:

At *B*: Dead load = 30 kN
         Imposed load = 180 kN
At *C*: Dead load = 60 kN
         Imposed load = 240 kN
At *D*: Dead load = 30 kN
         Imposed load = 60 kN

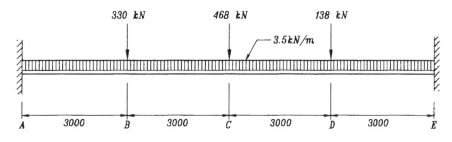

SK 7/33   Loading diagram.

Load combination at ultimate limit state = $1.4DL + 1.6IL$
Load factored self-weight of beam = $1.4 \times 2.5 = 3.5$ kN/m = w
Loads from secondary beams at ultimate limit state:

At *B*: $P_1 = (1.4 \times 30) + (1.6 \times 180) = 330$ kN
At *C*: $P_2 = (1.4 \times 60) + (1.6 \times 240) = 468$ kN
At *D*: $P_3 = (1.4 \times 30) + (1.6 \times 60) = 138$ kN

Fixed-end moment at $A = M_A$, given by:

$$M_A = -\left(\frac{wL^2}{12} + \frac{P_1 a_1 b_1^2}{L^2} + \frac{P_2 a_2 b_2^2}{L^2} + \frac{P_3 a_3 b_3^2}{L^2}\right)$$

$$= -\left(\frac{3.5 \times 12^2}{12} + \frac{330 \times 3 \times 9^2}{12^2} + \frac{468 \times 6 \times 6^2}{12^2} + \frac{138 \times 9 \times 3^2}{12^2}\right)$$

$$= -1378.5 \text{ kNm}$$

Fixed-end moment at $E = M_E$, given by:

$$M_E = -\left(\frac{wL^2}{12} + \frac{P_1 b_1 a_1^2}{L^2} + \frac{P_2 b_2 a_2^2}{L^2} + \frac{P_3 b_3 a_3^2}{L^2}\right) = -1162.5\,\text{kNm}$$

The reaction at $A = R_A$, given by:

$$R_A = \frac{1378.5 - 1162.5}{12} + \frac{330 \times 9}{12} + \frac{468}{2} + \frac{138 \times 3}{12} + \frac{3.5 \times 12}{2}$$

$$= 555\,\text{kN}$$

Similarly, reaction at $E = R_E = 423\,\text{kN}$
Bending moment at $B = M_B$, given by:

$$M_B = (555 \times 3) - 1378.5 - \frac{3.5 \times 3^2}{2} = +271\,\text{kNm}$$

Similarly $M_C = +898.5\,\text{kNm}$ and $M_D = +91\,\text{kNm}$

Note: Sagging moments are positive and hogging moments are negative.

Step 3 *Draw bending moment and shear force diagrams*

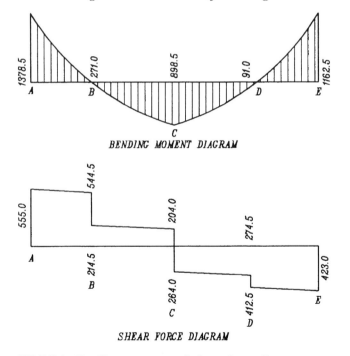

SK 7/34 Bending moment and shear force diagrams.

Step 4 *Section classification*
For a member with a destabilizing load between the restraints, m and n should be taken as equal to 1.0.

The effective length $L_E = 0.85L$ for a member with the compression flange laterally restrained, the beam fully restrained against torsion at the supports and both flanges fully restrained against rotation on plan at the supports. This is achieved by the full moment connection at the concrete wall.

$$L_E = 0.85 \times 12\,000 = 10\,200\text{ mm}$$

$$\text{Minimum plastic modulus required} = \frac{M_{max}}{p_y} = \frac{1378.5 \times 10^6}{355}$$

$$= 3883 \times 10^3 \text{ mm}^3$$

From steel section tables find a section with plastic modulus greater than 3883 cm³. This is found to be *UB 686 × 254 × 125 kg/m* with a minimum radius of gyration $r_y = 5.24$ cm and torsional index $x = 43.9$. Using these values, the slenderness ratio $\lambda = L_E/r_y = 10\,200/52.4 = 195$. Obtain $p_b = 73$ N/mm² from Table 19(d) of BS 5950: Part 1.

Revised plastic modulus required $= (1378.5 \times 10^6)/73$

$$= 18\,884 \times 10^3 \text{ mm}^3$$

This is a very conservative estimate, and a first trial section could be *UB 914 × 305 × 289 kg/m* with a plastic modulus equal to $12\,600 \times 10^3$ mm³.

For the chosen section the following dimensions are relevant:

$D = 926.6$ mm; $B = 307.8$ mm; $t = 19.6$ mm; $T = 32$ mm; $b/T = 4.81$; $d/t = 42.1$; $I_{xx} = 505 \times 10^6$ mm³; $r_{yy} = 6.51$ cm; $Z_{xx} = 109 \times 10^5$ mm³; $p_y = 345$ N/mm² for material thickness over 16 mm.

$$\varepsilon = \sqrt{\frac{275}{345}} = 0.89$$

$b/T = 4.81 < 8.5\varepsilon$ and $d/t = 42.1 < 79\varepsilon$

Therefore, the section is classified as plastic.

Step 5 **Section properties**
Area of section $= A = 369$ cm²
Shear area $= A_v = tD = 19.6 \times 926.6 = 18\,161$ mm²
$S_x = 12\,600$ cm³; buckling parameter $u = 0.867$; torsional index $x = 31.9$

Step 6 **Check shear capacity**
$P_v = 0.6 p_y A_v = 0.6 \times 345 \times 18\,161 \times 10^{-3} = 3759$ kN $> F_v = 555$ kN
Shear capacity is satisfied.

Steps 7 and 8 Not required.

Step 9 **Section moment capacity with low shear**
$0.6 P_v = 0.6 \times 3759 = 2256$ kN $> F_v$

The section moment capacity is based on low shear force.

$$M_{cx} = p_y S_x = 345 \times 12\,600 \times 10^{-3} = 4347\,\text{kNm}$$

$$k = \frac{\text{ultimate moment}}{\text{service moment}} = \frac{1378.5}{885} = 1.56$$

$$k p_y Z_x = 1.56 \times 345 \times 10\,900 \times 10^{-3} = 5866\,\text{kNm}$$

$$M < M_{cx} < k p_y Z_x \quad \text{(Satisfied)}$$

Steps 10 and 11   Not required.

Step 12   *Member moment capacity (lateral torsional buckling)*
$L_E = 10\,200$ mm (see Step 4)
$m = 1$ and $n = 1$ for member subjected to destabilizing loads

$$\lambda = \frac{L_E}{r_y} = \frac{10\,200}{65.1} = 157$$

$\lambda/x = 157/31.9 = 4.9$ and $N = 0.5$ for symmetrical section

From Table 14 of BS 5950: Part 1 $v = 0.82$

$\lambda_{LT} = nuv\lambda = 1 \times 0.867 \times 0.82 \times 157 = 112$
$p_b = 115\,\text{N/mm}^2$ (from Table 11 of BS 5950: Part 1 for $p_y = 345\,\text{N/mm}^2$)

Buckling resistance moment $M_b = p_b S_x = 115 \times 12\,600 \times 10^{-3}$
$= 1449\,\text{kNm}$
$M = 1378.5 < M_b = 1449\,\text{kNm}$ (Satisfied)

Step 13   *Web buckling of unstiffened web*
At the supports, full moment connection with the concrete walls will require end plates and hence the end reaction of the beam will not be transferred directly through the web. Web buckling at the supports need not be checked.

The secondary beams are resting on the top flange of the beam, and the concentrated load from the secondary beam is applied through the top flange. The top flange is assumed to be unrestrained from movement and rotation relative to web at the point of application of load. The flange with applied load is also not restrained from movement relative to the other flange.

The effective length of web carrying the applied load in compression is equal to the full depth of the section under these circumstances, which gives the slenderness ratio as follows:

$$\lambda = 3.5 \frac{d}{t} = 3.5 \times 42.1 = 147$$

Compressive strength $p_c = 75\,\text{N/mm}^2$ from Table 27(c) of BS 5950: Part 1, corresponding to $\lambda = 147$ and $p_y = 345\,\text{N/mm}^2$.

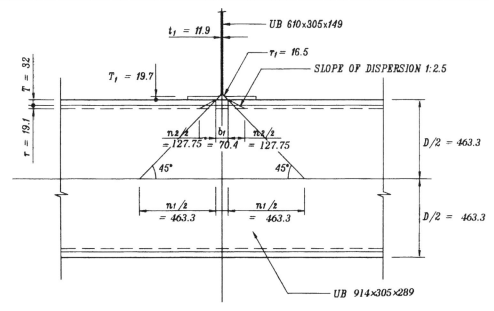

SK 7/35 Distribution of concentrated load from secondary beam through web of main beam.

Size of secondary beam = $610 \times 305 \times 149$ kg/m

$t_1 = 11.9$ mm; $T_1 = 19.7$ mm; $r_1 = 16.5$ mm
$b_1 = 2T_1 + t_1 + 1.16r_1 = (2 \times 19.7) + 11.9 + (1.16 \times 16.5) = 70.4$ mm
$n_1 = D = 926.6$ mm
$P_W = (b_1 + n_1)tp_c = (70.4 + 926.6) \times 19.6 \times 75 \times 10^{-3} = 1466$ kN
$P_W > F_v = 468$ kN (Satisfied)

Web stiffeners are not required.

Step 14 **Web bearing for unstiffened web**
$n_2 = 2 \times 2.5(T + r) = 5 \times (32 + 19.1) = 255.5$ mm
Local bearing capacity of the web P_{LW} is given by:

$$P_{LW} = (b_1 + n_2)tp_{yw}$$

$$= (70.4 + 255.5) \times 19.6 \times 345 \times 10^{-3}$$

$$= 2204\,\text{kN} > F_v \quad \text{(Satisfied)}$$

Bearing stiffeners are not required.

Steps 15 to 21 Not required.

Step 22 **Deflection due to imposed load**
Imposed loads at the serviceability limit state are as follows:

at $B = 180$ kN; at $C = 240$ kN; at $D = 60$ kN

This gives maximum deflection in span equal to 4.85 mm.
Maximum allowable deflection = $l/200 = 60$ mm (Satisfied)

7.4.4 Example 7.4: Design of a stiffened plate girder

The plate girder will be used in a sports stadium where the columns are spaced at 24 m centres. The secondary beams carrying the seats for the spectators are assumed to be resting on the top flange of the plate girders. The secondary beams are at 6 m centres. The girder is

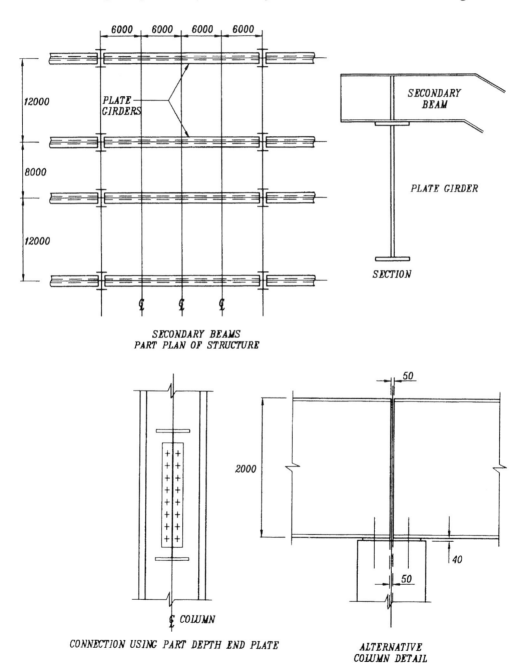

SK 7/36 Plan, section and connection detail of the plate girder structural arrangement.

connected to the web of the columns or, alternatively, placed on a capping plate on top of the column.

Step 1 **Determine design strength**
Select steel = Grade 50.

For thickness of material up to 16 mm $p_y = 355 \text{ N/mm}^2$
For thickness of material up to 40 mm $p_y = 345 \text{ N/mm}^2$

Step 2 **Carry out analysis**
The plate girder is assumed to be simply supported at the columns. The span of the plate girder is taken as 24 m, centre-to-centre of bearings. The self-weight of the girder is assumed to be 6 kN/m.

Factored ultimate self-weight of girder = $1.4 \times 6 = 8.4 \text{ kN/m}$
Concentrated dead load from the secondary beams
 = 300 kN per beam
Concentrated imposed load from the secondary beams
 = 300 kN per beam
Factored ultimate concentrated load from secondary beams
 = $(1.4 \times 360) + (1.6 \times 300) = 984 \text{ kN}$ per beam
Ultimate end reaction of the plate girder
 = $(8.4 \times 12) + (984 \times 1.5) = 1576.8 \text{ kN}$
Ultimate bending moment at B (see SK 7/37)

$$= (1576.8 \times 6) - \left(8.4 \times \frac{6^2}{2}\right)$$

$$= 9309.6 \text{ kNm}$$

Ultimate bending moment at C

$$= (1576.8 \times 12) - (984 \times 6) - \left(8.4 \times \frac{12^2}{2}\right)$$

$$= 12\,412.8 \text{ kNm}$$

Service bending moment at $C = 8352 \text{ kNm}$

Step 3 Draw bending moment and shear force diagrams

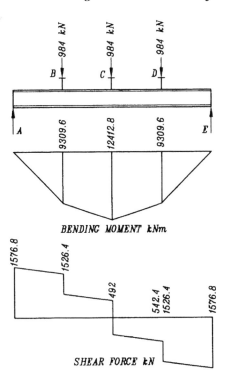

SK 7/37 Bending moment and shear force diagrams of plate girder.

Step 4 Choice of section and section classification

Select depth of girder $= D = \dfrac{L}{12} = \dfrac{24\,000}{12} = 2000\,\text{mm}$

Select thickness of web $= t = \dfrac{D}{150} = \dfrac{2000}{150} \approx 12\,\text{mm}$

Area of flange required $= A_f = \dfrac{M_{max}}{Dp_y} = \dfrac{12\,412.8 \times 10^6}{2000 \times 345} = 17\,990\,\text{mm}^2$

Select equal flange size $= 500 \times 40 = 20\,000\,\text{mm}^2$ each flange
Section geometry: $D = 2000\,\text{mm}$; $d = 1920\,\text{mm}$; $t = 12\,\text{mm}$; $T = 40\,\text{mm}$;
  $B = 500\,\text{mm}$; $b = 0.5(B - T) = 244\,\text{mm}$ (see SK 7/38)
Design strength for flange $= p_{yf} = 345\,\text{N/mm}^2$
Design strength for web $= p_{yw} = 355\,\text{N/mm}^2$

$$\varepsilon_f = \sqrt{\dfrac{275}{p_{yf}}} = 0.89 \quad \text{and} \quad \varepsilon_w = \sqrt{\dfrac{275}{p_{yw}}} = 0.88$$

$$\dfrac{b}{T} = \dfrac{244}{40} = 6.1 < 7.5\varepsilon_f = 6.675$$

$$\dfrac{d}{t} = \dfrac{1920}{12} = 160 > 120\varepsilon_w = 105.6$$

Flanges are classified as plastic but the web is classed as slender.

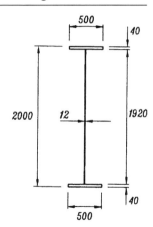

SK 7/38 Section of the plate girder selected.

Step 5 *Determine section properties*
Area of section $= A = (1920 \times 12) + (2 \times 500 \times 40) = 63\,040\,\text{mm}^2$
Shear area $A_v = td = 12 \times 1920 = 23\,040\,\text{mm}^2$

$$I_x = \left(\frac{1}{12} \times 12 \times 1920^3\right) + \left(2 \times \frac{1}{12} \times 500 \times 40^3\right)$$
$$+ (2 \times 500 \times 40 \times 980^2) = 4.55 \times 10^{10}\,\text{mm}^4$$

$$I_y = \left(\frac{1}{12} \times 1920 \times 12^3\right) + \left(2 \times \frac{1}{12} \times 40 \times 500^3\right) = 8.34 \times 10^8\,\text{mm}^4$$

$$r_x = \sqrt{\frac{I_x}{A}} = \sqrt{\frac{4.55 \times 10^{10}}{63\,040}} = 850\,\text{mm}$$

$$r_y = \sqrt{\frac{I_y}{A}} = \sqrt{\frac{8.34 \times 10^8}{63\,040}} = 115\,\text{mm}$$

$$Z_x = \frac{4.55 \times 10^{10}}{1000} = 4.55 \times 10^7\,\text{mm}^3$$

$$Z_y = \frac{8.34 \times 10^8}{250} = 3.336 \times 10^6\,\text{mm}^3$$

$$S_x = (500 \times 40 \times 1960) + (12 \times 1920^2 \div 4) = 5.026 \times 10^7\,\text{mm}^3$$

$$J = \frac{1}{3}(t_1^3 b_1 + t_2^3 b_2 + t_w^3 h_w) = \frac{1}{3}[(40^3 \times 500 \times 2) + (12^3 \times 1920)]$$

$$= 2.244 \times 10^7\,\text{mm}^4$$

$N = 0.5$ for equal flanged section

$$H = \frac{h_s^2 t_1 t_2 b_1^3 b_2^3}{12(t_1 b_1^3 + t_2 b_2^3)} = \frac{1960^2 \times 40 \times 40 \times 500^3 \times 500^3}{12 \times [(40 \times 500^3) + (40 \times 500^3)]} = 8 \times 10^{14}$$

$$\gamma = 1 - \frac{I_y}{I_x} = 1 - \frac{0.834 \times 10^9}{4.55 \times 10^{10}} = 0.982$$

$$u = \left(\frac{I_y S_x^2 \gamma}{A^2 H}\right)^{0.25} = 0.898$$

$$x = 1.132 \left(\frac{AH}{I_y J}\right)^{0.5} = 58.76$$

Step 6 Not required.

Step 7 **Check minimum web thickness**

$$\frac{d}{250} = \frac{1920}{250} = 7.68\,\text{m} \quad \text{and} \quad \frac{d}{250} \times \left(\frac{p_{yf}}{345}\right) = 7.68\,\text{mm}$$

$$t = 12\,\text{mm} > \frac{d}{250} \quad \text{and} \quad \frac{d}{250} \times \left(\frac{p_{yf}}{345}\right) \quad \text{(Satisfied)}$$

Step 8 **Check shear buckling of slender web**

The plate girder can be designed to resist shear by using four different methods.

Method 1: Without using tension-field action

In sections AB and DE of the plate girder the maximum shear force is 1576.8 kN (see SK 7/37).

$$f_v = \frac{F_v}{dt} = \frac{1576.8 \times 10^3}{1920 \times 12} = 68.4\,\text{N/mm}^2$$

$$\frac{d}{t} = 160, \quad p_{yw} = 355\,\text{N/mm}^2, \quad \text{select} \quad \frac{a}{d} = 0.95$$

This gives $q_{cr} = 73\,\text{N/mm}^2$ from Table 21(d) of BS 5950: Part 1, which is greater than $f_v = 68.4\,\text{N/mm}^2$.

For this design condition to be satisfied, use stiffeners at spacing $a = 0.95d = 0.95 \times 1920 = 1824\,\text{mm}$.

Use stiffeners at 1500 mm centres in sections AB and DE (SK 7/39).

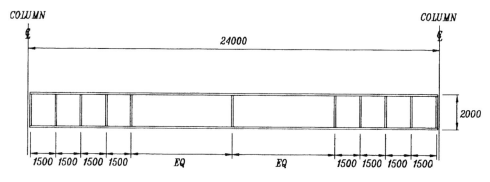

SK 7/39 Stiffener arrangement by first method of checking shear buckling.

In sections BC and CD of the plate girder, the maximum ultimate shear force is 542.4 kN.

Shear stress $f_v = \dfrac{F_v}{dt} = \dfrac{542.4 \times 10^3}{1920 \times 12} = 23.54 \, \text{N/mm}^2$

Assume $a/d = \infty$ which means that there are no stiffeners in sections BC and CD. From Table 21(d) of BS 5950: Part 1 $q_{cr} = 39 \, \text{N/mm}^2$, corresponding to $p_y = 355 \, \text{N/mm}^2$ and $a/d = \infty$. Because q_{cr} is greater than f_v, no stiffeners are required (see SK 7/40).

Method 2: Using partial tension-field action of the web
The first panel nearest to the bearing is designed without using tension-field action and the rest of the plate girder web is designed using tension-field action.

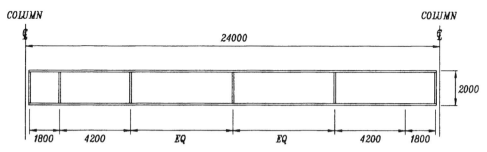

SK 7/40 Stiffener arrangement by second method of checking shear buckling.

It is assumed that under each secondary beam there is a stiffener. It has been found in the first method that a spacing of 1800 mm to the first stiffener from the bearing is adequate for web buckling without using tension-field action. The next stiffener is under the secondary beam 6 m from the bearing. The spacing between the first and the second stiffener in the web is $6 - 1.8 = 4.2 \, \text{m}$. This gives:

$\dfrac{a}{d} = \dfrac{4200}{1920} = 2.2$

The maximum ultimate shear force at 1800 mm from the support is:

$1576.8 - (1.8 \times 8.4) = 1561.7 \, \text{kN}$

$\therefore \quad f_v = \dfrac{1561.7 \times 10^3}{1920 \times 12} = 67.8 \, \text{N/mm}^2$

From Table 22(d) of BS 5950: Part 1 $q_b = 80 \, \text{N/mm}^2$, corresponding to $a/d = 2.2$, $p_y = 355 \, \text{N/mm}^2$ and $d/t = 160$. The second panel is adequate for shear buckling using tension-field action because $q_b > f_v$.

Method 3: Using full tension-field action and single bearing stiffeners
Assume stiffeners are at 3.0 m centres in sections AB and DE (see SK 7/41).

$\dfrac{a}{d} = \dfrac{3000}{1920} = 1.6$

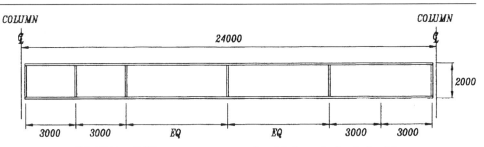

SK 7/41 Stiffener arrangement by third method of checking shear buckling.

From Table 22(d) of BS 5950: Part 1 $q_b = 94\,\text{N/mm}^2$, corresponding to $a/d = 1.6$, $p_y = 355\,\text{N/mm}^2$ and $d/t = 160$. The stiffener arrangement is adequate using tension-field action because q_b is greater than $f_v = 68.4\,\text{N/mm}^2$ (see method 1).

Method 4: Using full tension-field action and double stiffeners
See Step 18 for the design of stiffeners using this method.

Steps 9 and 10 Not required.

Step 11 *Section moment capacity with slender web where* **d/t > 63ε**
Plastic modulus of flanges only
$= S_f = A_f h_f = 500 \times 40 \times 1960 = 3.92 \times 10^7\,\text{mm}^3$
The flanges are classified as plastic (see Step 4).

$$M_c = p_y S_f = 345 \times 3.92 \times 10^7 \times 10^{-6}$$
$$= 13\,524\,\text{kNm} > 12\,412.8\,\text{kNm (Step 2)}$$

$$k = \frac{W_{\text{Ultimate}}}{W_{\text{Service}}} = \frac{12\,412.8}{8352} = 1.49 \text{ (Step 2)}$$

$Z_f =$ elastic modulus of flanges only

$$= \frac{1}{1000}\left[\left(\frac{1}{12} \times 500 \times 40^3\right) + (2 \times 500 \times 40 \times 980^2)\right]$$

$$= 3.842 \times 10^7\,\text{mm}^3$$

$$kp_y Z_f = 1.49 \times 3.842 \times 10^7 \times 345 \times 10^{-6}$$
$$= 19\,750\,\text{kNm} > 12\,412.8\,\text{kNm}$$

Section moment capacity is adequate.

Step 12 *Member moment capacity (lateral torsional buckling)*
Assume normal loading on the flange and that the secondary beams provide lateral restraint to the compression flange. At the support with the alternative detail of connection to the column, as shown in

SK 7/36, the compression flange is assumed to be laterally unrestrained, and both the flanges are free to rotate on plan. The girder is not fully restrained against torsion at the support.

Effective length for this condition at the support is given by:

$1.0L + 2D = 6000 + (2 \times 2000) = 10\,000\,\text{mm}$

Effective length of the girder between intermediate restraint
$= 1.0L = 6000\,\text{mm}$
Average effective length for sections AB and DE
$= \frac{1}{2}(10\,000 + 6000) = 8000\,\text{mm}$
Assume that the girder is not loaded between restraints, ignoring the self-weight of the girder, and use the equivalent moment method.

For sections AB and DE

$$n = 1.0; \quad \beta = \frac{0}{9309.6} = 0; \quad \text{and}$$

$m = 0.57$ (from Table 18 of BS 5950: Part 1)

$$\lambda = \frac{L_E}{r_y} = \frac{8000}{115} = 70 \text{ (see Step 5)}$$

$$\frac{\lambda}{x} = \frac{70}{58.76} = 1.2 \text{ (see Step 5)}$$

$N = 0.5$ (Step 5)
$v = 0.98$ (from Table 14 of BS 5950: Part 1)
$u = 0.898$ (from Step 5)
$\lambda_{LT} = nuv\lambda = 1 \times 0.898 \times 0.98 \times 70 = 62 \quad \therefore p_b = 204\,\text{N/mm}^2$
(from Table 12 of BS 5950: Part 1)
$S_x = 5.026 \times 10^7\,\text{mm}^3$ (from Step 5)
$M_b = S_x p_b = 5.026 \times 10^7 \times 204 \times 10^{-6} = 10\,253\,\text{kNm}$
$\bar{M} = mM = 0.57 \times 9309.6 = 5306\,\text{kNm} < M_b$ (Satisfied)

For sections BC and CD

$$\beta = \frac{9309.6}{12\,412.8} = 0.75 \text{ (see Step 3)}$$

$m = 0.875$ (from Table 18 of BS 5950: Part 1)

$$\lambda = \frac{L_E}{r_y} = \frac{6000}{115} = 52; \quad \frac{\lambda}{x} = \frac{52}{58.76} = 0.89$$

$N = 0.5$ (which gives $v = 0.99$ from Table 14 of BS 5950: Part 1)
$u = 0.898$ (from Step 5)
$\lambda_{LT} = nuv\lambda = 1 \times 0.898 \times 0.99 \times 52 = 46$
$\therefore p_b = 266\,\text{N/mm}^2$ (from Table 12 of BS 5950: Part 1)
$M_b = S_x p_b = 5.026 \times 10^7 \times 266 \times 10^{-6} = 13\,369\,\text{kNm}$
$\bar{M} = mM = 0.875 \times 12\,412.8 = 10\,861.2\,\text{kNm} < M_b$ (Satisfied)

The plate girder has adequate buckling resistance.

Step 13 Check web buckling of unstiffened web
Not required: see Step 17.

Step 14 Check web bearing of unstiffened web
Not required: see Step 17.

Step 15 Web check between stiffeners
P = concentrated load between stiffeners = 0
W = distributed load over shorter length than the panel length $a=0$
w = distributed load per unit length over the whole length of the panel
$a = 8.4\,\text{kN/m}$
$a = 6\,\text{m}$ in sections BC and CD (see Step 8)

$$\frac{a}{d} \approx 3$$

$$f_{ed} = \left[\frac{P+W}{d} + w\right]\frac{1}{t} = \frac{w}{t} = \frac{8.4\,\text{N/mm}}{12} = 0.7\,\text{N/mm}^2$$

$$p_{ed} = \left[1 + \frac{2}{(a/d)^2}\right]\frac{E}{(d/t)^2} = \left[1 + \frac{2}{3^2}\right] \times \frac{205 \times 10^3}{160^2}$$

$$= 9.79\,\text{N/mm}^2 > f_{ed} \text{ (Satisfied)}$$

Step 16 Intermediate transverse web stiffeners
The maximum allowable outstand of stiffener is given by:

$$19 t_s \sqrt{\frac{275}{p_y}} = 19 \times 12 \times \sqrt{\frac{275}{355}} = 200\,\text{mm}$$

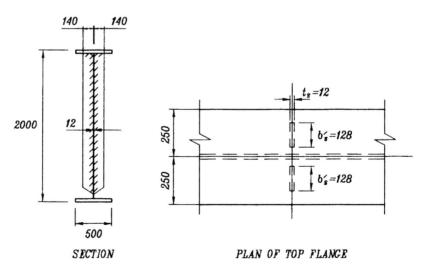

SK 7/42 Details of intermediate stiffeners.

Design of Beams

The maximum value of outstand to be considered in the design b_s is given by:

$$b_s = 13t_s\sqrt{\frac{275}{p_y}} = 137\,\text{mm}$$

Select 12 mm thick × 140 mm stiffeners (see SK 7/42).

Design of intermediate stiffener with load
Load on the intermediate stiffener from secondary beam $= F_x = 984\,\text{kN}$
Effective area of stiffener in contact with the flange allowing for 12 mm notch:

$$2b'_s t_s = 2 \times (140 - 12) \times 12 = 3072\,\text{mm}^2$$

where $b'_s = 140 - 12 = 128\,\text{mm}$
Required minimum area of contact for web bearing:

$$\frac{0.8F_x}{p_y} = \frac{0.8 \times 984\,000}{355} = 2217\,\text{mm}^2 < 3072\,\text{mm}^2$$

The web bearing condition is satisfied.

The applied load on the stiffener from the secondary beam is central on the web without any eccentricity and there is no lateral load on the stiffener at the level of the compression flange. Therefore:

$F = 0\,\text{kN}$; $M_s = 0\,\text{kNm}$
$b_s =$ outstand of stiffener $= 140\,\text{mm}$
$t_s =$ thickness of stiffener $= 12\,\text{mm}$
$t\ =$ thickness of web $= 12\,\text{mm}$
$I_s =$ moment of inertia of the stiffeners about the centre-line of the web

$$= \frac{t_s}{12}[(2b_s + t)^3 - t^3] = \frac{12}{12} \times \{[(2 \times 140) + 12]^3 - 12^3\}$$

$$= 24.9 \times 10^6\,\text{mm}^4$$

$0.75dt^3 = 0.75 \times 1920 \times 12^3 = 2.49 \times 10^6\,\text{mm}^4 < I_s$ (Satisfied)

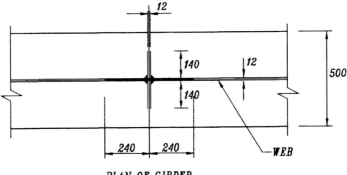

SK 7/43 Equivalent section of stiffener against buckling.

Take $b_s = 137$ mm instead of the actual 140 mm for the capacity calculations because of outstand limitations.

$$A_{sc} = 40t^2 + 2b_s t_s = (40 \times 12^2) + (2 \times 137 \times 12) = 9048 \text{ mm}^2$$

$$I_{sc} = \frac{1}{12} t_s(2b_s + t)^3 + \frac{1}{6} t^3 \left(20t - \frac{t_s}{2}\right) = 23.39 \times 10^6 \text{ mm}^4$$

$$r_{sc} = \sqrt{\frac{I_{sc}}{A_{sc}}} = \sqrt{\frac{23.39 \times 10^6}{9048}} = 50.84 \text{ mm}$$

The buckling resistance of an unloaded intermediate stiffener is given as follows:

Stiffener force $F_q = V - V_s \leq P_q$
$V = 1526.4$ kN (see Step 3) (applied maximum shear force)
$V_s = q_{cr} dt = 42 \times 1920 \times 12 \times 10^{-3} = 967.7$ kN (shear buckling resistance without tension-field action)
q_{cr} is obtained from Table 21(d) of BS 5950: Part 1, corresponding to $d/t = 160$, $a/d = 3$ and $p_y = 340$ N/mm^2
The slenderness ratio of an intermediate stiffener =

$$\lambda_q = \frac{0.7L}{r_{sc}} = \frac{0.7 \times 2000}{50.8} = 27.5$$

$$p_y = 355 - 20 = 335 \text{ N/mm}^2$$

Compressive strength $p_c = 312$ N/mm^2 (from Table 27(c) of BS 5950: Part 1)

$$P_q = p_c A_{sc} = 312 \times 9048 \times 10^{-3} = 2823 \text{ kN}$$

$$\therefore \quad V - V_s = 1526.4 - 967.7 = 558.7 < P_q \text{ (Satisfied)}$$

The buckling resistance of a loaded intermediate stiffener is given as follows.

Assume that the flange is restrained against rotation by the secondary beams.

$$\therefore \quad \lambda_x = \frac{0.7L_s}{r_{sc}} = \frac{0.7 \times 2000}{50.8} = 27.5$$

and $p_c = 312$ N/mm^2 (as before)

This gives $P_x = P_q = 2823$ kN

F_x = external load on the stiffener = 984 kN > F_q

The following condition has to be satisfied:

$$\frac{F_q - F_x}{P_q} + \frac{F_x}{P_x} + \frac{M_s}{M_{ys}} \leq 1 \quad (F_q - F_x = 0 \text{ and } M_s = 0)$$

$$\therefore \quad \frac{F_x}{P_x} = \frac{984}{2823} = 0.35 < 1 \quad \text{(Satisfied)}$$

Design of intermediate stiffener without load
Select 10×75 flats to be used as stiffeners.
The maximum allowable outstand is given by:

$$19t_s\sqrt{\frac{275}{p_y}} = 167\,\text{mm} > 75\,\text{mm}$$

The maximum outstand to be considered for strength is given by:

$$13t_s\sqrt{\frac{275}{p_y}} = 114\,\text{mm} > 75\,\text{mm}$$

$$I_s = \frac{t_s}{12}[(2b_s + t)^3 - t^3] = \frac{10}{12}[((2 \times 75) + 12)^3 - 12^3]$$

$$= 3.54 \times 10^6\,\text{mm}^4$$

$$0.75dt^3 = 0.75 \times 1920 \times 12^3 = 2.488 \times 10^6\,\text{mm}^4 < I_s\ \text{(Satisfied)}$$

$$A_{sc} = 40t^2 + 2b_s t_s = (40 \times 12^2) + (2 \times 75 \times 10) = 7260\,\text{mm}^2$$

$$I_{sc} = \frac{1}{12}t_s(2b_s + t)^3 + \frac{1}{6}t^3\left(20t - \frac{t_s}{2}\right) = 3.61 \times 10^6\,\text{mm}^4$$

$$r_{sc} = \sqrt{\frac{I_{sc}}{A_{sc}}} = 22.3\,\text{mm}$$

$$\lambda_q = \frac{0.7L_s}{r_{sc}} = \frac{0.7 \times 2000}{22.3} = 63$$

$$p_y = 355 - 20 = 335\,\text{N/mm}^2$$

$$p_{cq} = 227.5\,\text{N/mm}^2\ \text{(from Table 27(c) of BS 5950: Part 1)}$$

$$P_q = p_{cq}A_{sc} = 227.5 \times 7260 \times 10^{-3} = 1651.6\,\text{kN}$$

$$V_s = 967.7\,\text{kN and } V = 1551.6\,\text{kN}$$

$$F_q = V - V_s = 583.9\,\text{kN} < P_q\ \text{(Satisfied)}$$

Step 17 *Bearing stiffener without tension-field action*

Note: For a plate girder connected to the web of the column using flexible end plates or web cleats, the conventional bearing stiffener for the girder is not required. The alternative connection detail at the column shows that the plate girder rests on the bottom flange on a stiff bearing area (see SK 7/36).

Refer to Step 8 (Method 1).
Assume that the stiffeners in sections AB and DE are at 1500 mm centres.
Select 15 mm $\times$ 150 mm flats to be used as bearing stiffeners.
End reaction $= R_L = 1576.8$ kN
$t_s = 15$ mm; $b_s = 150$ mm; $t = 12$ mm; $a = 1500$ mm

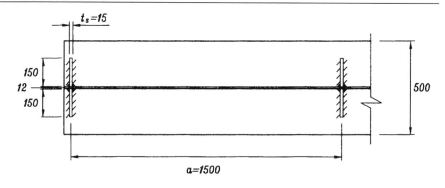

PLAN OF GIRDER

SK 7/44 Bearing stiffener arrangement of the plate girder.

A = area of stiffener in contact with the bottom flange
$= 2 \times 15 \times (150 - 12) = 4140 \, \text{mm}^2$ (allowing for 12 mm notch)

$$\frac{0.8 R_L}{p_y} = \frac{0.8 \times 1576.8 \times 10^3}{355} = 3553 \, \text{mm}^2 < A \text{ (Satisfied)}$$

M_s = moment due to eccentricity of end reaction $R_L = 0 \, \text{kNm}$
F = lateral load at the bearing = 0 kN

$$I_s = \frac{t_s}{12}[(2b_s + t)^3 - t^3] = 37.96 \times 10^6 \, \text{mm}^4$$

$a < \sqrt{2}d = 2715 \, \text{mm}$

$$\therefore \frac{1.5 d^3 t^3}{a^2} = \frac{1.5 \times 1920^3 \times 12^3}{1500^2} = 8.154 \times 10^6 \, \text{mm}^4 < I_s \text{ (Satisfied)}$$

$A_{sc} = 2 b_s t_s$
$= 4500 \, \text{mm}^2$ (ignoring the contribution of 20 times the thickness of web)

$$I_{sc} = \frac{1}{12} t_s (2b_s + t)^3$$

$= 38 \times 10^6 \, \text{mm}^4$ (ignoring the contribution of 20 times the thickness of web)

$$r_{sc} = \sqrt{\frac{I_{sc}}{A_{sc}}} = 91.90 \, \text{mm}$$

$$\lambda_x = \frac{k L_s}{r_{sc}} = \frac{0.7 \times 2000}{91.9} = 15$$

k is taken as 0.7 because the bottom flange is assumed to be restrained against rotation in the plane of the stiffeners by using holding-down bolts through capping plate.

$p_y = 355 - 20 = 335 \, \text{N/mm}^2$
$p_{cx} = 335 \, \text{N/mm}^2$ (from Table 27(c) of BS 5950: Part 1, corresponding to $\lambda = 15$)
$P_x = A_{sc} p_{cx} = 4500 \times 335 \times 10^{-3} = 1507.5 \, \text{kN} < 1576.8 = R_L$

(May be allowed)

Design of Beams 297

The buckling resistance P_x is slightly less than the applied reaction because, conservatively, the contribution of the web has been ignored.

Check direct bearing
The stiff bearing length is obtained by 45° dispersion from the edge of the column web to the flange of the plate girder through a 40 mm thick capping plate (see SK 7/36).
Assume width of stiff bearing $= b_1 = 40$ mm

$$n_2 = 2.5T = 2.5 \times 40 = 100 \text{ mm} \quad (T \text{ is thickness of bottom flange})$$

Web bearing resistance $= P_{LW} = (b_1 + n_2)tp_{yw} = 596.4$ kN
The total bearing resistance including the bearing stiffener is given by:

$$P_{LW} + Ap_y = 596.4 + 4140 \times 355 \times 10^{-3}$$
$$= 2066.1 \text{ kN} > R_L \text{ (Satisfied)}$$

A = area of stiffener in contact with the bottom flange.

Step 18 Bearing stiffener using tension-field action

Note: The plate girder in this example is connected to the web of the column using flexible end plates or web cleats, or alternatively is placed on top of a column capping plate. To illustrate the method of design of bearing stiffeners, it is assumed that the plate girder is placed on a stiff bearing area, as shown in SK 7/36.

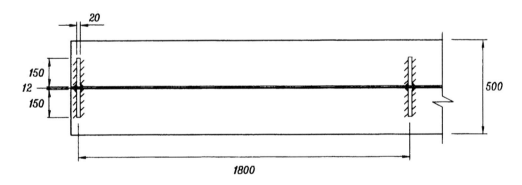

PLAN OF GIRDER

SK 7/45 Stiffener arrangement of the plate girder using tension-field action.

Refer to Step 8 (Method 2).
Place the first intermediate stiffener at 1800 mm from the centre-line of the bearing. The second web panel 4200 mm × 1920 mm is designed using tension-field action whereas the first panel is not.
End reaction $= R_L = 1576.8$ kN
Shear force at the first intermediate stiffener $= F_v = 1561.7$ kN

$$\frac{a}{d} = \frac{4200}{1920} = 2.19 \text{ for the first panel with tension-field action}$$

$d/t = 160$ and $p_y = 355\,\text{N/mm}^2$ which gives:

$q_{cr} = 45\,\text{N/mm}^2$ (from Table 21(d) of BS 5950: Part 1)

$q_b = 80\,\text{N/mm}^2$ (from Table 22(d) of BS 5950: Part 1)

Shear stress $= f_v = \dfrac{F_v}{dt} = \dfrac{1561.7 \times 10^3}{1920 \times 12} = 67.8\,\text{N/mm}^2 < q_b$

$$H_q = 0.75 dt p_y \left[1 - \dfrac{q_{cr}}{0.6 p_y}\right]^{\frac{1}{2}} \left[\dfrac{(f_v - q_{cr})}{(q_b - q_{cr})}\right]$$

$$= 0.75 \times 1920 \times 12 \times 355 \times \left[1 - \dfrac{45}{0.6 \times 355}\right]^{\frac{1}{2}} \times \left(\dfrac{67.8 - 45}{80 - 45}\right)$$

$$= 3549\,\text{kN}$$

$$R_{tf} = \dfrac{H_q}{2} = 1775\,\text{kN}$$

$$M_{tf} = \dfrac{H_q d}{10} = \dfrac{3549 \times 1.92}{10} = 681.4\,\text{kNm}$$

$a = 1800\,\text{mm}$ for the end panel

The shear capacity of this end panel subjected to a horizontal shear force equal to R_{tf} is calculated as follows.

Determine a/t ratio of the vertical beam: $1800/12 = 150 > 63\varepsilon$.

The web is classed as slender for this vertical beam. The spacing of the flanges of the plate girder is used to find the stiffener spacing ratio a/d for the vertical beam: $1920/1800 = 1.0$ approximately.

Therefore, from Table 21d of BS 5950: Part 1 the critical shear strength $q_{cr} = 78\,\text{N/mm}^2$.

The shear capacity

$$= V_{cr} = q_{cr} dt = 78 \times 1800 \times 12 \times 10^{-3} = 1685\,\text{kN} < R_{tf}$$

The 1800 mm wide end panel has not got adequate horizontal shear capacity. Use a horizontal stiffener at the centre of the vertical beam and no further calculations are necessary. Use two 10×75 flats as horizontal stiffeners based on the calculations in Step 16 in the section covering the design of an intermediate stiffener without load.

Additional vertical load on the bearing stiffener due to tension-field action is given by:

$$\dfrac{M_{tf}}{a} = \dfrac{681.4}{1.8} = 378.6\,\text{kN}$$

Total vertical load on the bearing stiffener

$$= F_x = R_L + 378.6 = 1955.4\,\text{kN}$$

Follow method in Step 17 to design the bearing stiffener.

Select 20 × 150 flats as bearing stiffeners.

A = area in contact with the flange = $2 \times 20 \times (150 - 12)$
= 5520 mm² (allowing for 12 mm notch)

Contact area required to satisfy bearing stress:

$$\frac{0.8 F_x}{p_y} = \frac{0.8 \times 1955.4 \times 10^3}{355}$$

$$= 4406.5 \text{ mm}^2 < A \text{ (Satisfied)}$$

$t_s = 20$ mm; $b_s = 150$ mm; $t = 12$ mm

$$I_s = \frac{t_s}{12}[(2b_s + t)^3 - t^3] = 50.62 \times 10^6 \text{ mm}^4$$

$a = 1800$ mm $< \sqrt{2}d = 2715$ mm

$$\therefore \quad I_s > \frac{1.5 d^3 t^3}{a^2} = \frac{1.5 \times 1920^3 \times 12^3}{1800^2} = 5.66 \times 10^6 \text{ mm}^4 \text{ (Satisfied)}$$

$A_{sc} = 2 b_s t_s = 6000$ mm²
  (ignoring contribution of 20 times the thickness of web)

$$I_{sc} = \frac{1}{12} t_s (2 b_s + t)^3 = 50.62 \times 10^6 \text{ mm}^4$$
  (ignoring contribution of web)

The maximum permitted outstand of stiffener b is given by:

$$13 t_s \sqrt{\frac{275}{p_y}} = 172 \text{ mm} > 150$$

$$r_{sc} = \sqrt{\frac{I_{sc}}{A_{sc}}} = 91.85 \text{ mm}$$

$$\lambda_x = \frac{kL_s}{r_{sc}} = \frac{0.7 \times 2000}{91.85} = 15$$

$p_y = 355 - 20 = 335$ N/mm²

From Table 27(c) of BS 5950: Part 1, $p_{cx} = 335$ N/mm²

$P_x = A_{sc} p_{cx} = 6000 \times 335 \times 10^{-3} = 2010$ kN $> F_x = 1955.4$ kN
(Satisfied)

Check bearing
$P_{LW} + A p_y = 596.4 + (5520 \times 355 \times 10^{-3}) = 2556$ kN > 1955.4 kN
(Satisfied)

See Step 17 for P_{LW}.

Step 19 Torsion stiffener

Note: The girder is not fully restrained torsionally at the support where it rests on the column capping plate. See Step 12 where the lateral torsional buckling has been checked without torsional restraint at the support. If part depth flexible end plate or web cleats are used to connect the girder to the column, then full end torsional restraint may be assumed.

Step 20 *Connection of stiffeners*
Maximum vertical load on the stiffener $= 1955.4\,\text{kN} = F$ (see Step 18)
$b_s = 150\,\text{mm}$; $t_s = 20\,\text{mm}$; $t = 12\,\text{mm}$
$A_{sc} = 6000\,\text{mm}^2$ (see Step 18)

$$w_s = \frac{t^2}{5b_s} + \frac{Fb_s t_s}{A_{sc} d} = \frac{12^2}{5 \times 150} + \frac{1955.4 \times 150 \times 20}{6000 \times 1920} = 0.70\,\text{kN/mm}$$

or $$w_s = \frac{b_s t_s p_y}{L_s} = \frac{150 \times 20 \times 355}{2000} = 0.53\,\text{kN/mm}$$

Two runs of 6 mm fillet weld are required.

Step 21 Not required.

Step 22 *Deflection due to imposed load*
Imposed concentrated load at the service limit from the secondary beams $= P = 300\,\text{kN}$
Moment of inertia $= I = 4.55 \times 10^{10}\,\text{mm}^4$ (see Step 4)

$$\text{Deflection} = \frac{19PL^3}{384EI} = \frac{19 \times 300 \times 10^3 \times 24\,000^3}{384 \times 205 \times 10^3 \times 4.55 \times 10^{10}}$$

$$= 22\,\text{mm} < \frac{l}{200} = 120\,\text{mm}$$

Step 23 *Check for local stresses*
Not required.

Step 24 *Connection of flange to web*
$A_f = 40 \times 500 = 20\,000\,\text{mm}^2$
$I_x = 4.55 \times 10^{10}\,\text{mm}^4$
$h_t = h_b = 980\,\text{mm}$
$Q_{xt} = Q_{xb} = A_f h_t = 20\,000 \times 980 = 196 \times 10^5\,\text{mm}^3$
$V_{max} = 1576.8\,\text{kN}$

Shear flow per unit length

$$= v_{xt} = v_{xb} = \frac{VQ_x}{I_x} = \frac{1576.8 \times 196 \times 10^5}{4.55 \times 10^{10}} = 0.68\,\text{kN/mm}$$

Use two continuous runs of 6 mm fillet weld $= 1.34\,\text{kN/mm}$

7.4.5 Example 7.5: Design of a gantry girder for an electric overhead travelling crane

Basic data
Heavy duty overhead travelling crane selected for the maintenance bay in a workshop.

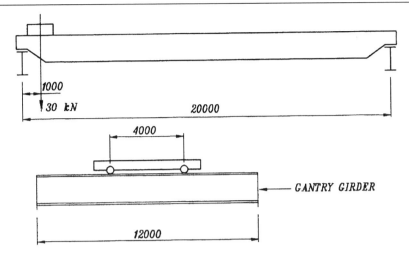

SK 7/46 General arrangement of the crane on the girders.

Simply supported gantry girder has a span of 12 m.
Span of crane bridge = L_c = 20 m
Capacity of crane = W_{cap} = 200 kN
Weight of crab = W_{cb} = 30 kN
Weight of crane bridge = W_c = 150 kN
Wheel spacing = a_w = 4 m
Minimum hook approach = a_h = 1 m
Total number of wheels = N = 4
Class of utilisation = U_6 = 1×10^6 cycles (from BS 2573: Part 1)
State of loading = Q_3 (from Table 2 of BS 2573: Part 1)
Nominal load spectrum factor = K_p = 0.8 (from Table 2 of BS 2573: Part 1)
Duty factor = 0.95 (from Table 4(a) of BS 2573: Part 1)
Impact factor = I = 1.3 (from Table 4(a) of BS 2573: Part 1)

The maximum unfactored static wheel load W_{us} is given by:

$$W_{us} = \frac{2}{N}\left[\frac{W_c}{2} + (W_{cb} + W_{cap})\left(\frac{L_c - a_h}{L_c}\right)\right]$$

$$= \frac{2}{4} \times \left[\frac{150}{2} + (30 + 200) \times \frac{(20-1)}{20}\right] = 146.75 \text{ kN}$$

The vertical wheel load from the crane CLV is given by:

$$CLV = IW_{us} = 1.3 \times 146.75 = 191 \text{ kN}$$

The horizontal transverse surge load from the crane CLH_s is given by:

$$CLH_s = \frac{0.1(W_{cb} + W_{cap})}{N} = 5.75 \text{ kN per wheel}$$

The horizontal transverse wheel load due to crabbing CLH_c is given by:

$$CLH_c = \frac{L_c W_{us}}{40 a_w} \geq \frac{W_{us}}{20} = \frac{20 \times 146.75}{40 \times 4} \geq \frac{146.75}{20}$$

$$= 18.35 \text{ kN per wheel}$$

Step 1 **Determine design strength**

Selected design grade of steel = Grade 43. This selection is made simply for the purpose of illustrating the method of analysis and design and is not based on considerations of overall economy of construction. Generally, the higher grades of steel will provide a more cost-effective solution.

$p_y = 275 \text{ N/mm}^2$ up to 16 mm thickness of material
$\quad = 265 \text{ N/mm}^2$ up to 40 mm thickness of material
$\quad = 255 \text{ N/mm}^2$ up to 63 mm thickness of material

Step 2 **Carry out analysis**

Assume self-weight of girder = 5 kN/m
By inspection only three load cases have to be examined as stated below (refer to Table 4.1):

Load case 27: $1.4DL + 1.4CLV + 1.4CLH$
Load case 31: $1.4DL + 1.6CLV$
Load case 33: $1.4DL + 1.6CLH$

Load case 27

Simply supported girder of span = 12 m
Ultimate dead load = $1.4 \times 5 = 7.0$ kN/m
Total ultimate dead load on span = $W_G = 7 \times 12 = 84$ kN
Ultimate vertical crane wheel load = $W_V = 1.4 \times 191 = 267.4$ kN
The reaction at end A is given by (see SK 7/47):

$$R_A = \frac{2W_V}{L}\left[\frac{L}{2} - \frac{a_w}{4}\right] + \frac{W_G}{2} = \frac{2 \times 267.4}{12} \times \left[\frac{12}{2} - \frac{4}{4}\right] + \frac{84}{2}$$

$$= 264.8 \text{ kN}$$

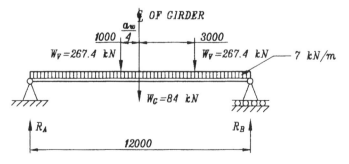

SK 7/47 Location and direction of vertical and horizontal loads on crane girder to produce maximum bending moment in the girder.

The reaction at end B is given by:

$$R_B = (2 \times 267.4) + (84 - 264.8) = 354.0 \text{ kN}$$

The maximum ultimate limit moment due to vertical loading is given by (see SK 7/47):

$$M_V = R_A \left[\frac{L}{2} - \frac{a_w}{4} \right] - \frac{W_G}{2L} \left[\frac{L}{2} - \frac{a_w}{4} \right]^2$$

$$= 264.8 \times \left[\frac{12}{2} - \frac{4}{4} \right] - \frac{84}{2 \times 12} \times \left[\frac{12}{2} - \frac{4}{4} \right]^2 = 1236.5 \text{ kNm}$$

Service limit moment $= 1236.5/1.4 = 883.2 \text{ kNm}$.

The maximum shear force due to ultimate vertical loading is given by:

$$F_{v,\max} = W_V + W_V \left[\frac{L - a_w}{L} \right] + \frac{W_G}{2}$$

$$= 267.4 + \left(267.4 \times \left[\frac{12 - 4}{12} \right] \right) + \frac{84}{2}$$

$$= 487.7 \text{ kN}$$

Ultimate horizontal crane wheel load due to surge $= W_{HS}$

$$W_{HS} = 1.4 C L H_S = 1.4 \times 5.75 = 8.05 \text{ kN}$$

Horizontal end reaction at A is given by:

$$R_A = \frac{2 W_{HS}}{L} \left[\frac{L}{2} - \frac{a_w}{4} \right] = \frac{2 \times 8.05}{12} \times \left[\frac{12}{2} - \frac{4}{4} \right] = 6.708 \text{ kN}$$

$$R_B = (2 \times 8.05) - 6.708 = 9.392 \text{ kN}$$

Maximum horizontal bending moment due to horizontal surge load is given by:

$$M_{HS} = R_A \left[\frac{L}{2} - \frac{a_w}{4} \right] = 33.54 \text{ kNm}$$

Ultimate horizontal crane wheel load due to crabbing $= W_{HC}$

$$W_{HC} = 1.4 C L H_C = 1.4 \times 18.35 = 25.69 \text{ kN}$$

$$R_A = \frac{W_{HC} a_w}{L} = \frac{25.69 \times 4}{12} = 8.56 \text{ kN}$$

Maximum horizontal bending moment due to crabbing is given by:

$$M_{HC} = R_A (L - a_w) = 8.56 \times (12 - 4) = 68.5 \text{ kNm}$$

Load case 31
The analysis is similar to load case 27.

$W_V = 1.6 CLV = 1.6 \times 191 = 305.6 \, \text{kN}$
$R_A = 296.7 \, \text{kN}$
$R_B = 398.5 \, \text{kN}$
$M_V = 1396 \, \text{kNm}$
$F_{v,\text{max}} = 551.3 \, \text{kN}$

Load case 33
$W_{HC} = 1.6 CLH_C = 1.6 \times 18.35 = 29.36 \, \text{kN}$
$R_A = 9.79 \, \text{kN}$
$M_{HC} = 78.29 \, \text{kNm}$

Step 3 *Draw bending moment and shear force diagram*
May be omitted.

Step 4 *Section classification*
Maximum moment due to vertical loading $= 1396 \, \text{kNm}$
Plastic modulus required for $p_y = 275 \, \text{N/mm}^2$:

$$\frac{1396 \times 10^6}{275} = 5.076 \times 10^6 \, \text{mm}^3$$

From steel section tables it is observed that a minimum section size required will be $762 \times 267 \times 147 \, \text{kg/m}$ UB. This section has a minimum radius of gyration equal to 5.39 cm and torsional index equal to 45.1. Using this radius of gyration, the slenderness ratio for the effective length of 12 m is 223. From Table 19 of BS 5950: Part 1, the bending strength p_b is $60 \, \text{N/mm}^2$ for a λ of 223 and a torsional index of 45.

Assume bending strength equal to $100 \, \text{N/mm}^2$ because the bigger section size will have a bigger radius of gyration. With p_b equal to $100 \, \text{N/mm}^2$, the plastic modulus required is $13\,960 \, \text{mm}^3$. Select UB $914 \times 419 \times 343 \, \text{kg/m}$ with a plastic modulus of $15\,500 \, \text{mm}^3$.

The maximum horizontal bending moment is 78.29 kNm. Therefore, the section modulus required is:

$$\frac{M}{p_y} = \frac{78.29 \times 10^6}{275} = 284\,691 \, \text{mm}^3$$

Assume the width of the top plate is 450 mm and the thickness of plate is 10 mm. Therefore, the section modulus of this plate for horizontal bending is:

$$\frac{bt^2}{6} = \frac{10 \times 450^2}{6} = 337\,500 \, \text{mm}^3 > 284\,691 \, \text{mm}^3$$

Trial size: UB $914 \times 419 \times 343 \, kg/m$ with a welded top plate 450×10

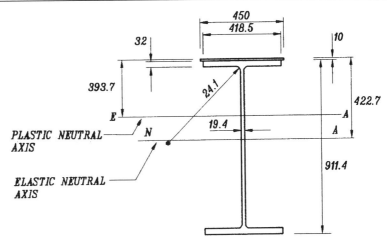

SK 7/48 Trial section of the gantry girder.

Section classification of UB
p_y is 265 N/mm² for the flange and the web because the thickness of material is greater than 16 mm but less than 40 mm.

$$\varepsilon = \sqrt{\frac{275}{p_y}} = 1.02$$

$$\frac{b}{T} = 6.54 \text{ (classification plastic)}$$

$$\frac{d}{t} = 41.2 < \frac{79\varepsilon}{0.4 + 0.6\alpha} \text{ (classification plastic)}$$

because α is less than 1.

Section classification of the compound section
Two checks are necessary for the top plate for classification purposes.

Check 1
Treat the top plate between points of attachment to the flange of the UB as an internal element of the compression flange.

The width of top plate between points of attachment to the UB, b, is 418.5 mm. So:

$$\frac{b}{T} = \frac{418.5}{10} = 41.85 > 28\varepsilon$$

Therefore, the top plate is classified as slender as per Table 7 of BS 5950: Part 1.

The strength reduction factor for slender elements as per Table 8 of BS 5950: Part 1 is given by:

$$\frac{21}{(b/T\varepsilon) - 7} = \frac{21}{(41.85/1) - 7} = 0.60$$

∴ Design compressive strength for the top plate
$= p_y = 275 \times 0.60 = 165 \text{ N/mm}^2$

Check 2
Projection of top plate beyond the UB, b, has to be checked for classification ($b = 15.75$ mm).

$$\frac{b}{T} = 1.575 < 7.5\varepsilon$$

The projected part of the top plate is classified as plastic. Check 1 dictates that the top plate should be treated as slender.
Depth of elastic neutral axis from the top surface of the compound girder $= Y_{NA}$
Sectional area of the UB $= 437$ cm^2
Sectional area of the top plate $= 45$ cm^2

$$Y_{NA} = \frac{\left(\frac{91.14}{2} + 1\right) \times 437 + (45 \times 0.5)}{(437 + 45)} = 42.27 \text{ cm}$$

$D_c =$ total depth of the compound section $= 91.14 + 1 = 92.14$ cm
Depth of plastic neutral axis (equivalent equal area axis) from the top surface of the compound section $= Y_{PL}$

$$\text{Equivalent area of top plate} = 45 \times \frac{p_{y,\text{top plate}}}{p_{y,\text{UB}}} = 45 \times \frac{165}{265} = 28 \text{ cm}^2$$

The shift of the plastic neutral axis from the centre of the UB towards the top flange of the compound girder is denoted x. The equivalent area of the top plate will be equally divided between the compression and the tension zone about the plastic neutral axis.

$$\therefore \quad x = \frac{\text{equivalent area of top plate}}{2 \times \text{thickness of web}} = \frac{28}{2 \times 1.94} = 7.2 \text{ cm}$$

$$Y_{PL} = \left(\frac{91.14}{2} + 1\right) - 7.2 = 39.37 \text{ cm}$$

Y_c is the distance from the plastic neutral axis to the edge of the web connected to the compression flange.

$$Y_c = Y_{PL} - T_{UB} - T_{\text{top plate}} - \text{root radius of UB}$$
$$= 39.37 - 3.2 - 1.0 - 2.41 = 32.76 \text{ cm}$$

$$a = \frac{2Y_c}{d} = \frac{2 \times 32.76}{79.91} = 0.82$$

Refer to Table 7 of BS 5950: Part 1.

$$\frac{79\varepsilon}{0.4 + 0.6\alpha} = \frac{79 \times 1.02}{0.4 + (0.6 \times 0.82)} = 90.3$$

$$\frac{d}{t} = 41.2 < 90.3$$

The web is classified as plastic.

Step 5 *Section properties*

Section properties of UB
$A = 437\,\text{cm}^2$; $I_{xx} = 625\,000\,\text{cm}^4$; $I_{yy} = 39\,200\,\text{cm}^4$; $r_x = 37.8\,\text{cm}$;
$r_y = 9.46\,\text{cm}$; $Z_x = 13\,700\,\text{cm}^3$; $Z_y = 1870\,\text{cm}^3$; $S_x = 15\,500\,\text{cm}^3$;
$S_y = 2890\,\text{cm}^3$; $J = 1190\,\text{cm}^4$

Section properties of top plate
$A = 45\,\text{cm}^2$; $I_{xx} = 45 \times 1^3 \div 12 = 3.75\,\text{cm}^4$; $I_{yy} = 45^3 \div 12 = 7594\,\text{cm}^4$;
$J = 15\,\text{cm}^4$; $Z_y = 337.5\,\text{cm}^3$; $S_y = 506.2\,\text{cm}^3$

Section properties of compound section
$D_c = 92.14\,\text{cm}$; $A = 437 + 45 = 482\,\text{cm}^2$; $Y_{NA} = 42.27\,\text{cm}$;
$Y_{PL} = 39.37\,\text{cm}$

$$I_{xx} = [625\,000 + 3.75 + 437 \times (46.57 - 42.27)^2]$$
$$+ [45 \times (42.27 - 0.5)^2] = 711\,597\,\text{cm}^4$$

$$I_{yy} = 39\,200 + 7594 = 46\,794\,\text{cm}^4$$

$$r_x = \left(\frac{I_{xx}}{A}\right)^{0.5} = \left(\frac{711\,597}{482}\right)^{0.5} = 38.42\,\text{cm}$$

$$r_y = \left(\frac{I_{yy}}{A}\right)^{0.5} = 9.85\,\text{cm}$$

$$Z_{x,\text{top}} = \frac{I_{xx}}{Y_{NA}} = \frac{711\,597}{42.27} = 16\,834.6\,\text{cm}^3$$

$$Z_{x,\text{bottom}} = \frac{I_{xx}}{(D_c - Y_{NA})} = 14\,269\,\text{cm}^3$$

$$Z_y = \frac{2 I_{yy}}{B} = \frac{2 \times 46\,794}{45} = 2080\,\text{cm}^3$$

The plastic modulus of the compound section, ignoring fillets, is given by:

$$S_{xc} = A_{\text{plate}}\left(Y_{PL} - \frac{t_{\text{plate}}}{2}\right) + BT\left(Y_{PL} - t_{\text{plate}} - \frac{T}{2}\right)$$
$$+ \frac{t}{2}(Y_{PL} - t_{\text{plate}} - T)^2$$
$$+ \frac{t}{2}(D_c - Y_{PL} - T)^2 + BT\left(D_c - Y_{PL} - \frac{T}{2}\right)$$
$$= [28 \times (39.37 - 0.5)] + [41.85 \times 3.2 \times (39.37 - 1 - 1.6)]$$
$$+ \left[\frac{1.94}{2} \times (39.37 - 1 - 3.2)^2\right]$$
$$+ \left[\frac{1.94}{2} \times (92.14 - 39.37 - 3.2)^2\right]$$
$$+ [41.85 \times 3.2 \times (92.14 - 39.37 - 1.6)]$$
$$= 16\,448\,\text{cm}^3$$

$$J = J_B + J_P = 1190 + 15 = 1205 \text{ cm}^4$$

$$h_s = D - \frac{T}{2} \text{ (approximately)}$$

$$= 91.14 - 1.6 = 89.54 \text{ cm}$$

$$x = 0.566 h_s \sqrt{\frac{A}{J}} = 0.566 \times 89.54 \times \sqrt{\frac{482}{1205}} = 32$$

$$\gamma = 1 - \frac{I_y}{I_x} = 1 - \frac{46\,794}{711\,597} = 0.93$$

$$u = \sqrt[4]{\left(\frac{4S_x^2 \gamma}{A^2 h_s^2}\right)} = \sqrt[4]{\frac{4 \times 16\,530^2 \times 0.93}{482^2 \times 89.54^2}} = 0.88$$

$$I_{cf} = \frac{T_{plate} B_{plate}^3}{12} + \frac{TB^3}{12} = \frac{1 \times 45^3}{12} + \frac{3.2 \times 41.85^3}{12} = 27\,140 \text{ cm}^4$$

$$I_{tf} = \frac{TB^3}{12} = \frac{3.2 \times 41.85^3}{12} = 19\,546 \text{ cm}^4$$

$$N = \frac{I_{cf}}{I_{cf} + I_{tf}} = 0.58$$

Step 6 *Check shear capacity*
$A_v = Dt = 911.4 \times 19.4 = 17\,681 \text{ mm}^2$
$P_v = 0.6 p_y A_v = 0.6 \times 265 \times 17\,681 \times 10^{-3} = 2811 \text{ kN}$
$F_{v,max} = 551.3 \text{ kN (load case 31)} < P_v$ (Satisfied)

Steps 7 and 8 Not required.

Step 9 *Moment capacity with low shear*

$0.6 P_v = 0.6 \times 2811 = 1686.6 \text{ kN} > F_{v,max}$

Low shear prevails for all loading conditions.

Vertical loading
Maximum ultimate moment = 1396 kNm (load case 31)
Service limit moment = 883.2 kNm

$$k = \frac{1396}{883.2} = 1.58$$

Compound section moment capacity $= S_{xc} p_y = 16\,448 \times 265 \times 10^{-3}$
$= 4359 \text{ kNm}$
$k p_y Z_{xb} = 1.58 \times 265 \times 14\,269 \times 10^{-3} = 5974 \text{ kNm} > 4359$ (Satisfied)

Horizontal loading
Maximum ultimate horizontal loading moment = 78.29 kNm
Consider top plate only.
Section moment capacity
$$= S_{y,plate} p_y = 506.2 \times 275 \times 10^{-3} = 139.2 \text{ kNm}$$

Check $\dfrac{M_x}{M_{cx}} + \dfrac{M_y}{M_{cy}} \leq 1$ or $\dfrac{1396}{4380} + \dfrac{78.29}{139.2} = 0.88 < 1$

Section moment capacity is adequate.

Steps 10 and 11 Not required.

Step 12 *Member moment capacity (lateral torsional buckling)*
As per Clause 4.3.3 of BS 5950: Part 1, the gantry girder may be regarded as torsionally restrained about its longitudinal axis because, at the columns, both flanges are effectively held in position relative to each other (see Step 19). Assume, conservatively, that both the flanges are free to rotate on plan.

The loading on the gantry girder is considered destabilizing. The compression flange is laterally restrained at the columns.
The effective length $L_E = 1.2L = 14400$ mm

$$\text{Slenderness ratio} = \lambda = \frac{L_E}{r_y} = \frac{14\,400}{98.5} = 146 \text{ (see Step 5)}$$

$\lambda/x = 146/32 = 4.6$ (see Step 5)
$N = 0.58$ (see Step 5)
$v = 0.8$ (from Table 14 of BS 5950: Part 1)
$u = 0.88$ (see Step 5)
$m = n = 1.0$ for gantry girders, as per BS 5950: Part 1: Table 13
$\lambda_{LT} = uv\lambda = 0.88 \times 0.8 \times 146 = 103$
$p_b = 117 \text{ N/mm}^2$ (from Table 12 of BS 5950: Part 1)
$M_b = S_{xc} p_b = 16\,448 \times 117 \times 10^{-3} = 1924 \text{ kNm}$

Check $\dfrac{mM_x}{M_b} + \dfrac{mM_y}{p_y Z_y} \leq 1$ or $\dfrac{1396}{1924} + \dfrac{78.29}{275 \times 2080 \times 10^{-3}} = 0.86$

Member capacity check is satisfied.

Note: Z_y for the compound section has been considered.

Step 13 *Web buckling of unstiffened web*
At the bearing, assume that the bottom flange has a minimum stiff bearing length b_1 equal to 20 mm. The spread of reaction load up to the half depth of the girder is $n_1 = D/2 = 455.7$ mm. The bottom flange of the girder at the support is assumed to be restrained against translation and rotation. The web, with a width of $(b_1 + n_1)$, is treated as a column.

$$\text{Slenderness ratio } \lambda = 2.5 \frac{d}{t} = 2.5 \times 41.2 = 103$$

310 Structural Steelwork

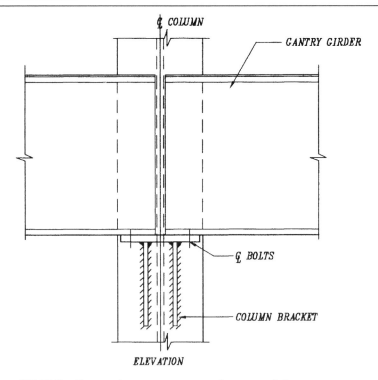

SK 7/49 Supporting arrangement of gantry girder.

Corresponding to this slenderness ratio, the compressive strength $p_c = 118\,\text{N/mm}^2$ for $p_y = 265\,\text{N/mm}^2$ from Table 27(c) of BS 5950: Part 1.

Web buckling capacity P_w

$$= (b_1 + n_1)tp_c = (20 + 455.7) \times 19.4 \times 118\,\text{N}$$

$$= 1089\,\text{kN} > F_{v,\text{max}} = 551.3\,\text{kN} \text{ (Satisfied)}$$

Bearing stiffeners are not required.

Web buckling under wheel load
Assume height of rail $H_R = 100\,\text{mm}$
The stiff bearing length $b_1 = 2(100 + 10) = 220\,\text{mm}$ up to top of flange
The total length of dispersion $n_1 = D = 911.4\,\text{mm}$

$$\lambda = 3.5\frac{d}{t} = 144$$

$p_c = 72\,\text{N/mm}^2$ from Table 27(c) of BS 5950
$P_w = (b_1 + n_1)tp_c = 1580\,\text{kN} >$ factored wheel load
$\quad = 305.6\,\text{kN}$ (Satisfied)

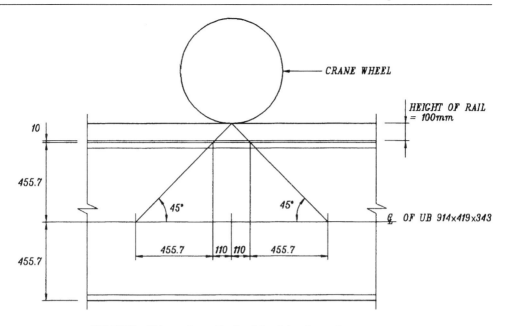

SK 7/50 Dispersion of wheel load in the web.

Step 14 *Web bearing of unstiffened web*
$b_1 = 20 \, \text{mm}$
$n_2 = 2.5(T+r) = 2.5 \times (32 + 24.1) = 140 \, \text{mm}$
$P_{\text{LW}} = (b_1 + n_2)tp_{\text{yw}} = (20 + 140) \times 19.4 \times 265 \times 10^{-3}$
$\quad\quad = 822 \, \text{kN} > 551.3 \, \text{kN}$

Web bearing strength is satisfied.

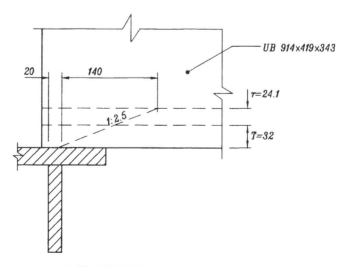

PART ELEVATION AT SUPPORT

SK 7/51 Local bearing stress in the web at the support.

312 Structural Steelwork

Steps 15 to 18 Not required.

Step 19 **Torsional restraint at the support**

Since bearing stiffeners are not required, as established in Step 14, the torsional restraint at the support may be provided by effectively holding both flanges laterally in position by external means. This is achieved by connecting the top flange to the roof leg of the stanchion. The top flange has to be restrained laterally to resist the horizontal end reaction due to surge or crabbing. The maximum horizontal reaction at the support due to horizontal loading on the girder from the crane should be increased by a force equal to 1% of the maximum flange force due to vertical loading to become effective in restraining torsion at the support. The bottom flange is held in position by direct bolted connection to the bracket on the column as shown in SK 7/52 (see Clause 4.3.3 of BS 5950: Part 1).

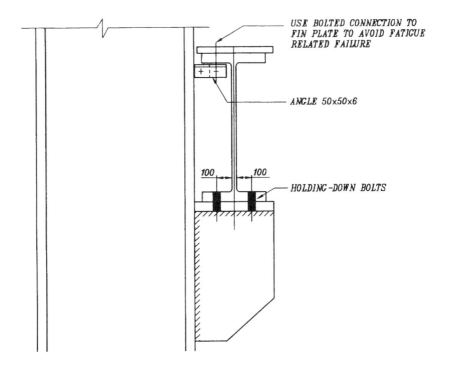

SK 7/52   Connection detail of gantry girder to column.

See Step 2. The maximum ultimate horizontal reaction on the column due to crane crabbing action is 9.79 kN.

The maximum flange force due to vertical loading is given approximately by:

$$\frac{M}{D} = \frac{1396}{0.9114} = 1532 \, \text{kN}$$

Design of Beams

The connection between the top flange of the girder and the roof leg of the column should be designed for a compression or tension load given by:

$$P = 9.79 + 15.32 = 25.11 \text{ kN}$$

The effective length of this connecting member is less than 1 m. An angle $50 \times 50 \times 6$ will be adequate to carry this loading.

Steps 20 and 21 Not required.

Step 22 *Deflection due to imposed loads*

Deflection limit due to vertical crane wheel loads

$$= \frac{l}{600} = \frac{12\,000}{600} = 20 \text{ mm}$$

Deflection limit due to horizontal crane surge

$$= \frac{l}{500} = \frac{12\,000}{500} = 24 \text{ mm}$$

Unfactored service limit static vertical wheel load $= W_{us} = 146.75 \text{ kN}$
The maximum vertical deflection is given by:

$$\delta = \frac{W_{us} L^3}{6EI_x} \left[\frac{3}{8}\left(1 - \frac{a_w}{L}\right) - \left(\frac{L - a_w}{2L}\right)^3 \right]$$

$$= \frac{146.75 \times 12\,000^3}{6 \times 205 \times 711\,597 \times 10^4} \times \left[\frac{3}{8}\left(1 - \frac{4}{12}\right) - \left(\frac{12 - 4}{2 \times 12}\right)^3 \right]$$

$$= 6.17 \text{ mm} < 20 \text{ mm}$$

Wheel load due to crane surge $= 5.75 \text{ kN}$

The maximum horizontal deflection due to crane surge, assuming the load is carried by the top flange of the girder only, is given by:

$$\delta_h = \frac{5.75 \times 12\,000^3}{6 \times 205 \times 7594 \times 10^4} \times \left[\frac{3}{8}\left(1 - \frac{4}{12}\right) - \left(\frac{12 - 4}{2 \times 12}\right)^3 \right]$$

$$= 22.65 \text{ mm} < 24 \text{ mm}$$

The maximum horizontal deflection due to crabbing load is given by:

$$\delta_{hc} = \frac{PL^3}{48EI_y} \left[1 - \frac{3a}{L} + 4\left(\frac{a}{L}\right)^3 \right]$$

$$= \frac{18.35 \times 12\,000^3}{48 \times 205 \times 7594 \times 10^4} \times \left[1 - \frac{3 \times 2}{12} + 4 \times \left(\frac{2}{12}\right)^3 \right]$$

$$= 22.0 \text{ mm} < 24 \text{ mm}$$

All deflection limits are satisfied.

Step 23 *Local stresses*

Allow 45° dispersion of wheel load through the rail and the flange plates.

$X_R = 2(H_R + T + T_1) = 2(100 + 32 + 10) = 284$ mm
H_R = height of rail, assumed to be 100 mm
T = thickness of top flange of the UB = 32 mm
T_1 = thickness of additional top plate = 10 mm

The rail should not be continuously welded at the joints of the simply supported gantry girders and the columns because the end rotations of large span gantry girders may give rise to fatigue related cracking of the rail.

The local compression on the web of the girder is most pronounced when the wheel is at the end of the girder. The load distribution may be over $X_R/2 = 142$ mm only.

The maximum local compressive stress is given by:

$$f_w = \frac{W_v}{X_R t} = \frac{305.6 \times 10^3}{142 \times 19.4} = 111 \text{ N/mm}^2 < p_y$$

$$= 265 \text{ N/mm}^2 \text{ (Satisfied)}$$

Step 24 *Connection of flange plate to UB*

The top plate is connected to the top flange of the UB by fillet welds.
Maximum ultimate vertical shear force $F_v = 551.3$ kN (load case 31)
Area of top plate $A_p = 45$ cm^2
Moment of inertia of compound girder section $I_x = 711\,597$ cm^4
$Q_x = A_p (Y_{NA} - 0.5T_1) = 45 \times (42.27 - 0.5) = 1880$ cm^3
Shear flow per unit length along the length of the girder at the point of maximum shear is given by:

$$v_x = \frac{F_v Q_x}{I_x} = \frac{551.3 \times 1880 \times 10^3}{711\,597 \times 10^4}$$

$$= 0.14 \text{ kN/mm (due to vertical shear force)}$$

Note: If the top plate was directly connected by weld to the web, as in the case of a built-up girder, the connection of the plate to the web would have been required to transmit the local wheel load by vertical shear through the weld.

The horizontal wheel load due to crane surge or crabbing action is transmitted by the rail to the top plate. The top plate carries a portion of this horizontal wheel load depending on its relative stiffness with respect to the top flange of the UB. The remaining horizontal load is transmitted to the top flange of the UB by means of the welded connection.

Shear in the welded connection due to local horizontal wheel load
Moment of inertia of top plate for horizontal loading
  $= 75.93 \times 10^6$ mm^4
Moment of inertia of the top flange of UB for horizontal loading
  $= 195.45 \times 10^6$ mm^4

The top plate and the top flange of the UB carry the horizontal loading in the proportion $1 : (195.45 \div 75.93) = 1 : 2.6$.
Maximum ultimate horizontal wheel load due to crabbing action $= 29.36\,\text{kN}$

Assume that this load is transmitted from the rail to the top plate by means of a 75 mm wide rail fastener. Therefore, the local load on the fillet weld connecting the top plate to the top flange of the UB is:

$$\frac{29.36}{75} \times \frac{2.6}{2.6+1} = 0.28\,\text{kN/mm (acting transversely to the girder)}$$

Therefore, the combined horizontal shear on the connecting weld is:

$$\sqrt{(0.14^2 + 0.28^2)} = 0.31\,\text{kN/mm}$$

Use two continuous runs of 5 mm fillet weld to connect the top plate to the top flange of the UB $= 0.80\,\text{kN/mm}$.

Step 25 *Fatigue of gantry girder*
Static load per wheel $W_{US} = 146.75\,\text{kN}$
$W_{FT} = K_p I W_{US} = 0.8 \times 1.3 \times 146.75 = 152.6\,\text{kN}$ (see basic data at the beginning of this example)
$CLH_c = 18.35\,\text{kN}$ due to crabbing
Dead load $= 5\,\text{kN/m}$
Maximum bending moment due to dead load $= 5 \times 12^2 \div 8 = 90\,\text{kNm}$
The maximum bending moment due to vertical wheel load is given by:

$$\frac{2W_{FT}}{L}\left(\frac{L}{2} - \frac{a_w}{4}\right)^2 = \frac{2 \times 152.6}{12} \times \left(\frac{12}{2} - \frac{4}{4}\right)^2 = 635.8\,\text{kNm}$$

The corresponding bending moment due to horizontal crabbing load is given by:

$$CLH_c \frac{a_w}{L}\left(\frac{L}{2} - \frac{a_w}{4}\right) = 18.35 \times \frac{4}{12} \times 5 = 30.6\,\text{kNm}$$

The maximum bending moment due to horizontal crabbing load is:

$$\frac{18.35 \times 4 \times 8}{12} = 48.93\,\text{kNm}$$

The corresponding bending moment due to vertical wheel load is:

$$\frac{152.6 \times 4 \times 8}{12} = 406.9\,\text{kNm}$$

$$Z_{xt} = 16\,834.6\,\text{cm}^3 \text{ (see Step 5)}$$

$$Z_{xb} = 14\,269\,\text{cm}^3 \text{ (see Step 5)}$$

$$Z_{\text{top flange}} = \frac{I_{cf}}{0.5B} = \frac{27\,140}{22.5} = 1206\,\text{cm}^3 \text{ (see Step 5)}$$

Dead load stresses

$$p_{cd} = \frac{90 \times 10^6}{16\,834.6 \times 10^3} = +5.3\,\text{N/mm}^2$$

$$p_{td} = \frac{90 \times 10^6}{14\,269 \times 10^3} = -6.3\,\text{N/mm}^2$$

Stresses due to vertical wheel load

$$p_{cv,max} = \frac{635.8 \times 10^6}{16\,834.6 \times 10^3} = +37.8\,\text{N/mm}^2$$

$$p_{tv,max} = \frac{635.8 \times 10^6}{14\,269 \times 10^3} = -44.6\,\text{N/mm}^2$$

Corresponding stresses due to horizontal crabbing load

$$p_{cc} = \frac{\pm 30.6 \times 10^6}{1206 \times 10^3} = \pm 25.4\,\text{N/mm}^2$$

Similarly:
Maximum stresses due to horizontal crabbing load

$$p_{cc} = \pm 40.6\,\text{N/mm}^2$$

Corresponding stresses due to vertical wheel load

$$p_{cv} = +24.17\,\text{N/mm}^2$$
$$p_{ct} = -28.52\,\text{N/mm}^2$$

Variation of stresses in the top flange
$f_{max} = 5.3 + 37.8 + 25.4 = +68.5\,\text{N/mm}^2$
or $= 5.3 + 24.2 + 40.6 = +70.1\,\text{N/mm}^2$
$f_{min} = 5.3 + 37.8 - 25.4 = +17.7\,\text{N/mm}^2$
or $= 5.3 + 24.2 - 40.6 = -11.1\,\text{N/mm}^2$

$$\frac{f_{min}}{f_{max}} = \frac{-11.1}{70.1} = -0.16$$

Refer to Clause 8.6 of BS 2573: Part 1. Assume class C construction. From Table 24 of BS 2573, $p_{c,max} = 210\,\text{N/mm}^2$ for $N = 1 \times 10^6$ cycles. This stress limit is not exceeded.

Variation of stresses in bottom flange
$f_{max} = -6.3\,\text{N/mm}^2$
$f_{min} = -6.3 - 44.6 = -50.9\,\text{N/mm}^2$

$$\frac{f_{max}}{f_{min}} = \frac{6.3}{50.9} = 0.12$$

From Table 24 of BS 2573, $p_{t,max} = 174\,\text{N/mm}^2$ for $N = 1 \times 10^6$ cycles. This stress limit is not exceeded.

Design of Beams 317

7.5 BEAMS SUBJECT TO TORSION

General

If the resultant shear force in a section passes through the shear centre of the section, then there is no torsion induced. A section symmetrical about both of the principal axes has the centroid and the shear centre coincidental. The shear centre of a channel section lies on the X–X axis about which the section is symmetrical.

7.5.1 Torsional resistance

The torsion is resisted by the section in two different ways, i.e. uniform torsion and warping torsion. When the rate of change of angle of twist is constant along the length of a member, then uniform torsion occurs. Warping results in bending stresses in the flanges of an open section like a universal beam. Warping in that section gives rise to equal and opposite shear parallel to the two flanges of the UB. The section resists torsion by the sum of the resistances of uniform and warping torsion.

The torsional rigidity relating to uniform torsion is GJ; the warping rigidity is EH.

E = elastic modulus
H = warping constant
G = shear modulus
J = torsional constant

Closed sections have a very large torsional rigidity GJ compared to warping rigidity EH and hence they resist torque essentially by uniform torsion. The total torsional resistance at any point in a beam T_r is given by:

$$T_r = T_P + T_W$$

where T_P = resistance by uniform torsion = $GJ \dfrac{d\phi}{dx} = GJ\phi'$

T_W = resistance by warping = $EH \dfrac{d^3\phi}{dx^3} = EH\phi'''$

x = length along the longitudinal axis of the member
ϕ = rotation of the member about the longitudinal x-axis

Equating the applied torque T_q to the resistance T_r gives the following relationship:

$$\frac{T_q}{GJ} = \phi' - a^2 \phi'''$$

where a = torsional bending constant = $\sqrt{\dfrac{EH}{GJ}}$

7.5.2 Stresses in closed sections

Assume uniform torsion for closed sections, as in rectangular and circular hollow structural sections. The total angle of twist ϕ is given by:

$$\phi = \frac{T_q x}{GJ}$$

where $J = \dfrac{4A_h^2}{\sum\left(\dfrac{s}{t}\right)}$

A_h is the area enclosed by the mean perimeter of the closed section and $\sum(s/t)$ is the summation of individual component lengths of the closed section along the perimeter divided by the thickness of the component. For a section with uniform thickness along its perimeter of length S this value is S/t.

For a closed thin-walled section of uniform thickness t, the total angle of twist is given by:

$$\phi = \frac{T_q x S}{4A_h^2 G t}$$

τ_t, the shear stress in a closed section subjected to torque T_q, is given by:

$$\tau_t = \frac{T_q}{2A_h t} \quad \text{for thin-walled sections}$$

$$\tau_t = \frac{T_q}{C} \quad \text{for thick-walled sections}$$

C is the torsional modulus constant and can be found from published tables (see *Design of Members Subject to Combined Bending and Torsion*, The Steel Construction Institute, 1989).

7.5.3 Stresses in open sections

Uniform torsion
The total angle of twist is given by:

$$\phi = \frac{T_q x}{GJ} = Gt\phi'$$

The maximum shear stress in a component of the open section with thickness t is given by:

$$\tau_t = Gt\frac{d\phi}{dx}$$

Warping torsion
The torsional warping couple acts as equal and opposite forces in the flanges at a lever arm distance equal to the centroidal distance between the flanges. These forces give rise to bending in the plane of

the flanges. The tensile and compressive bending stress σ_w in each flange due to this bending is given by:

$$\sigma_w = EW_{ns}\frac{d^2\phi}{dx^2} = EW_{ns}\phi''$$

where W_{ns} = the normalized warping constant

$$W_{ns} = \frac{hB}{4} \text{ for symmetrical I- and H-sections}$$

$$= \frac{\left(B - \frac{t}{2} - c\right)h}{2} \text{ at tip of flange of channel sections}$$

$$= \frac{ch}{2} \text{ at the back end of flange of channel sections}$$

The distance of the shear centre from the centre of web is given by:

$$c = \frac{\left(B - \frac{t}{2}\right)^2 T}{2\left(B - \frac{t}{2}\right)T + h\frac{t}{3}}$$

where h = distance between centroids of flanges
B = width of flange
t = thickness of web
T = thickness of flange

The warping shear stress in the flanges is given by:

$$\tau_w = -\frac{ES_{ws}\phi'''}{t}$$

where S_{ws} = warping statical moment

$$= \frac{hB^2 T}{16} \text{ for symmetrical I- and H-sections}$$

$$= \frac{\left(B - \frac{t}{2} - c\right)^2 hT}{4} \text{ for channel sections}$$

The angle of twist ϕ due to torsion is amplified by the presence of bending moment. This induces additional warping moments in the flanges and additional torsional shears. Additional minor axis bending moment is also induced due to the rotation of the member about its longitudinal axis. As the beam rotates, some of the loading is transferred to minor axis bending. This additional minor axis bending M_{yt} is proportional to the angle of twist ϕ.

Stresses from elastic analyses of section

Major axis bending stress $= \sigma_{bx} = \dfrac{M_x}{Z_x}$

Minor axis bending stress $= \sigma_{by} = \dfrac{M_y}{Z_y}$

Additional minor axis bending stress $= \sigma_{byt} = \dfrac{M_{yt}}{Z_y} = \dfrac{\phi M_x}{Z_y}$

Warping bending stress in flanges $= \sigma_w = E W_{ns} \phi''$

M_x and M_y are applied moments about the X–X and Y–Y axes respectively.

7.5.4 Checks for capacity

The following checks are necessary.

Section capacity check

$$\sigma_{bx} + \sigma_{by} + \sigma_{byt} + \sigma_w \leq p_y$$

Member buckling check

$$\dfrac{\bar{M}_x}{M_b} + \dfrac{\bar{M}_y}{p_y Z_y} + \dfrac{(\sigma_{byt} + \sigma_w)}{p_y}\left[1 + 0.5\,\dfrac{\bar{M}_x}{M_b}\right] \leq 1.0$$

where $\bar{M}_x$ = equivalent moment about X-axis $= m_x M_x$
$\bar{M}_y$ = equivalent moment about Y-axis $= m_y M_y$

Torsional shear stresses are also amplified by the bending effect. The amplified final torsional shear stress τ_{vt} is given by:

$$\tau_{vt} = (\tau_t + \tau_w)\left(1 + 0.5\,\dfrac{\bar{M}_x}{M_b}\right)$$

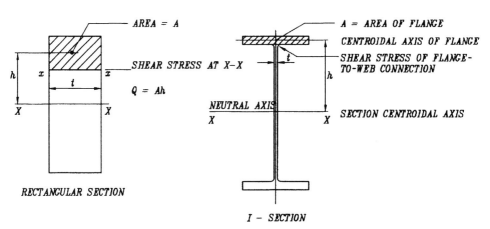

SK 7/53 First moment of area for the determination of flexural shear stress.

This shear stress due to torsion should be added to the flexural shear stress due to bending τ_b, which is is given by:

$$\tau_b = \frac{F_v Q}{It}$$

where F_v = applied shear force at the section
Q = first moment of area (see SK 7/53) about the neutral axis
I = second moment of area about the neutral axis
t = thickness of element in which shear stress is determined

Values of angle of twist ϕ and its derivatives for a selected number of commonly occurring cases

Case 1: Beam with concentrated torque T_q at midspan of beam
The beam-end conditions are torsionally fixed but the flanges are free to warp. The end condition for flexure may be either simply supported or continuous. The angle of twist and its derivatives may be determined at any point along the length x of the beam by using the following formulae. Assume $x = 0$ at the left-hand end of the beam.

$$\phi = \frac{T_q a}{GJ}\left[\frac{x}{2a} + \left\{\frac{\sinh(L/2a)}{\tanh(L/a)} - \cosh(L/2a)\right\}\sinh(x/a)\right]$$

$$\frac{d\phi}{dx} = \frac{T_q}{GJ}\left[0.5 + \left\{\frac{\sinh(L/2a)}{\tanh(L/a)} - \cosh(L/2a)\right\}\cosh(x/a)\right]$$

$$\frac{d^2\phi}{dx^2} = \frac{T_q}{GJa}\left[\left\{\frac{\sinh(L/2a)}{\tanh(L/a)} - \cosh(L/2a)\right\}\sinh(x/a)\right]$$

$$\frac{d^3\phi}{dx^3} = \frac{T_q}{GJa^2}\left[\left\{\frac{\sinh(L/2a)}{\tanh(L/a)} - \cosh(L/2a)\right\}\cosh(x/a)\right]$$

Case 2: Beam with distributed uniform torque T_q/L per unit length giving a total torque T_q on the beam
The beam-end condition is torsionally fixed and free to warp. The end condition for bending may be either simply supported or continuous. The angle of twist ϕ and its derivatives may be found anywhere along the length x of the beam by using the following formulae.

$$\phi = \frac{T_q a}{GJ}$$

$$\times \left(\frac{a}{L}\right)\left[\frac{L^2}{2a^2}\left(\frac{x}{L} - \frac{x^2}{L^2}\right) + \cosh\left(\frac{x}{a}\right) - \tanh\left(\frac{L}{2a}\right)\sinh\left(\frac{x}{a}\right) - 1\right]$$

$$\frac{d\phi}{dx} = \frac{T_q}{GJ}\left(\frac{a}{L}\right)\left[\frac{L^2}{2a^2}\left(\frac{1}{L} - \frac{2x}{L^2}\right) + \sinh\left(\frac{x}{a}\right) - \tanh\left(\frac{L}{2a}\right)\cosh\left(\frac{x}{a}\right)\right]$$

$$\frac{d^2\phi}{dx^2} = \frac{T_q}{GJ}\left(\frac{1}{L}\right)\left[-1+\cosh\left(\frac{x}{a}\right)-\tanh\left(\frac{L}{2a}\right)\sinh\left(\frac{x}{a}\right)\right]$$

$$\frac{d^3\phi}{dx^3} = \frac{T_q}{GJa}\left(\frac{1}{L}\right)\left[\sinh\left(\frac{x}{a}\right)-\tanh\left(\frac{L}{2a}\right)\cosh\left(\frac{x}{a}\right)\right]$$

For all other cases of loading refer to *Design of Members Subject to Combined Bending and Torsion*, The Steel Construction Institute, 1989.

7.6 EXAMPLE 7.6: DESIGN OF A BEAM WITH CONCENTRATED APPLIED TORQUE AT THE CENTRE OF SPAN

Span of main beam = 6000 mm
Vertical ultimate load from the secondary beam = 100 kN
Eccentricity of loading = 75 mm from the face of web of main beam
Size of secondary beam = 305 × 165 × 46 kg/m
Assume self-weight of main beam = 1 kN/m

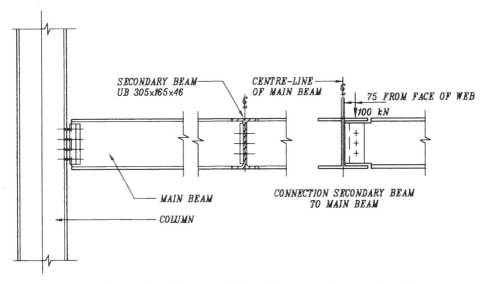

SK 7/54 Simply supported main beam with a secondary beam connected by a fin plate to the centre of span.

Step 1 **Determine design strength p_y**
Select steel = Grade 43

$p_y = 275$ N/mm² for thickness of material up to 16 mm
$p_y = 265$ N/mm² for thickness of material up to 40 mm

Step 2 **Carry out analysis**
Span of beam = 6000 mm
Ultimate limit state bending moment at midspan

$$= \frac{100 \times 6}{4} + \frac{1.4 \times 1 \times 6^2}{8}$$

$$= 156.3 \text{ kNm}$$

Support shear = $50 + 1.4 \times 1 \times 3 = 54.2$ kN
Shear at the connection of secondary beam = 50 kN
Torque = $100 \times (75 + t/2) \div 1000 = 8.0$ kNm

Step 3 *Choose trial section and section classification*
Maximum ultimate bending moment = 156.3 kNm

$$\text{Required minimum plastic modulus} = \frac{156.3 \times 10^6}{275} = 568 \times 10^3 \text{ mm}^3$$

Choose a section which is at least $L/20$ (= 300 mm) in overall depth, a wide flange with a large torsion constant J, and with sufficient margin in plastic modulus to cater for lateral torsional buckling and torsion effects. A column section UC has more torsional stiffness and should be chosen where large torsional stresses are expected.

Choose a section which is at least of the same depth as the secondary beam (or deeper) to simplify the connection geometry. Also make sure that the section chosen has not got a slender web.

Chosen section is $305 \times 305 \times 97$ kg/m UC which has the following section properties:

$D = 307.8$ mm; $B = 304.8$ mm; $t = 9.9$ mm; $T = 15.4$ mm; $d = 246.6$ mm; $b/T = 9.9$; $d/t = 24.9$; $I_{xx} = 22\,200$ cm^4; $I_{yy} = 7272$ cm^4; $r_x = 13.4$ cm; $r_y = 7.68$ cm; $Z_x = 1443$ cm^3; $Z_y = 477$ cm^3; $S_x = 1589$ cm^3; $u = 0.850$; $x = 19.3$; $H = 1.55$ dm^6; $J = 91.1$ cm^4; $A = 123$ cm^2

The section classification is semi-compact for the flanges and plastic for the web, as per BS 5950: Part 1.

Step 4 *Check shear capacity*
$P_v = 0.6 p_y A_v$
A_v = shear area of section = $tD = 9.9 \times 307.8 = 3047$ mm^2

$\therefore \quad P_v = 0.6 \times 275 \times 3047 \times 10^{-3} = 502.8$ kN

$0.6 P_v = 0.6 \times 502.8 = 302$ kN $> F_v = 54.2$ kN

There is very little interaction between bending and shear stress in the web and the section will be checked for capacity with low shear.

Step 5 *Determine the section moment capacity with low shear*
Ultimate section moment capacity about the major axis for a semi-compact section is given as:

$M_{cx} = p_y Z_x = 275 \times 1443 \times 10^3 \times 10^{-6}$

$= 396.8$ kNm > 156.3 (Satisfied)

Step 6 *Check lateral torsional buckling*
Assume that lateral restraint to the compression flange is provided at the connection with the secondary beam. Lateral and torsional restraint is provided at the support. At the supports assume that the compression flange is laterally restrained and the beam is fully restrained against torsion, but the flanges are free to rotate on plan. Effective length between restraints = $1.0L = 3000$ mm

Slenderness ratio $= \dfrac{L_E}{r_y} = \dfrac{3000}{76.8} = 39.1$

$\dfrac{\lambda}{x} = \dfrac{39.1}{19.3} = 2$

$N = 0.5$ for symmetrical sections
$v = 0.96$ (from Table 14 of BS 5850: Part 1)
β = ratio of minimum moment over maximum moment in the length of the beam between restraints $= 0$

Equivalent uniform moment factor $m = 0.57$ (from Table 18 of BS 5950: Part 1)
Slenderness correction factor $= n = 1.0$

$\lambda_{LT} = nuv\lambda = 1 \times 0.850 \times 0.96 \times 39.1 = 32$

Bending strength $p_b = 274\,\text{N/mm}^2$ for $\lambda_{LT} = 32$ and $p_y = 275\,\text{N/mm}^2$ (from Table 11 of BS 5950: Part 1).

$$\therefore\ M_b = p_b S_x = 274 \times 1589 \times 10^3 \times 10^{-6}$$
$$= 435.4\,\text{kNm} > M_{cx}\ (\text{use } M_{cx})$$

Equivalent moment $\bar{M}_x = mM_x = 0.57 \times 156.3$
$$= 89.1\,\text{kNm} < 396.8\,\text{kNm}\ (\text{Satisfied})$$

Step 7   Check combined bending and torsion

$$\dfrac{\bar{M}_x}{M_b} + \dfrac{\sigma_{byt} + \sigma_w}{p_y}\left[1 + 0.5\dfrac{\bar{M}_x}{M_b}\right] \le 1.0$$

$$\dfrac{E}{G} = 2(1+\mu) = 2.6\quad \text{for steel}$$

$$a = \sqrt{\dfrac{EH}{GJ}} = \sqrt{\dfrac{2.6 \times 1.55 \times 10^6}{91.1}} = 210.3\,\text{cm}$$

With reference to case 1 (beam with concentrated torque T_q at the centre of span), the angle of twist ϕ at midspan, where $x = 0.5L = 3000$ mm, is given by:

$$\phi = \dfrac{T_q a}{GJ}\left[\dfrac{x}{2a} + \left\{\dfrac{\sinh(L/2a)}{\tanh(L/a)} - \cosh(L/2a)\right\}\sinh(x/a)\right]$$

$$= \dfrac{8.0 \times 10^6 \times 2103}{79\,000 \times 91.1 \times 10^4}$$

$$\times \left[\dfrac{3000}{2 \times 2103} + \left\{\dfrac{\sinh(6000/[2 \times 2103])}{\tanh(6000/2103)}\right.\right.$$

$$\left.\left. - \cosh(6000/[2 \times 2103])\right\}\sinh(3000/2103)\right]$$

$$= 0.0626\ \text{radian}$$

Design of Beams

Determine stresses about the Y–Y axis due to rotation of the beam
$M_{yt} = M_x \phi = 156.3 \times 0.0626 = 9.78 \text{ kNm}$

$$\sigma_{byt} = \frac{M_{yt}}{Z_y} = \frac{9.78 \times 10^6}{477 \times 10^3} = 20.5 \text{ N/mm}^2$$

Determine flange stresses due to warping
$\sigma_w = E W_{ns} \phi''$
For a symmetrical I- and H-section

$$W_{ns} = \frac{hB}{4} = \frac{292.4 \times 304.8}{4} = 22\,281 \text{ mm}^2$$

h = centre-to-centre distance between the flanges
$= 307.8 - 5.4 = 292.4 \text{ mm}$

$$\phi'' = \frac{T_q}{GJa} \left[\frac{\sinh(L/2a)}{\tanh(L/a)} - \cosh(L/2a) \right] \sinh(x/a)$$
$$= -2.355 \times 10^{-8}$$

$\therefore \quad \sigma_w = 205\,000 \times 22\,281 \times 2.355 \times 10^{-8} = 108 \text{ N/mm}^2$

Check interaction formula

$$\frac{89.1}{396.8} + \frac{(20.5 + 108)}{275} \times \left[1 + 0.5 \times \frac{89.1}{396.8} \right] = 0.74 < 1.0 \text{ (Satisfied)}$$

Section capacity check for maximum stress

$$\sigma_{bx} = \frac{M_x}{Z_x} = \frac{156.3 \times 10^6}{1443 \times 10^3} = 108.3 \text{ N/mm}^2$$

$$\sigma_{by} = \frac{M_y}{Z_y} = 0$$

$\sigma_{bx} + \sigma_{by} + \sigma_{byt} + \sigma_w = 108.3 + 20.5 + 108$
$\phantom{\sigma_{bx} + \sigma_{by} + \sigma_{byt} + \sigma_w} = 236.8 \text{ N/mm}^2 < p_y = 275 \text{ N/mm}^2$

The stresses in the section are within permissible limits.

Step 8 **Check combined shear stress**
Check shear stress at the support where it will be critical.

Flexural shear stress
A_1 = area of the flange = $BT = 304.8 \times 15.4 = 4694 \text{ mm}^2$
A_2 = area of web above neutral axis = $\frac{1}{2}(D - 2T)t$
$ = 0.5 \times [307.8 - (2 \times 15.4)] \times 9.9 = 1371 \text{ mm}^2$
y_1 = distance of centre of flange from X–X axis = 146.2 mm
y_2 = distance of centre of area A_2 from X–X axis = 69.3 mm
A_3 = area of half of flange from the edge of web = 147.5×15.4
$ = 2272 \text{ mm}^2$
Q_w = first moment of area for web shear stress = $A_1 y_1 + A_2 y_2$
$ = 781\,273 \text{ mm}^3$
Q_f = first moment of area for flange shear stress = $A_3 y_1$
$ = 332\,166 \text{ mm}^3$

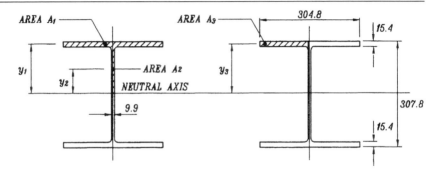

SECTIONS OF UC 305×305×97

SK 7/55 Determination of first moment of area to find flexural shear stress.

More accurate values of Q_w and Q_f can be obtained from tables in *Design of Members Subject to Combined Bending and Torsion*, The Steel Construction Institute, 1989.

The flexural shear stress in the web is given by:

$$\tau_{bw} = \frac{F_v Q_w}{I_{xx} t} = \frac{54.2 \times 10^3 \times 781\,233}{22\,200 \times 10^4 \times 9.9} = 19.3\,\text{N/mm}^2$$

The flexural shear stress in the flange is given by:

$$\tau_{bf} = \frac{F_v Q_f}{I_{xx} T} = 5.3\,\text{N/mm}^2$$

Torsional shear stress (uniform torsion)

$$\tau_t = G t \phi'$$

$$\phi' = \frac{T_q}{GJ}\left[0.5 + \left(\frac{\sinh(L/2a)}{\tanh(L/a)} - \cosh(L/2a)\right)\cosh\left(\frac{x}{a}\right)\right]$$

$$= 3.0357 \times 10^{-5} \quad \text{at } x = 0$$

$$\phi''' = \frac{T_q}{GJa^2}\left[\frac{\sinh(L/2a)}{\tanh(L/a)} - \cosh\left(\frac{L}{2a}\right)\right]\cosh\left(\frac{x}{a}\right)$$

$$= -5.705 \times 10^{-12} \quad \text{at } x = 0$$

The stress due to uniform torsion in the web is given by:

$$\tau_{tw} = G t \phi' = 79\,000 \times 9.9 \times 3.0357 \times 10^{-5}$$

$$= 23.7\,\text{N/mm}^2$$

The stress due to uniform torsion in the flange is given by:

$$\tau_{tf} = G T \phi' = 36.9\,\text{N/mm}^2$$

Torsional shear stress (warping)
The warping statical moment is given by:

$$S_{ws} = \frac{hB^2T}{16} = \frac{292.4 \times 304.8^2 \times 15.4}{16} = 26.15 \times 10^6 \text{ mm}^4$$

The shear stress due to warping in the flange is given by:

$$\tau_{wf} = \frac{-ES_{ws}\phi'''}{T}$$

$$= \frac{205\,000 \times 26.15 \times 10^6 \times 5.705 \times 10^{-12}}{15.4} = 2 \text{ N/mm}^2$$

The amplified torsional shear stress due to bending effect τ_{vt} is given by:

$$\tau_{vt} = (\tau_t + \tau_w)\left(1 + \frac{0.5\bar{M}_x}{M_b}\right)$$

$$= (23.7 + 0) \times \left(1 + \frac{0.5 \times 89.1}{396.8}\right) = 26.4 \text{ N/mm}^2 \text{ for the web}$$

$$= (36.9 + 2) \times \left(1 + \frac{0.5 \times 89.1}{396.8}\right) = 43.3 \text{ N/mm}^2 \text{ for the flange}$$

Final maximum shear stress in web
= flexural shear stress + torsional shear stress
= 19.3 + 26.4 = 45.7 N/mm²
Final maximum shear stress in flange = 5.3 + 43.3 = 48.6 N/mm²
Allowable maximum shear stress = $0.6p_y$ = 0.6 × 275 = 165 N/mm²
Shear stresses are within limits.

Chapter 8
Design of Composite Beams and Columns

8.1 COMPOSITE BEAMS

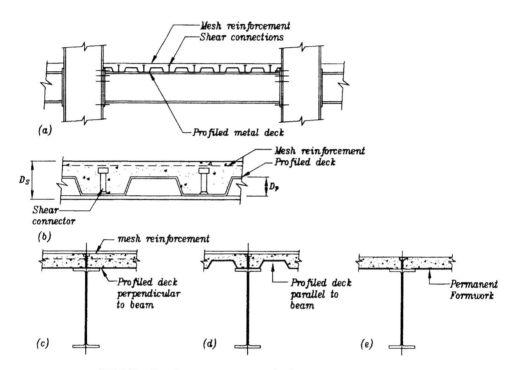

SK 8/1 Steel–concrete composite beam.

8.1.1 Principal issues

The principal issues governing the design of composite beams are as follows.

(1) The interface between the concrete and the steel beam is capable of transmitting shear such that the concrete and the steel beam act compositely as one section. The shear connectors have to be designed in such a manner that this condition is achieved.
(2) At the construction stage, the steel beams should be strong enough to carry the weight of wet concrete, construction-imposed load and some concentrated plant load, including any shuttering or permanent formwork used. The steel beams may be propped during the construction stage to avoid carrying these loads.

(3) The composite plastic stress state will be dependent on the location of the plastic neutral axis. This may necessitate some lengthy computations depending on whether the neutral axis is in the concrete, in the flange, or in the web of the steel beam.
(4) The compression flange of the steel beam may be considered as effectively restrained in a simply supported beam by the hardened concrete, but adequate measures have to be considered at the construction stage when the concrete is wet. The horizontal liquid thrust of the concrete on the side shutters could be quite considerable.
(5) The effective width of the concrete slab in compression acting compositely with the steel beam requires consideration.
(6) The mechanism of shear transfer between the concrete and the steel beam, and also within the concrete slab itself (longitudinally along the beam), should be investigated.
(7) Care should be taken to limit the d/t ratio of the web in such a fashion as to get a plastic section for the development of full plastic moment of resistance of the composite cross-section. The classification of the web will depend on the location of the neutral axis.
(8) At the serviceability limit state the calculations of deflection will require the considerations of propped or unpropped construction, cracked or uncracked moment of inertia, shrinkage and creep of the concrete, and full or partial interaction of shear studs.
(9) The method of analysis of continuous beams dictates the percentage of redistribution of support moment to the span. This percentage will depend on whether a gross uncracked section or a cracked section is used in the analysis, and also whether the compactness of the steel section would allow ductile plastic behaviour.

8.1.2 Design basis

Elastic analysis of simple beams and continuous beams is carried out with cracked or uncracked moment of inertia of the section. The elastic analysis results are used to determine deflection, vibration characteristics and concrete serviceability limit states, such as crack widths. During construction the unpropped steel section will carry the weight of the concrete and the locked-in stresses influence the serviceability limit state. Creep and shrinkage also affect the serviceability limit state.

To carry out the analysis it is necessary to find the section properties. For a continuous beam, the section properties change quite dramatically from midspan, where the concrete flange is in compression, to the supports, where the concrete flange is in tension. At the supports, the concrete flange in tension is ignored.

8.1.3 Effective breadth

The concept of effective breadth of the concrete slab in compression comes from the recognition of 'shear lag' effects. Beyond a certain

Design of Composite Beams and Columns 331

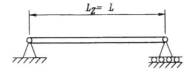

SIMPLY SUPPORTED

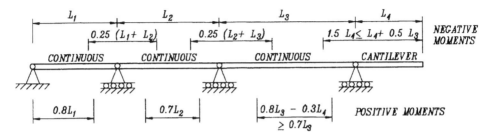

SK 8/2 Effective breadth of compression flange.

distance from the web of a beam, the compressive flange is considered ineffective and hence is ignored in the calculations. The determination of the effective width is totally empirical and has no theoretical basis. The effective breadth is dependent on the type of loading, the support conditions and the location of the section in relation to the supports.

An approximate method of finding this value is given in BS 5950: Part 3.1. The effective breadth in the code is a factor of the length L_Z between adjacent points of zero moment or points of contraflexure. For a simply supported beam L_Z is equal to the span; for a continuous beam the proportions of span to be used for finding the effective breadth are given in SK 8/2.

8.1.4 Modular ratio

This is another important factor which is required in order to determine the condition of stresses in a composite beam under the serviceability limit state. The modular ratio is dependent on several factors, including the age of concrete at loading, stress state, length of time of applied loading etc. The most important influence on modular ratio is from creep of the concrete under sustained loading, and, therefore, the British Standard code formula allows for a correction of the modular ratio, depending on the ratio of the short- and long-term loads.

8.1.5 Transformed section: elastic section properties

In elastic analysis of the section properties the concrete is converted into equivalent areas of steel by dividing the area of the concrete by an appropriate modular ratio. The section then becomes a compound steel section. The neutral axis will lie at the centre of gravity of the compound section where it can also be proved that the tensile and the compressive forces on either side of the neutral axis will be equal when a bending moment is applied, assuming that the plane section remains plane before and after bending. The moment of inertia of the compound section is the sum of moments of inertia of the component parts about their own centroidal axis plus their areas multiplied by the square of the distance of their centroidal axes from the compound section neutral axis. There can be two different cases of section stress distribution under elastic conditions. The neutral axis may lie in the steel section when the section properties are classed as uncracked. When the neutral axis lies in the concrete flange, the concrete below the neutral axis is deemed cracked and is ignored in the calculation of section properties. The derivation of the formulae for the neutral axis and moment of inertia is shown below. Tables 8.1 and 8.2 give values of the section properties for different locations of the neutral axis.

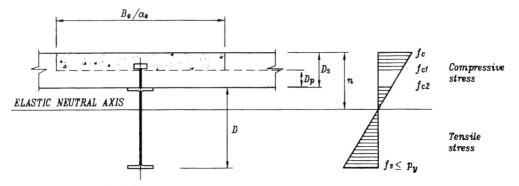

SK 8/3 Elastic stress distribution and section properties.

A = area of steel section
B_e = total effective breadth of concrete flange
D = depth of steel section
D_p = overall depth of profiled sheet
D_s = overall depth of concrete slab
n = depth of elastic neutral axis from concrete flange
α_e = modular ratio
f_c = stress in concrete fibre
f_s = stress in steel beam

The stress diagram shown in SK 8/3 is due to a bending moment M. The concrete flange is totally in compression and uncracked.

$$f_{c1} = \frac{f_c}{n}[n - (D_s - D_p)] = f_c\left[1 - \frac{(D_s - D_p)}{n}\right]$$

$$f_{c2} = \frac{f_c}{n}(n - D_s)$$

The total compressive force in the section is given by:

$$C = \frac{1}{2}\left(\frac{B_e}{\alpha_e}\right)\left[f_c + f_c\left(1 - \frac{(D_s - D_p)}{n}\right)\right](D_s - D_p)$$

The total tensile force in the section is given by:

$$T = \frac{A}{2}(f_s + f_{c2}) - Af_{c2}$$

Equating tension T and compression C in the section and simplifying:

$$n = \frac{\frac{1}{2}(D_s - D_p) + \alpha_e r(\frac{1}{2}D + D_s)}{1 + \alpha_e r}$$

where $r = A/(D_s - D_p)B_e$

8.1.6 Second moment of area for elastic analysis

I_g = second moment of area of uncracked composite section
I_p = second moment of area of cracked composite section for positive moments
I_n = second moment of area of cracked composite section for negative moments
I_x = second moment of area of steel beam about major axis

$$I_g = A\left(\frac{D}{2} + D_s - n\right)^2 + I_x + \frac{1}{12}\frac{B_e}{\alpha_e}(D_s - D_p)^3$$

$$+ \frac{B_e}{\alpha_e}(D_s - D_p)\left(n - \frac{(D_s - D_p)}{2}\right)^2$$

$$= I_x + \frac{B_e(D_s - D_p)^3}{12\alpha_e} + \frac{A(D + D_s + D_p)^2}{4(1 + \alpha_e r)}$$

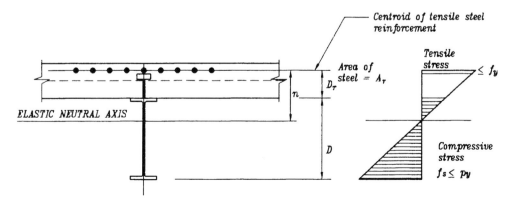

SK 8/4 Second moment of area of cracked section for negative moment.

334 Structural Steelwork

With reference to SK 8/4, D_r is the distance from the top of the steel beam to the centroid of tensile reinforcement and A_r is the area of tensile reinforcement in the concrete slab in the effective section.

The neutral axis in an elastic stress state can be found easily by taking moments of the areas about the centroid of the steel reinforcement.

$$n(A + A_r) = A(\tfrac{1}{2}D + D_r)$$

$$n = \frac{A(\tfrac{1}{2}D + D_r)}{(A + A_r)}$$

$$I_n = I_x + A(\tfrac{1}{2}D + D_r - n)^2 + A_r n^2 = I_x + \frac{AA_r(D + 2D_r)^2}{4(A + A_r)}$$

Table 8.1 Summary of section properties.

Section parameter	Value of section property
Uncracked section: depth of neutral axis for positive moment	$n = \dfrac{\tfrac{1}{2}(D_s - D_p) + \alpha_e r(\tfrac{1}{2}D + D_s)}{(1 + \alpha_e r)}$
Cracked section: depth of neutral axis for positive moment	$n = \dfrac{(D + 2D_s)}{1 + \left[1 + \dfrac{B_e}{A\alpha_e}(D + 2D_s)\right]^{\tfrac{1}{2}}}$
Cracked section: depth of neutral axis for negative moment	$n = \dfrac{A(\tfrac{1}{2}D + D_r)}{(A + A_r)}$
Uncracked section: second moment of area for positive moment	$I_g = I_x + \dfrac{B_e(D_s - D_p)^3}{12\alpha_e} + \dfrac{A(D + D_s + D_p)^2}{4(1 + \alpha_e r)}$
Cracked section: second moment of area for positive moment	$I_p = I_x + \dfrac{B_e n^3}{3\alpha_e} + A(\tfrac{1}{2}D + D_s - n)^2$
Cracked section: second moment of area for negative moment	$I_n = I_x + \dfrac{AA_r(D + 2D_r)^2}{4(A + A_r)}$

Table 8.2 Elastic section modulus.

Section	Moment	Condition	Neutral axis	Modulus concrete	Modulus steel
Cracked	Positive	$A < \dfrac{(D_s - D_p)^2 B_e}{(D + 2D_p)\alpha_e}$	See Table 8.1	$Z_p = \dfrac{I_p \alpha_e}{n}$	$Z_s = \dfrac{I_p}{(D + D_s - n)}$
Uncracked	Positive	$A \geq \dfrac{(D_s - D_p)^2 B_e}{(D + 2D_p)\alpha_e}$	See Table 8.1	$Z_g = \dfrac{I_g \alpha_e}{n}$	$Z_g = \dfrac{I_g}{(D + D_s - n)}$
Cracked	Negative		See Table 8.1	$Z_r = I_n/n$	$Z_s = I_n/(D + D_r - n)$

8.1.7 Plastic section properties

The fundamental assumption in plastic analysis is that the strain in the steel beam is high enough to allow full plastic capacity to develop, i.e. the stress block is rectangular with design strength p_y. But this depends on the compactness of the web. If the web of the steel beam is non-compact then the stress block is not rectangular because web buckling at much lower rotation of the section prevents the design strength from being achieved in all parts of the section. There are two different methods of finding the plastic section properties, one with a full plastic rectangular stress block for compact webs and another with triangular stress distribution, as in elastic analysis for non-compact webs. The concrete strain is limited to 0.0035 which results in a rectangular stress block with an average stress of $0.45f_{cu}$. The plastic analysis is also carried out for two loading conditions, i.e. sagging and hogging moments. For hogging moments the concrete is totally ignored, as in a reinforced concrete analysis, but any tensile reinforcement provided for the purpose of the design of the composite beam may be taken into account provided it lies within the effective width of the flange. The compressive reinforcement in the concrete should generally be ignored in the computations because they are not normally tied in with the steel beam by links going round them to prevent buckling under load.

It should be borne in mind that the plastic analysis is independent of the construction sequence (i.e. propped or unpropped) because at the plastic state the total load is resisted by a total internal assumed stress block, and the effects of shrinkage, creep and locked-in stresses due to the construction sequence do not influence this analysis.

In composite construction, steel decking is generally used. The depth of this decking, when running perpendicular to the beam, is ignored in the calculation. For a solid slab or a slab with participating precast concrete permanent formwork, the total depth of the concrete construction should be taken into consideration. The following derivation of the plastic moment of resistance for different conditions of loading illustrates the principle. For the benefit of designers the formulae are presented in a tabular form for easy reference and application – see Tables 8.3 and 8.4 for plastic moment capacities under different conditions of loading, geometric proportions and slenderness of the elements.

A = area of the steel beam
A_r = area of the tensile reinforcement in the effective cross-section
B = the width of the flange of the steel beam
B_e = effective width of the concrete flange
D_p = depth of the profiled sheet
D_r = distance from the top of steel beam to centroid of tensile reinforcement
D_s = overall depth of the concrete flange
d = clear depth of web of the steel beam
f_{cu} = characteristic 28-day cube strength of concrete

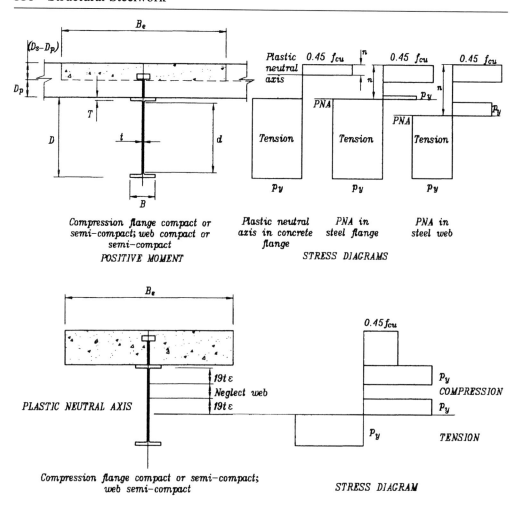

SK 8/5 Plastic section properties for positive moments.

f_y = characteristic yield strength of steel reinforcement
M_c = plastic moment capacity of the composite section
M_s = plastic moment capacity of the steel beam
N = actual number of shear connectors for positive or negative moment as relevant
n = depth of plastic neutral axis (PNA) from top of the concrete flange
p_y = design strength of steel beam
Q = capacity of shear connectors for positive or negative moment as relevant
R_c = resistance of the concrete flange = $0.45 f_{cu} B_e (D_s - D_p)$
R_f = resistance of the flange of the steel beam = $BT p_y$
R_n = resistance of slender steel beam = $R_s - R_v + R_o$
R_o = resistance of slender web of the steel beam = $38\varepsilon t^2 p_y$

Design of Composite Beams and Columns

R_q = resistance of shear connection = NQ
R_r = resistance of steel tensile reinforcement = $0.87 f_y A_r$
R_s = resistance of steel beam = $A p_y$
R_v = resistance of clear web depth = $dt p_y$
R_w = resistance of overall web depth = $R_s - 2R_f$
T = thickness of the flange of steel beam
t = thickness of the web of steel beam

$$\varepsilon = \sqrt{\frac{275}{p_y}}$$

Equating tensile and compressive internal force in the section:

$$0.45 f_{cu} B_e n = A p_y$$

$$\therefore \quad n = \frac{A p_y}{0.45 f_{cu} B_e} = \frac{R_s}{R_c}(D_s - D_p)$$

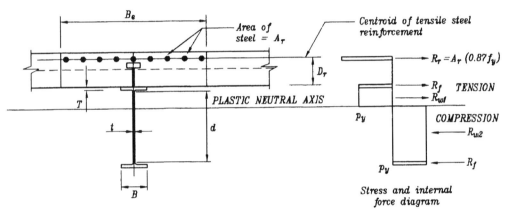

Compression flange compact or semi-compact;
web compact or semi-compact

NEGATIVE MOMENT

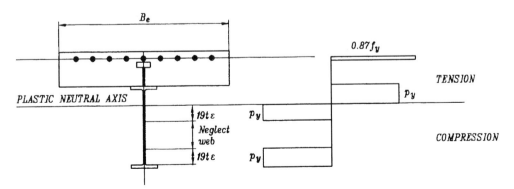

Compression flange compact or semi-compact;
web semi-compact

STRESS DIAGRAM

NEGATIVE MOMENT

SK 8/6 Plastic section properties for negative moments.

Table 8.3 Positive plastic moment capacities of composite sections.

Location of neutral axis	Relationship of resistances	Classification of web	Positive plastic moment capacity with full shear connection	Positive plastic moment capacity with partial shear connection
Neutral axis in web of steel beam	$R_c < R_w$ or $R_q < R_w$	Compact $d/t \leq 76\varepsilon$ or $\dfrac{d}{t} \leq \dfrac{76\varepsilon}{1-(R_c/R_v)}$ or $\dfrac{d}{t} \leq \dfrac{76\varepsilon}{1-(R_q/R_v)}$	$M_c = M_s + R_c\left[\dfrac{D + D_s + D_p}{2}\right]$ $-\dfrac{R_c^2}{R_v}\dfrac{d}{4}$	$M_c = M_s + R_q\left[\dfrac{D}{2} + D_s - \dfrac{R_q}{R_c}\dfrac{(D_s - D_p)}{2}\right]$ $-\dfrac{R_q^2}{R_v}\dfrac{d}{4}$
Neutral axis in web of steel beam	$R_c < R_w$ or $R_q < R_w$	Not compact $\dfrac{d}{t} > \dfrac{76\varepsilon}{1-(R_c/R_v)}$ or $\dfrac{d}{t} > \dfrac{76\varepsilon}{1-(R_q/R_v)}$	$M_c = M_s + R_c\left[\dfrac{D + D_s + D_p}{2}\right]$ $-\left[\dfrac{R_c^2 + (R_v - R_c)(R_v - R_c - 2R_o)}{R_v}\right]\dfrac{d}{4}$	$M_c = M_s + R_q\left[\dfrac{D}{2} + D_s - \dfrac{R_q}{R_c}\dfrac{(D_s - D_p)}{2}\right]$ $-\left[\dfrac{R_q^2 + (R_v - R_q)(R_v - R_q - 2R_o)}{R_v}\right]\dfrac{d}{4}$
Neutral axis in flange of steel beam	$R_c \geq R_w$ $R_s > R_c$ or $R_q \geq R_w$	Not required	$M_c = R_s\dfrac{D}{2} + R_c\left(\dfrac{D_s + D_p}{2}\right)$ $-\dfrac{(R_s - R_c)^2}{R_f}\dfrac{T}{4}$	$M_c = R_s\dfrac{D}{2} + R_q\left[D_s - \dfrac{R_q}{R_c}\left(\dfrac{D_s - D_p}{2}\right)\right]$ $-\dfrac{(R_s - R_q)^2}{R_f}\dfrac{T}{4}$
Neutral axis in concrete flange	$R_c \geq R_w$ $R_s \leq R_c$ $R_q < R_s$	Not required	$M_c = R_s\left[\dfrac{D}{2} + D_s - \dfrac{R_s}{R_c}\dfrac{(D_s - D_p)}{2}\right]$	$M_c = R_q\left[\dfrac{D}{2} + D_s - \dfrac{R_q}{R_c}\dfrac{(D_s - D_p)}{2}\right]$

Table 8.4 Negative plastic moment capacities of composite sections.

Location of neutral axis	Relationship of resistances	Classification of web	Negative moment capacity with full shear connection $N_N = F_N/Q_N$
Neutral axis in web of steel beam	$R_r < R_w$	Compact web $d/t \leq 38\varepsilon$ or $\dfrac{d}{t} \leq \dfrac{76\varepsilon}{1+(R_r/R_v)}$	$M_c = M_s + R_r\left(\dfrac{D}{2} + D_r\right) - \dfrac{R_r^2}{R_v}\dfrac{d}{4}$
	$R_r < R_w \quad R_r < R_o$	Web not compact $d/t > 38\varepsilon$ or $\dfrac{d}{t} > \dfrac{76\varepsilon}{1+(R_r/R_v)}$	$M_c = M_s + R_r\left(\dfrac{D}{2} + D_r\right)$ $- \left[\dfrac{R_r^2 + (R_v + R_r)(R_v + R_r - 2R_o)}{R_v}\right]\dfrac{d}{4}$
Neutral axis in flange	$R_r \geq R_w; \quad R_r < R_s$	Compact web $d/t \leq 38\varepsilon$	$M_c = R_s\dfrac{D}{2} + R_rD_r - \dfrac{(R_s - R_r)^2}{R_f}\dfrac{T}{4}$
	$R_r \geq R_w; \quad R_r \geq R_s$	$d/t \leq 38\varepsilon$	$M_c = R_s\left(\dfrac{D}{2} + D_r\right)$
Neutral axis in flange	$R_r \geq R_o; \quad R_r < R_n$	Web not compact $d/t > 38\varepsilon$	$M_c = R_n\dfrac{D}{2} + R_rD_r - \dfrac{(R_n - R_r)^2}{R_f}\dfrac{T}{4}$
	$R_r \geq R_o; \quad R_r \geq R_n$	$d/t > 38\varepsilon$	$M_c = R_n\left(\dfrac{D}{2} + D_r\right)$

Taking moments about the top of the concrete flange:

$$M_c = Ap_y\left(D_s + \frac{D}{2}\right) - 0.45f_{cu}B_e\frac{n^2}{2} = R_s\left[D_s + \frac{D}{2} - \frac{R_s}{R_c}\left(\frac{D_s - D_p}{2}\right)\right]$$

Equating tensile and compressive internal forces (see SK 8/6):

$$R_r + R_f + R_{w1} = R_f + R_{w2}$$

$$R_r = R_{w2} - R_{w1} = (d - n - n)tp_y = (d - 2n)tp_y$$

$$\therefore \quad n = \left(d - \frac{R_r}{tp_y}\right)\frac{1}{2} = \frac{R_v - R_r}{2tp_y}$$

Taking moments about the bottom of steel beam:

$$M_c = R_r(D_r + T) + R_f\left(\frac{T}{2}\right) - 2R_{w1}\frac{n}{2} + R_v\frac{d}{2} + R_f\left(d + \frac{T}{2}\right)$$

Substituting

$$R_{w1}n = n^2 tp_y = R_v\left(\frac{d}{4}\right) + \left(\frac{R_r^2}{R_v}\frac{d}{4}\right) - \left(\frac{2R_r d}{4}\right)$$

and

$$M_s = R_f(d + t) + R_v\frac{d}{4}$$

The simplified form of M_c is given by:

$$M_c = R_r\left(D_r + \frac{D}{2}\right) + M_s - \frac{R_r^2}{R_v}\frac{d}{4}$$

See Tables 8.3 and 8.4 for positive and negative moment capacities of composite sections at the ultimate limit state for steel beams with equal flanges.

8.1.8 Redistribution of support moments

If a frame or a continuous beam analysis is carried out assuming a uniform uncracked moment of inertia based on the section at the midspan of the beam, there is a large overestimation of stiffness at the supports and consequently the support negative moments are large. The redistribution of moment is allowed from the support to the midspan but any redistribution means that a plastic hinge is allowed to form and this hinge must have adequate rotational ductility. For slender sections the rotational ductility is absent; for plastic sections it is assured. Therefore, depending on the initial method of analysis using a cracked or uncracked moment of inertia, and also depending on the classification of the steel beam, various maximum percentages of redistribution may be allowed. Redistribution of moments will inevitably mean a reduction of robustness for serviceability and the deflection or crack control criteria may govern the design. Up to 50% redistribution may be allowed for a special category of composite

Design of Composite Beams and Columns

Table 8.5 Maximum permissible percentage of redistribution of support moments.

Type of analysis	Classification of compression flange at the supports of continuous beam				
	Class 4 slender	Class 3 semi-compact	Class 2 compact	Class 1 plastic (general)	Class 1 plastic (unreinforced)
Elastic analysis using gross uncracked section properties	10%	20%	30%	40%	50%
Elastic analysis using cracked section properties	0%	10%	20%	30%	40%

beam where only nominal reinforcement is provided in the concrete slab and this reinforcement is not taken into account for the determination of plastic moment capacity for negative moments. In this case, the steel beam should be class 1 plastic. Table 8.5 gives the maximum redistribution percentages.

8.1.9 Shear connection

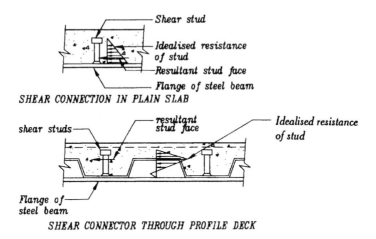

SK 8/7 Shear connectors in plain slabs and through profiled sheeting.

Shear connection is generally by studs. These studs may come ready-welded on the beam from the shops or they can be shot fired after erection of the beams. The most modern method is to attach the studs by through-deck welding. The characteristic strengths of standard shear connectors are given in Table 5 of BS 5950: Part 3.1. The strength is dependent on the strength of concrete and the diameter of the stud, assuming a nominal length of the stud. The head of the stud is required to prevent uplift.

Failure by shear is essentially brittle in nature. To allow for a level of ductility and also to include a measure of the material factor, the design strength of the shear connector is taken as 80% when the concrete is in compression and 60% when the concrete is in tension. The adequacy of shear connectors must be checked at locations of high shear under concentrated loads, at the points of sudden change in the cross-section of the beam, at numerous points on the tapered section of the beam and also frequently on the span if the concrete flange is large. Normally, however, only two checks are made for the adequacy of shear connectors – one at midspan for positive moment and one at the support for negative moment.

Partial shear connection is sometimes used to economise on construction costs. This involves reducing the number of shear connectors required for positive moments only. Table 8.3 gives the plastic moment capacities for partial shear connection where the whole compressive flange load of the concrete cannot be transferred by the connectors, and the shear connector resistance R_q replaces the compressive flange force R_c or steel beam tensile force R_s, whichever is the smaller. The number of shear connectors required to transfer negative moment must not be reduced.

8.1.10 Longitudinal shear transfer

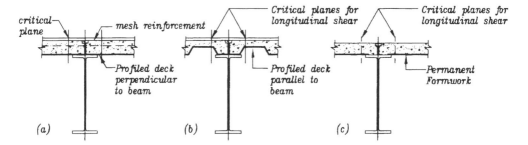

SK 8/8 Failure planes in slab in longitudinal shear.

The shear connectors transfer shear to the concrete, and the concrete section resisting this shear should therefore be checked for cracking or splitting. The concrete slab resists the longitudinal shear along the length of the beam by shear resistance along the critical shear crack planes. In an internal beam there are many such critical sections on either side of the beam.

The profile sheeting and the transverse reinforcement contribute in carrying this longitudinal shear by dowel action. The British Standard code allows the inclusion of these components in the calculation of total shear resistance along the length of the beam.

8.1.11 Reduction of plastic moment capacity due to high shear

As in the case of ordinary steel beams, the composite beam has to be checked with respect to the effect of high shear load and bending moment acting simultaneously. The plastic moment capacity of the section reduces with shear after the shear in the web of the steel section reaches a value greater than $0.5P_v$, where P_v is the shear capacity of the web taking into account any reduction due to shear buckling effects. The total shear in a composite section must be taken by the steel section. The reduction in capacity is due to progressive ineffectiveness of the web in carrying bending stresses when the shear stress gets higher and higher and reaches the maximum value P_v. When applied shear F_v reaches P_v of the section, the contribution of the web towards the plastic moment of resistance of the composite section vanishes.

8.1.12 Deflections

The limit of deflection of beams for imposed loads as per BS 5950: Part 1 is generally $l/200$. To determine the deflection of a simply supported composite beam, the gross uncracked section properties may be used. For further refinement, if the neutral axis happens to be in the concrete flange, cracked section properties may be used for a part of the beam near the middle. For a continuous beam, the best way to find the deflection due to imposed load is to carry out computer analysis using the incremental loading method. Firstly, put all the dead load on the steel beam for unpropped construction. At this stage the whole beam should be fully elastic. For the composite section find section properties for negative moment and positive moment. Model the beam with assumed points of contraflexure and beam section properties corresponding to positive or negative moment. Progressively load the beam, firstly with permanent imposed loading and then, stage by stage, with further imposed loading in a pattern loading format to cause maximum support bending moment. The first plastic hinges will form at the supports when plastic capacities are reached. Remodel the beam with hinges at the supports until the span becomes simply supported by the formation of hinges at the two ends of the beam. Load this beam until all of the imposed service load is used up. Add the progressive deflection at every stage of loading from the end of the dead load application and this will give the total deflection due to the imposed loading.

The British Standard code allows the use of an approximate formula for finding this imposed load deflection. (Note that plastic hinges at service load for deflection calculations form only in cases of high levels of redistribution of moments.)

8.1.13 Effect of dead load

In a simply supported beam, at service load the maximum fibre stresses can be calculated using the transferred section properties. These stresses must be less than p_y for steel and $0.5f_{cu}$ for concrete. When unpropped construction is used, the stresses should be found in three stages and the effects added together. In the first stage for the dead load of wet concrete calculate the stresses in the steel beam, taking the section properties of the steel beam only.

8.1.14 Effect of shrinkage and creep

The second stage is the calculation of stresses in the steel beam and the concrete flange due to shrinkage of the concrete. The concrete flange will try to shrink but the steel beam will resist this shrinkage. This will create a compressive load at the level of the flange of the steel beams. The compressive load on the top flange creates a bending of the composite beam, increasing the tension in the bottom flange of the steel beam and compression in the concrete flange. The shrinkage strain for an internal member in a heated building may be of the order of 300×10^{-6} and for an external element it could be of the order of 100×10^{-6}. This shrinkage strain may be reduced by a factor of 0.5 to take into account the effect of creep.

The curvature of a simply supported composite beam due to the action of concrete shrinkage may be expressed as:

$$\frac{1}{R} = \frac{\varepsilon_s(D + D_s + D_p)A}{2(1 + \alpha_e r)I_c}$$

where ε_s = effective shrinkage strain in concrete
D = depth of steel beam
α_e = modular ratio
D_s = depth of concrete slab
D_p = depth of profile sheeting
A = cross-sectional area of the steel beam

$$r = \frac{A}{(D_s - D_p)B_e}$$

B_e = effective width of flange of the composite beam.

From the expression $M/EI = 1/R$ the bending moment in the composite section may be found. This bending moment may be used to calculate the stresses in the composite section due to shrinkage. Normally, these calculations are omitted for composite beams in buildings.

The deflection at the centre of the span of the composite beam due to shrinkage is given by:

$$\delta = 0.125 \frac{1}{R} L^2 = \frac{0.125 ML^2}{EI}$$

8.1.15 Effect of imposed loads

In the third stage, the stresses in the composite section due to imposed loads and any wind loads should be calculated using the transferred section properties of the composite section.

For composite beams one should use the incremental plastic analysis and the structural model should be changed as the hinges form. The stress checks are only to be carried out at the centre of the spans. For continuous beams where redistribution of moments have been allowed, the objective is to prevent any irreversible permanent

Design of Composite Beams and Columns 345

deformation/rotation at the centres of spans. The cracking or rotation over the supports may be ignored. The stresses at service load at midspan must be kept below p_y at extreme fibres of steel beam and at a maximum of $0.5f_{cu}$ in the concrete flange. The stress distribution at service load in the composite section is assumed to be triangular. For unpropped construction, the same three-stage investigation of stresses should be carried out and the total stress at the point of interest is the sum of the stresses from all three stages.

8.1.16 Step-by-step design procedure for composite beams

Step 1 *Select type of beam and slab construction*
There are three different elements in the construction of the composite beam. The type, depth, material properties and the method of connection should be decided at the onset of the design process for all three elements. These elements are as follows.

Reinforced concrete slab: The depth, grade of concrete and amount of reinforcement will depend on the span of the slab, the imposed loads and also whether the slab is designed as continuous or simply supported. Guidance should be sought from books on reinforced concrete design and construction and BS 8110: Part 1.

Profile steel sheeting: If used, the guidance on depth, thickness, material properties and connection system should be obtained from the manufacturer's catalogue. The geometry of the profile sheeting will depend on the span, the loading and the construction method (propped or unpropped).

Steel beam: Generally, equal flanged steel UB sections will be most cost-effective. Grade 50 steel should be used whenever possible. Sometimes for very large span beams castellated sections are used. It should be borne in mind that additional fabrication costs can wipe out the savings in material costs. For unpropped construction in simply supported beams, a span to overall depth ratio of 20 is generally suitable whereas in continuous construction the same ratio could be 25.

Step 2 *Determine loading*
The loading on the beam has to be found for the construction stage and for the completed structure. Determine the loading for the construction stage at the serviceability and ultimate limit state. Similarly, determine the loading for the composite stage at the serviceability and ultimate limit state. The serviceability loading at the composite stage for unpropped construction is an additional loading to the stress state of the steel beam after construction. This means that the stress state at the service limit state for unpropped construction should be found by first finding the stress in the steel beam due to construction dead load only (do not consider temporary construction live load) and then adding to this stress the stresses due to other permanent and imposed loads at composite stage.

Step 3 Determine material and section properties

Modular ratio α_e

$$\alpha_e = \alpha_s + \rho_l(\alpha_l - \alpha_s)$$

α_l = the modular ratio for long-term loading taking into account creep of concrete
ρ_l = the proportion of total loading which is long term.
α_s = the modular ratio for short-term loading

Typical values of α_s and α_l are given below:
  Normal weight concrete: $\alpha_s = 6$; $\alpha_l = 18$
  Lightweight concrete: $\alpha_s = 10$; $\alpha_l = 25$

Select design strength p_y for structural steel section and f_y for steel reinforcement in the concrete.

Effective breadth of concrete flange B_e

b = actual breadth of flange on each side of centre-line of beam measured as half the distance to the adjacent beam

$b_e = \dfrac{L_z}{8} \leq b$ for slab spanning perpendicular to the beam.

$b_e = \dfrac{L_z}{8} \leq 0.8b$ for slab spanning parallel to the beam

B_e = total effective breadth = sum of the effective breadth b_e from each side of the centre-line of the beam
L_Z = effective span L for simply supported beam
     = distance between points of zero moment for continuous beam
For a continuous beam the distance L_Z may be taken as in SK 8/2.

Steel reinforcement in the concrete flange for negative bending over support may not be necessary after redistribution of support moments. Before carrying out the cracked section analysis of the continuous composite beam, a check should be made to determine if steel reinforcement is required to resist support moment after the maximum allowable percentage of redistribution. If the bending moment over the support after the maximum allowable redistribution is greater than the plastic moment capacity of the steel beam, then only the steel reinforcement becomes necessary.

Trial area of reinforcement A_r for negative bending moment, when necessary, may be obtained from:

$$A_r \leq \dfrac{dtp_y}{0.87f_y}\left[76\varepsilon\left(\dfrac{t}{d}\right) - 1\right]$$

The area of tensile reinforcement should not be so high as to shift the plastic neutral axis far towards the reinforcement, resulting in a large depth of the web being in compression. This may cause problems of compactness of the web, resulting in a decrease of plastic moment capacity.

Determine the location of the neutral axis and the second moment of area. See Tables 8.1 to 8.4 for section properties. Determine section properties for the transformed composite section.

Step 4 **Carry out analysis**

There are three basic decisions to be taken regarding the analysis:

(1) Type of analysis – elastic or plastic.
(2) Redistribution of moments in continuous construction (percentage).
(3) Type of shear connection – full or partial.

The bending moments and shear should be determined for three stages as follows.

(1) *Construction stage:* For unpropped construction, the bending moments should be found separately for dead load of wet concrete and for construction-imposed loading, including any stacking, heaping, dumping, plant and machinery etc. The steel beams have to be checked for construction loads, including imposed loads, but locked-in stresses for dead loads of wet concrete alone should be used to check the service limit state stress distribution in the steel beam.
(2) *Service limit state:* Analysis must be carried out to find the bending moment due to imposed loads after the concrete has hardened, when the beam behaves compositely with the concrete slab. For propped construction, the construction stage need not be investigated, and the service limit state should consider the combined dead and imposed load on the composite beam after the concrete has hardened and the props have been removed.
(3) *Ultimate limit state:* The full combined dead and imposed loads multiplied by appropriate load factors should be used to find the ultimate bending moments and shears. The composite section should be used to check the resistance.

Note: Recommended procedure for continuous composite beam is to find, by the use of recognised computer software, the bending moment, shear and deflections for unit distributed loading on single spans separately and for all spans under loaded conditions. The model of the beam should include cracked section properties for a distance of $0.15L$ on either side of the supports. By progressive loading of this model, the pre-determined limits of plastic hinges on the composite beam will be reached. When a hinge has formed, the model should be altered to allow for this hinge in the beam and subsequent loads should be applied on this altered model to find the final bending moments, shears etc. The pre-determined level of the maximum bending moment depends on the percentage of redistribution of support moment allowed.

Step 5 **Check section classification**

Steel beam: at construction stage without composite action
Check section classification as per Clause 3.5.2 and Table 7 of BS 5950: Part 1. See Chapter 7.

Composite beam: web
Check web classification as per Table 8.6 based on BS 5950: Part 3.1.

348 Structural Steelwork

Table 8.6 Limiting depth to thickness ratios of steel beam webs in composite construction.

Condition of web	Class of section		
	Class 1 plastic	Class 2 compact	Class 3 semi-compact
Web with neutral axis at mid-depth	$\dfrac{d}{t} \leq 64\varepsilon$	$\dfrac{d}{t} \leq 76\varepsilon$	$\dfrac{d}{t} \leq 114\varepsilon$
Web with neutral axis generally elsewhere	$\dfrac{d}{t} \leq \dfrac{64\varepsilon}{1+r}$	$\dfrac{d}{t} \leq \dfrac{76\varepsilon}{1+r}$	For rolled sections $\dfrac{d}{t} \leq \dfrac{114\varepsilon}{1+2r}$ when $r \geq 0.66$ For welded sections $\dfrac{d}{t} \leq \left(\dfrac{41}{r} - 13\right)\varepsilon$ when $r \geq 0.66$ For rolled and welded sections $\dfrac{d}{t} \leq \dfrac{114\varepsilon}{1+2r}$ when $0.66 > r \geq 0$ For rolled and welded sections $\dfrac{d}{t} \leq \dfrac{114\varepsilon(1+r)}{(1+2r)^{\frac{1}{2}}}$ when $r < 0$

r is called the web stress ratio. It is the ratio of the mean longitudinal stress in the web to the design strength p_y. Compressive stress is taken as positive and tensile stress is taken as negative.

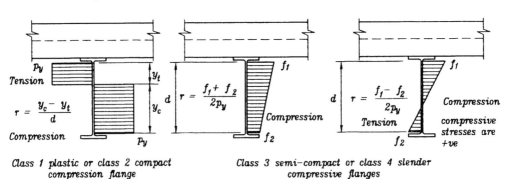

SK 8/9 Formulae for the determination of web stress ratio r for beams with unequal flanges.

General rules for flanges of steel beams in compression
(1) A steel compression flange of a beam connected by shear connectors to a solid concrete slab may be assumed to be class 1 plastic.
(2) A steel compression flange of a beam connected by shear connectors to a ribbed slab with profiled sheets, where either the ribs are at a maximum inclination of 45° to the axis of the beam or the minimum width of rib is not less than half the width of the flange of the steel beam, may be classified as plastic if it is plastic

or compact according to BS 5950: Part 1, and may be classified as compact if it is semi-compact according to BS 5950: Part 1.
(3) Use rectangular plastic stress distribution in the steel compressive flange if it is classified as compact or plastic.
(4) When the steel compressive flange is semi-compact or slender, assume triangular stress distribution, as in elastic analysis.
(5) If the web or the steel compressive flange is slender, use the stress reduction factors in Table 8 of BS 5950: Part 1.

General rules for web
(1) Calculate the web stress ratio r using the formulae given below for steel beams with equal flanges.

Elastic stress distribution
With elastic stress distribution, when flanges are semi-compact or slender or the web is slender, the following formulae may be used to find r for steel beams with equal flanges.

For positive moments: $\quad r = -\dfrac{F_c}{R_s} \geq -1$

For negative moments: $\quad r = \dfrac{F_r}{R_s} \leq 1$

where F_c = compressive force in the concrete flange
F_r = tensile force in steel reinforcement
R_s = resistance of steel beam = Ap_y

Plastic stress distribution
With plastic stress distribution, when steel beam flanges are equal and the steel compression flange is classed as compact or plastic and the web is not slender, the following formulae may be used to determine r.

For positive moments: $\quad r = -\dfrac{F_c}{R_v} \geq -1$

For negative moments: $\quad r = \dfrac{R_r}{R_v} \leq 1$

where F_c = compressive force in the concrete flange
$= R_c = 0.45 f_{cu} B_e (D_s - D_p) \leq R_q$
R_q = resistance of shear connectors = NQ
$R_v = dt p_y$ for the steel beam
$R_r = 0.87 f_y A_r$
A_r = area of tensile steel reinforcement

(2) Determine classification of web as per Table 2 of BS 5950: Part 3.1.
(3) Use the formulae in Tables 8.3 and 8.4 according to whether the web is plastic, compact or semi-compact.
(4) When the web is slender use the design stress reduction factor as in Clause 3.6 and Table 8 of BS 5950: Part 1.

Step 6 Determine ultimate plastic moment capacities and check shear capacity

Use the formulae in Tables 8.3 and 8.4 according to positive moment or negative moment capacity. Check that the section capacities are adequate to resist the effects of the applied loading.

Determine section shear capacity P_v as per BS 5950: Part 1.

Note: P_v is the lesser of shear capacity and shear buckling resistance as per BS 5950: Part 1.

Shear capacity $P_v = 0.6 p_y A_v$

A_v is given by:

For rolled I-, H- and channel sections with loads parallel to the web $= tD$

For built-up sections and boxes with loads parallel to the web $= td$

For solid bars and plates $= 0.9A$

For rectangular hollow sections with loads parallel to the webs

$$= \left(\frac{D}{D+B}\right) A$$

For circular hollow sections $= 0.6A$

For any other case $= 0.9 A_0$

where t = total web thickness
 B = breadth of section
 D = overall depth of section
 d = depth of the web
 A = area of the section
 A_0 = area of the rectilinear element of the section which has the largest dimension parallel to the direction of the load

The shear capacity is governed by shear buckling when the ratio d/t exceeds 63ε.

Shear buckling resistance $= V_{cr} = q_{cr} dt$

The critical shear strength q_{cr} can be obtained from Tables 21(a) to 21(d), as appropriate, in BS 5950: Part 1. The stiffener spacing ratio a/d may be taken as infinity for steel beams without stiffeners, where a is the spacing of stiffeners.

Reduction of plastic moment capacities due to high shear

If the applied shear F_v exceeds $0.5 P_v$, then the reduced plastic moment capacity M_{cv} is given by:

$$M_{cv} = M_c - (M_c - M_f)\left(\frac{2F_v}{P_v} - 1\right)^2$$

where M_c = plastic moment capacity of the composite section
 M_f = plastic moment capacity of the section remaining after deduction of shear area A_v

For sections with a compression flange which is semi-compact, or a web or a compression flange which is slender, M_{cv} should not be taken as greater than the elastic moment capacity.

Step 7 ***Check stability of compression flange (lateral torsional buckling)***
(During construction and as composite beams)
The following rules apply:

(1) The compression flange in a simply supported beam will be deemed to be fully restrained by the concrete slab or by the profiled steel sheet, if used, at the hardened construction stage. Otherwise, during construction, appropriate restraint of the compression flange has to be provided by temporary bracing.
(2) The compression top flange at the centre of a continuous beam may be regarded as fully restrained by the hardened concrete slab or by the profiled steel sheeting during construction, if used as permanent formwork.
(3) The compression bottom flange at the support of a continuous beam must be laterally and torsionally restrained at the column where plastic hinges may be allowed to form due to redistribution of moments.
(4) It is assumed that the hardened concrete slab or the profiled steel sheeting will offer the overall buckling resistance of the bottom compression flange of the continuous beam by providing the lateral restraint at the top flange. The check of this stability should be carried out using the method in BS 5950 Part 1: Appendix G: Design of restrained members with unrestrained compression flange.

Steps to follow to carry out stability check as per BS 5950: Part 1: Appendix G

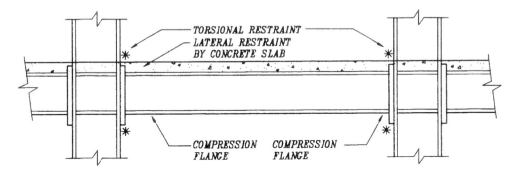

SK 8/10 Stability check as per BS 5950: Part 1: Appendix G.

(1) Take L to be the distance between the points where both flanges are torsionally restrained — usually at the columns for the continuous beam. Additional torsional stiffeners can be provided in the span if required.
(2) Select r_y, which is the minor axis radius of gyration of the steel beam only.
(3) Determine $\lambda = L/r_y$.
(4) Determine a, the distance between the centre of the steel beam and the centre of the concrete slab (or the centre of the profiled

sheet during construction), which is the distance between the reference axis of rotation and the restraint axis.

(5) Determine h_s, the distance between the centre of the flanges of the steel beam, which is the distance between the shear centre of the flanges.

(6) Determine x, the torsional index of the steel beam, from published tables or by using the formulae given in Chapter 7.

(7) Take u as 0.9 for rolled I-, H- and channel sections, or as 1.0 for other sections.

(8) Determine $Y = \left[\dfrac{1 + \left(\dfrac{2a}{h_s}\right)^2}{1 + \left(\dfrac{2a}{h_s}\right)^2 + \dfrac{1}{20}\left(\dfrac{\lambda}{x}\right)^2} \right]^{\frac{1}{2}}$

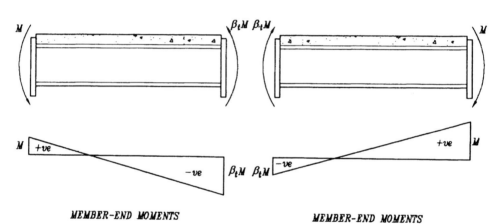

MEMBER-END MOMENTS   MEMBER-END MOMENTS

MOMENT PRODUCING TENSION AT THE TOP OF BEAM IS +ve

SK 8/11 Determination of β_t.

(9) Determine $\beta_t = \dfrac{\text{algebraically smaller end moment}}{\text{algebraically higher end moment}} \geq -1$

Note that in the determination of the algebraically smaller end moment, a negative moment is smaller than zero moment.

(10) Determine the equivalent uniform moment factor m_t from Table 39 of BS 5950: Part 1: Appendix G using values of Y and β_t.

(11) Draw bending moment diagram for the beam and determine:

$$n_t = \left[\dfrac{1}{12} \left\{ \dfrac{N_1}{M_1} + \dfrac{3N_2}{M_2} + \dfrac{4N_3}{M_3} + \dfrac{3N_4}{M_4} + \dfrac{N_5}{M_5} + 2\left(\dfrac{N_S}{M_S} - \dfrac{N_E}{M_E}\right) \right\} \right]^{\frac{1}{2}}$$

where N_1 to N_5 are bending moments at the quarter points starting from end 1. Only positive values of these moments should be considered in the determination of n_t, positive moments being those which cause compression in the unrestrained flange. The beam is divided into four equal parts

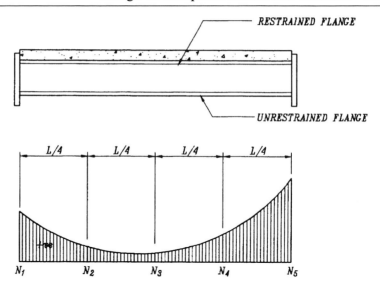

SK 8/12 **Bending moments at intermediate points on the beam.**

between torsional restrains to both flanges. The node numbers of these parts are 1 to 5 and the bending moments N_1 to N_5 correspond to these node numbers. N_5 is the bending moment at end 2 of the beam. The plastic moment capacities at the nodes 1 to 5 are designated as M_1 to M_5 respectively.

N_S/M_S is the greatest of all the span moment ratios (i.e. the greater of N_2/M_2, N_3/M_3 and N_4/M_4).

N_E/M_E is the greatest of the end moment ratios (i.e. the greater of N_1/M_1 and N_5/M_5).

Only positive values of the expression $[(N_S/M_S) - (N_E/M_E)]$ should be included in the calculation of n_t.

For sections where an elastic linear stress distribution is assumed, the moment capacity is given by $M = p_y Z_x$, where Z_x is the elastic section modulus.

For sections where full plastic stress distribution can take place, the moment capacity is given by $M = p_y S_x$, where S_x is the plastic section modulus.

Note: For most applications the factor n_t may be conservatively assumed to be 1. n_t is also taken as 1 where there are no intermediate loads between restraints.

(12) Determine c (which is taken as 1.0 for uniform members). For tapered members c is given by:

$$c = 1 + \frac{3}{x-9}(R-1)^{\frac{2}{3}} q^{\frac{1}{2}}$$

where R is the ratio of the greater depth to the lesser depth between effective torsional restraint and q is the ratio of the tapered length to the total length between effective torsional restraints.

(13) Determine $\lambda_{TB} = n_t u v_t c \lambda$

$$v_t = \left[\frac{\dfrac{4a}{h_s}}{1 + \left(\dfrac{2a}{h_s}\right)^2 + \dfrac{1}{20}\left(\dfrac{\lambda}{x}\right)^2} \right]^{\frac{1}{2}}$$

(14) Using the values of λ_{TB} and p_y determine p_b from Table 11 or Table 12 of BS 5950: Part 1.

(15) Determine equivalent uniform moment $\bar{M}$ given by:

$$\bar{M} = m_t M_A$$

where M_A is the maximum moment in the length of the member under consideration.

(16) Determine buckling moment of resistance M_b given by:

$$M_b = p_b S_x \leq p_y Z$$

(17) Check $\bar{M}/M_b \leq 1$.

Step 8 Shear connection

The characteristic strength of shear connectors commonly used in the form of headed studs is given in Table 8.7 for different grades of concrete. Use the strength for 40 N/mm² for concrete of higher grades. This is according to BS 5950: Part 3.1.

Determine the capacity of the shear connector in the positive moment region, which is given by:

$$Q_p = 0.8 Q_K$$

Determine the capacity of the shear connector in the negative moment region, which is given by:

$$Q_n = 0.6 Q_K$$

When the design is based on plastic moment capacity of the composite section, determine F_P, the longitudinal compressive force in the concrete slab due to maximum positive moment:

$$F_P = 0.45 f_{cu} B_e (D_s - D_p) \quad \text{or} \quad A p_y \quad \text{(whichever is the lower)}$$

Table 8.7 Characteristic strength Q_K of headed studs (kN).

Dimensions of stud shear connector		Strength of normal weight concrete			
Flange diameter	Nominal height	25 N/mm²	30 N/mm²	35 N/mm²	40 N/mm²
25	100	146	154	161	168
22	100	119	126	132	139
19	100	95	100	104	109
19	75	82	87	91	96
16	75	70	74	78	82
13	65	44	47	49	52

Note: Use 90% of the values in Table 8.7 for lightweight aggregate concrete.

For partial shear connection this force may be taken as equal to $R_q < R_c$ and R_s. (See Step 4 for an explanation of the notations.)

When the design is based on elastic moment capacity of the composite section, determine the longitudinal compressive force in the concrete slab using the trapezoidal stress distribution and using the effective breadth B_e.

$$F_P = 0.5(f_1 + f_2)B_e D_c$$

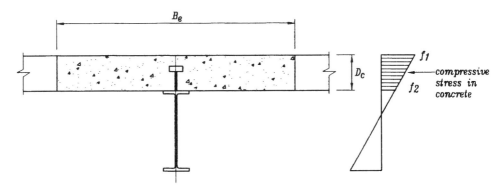

SK 8/13 Compressive force in concrete flange.

where f_1 and f_2 are the compressive stresses at the top and bottom of the concrete section in compression and D_c is the depth of the concrete section in compression.

Determine the longitudinal tensile force in the reinforcement within the negative moment region:

$$F_n = 0.87 f_y A_r$$

Determine the number of shear connectors required due to positive moment, which is given by:

$$N_p = F_p / Q_p$$

Determine the number of shear connectors required due to negative moment, which is given by:

$$N_n = F_n / Q_n$$

The total number of shear connectors required over a length from the point of maximum positive bending moment to each adjacent support is:

$$N = N_p + N_n$$

For partial shear connection, the number of shear connectors required to resist negative moment N_n should not be reduced.

Spacing of shear connectors: Method 1
The following steps may be followed.

(1) Find the points on the beam where bending moment becomes zero. These points are at the supports for simply supported

beams and at the point of contraflexure for a continuous beam. The point of contraflexure should be determined after redistribution of moments.

(2) Find the distance l_n from the point of contraflexure to the adjacent support and l_p from the point of contraflexure to the point of maximum positive moment in the span.

(3) Find the uniform spacing for negative moment, which is given by:

$$S_n = \frac{l_n}{\dfrac{N_n}{J} - 1}$$

where J is the number of shear connectors in a row perpendicular to the longitudinal axis of the beam.

(4) Find the uniform spacing for positive moment, which is given by:

$$S_p = \frac{l_p}{\dfrac{N_p}{J} - 1}$$

(5) Find the uniform spacing for combined positive and negative moment, which is given by:

$$S_b = \frac{l_p + l_n}{\dfrac{N_p + N_n}{J} - 1}$$

(6) The uniform spacing S is the smallest of S_n, S_p or S_b. Adopt shear connectors at uniform spacing S on the beam with J numbers on each row. This method is very conservative and no more additional checks are necessary. It is assumed that full shear connection is required. Uniform spacing of shear connectors improves buildability and reduces the overall time and cost, especially if they are shop-welded by semi-automatic machines.

Spacing of shear connectors: Method 2
The following steps may be followed.

(1) Find the points on the beam where the negative steel over the support can be fully curtailed, taking into account the recommendations of BS 8110 regarding curtailment. The bars have to extend at least 12 diameters beyond the point where they are not required, provided the full tension anchorage length has been provided from the point of maximum negative moment.

(2) Find the distance of the point of curtailment from the adjacent support l_n and find the spacing S_n, which is given by:

$$S_n = \frac{l_n}{\left(\dfrac{N_n}{J}\right)}$$

Use this spacing over the support and into the span up to a length $l_n - S_n$ from the support.

(3) Find l_p, the distance from the point $l_n - S_n$ from support to the point where the positive moment is maximum. Find the spacing S_p, which is given by:

$$S_p = \frac{l_p}{\dfrac{N_p}{J} - 1}$$

Use spacing S_p over the positive moment region in the span.

Spacing of shear connectors: Method 3
After selecting a variable zonal spacing of shear connectors, the beam may be divided into zones. Within each zone, the spacing of shear connectors is kept constant. At all changes of spacing of the shear connectors, the following conditions should be satisfied. (N_j is the total number of shear connectors from the point of interest to the nearest support.)

For positive moment zones: $N_j = N_p \dfrac{M - M_s}{M_c - M_s} + N_n$ but $N_j \geq N_n$

For negative moment zones: $N_j = N_n \dfrac{M_c - M}{M_c - M_s}$ but $N_j \leq N_n$

M is the moment at the point under consideration
M_c is the positive or negative moment capacity of the composite section, as appropriate
M_s is the moment capacity of the steel section

Additional check of shear connectors
Additional checks at points on the span are necessary following method 3. These are needed at points where (1) a heavy concentrated load is applied in the span, (2) a sudden change of cross-section occurs, (3) at the point of tapering of the beam and (4) where the concrete flange is so large that M_c is more than 2.5 times M_s.

Reduction of stud capacity in profiled sheets
Ribs perpendicular to beam
The stud capacities Q_p and Q_n have to be reduced by the following factors when profiled steel sheets are used as permanent formwork. The reduction factor is given by:

For one stud per rib: $k = 0.85 \left(\dfrac{b_r}{D_p}\right)\left(\dfrac{h}{D_p} - 1\right) \leq 1$

For two studs per rib: $k = 0.60 \left(\dfrac{b_r}{D_p}\right)\left(\dfrac{h}{D_p} - 1\right) \leq 0.8$

For three or more studs per rib: $k = 0.50 \left(\dfrac{b_r}{D_p}\right)\left(\dfrac{h}{D_p} - 1\right) \leq 0.6$

where b_r = breadth of the concrete rib (see SK 8/14).
   D_p = overall depth of profiled sheet
   h = overall height of stud (not to be taken more than equal to $2D_p$ or $D_p + 75$, whichever is less)

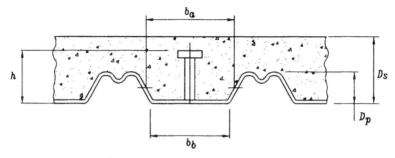

(a) OPEN TROUGH PROFILE $b_r = b_a$

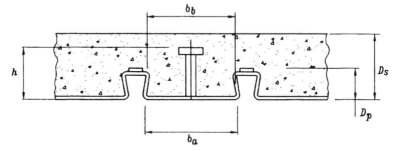

(b) RE-ENTRANT TROUGH PROFILE $b_r = b_b$

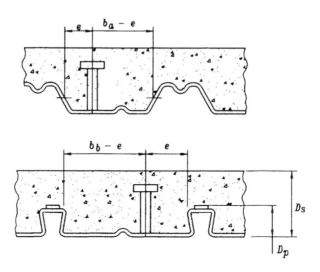

(c) NON-CENTRAL STUD

SK 8/14 Breadth of concrete rib b_r.

Ribs parallel to beam

$$k = 1 \quad \text{for} \quad \frac{b_r}{D_p} \geq 1.5$$

$$k = 0.6 \left(\frac{b_r}{D_p}\right)\left(\frac{h}{D_p} - 1\right) \quad \text{for} \quad \frac{b_r}{D_p} < 1.5 \quad \text{but } k \leq 1$$

Design of Composite Beams and Columns

Ribs at an angle θ to beam longitudinal axis

$$k = k_1 \sin^2 \theta + k_2 \cos^2 \theta$$

where k_1 = the reduction factor for ribs perpendicular to the beam
k_2 = the reduction factor for ribs parallel to the beam

Dimensions of studs and spacing
See SK 8/15.
Diameter of shank = d
Diameter of head of stud = $1.5d$
Depth of head of stud = $0.4d$
Ultimate tensile strength of stud material = $450 \, \text{N/mm}^2$
Maximum spacing of studs = 600 mm or $4D_s$ (whichever is less)
D_s = overall depth of slab
Minimum clear edge distance from the shank to the edge of flange of steel beam = 20 mm
Minimum longitudinal spacing along the beam = $5d$ centre-to-centre of transverse lines of connectors
Minimum transverse spacing across the beam = $4d$ centre-to-centre of longitudinal lines of connectors
Minimum transverse spacing for staggered connectors = $3d$ centre-to-centre of longitudinal lines
Maximum diameter of stud shear connector = 2.5 times the flange thickness of the steel beam

Partial shear connection for positive moment
The actual number of shear connectors used for positive moment connection is N_a, which may be less than N_p.
For beams up to 10 m span: $N_a \geq 0.4 N_p$
For beams over 16 m span: $N_a = N_p$
For beams between 10 and 16 m span: $\dfrac{N_a}{N_p} = \dfrac{L-6}{10} \geq 0.4$

where L = span in metres
The resistance of partial shear connector = $R_q = N_a Q_p$

Step 9 Check longitudinal shear

Applied longitudinal shear
Applied longitudinal shear on the concrete flange per unit length is given by:

$$v = \frac{NQ}{S}$$

where N = total number of shear connectors in a spacing S
$Q = Q_p$ or Q_n depending on whether positive or negative moment is connected.

Firstly, select minimum spacing in the positive moment zone and find v from the above formula using Q_p. Secondly, select the minimum spacing in the negative moment zone and find v from the same formula using Q_n. Choose v as the greater of these two values.

360 Structural Steelwork

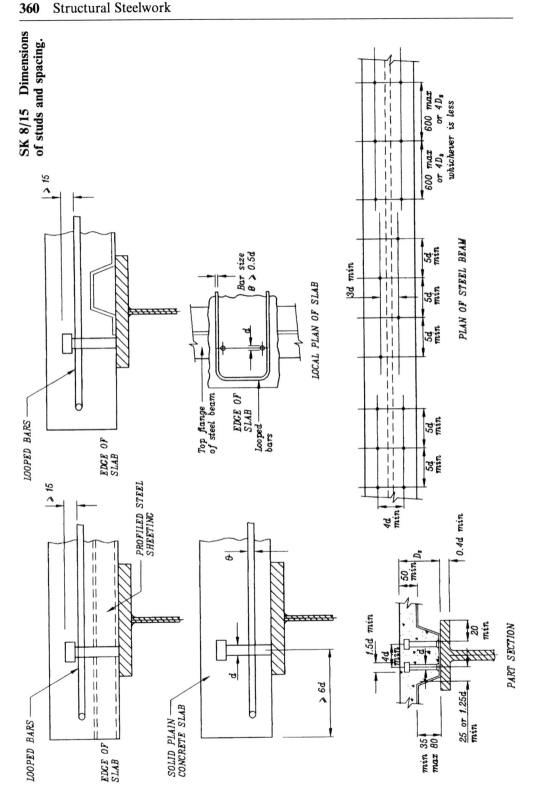

SK 8/15 Dimensions of studs and spacing.

Resistance to longitudinal shear

The resistance to longitudinal shear comes from the potential concrete shear surface (see SK 8/16), the dowel action of any anchored transverse reinforcement in the concrete slab and the contribution from the profiled steel sheeting.

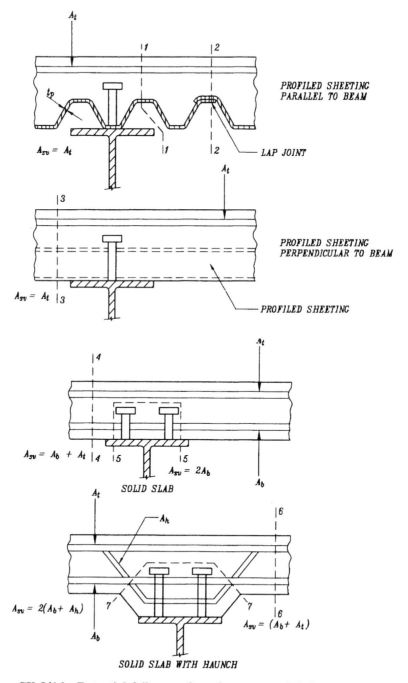

SK 8/16 Potential failure surfaces in concrete slab due to longitudinal shear.

The resistance per unit length of beam is given by:

$$v_r = 0.7 A_{sv} f_y + 0.03 \eta A_{cv} f_{cu} + v_p \leq 0.8 \eta A_{cv} \sqrt{f_{cu}} + v_p$$

where f_{cu} = characteristic strength of concrete
η = 1.0 for normal weight concrete
= 0.8 for lightweight aggregate concrete
A_{cv} = mean area of the concrete shear surface per unit length of beam for potential shear failure plane (see SK 8/16)
A_{sv} = cross-sectional area of the total steel reinforcement crossing the potential shear failure plane
v_p = contribution of the profiled steel sheeting, if applicable

For profiled sheets with ribs perpendicular to, and continuous over, the beam, v_p is given by:

$$v_p = t_p p_{yp}$$

where t_p = thickness of profiled steel sheet
p_{yp} = design strength of profiled steel sheet

For profiled sheets with discontinuity at the top of beams and stud shear connectors welded through the sheet to connect to the beam, v_p is given by:

$$v_p \left(\frac{N}{s}\right)(n\, d t_p p_{yp}) \leq t_p p_{yp}$$

where d = nominal shank diameter of the studs
n = 4 for most cases
N = number of shear connectors in a group
s = longitudinal spacing of groups of shear connectors

Check that $v_r \geq v$.

Where the ribs of the profiled sheeting run at an angle θ to the longitudinal axis of the beam, the resistance is given by:

$$v_r = v_1 \sin^2 \theta + v_2 \cos^2 \theta$$

where v_1 = value of v_r for ribs perpendicular to axis of beam
v_2 = value of v_r for ribs parallel to axis of beam

Step 10 **Check serviceability limit state**
Check deflection and stresses at service loads using the elastic properties of the composite section. For unpropped construction the full stress history has to be considered to include stage construction sequence.

Deflection
The deflections are to be determined for the serviceability limit state of loading only. For unpropped construction, the deflection due to imposed load is based on the properties of the composite section, but the deflection due to dead load of the concrete slab is based on the properties of the steel beam only. For propped construction all deflections are calculated using the properties of the composite section.

Deflection of composite simply supported beams
Deflection due to imposed load may be calculated using the properties of a composite beam with a gross uncracked section.

Deflection of composite continuous beams

$$\delta_c = \delta_0 \left[1 - 0.6 \frac{M_1 + M_2}{M_0}\right]$$

where δ_c = the deflection of the continuous composite beam at the middle of one of the spans due to imposed loading
δ_0 = the deflection of the same span due to same loading as a simply supported beam
M_0 = maximum span bending moment in the same span due to the same loading as a simply supported beam
M_1 and M_2 = the bending moments at the adjacent supports of the span due to same loading as a continuous beam (modified for patterned loading and shake-down effects)

Determination of M_1 and M_2 taking into account pattern loading and shake-down effects

Allowance for pattern loading
Load all spans of the continuous beam with unfactored imposed load and find M_1 and M_2 for the span under consideration. Reduce these moments by 30% for normal loading and 50% for storage loading (except adjacent to cantilevers) to account for pattern loading.

Allowance for shake-down effects
Shake-down effects are only to be considered if:

- plastic global analysis is used
- elastic global analysis is used with uncracked section properties and more than 40% redistribution of support moments
- elastic global analysis is used with cracked section properties and more than 20% redistribution of support moments

Determine M_1 and M_2 for unfactored dead load plus 80% of imposed loads for normal loading or determine M_1 and M_2 for unfactored dead load plus 100% of imposed loads for storage loads.

If moments M_1 and M_2 exceed the plastic moment capacity of the composite beam at supports 1 and 2, respectively, then the excess moments are called the shake-down moments.

Moments M_1 and M_2 found after the pattern loading correction should be further reduced by these shake-down moments. These finally reduced support moments should be used in the formula for deflection.

Correction of deflection due to partial shear connection

For propped construction: $\delta = \delta_c + 0.5\left(1 - \dfrac{N_a}{N_p}\right)(\delta_s - \delta_c)$

For unpropped construction: $\delta = \delta_c + 0.3\left(1 - \dfrac{N_a}{N_p}\right)(\delta_s - \delta_c)$

where δ = corrected final deflection of the composite beam allowing for partial shear connection
δ_c = deflection of the composite beam with full shear connection
δ_s = deflection of the steel beam alone subjected to the same loading
N_a = actual number of shear connectors used for positive moment, which should not be less than $0.4N_p$
N_p = the number of shear connectors required for full shear connection to develop positive moment capacity of the section

Step 11 Check pre-cambering requirement

Check the requirement of pre-cambering to counteract the deflection due to dead load of the wet concrete on the steel beam for unpropped construction. The construction sequence must be taken into consideration for continuous beams. All spans may not be poured at the same time and adjacent spans will be affected by stress and deflection according to the construction sequence followed.

Step 12 Check vibration

The natural frequency of the composite beam used in floor construction should be checked to prevent vibration induced oscillations. The frequency of the floor should be kept above 4 Hz to avoid susceptibility to vibration.

The natural frequency of a single degree of freedom system is given by:

$$f_n = \frac{1}{2\pi}\sqrt{\frac{K}{M}} \quad \text{(Hz)}$$

where K = stiffness of the system in N/m
M = mass of the system in kg

This expression can be rewritten as:

$$f_n = \frac{\sqrt{g}}{2\pi}\sqrt{\frac{K}{Mg}} = \frac{\sqrt{g}}{2\pi\sqrt{\delta_{st}}} = \frac{15.76}{\sqrt{\delta_{st}}}$$

where

$\delta_{st} = Mg/K$

= static deflection of the system due to gravity load Mg

The natural frequency of a composite beam may be found approximately by finding the deflection δ_{st} at midspan due to the service limit dead load, including all permanent loads plus an allowance for imposed load (10–50%).

Check $f_n = \dfrac{18}{\sqrt{\delta_{st}}} \geq 4\,\text{Hz}$

8.2 EXAMPLE 8.1: COMPOSITE BEAM

Design a four-span continuous composite beam to support the floor of a library. The stacking height of books should be taken as 3 m. The live loading on the floor is 2.4 kN/m² for every metre of stacking height. The beams are at 3 m centres. All four spans of the beam are 12 m. The construction of the floor will be unpropped.

Design parameters
Use trapezoidal profiled sheets for deck construction.

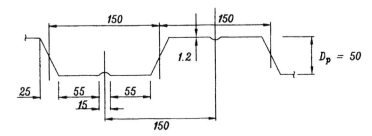

SK 8/17 Profiled sheet to be used for deck construction.

Height of profile = D_p = 50 mm
Spacing of trough = 300 mm
Width of trough = 150 mm
Thickness of steel = 1.2 mm
Self-weight of steel trough = 0.125 kN/m²
Shear connectors: 19 mm nominal diameter; 100 mm nominal height
Concrete deck slab:
  Overall thickness = D_s = 125 mm
  Density of concrete = 24 kN/m³
  Grade of concrete = 40 N/mm² at 28 days
Loading:
  Imposed loads = 7.2 kN/m²
  False ceiling under floor = 0.5 kN/m²
  Services under floor = 0.25 kN/m²
  Construction live load = 1.5 kN/m²

Step 1 *Select type of beam and slab construction*
Choose equal flanged steel beam from standard UB sections. Use Grade 50 steel for overall economy. Choose span-to-depth ratio of 25. Choose elastic analysis method using cracked section properties as appropriate for the composite beam. Use the maximum allowable percentage of support moment redistribution depending on the classification of the compression flange of the steel beam at the support. Prepare alternative designs for both partial and full shear connection.

Choose depth of section = 12 000/25 = 480 mm

Try UB 457 × 152 × 74 kg/m

Step 2 *Determine loading on the beam*
Construction stage: unpropped
Average depth of concrete slab $= 75 + 25 = 100$ mm
Self-weight of concrete slab $= 2.4$ kN/m^2
Self-weight of profiled decking $= 0.125$ kN/m^2
Total weight of concrete slab and decking $= 2.525$ kN/m^2
Self-weight of steel beam $= 0.74$ kN/m
Serviceability limit state loading on the beam at construction stage
$= [3 \times (2.525 + 1.5)] + 0.74 = 12.82$ kN/m
Serviceability limit state loading with dead load of concrete only
$= (3 \times 2.525) + 0.74 = 8.32$ kN/m
Ultimate limit state loading on the beam at construction stage
$= (1.4 \times 8.32) + (3 \times 1.6 \times 1.5) = 18.85$ kN/m (maximum)
$= (3 \times 0.125) + 0.74 = 1.12$ kN/m (minimum)

Composite stage
Serviceability limit state loading additional to construction stage dead load only
$= 3 \times (0.5 + 0.25 + 7.2) = 23.85$ kN/m (maximum)
$= 3 \times (0.5 + 0.25) = 2.25$ kN/m (minimum)
Ultimate limit state loading on the composite beam
$= 3 \times ([1.4 \times 2.525] + [1.4 \times 0.75] + [1.6 \times 7.2]) + (1.4 \times 0.74)$
$= 49.35$ kN/m (maximum)
$= 3 \times (2.525 + 0.75) + 0.74 = 10.57$ kN/m (minimum)

Step 3 *Determine material and section properties*

Modular ratio

$$\alpha_e = \alpha_s + \rho_l(\alpha_l - \alpha_s)$$

where $\alpha_l =$ modular ratio for long-term loading
$$= \frac{E_s}{E_c} = \frac{205}{28} = 7.3$$
$\rho_l =$ proportion of long-term loading
$$= \frac{3 \times (2.525 + 0.75) + 0.74}{8.32 + 23.85} = 0.33$$
$\alpha_l =$ modular ratio for long-term loading
$= 14.6$ (assumed)
$\therefore \quad \alpha_e = 7.3 + 0.33 \times (14.6 - 7.3) = 9.71$ (≈ 10, say)

Effective breadth of compression flange B_e
The slab is spanning perpendicular to the beam. Effective breadth b_e of the flange is found for both sides of the beam centre-line.

$b =$ half the distance to adjacent beam $= 1.5$ m
$$b_e - \frac{L_z}{8} \leq b$$

Determine L_Z for continuous beam from SK 8/2.
Positive moments, end spans: $L_Z = 0.8L = 9.6\,\text{m}$ ∴ $b_e = 1.2\,\text{m}$
Positive moments, central spans: $L_Z = 0.7L = 8.4\,\text{m}$ ∴ $b_e = 1.05\,\text{m}$
Negative moments, over supports: $L_Z = 0.5L = 6.0\,\text{m}$ ∴ $b_e = 0.75\,\text{m}$
B_e = total effective breadth = $2b_e$

Section properties of transformed area
First, check if tensile steel reinforcement is required in the concrete flange to resist negative bending moment.
Determine the maximum ultimate moment $M_{u,\text{max}}$ over the supports by elastic analysis, using constant moment of inertia and the appropriate coefficients from the tables in Chapter 11.

$$M_{u,\text{max}} = -0.107 \times \text{dead load} \times L^2 - 0.121 \times \text{live load} \times L^2$$
$$= [-0.107 \times (49.35 - [3 \times 7.2 \times 1.6]) \times 12^2]$$
$$- 0.121 \times (3 \times 7.2 \times 1.6) \times 12^2$$
$$= -830\,\text{kNm}$$

The plastic moment capacity M_s of the steel beam is given by:
$$M_s = S_x p_y = 1624 \times 10^3 \times 345 \times 10^{-6} = 560\,\text{kNm}$$

$560 \div 830 = 0.67$. This means that with 33% redistribution, the steel beam alone can resist the support moment. There is no need for additional tensile steel reinforcement because the level of redistribution required is less than the allowable level (50%) when elastic global analysis is carried out using uncracked section properties.

The continuous composite beam is designed as unreinforced, and plastic global analysis is used with cracked section properties. The conditions for plastic analysis, as described in Appendix D.3 of BS 5950: Part 3: Section 3.1, are satisfied because all spans are equal, there are no concentrated loads on the spans, and the compression flange and the web of the steel beam at the support are classified as class 1 plastic.

Section properties for positive moments
A = area of steel beam = $9510\,\text{mm}^2$

$$\frac{(D_s - D_p)^2 B_e}{(D + 2D_p)\alpha_e} = \frac{(125 - 50)^2 \times 2100}{[461.3 + (2 \times 50)] \times 10} = 2104\,\text{mm}^2$$

$$\therefore\quad A > \frac{(D_s - D_p)^2 B_e}{(D + 2D_p)\alpha_e}$$

The elastic neutral axis is in the steel beam for positive moments. The concrete is uncracked and the gross uncracked section properties should be used.
(See Table 8.1.) The depth of the neutral axis below the top of the concrete slab is given by:

$$n = \frac{\frac{1}{2}(D_s - D_p) + \alpha_e r(\frac{1}{2}D + D_s)}{1 + \alpha_e r}$$

End spans: $r = \dfrac{A}{(D_s - D_p)B_e} = \dfrac{9510}{75 \times 2400} = 0.0528$

Central spans: $r = \dfrac{9510}{75 \times 2100} = 0.0604$

End spans: $n = \dfrac{\frac{1}{2}(125 - 50) + (10 \times 0.0528 \times [\frac{1}{2}(461.3) + 125])}{1 + (10 \times 0.0528)}$

$= 147.4\,\text{mm}$

Central spans: $n = 180.7\,\text{mm}$

Moment of inertia of composite section
I_x = moment of inertia of steel beam about major axis = $32\,470\,\text{cm}^4$

Positive moments

$$I_g = I_x + \dfrac{B_e(D_s - D_p)^3}{12\alpha_e} + \dfrac{A(D + D_s + D_p)^2}{4(1 + \alpha_e r)}$$

End spans: $I_y = 3.247 \times 10^8 + \dfrac{2400 \times 75^3}{12 \times 10} + \dfrac{9510 \times (461.3 + 125 + 50)^2}{4 \times (1 + 10 \times 0.0528)}$

$= 9.631 \times 10^8\,\text{mm}^4$

Central spans: $I_g = 9.322 \times 10^8\,\text{mm}^4$

Negative moments
Moment of inertia is given by:

$I_n = I_x$ (because $A_r = 0$) $= 3.2470 \times 10^8\,\text{mm}^4$

Step 4 **Carry out analysis**
The analysis, using computer software, is carried out for unit uniformly distributed load on all spans and on each span separately. The model is created with cracked section properties over the supports for a distance of $0.15L$ on either side of the supports.

It can be advantageous to tabulate the unit load analysis of the composite beam, as shown in Table 8.8.

Maximum shear coefficients at support B for span 1:
 Span 1 loaded: $V_{B1} = 6.532$
 Span 2 loaded: $V_{B2} = 0.453$
 Span 3 loaded: $V_{B3} = -0.084$
 Span 4 loaded: $V_{B4} = 0.017$
 All spans loaded: $V_B = 6.919$

Determine the maximum ultimate bending moment over support B as follows (refer to Table 8.8).
 Condition: Spans 1, 2 and 4 loaded with maximum ultimate load of $49.35\,\text{kN/m}$ and span 3 loaded with minimum ultimate load $10.57\,\text{kN/m}$. Load all spans with $10.57\,\text{kN/m}$ and spans 1, 2 and 4 with $49.35 - 10.57 = 38.78\,\text{kN/m}$.
 $M_{B,\text{max}} = 38.78 \times (6.38 + 5.44 + 0.21) + (10.57 \times 11.03) = 583.1\,\text{kNm}$

The ultimate bending moment capacity at the support is the plastic moment capacity M_s of the steel beam, which is $560\,\text{kNm}$. Therefore,

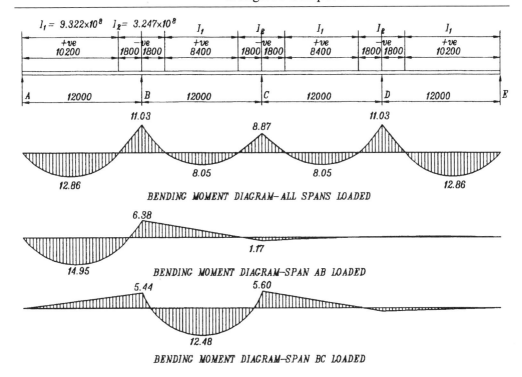

SK 8/18 Computer model of the continuous beam and results.

the redistribution of support moment required $= 560/583.1 = 0.96$ (or 4%) only.

Global ultimate load analysis of span 1
All spans are loaded with minimum ultimate load $= 10.57\,\text{kN/m}$
Bending moment at support $B = -11.03 \times 10.57 = -116.60\,\text{kNm}$
Difference between this moment and the ultimate moment at support
  B to form a plastic hinge $= 560 - 116.6 = 443.4\,\text{kNm}$
The loading on spans 1, 2 and 4 required to form a hinge at support B is given by:

$$\frac{443.4}{6.38 + 5.44 + 0.21} = 36.86\,\text{kN/m}$$

The remainder of the ultimate loading $= 38.78 - 36.86 = 1.92\,\text{kN/m}$ should be applied on span 1 which behaves as simply supported for this loading.
Positive bending moment in span 1 at $0.45L = 5.4\,\text{m}$ is given by:

$$M_{\text{UP}} = (10.57 \times 12.86) + (36.86 \times [14.95 - 2.45 - 0.09])$$
$$+ (1.92 \times [(6 \times 5.4) - (5.4^2 \div 2)])$$
$$= 627.6\,\text{kNm}$$

Similarly, global analysis of span 2 can be carried out with progressive hinge formation at support B and support C.

Table 8.8 Tabulated results of unit load analysis.

	Location	Span 1 loaded		Span 2 loaded		Span 3 loaded		Span 4 loaded		All spans loaded	
		Bending moment	Deflection	Bending moment	Deflection	Bending moment	Deflection	Bending moment	Deflection	Bending moment	Deflection
Span 1	0.40L	14.73	1.046	−2.17	−0.254	0.60	0.056	−0.13	−0.012	12.87	0.829
	0.45L	14.95	1.078	−2.45	−0.273	0.55	0.055	−0.12	−0.011	12.86	0.845
	0.50L	14.81	1.082	−2.72	−0.287	0.50	0.053	−0.10	−0.011	12.49	0.836
	0.55L	14.31	1.059	−2.99	−0.297	0.45	0.050	−0.09	−0.011	11.76	0.806
	0.60L	13.45	1.011	−3.26	−0.301	0.40	0.047	−0.08	−0.010	10.66	0.753
Support B	1.00L	−6.38	0	−5.44	0	1.00	0	−0.21	0	−11.03	0
Span 2	1.40L	−3.36	−0.309	11.78	0.861	−2.96	−0.273	0.62	0.057	7.12	0.380
	1.45L	−2.98	−0.301	12.31	0.895	−2.63	−0.265	0.55	0.055	7.77	0.407
	1.50L	−2.61	−0.287	12.48	0.907	−2.30	−0.253	0.48	0.053	8.05	0.420
	1.55L	−2.23	−0.268	12.29	0.894	−1.97	−0.237	0.41	0.049	7.98	0.417
	1.60L	−1.85	−0.245	11.74	0.859	−1.64	−0.216	0.34	0.045	7.55	0.399
Support C	2.00L	1.17	0	−5.60	0	−5.60	0	1.17	0	−8.87	0

Design of Composite Beams and Columns

Ultimate load analysis at construction stage
Steel beam only; unpropped; use coefficients from Table 11.1.

Condition 1: Span 1 (AB) only loaded with wet concrete
Ultimate dead load of beam and profiled sheet
$= 1.4 \times 1.12 = 1.568$ kN/m
Total ultimate dead and live load at construction stage
$= 18.85$ kN/m (see Step 2)
Ultimate load of wet concrete alone $= 18.85 - 1.568 = 17.282$ kN/m
Bending moments over supports using coefficients from Table 11.1
(Use inertia ratio equal to 1.0):

$M_B = -([1.568 \times 0.1071] + [17.282 \times 0.0670]) \times 12^2$
$\quad = -190.92$ kNm
$M_C = -([1.568 \times 0.0714] - [17.282 \times 0.0179]) \times 12^2$
$\quad = +28.42$ kNm
$M_D = -([1.568 \times 0.1071] + [17.282 \times 0.0045]) \times 12^2$
$\quad = -35.38$ kNm

Condition 2: Spans 1 and 2 loaded with wet concrete
$M_B = -([1.568 \times 0.1071] + [17.282 \times 0.1161]) \times 12^2$
$\quad = -313.11$ kNm
$M_C = -([1.568 \times 0.0714] + [17.282 \times 0.0384]) \times 12^2$
$\quad = -111.7$ kNm
$M_D = -([1.568 \times 0.1071] - [17.282 \times 0.0089]) \times 12^2$
$\quad = -2.00$ kNm

Service load analysis at construction stage

Spans 1 (AB) and 2 (BC) loaded with wet concrete and construction live load
Support moment at $B = -0.116 \times 12.82 \times 12^2 = -214.1$ kNm

All spans loaded with dead load of concrete only
Support moment at $B = -0.107 \times 8.32 \times 12^2 = -128.2$ kNm
Span moment in span $1 = 0.077 \times 8.32 \times 12^2 = 92.3$ kNm
The maximum ultimate shear force in span 1 at support B is given by:

$F_v = (10.57 \times 6.919) + (36.86 \times ([6.532 + 0.453 + 0.017])$
$\quad + (1.92 \times 12 \times 0.5) = 342.7$ kN

Step 5 *Check section classification*
Construction stage; steel beam only; unpropped.
UB $457 \times 152 \times 74$ kg/m; $d/t = 41.1$; $b/T = 4.49$; $T = 17$ mm; $t = 9.9$ mm
Check section classification as per Table 7 of BS 5950: Part1.

For flange: $\varepsilon = \sqrt{(275/345)} = 0.89$

Flange is class 1 plastic because $b/T < 8.5\varepsilon = 7.56$
For web with neutral axis at mid-depth: web is class 1 plastic because $d/t < 79\varepsilon$

Composite beam
The ribs of the profiled steel sheet run at right angles to the beam, and the flange is classed as plastic. Therefore, rectangular plastic stress distribution in the composite section is assumed.

Concrete resistance $= R_c = 0.45 f_{cu} B_e (D_s - D_p)$
$= 0.45 \times 40 \times 2100 \times 75 \times 10^{-3} = 2835 \text{ kN}$
Web resistance $= R_v = dt p_y = 406.9 \times 9.9 \times 355 \times 10^{-3} = 1430 \text{ kN}$
Steel beam resistance $= R_s = A p_y = 9510 \times 345 \times 10^{-3} = 3281 \text{ kN}$

For positive moments, since R_s is greater than R_c, the neutral axis lies in the compression flange, and the whole web is in tension at the ultimate limit state. The web classification for buckling in compression is not necessary.

For negative moment the steel beam only is effective, and the web classification as in the construction stage is class 1 plastic.

Step 6 *Determine ultimate plastic moment capacities*
Use the formulae in Tables 8.3 and 8.4.

Span 1
Resistance of concrete flange $= R_c = 0.45 f_{cu} B_e (D_s - D_p)$
$= 0.45 \times 40 \times 2400$
$\times (125 - 50) \times 10^{-3}$
$= 3240 \text{ kN}$
Resistance of steel flange $= R_f = BT p_y = 152.7 \times 17 \times 345 \times 10^{-3}$
$= 895.6 \text{ kN}$
Resistance of steel beam $= R_s = A p_y = 9510 \times 345 \times 10^{-3} = 3281 \text{ kN}$
Resistance of clear web depth $= R_v = dt p_y = 406.9 \times 9.9 \times 355 \times 10^{-3}$
$= 1430 \text{ kN}$
Resistance of overall web depth $= R_w = R_s - 2R_f = 1489.8 \text{ kN}$

where $A =$ area of steel beam $= 9510 \text{ mm}^2$
$B =$ breadth of flange of steel beam $= 152.7 \text{ mm}$
$B_e =$ effective breadth of concrete flange $= 2400 \text{ mm}$
$d =$ clear depth of web $= 406.9 \text{ mm}$
$f_{cu} =$ grade of concrete $= 40 \text{ N/mm}^2$
$D_s =$ overall depth of slab $= 125 \text{ mm}$
$D_p =$ depth of profiled steel sheet $= 50 \text{ mm}$
$T =$ thickness of steel flange $= 17 \text{ mm}$
$t =$ thickness of web of steel beam $= 9.9 \text{ mm}$
$D =$ overall depth of steel beam $= 461.3 \text{ mm}$

Positive moments: full shear connection
$R_c > R_w$ and $R_s > R_c$
The positive moment plastic neutral axis lies in the steel flange. The plastic moment capacity of the composite beam M_c is given by:

$$M_c = R_s \frac{D}{2} + R_c \frac{D_s + D_p}{2} - \frac{(R_s - R_c)^2}{R_f} \frac{T}{4}$$

$$= \left(3281 \times \frac{0.4613}{2}\right) + \left(3240 \times \frac{0.175}{2}\right)$$

$$- \left(\frac{(3281 - 3240)^2}{895.6} \times \frac{0.017}{4}\right) = 1040.3 \text{ kNm}$$

Design of Composite Beams and Columns

Partial shear connection
N_p = number of shear connectors required for full shear connection
N_a = actual number of shear connectors used.
The allowable minimum is given by:

$$\frac{N_a}{N_p} = \frac{(L-6)}{10} = \frac{12-6}{10} = 0.6$$

where L is the span in metres. Therefore:

$$R_q = 0.6 R_c \quad \text{or} \quad 0.6 R_s \text{ (whichever is the lower)}$$

or $\quad R_q = 0.6 \times 3240 = 1944\,\text{kN}$

where R_q is the resistance of shear connection.

Positive moments partial shear connection
$R_q > R_w$ (∴ plastic neutral axis is in the flange)

$$M_c = R_s \frac{D}{2} + R_q \left\{ D_s - \frac{R_q}{R_c} \frac{D_s - D_p}{2} \right\} - \frac{(R_s - R_q)^2}{R_f} \frac{T}{4}$$

$$= \left(\frac{3281 \times 0.4613}{2}\right) + \left(1944\left\{0.125 - 0.6 \times \frac{0.075}{2}\right\}\right)$$

$$- \left(\frac{(3281 - 1944)^2}{895.6} \times \frac{0.017}{4}\right)$$

$$= 947.5\,\text{kNm} > 755\,\text{kNm} \text{ (see Step 7)}$$

Check shear capacity
Shear capacity $= P_v = 0.6 t D p_y = 0.6 \times 9.9 \times 461.3 \times 355 \times 10^{-3}$
$= 972.7\,\text{kN}$
Maximum ultimate shear force $= F_v = 342.7\,\text{kN} < 0.5 P_v = 486.4\,\text{kN}$
No reduction of moment capacity is needed for high shear load.
Shear buckling resistance $= V_{cr} = q_{cr} dt$
$d/t = 41.1$, which gives $q_{cr} = 213\,\text{N/mm}^2$ from Table 21(d) of BS 5950: Part 1
$V_{cr} = 213 \times 406.9 \times 9.9 \times 10^{-3} = 858\,\text{kN} > F_v$

Step 7 *Check stability of compression flange*

Construction stage
Span 1 only is loaded with wet concrete
Bending moment at B when span 1 is loaded with wet concrete $= -190.92\,\text{kNm}$ (see Step 4) at the ultimate limit state. It is assumed that the profiled sheet will give lateral restraint to the top flange of the steel beam.
Check span 2 for lateral stability of the compression bottom flange, which is unrestrained.

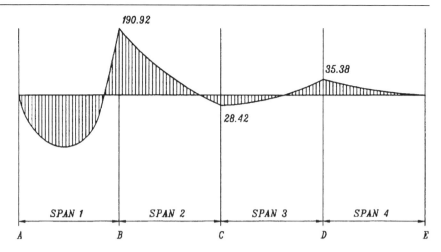

SK 8/19 Bending moment diagram of span 1 loaded with wet concrete.

L = distance between the points where both flanges are torsionally restrained
$= 12\,000$ mm
r_y = minor axis radius of gyration $= 3.26$ cm

$$\lambda = \frac{L}{r_y} = \frac{12\,000}{32.6} = 368$$

a = distance between the centre of beam and the centre of restraint axis (i.e. the profiled sheet)

$$= \frac{D}{2} + \frac{D_p}{2} = 255 \text{ mm}$$

h_s = distance between the centre of the flanges of the steel beam
$= D - T = 444$ mm
x = torsional index of the steel beam from published tables $= 30$
u = buckling parameter from published tables $= 0.87$

$$Y = \left[\frac{1 + \left(\frac{2a}{h_s}\right)^2}{1 + \left(\frac{2a}{h_s}\right)^2 + \frac{1}{20}\left(\frac{\lambda}{x}\right)^2} \right]^{\frac{1}{2}} = \left[\frac{1 + \left(\frac{2 \times 255}{444}\right)^2}{1 + \left(\frac{2 \times 255}{444}\right)^2 + \frac{1}{20}\left(\frac{368}{30}\right)^2} \right]^{\frac{1}{2}}$$

$= 0.485$

$$\beta_t = \frac{\text{algebraically smaller end moment}}{\text{algebraically higher end moment}} = \frac{-28.42}{190.92} = -0.15$$

Note: Moment which produces compression at the unrestrained flange is taken as positive.

Using $\beta_t = -0.15$ and $Y = 0.485$, determine the equivalent moment factor m_t from Table 39 of BS 5950: Part 1:

$m_t = 0.55$

Design of Composite Beams and Columns 375

The slenderness correction factor $n_t = 1.0$ where there are no intermediate loads between restraints (ignoring the self-weight of the profiled sheet and the beam).
$c = 1.0$ for uniform members

$$v_t = \left[\frac{\dfrac{4a}{h_s}}{1 + \left(\dfrac{2a}{h_s}\right)^2 + \dfrac{1}{20}\left(\dfrac{\lambda}{x}\right)^2} \right]^{\frac{1}{2}} = \left[\frac{\dfrac{4 \times 255}{444}}{1 + \left(\dfrac{2 \times 255}{444}\right)^2 + \dfrac{1}{20}\left(\dfrac{368}{30}\right)^2} \right]^{\frac{1}{2}}$$

$$= 0.483$$

The minor axis slenderness ratio λ_{TB} is given by:

$$\lambda_{TB} = n_t u v_t c \lambda = 1.0 \times 0.87 \times 0.483 \times 1.0 \times 368 = 155$$

Bending strength from Table 11 of BS 5950: Part 1 corresponding to $\lambda_{TB} = 155$ and $p_y = 355\,\text{N/mm}^2$ is $p_b = 67\,\text{N/mm}^2$.
Buckling moment of resistance:

$$M_b = S_x p_b = 1624 \times 10^3 \times 67 \times 10^{-6} = 109\,\text{kNm}$$

The equivalent uniform moment $\bar{M}$ is given by:

$$\bar{M} = m_t M_A = 0.55 \times 190.92 = 105\,\text{kNm}$$

$$\frac{\bar{M}}{M_b} = 0.963 < 1.0$$

Therefore, the beam selected is stable during construction.

Spans 1 and 2 are loaded with wet concrete
Check span 1 for stability of bottom flange at the support B.

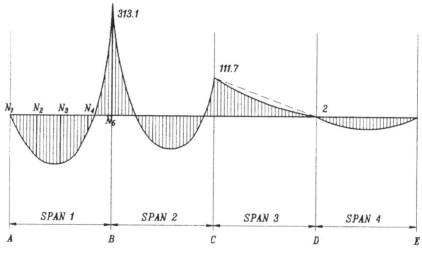

SK 8/20 Bending moment diagram when spans 1 and 2 are loaded.

Maximum support moment at $B = -313.1$ kNm (See Step 4)
$\beta_t = 0/313.1 = 0$
$\therefore m_t = 0.6$ from Table 39 of BS 5950: Part 1
$\lambda = 368$, $u = 0.87$, $m_t = 0.6$, $v_t = 0.483$ and $c = 1.0$
The slenderness correction factor n_t is different, being dependent on the shape of bending moment diagram.

Determination of n_t: Consider only positive values of applied moments at quarter points. Positive moments are those which produce compression in the unrestrained flange.

$N_1 = 0$, $N_2 = 0$, $N_3 = 0$, $N_4 = 0$ and $N_5 = 313.1$

$N_s = 0$ and $\dfrac{N_E}{M_E} = \dfrac{313.1}{560} = 0.56$ where $M_E = M_s$ (plastic capacity)

See SK 8/20 for an explanation of the notation.

$$n_t = \left[\frac{1}{12}\left\{\frac{N_5}{M_5} + 2\left(\frac{N_s}{M_s} - \frac{N_E}{M_E}\right)\right\}\right]^{\frac{1}{2}}$$

$$= \left[\frac{1}{12} \times 0.56\right]^{\frac{1}{2}}$$

$= 0.216$ taking only positive values of the expression $\dfrac{N_s}{M_s} - \dfrac{N_E}{M_E}$

$\lambda_{TB} = n_t u v_t c \lambda = 0.216 \times 0.87 \times 0.483 \times 1.0 \times 368 = 33$

Bending strength from Table 11 of BS 5950: Part 1 is $p_b = 333$ N/mm^2

$$\begin{aligned}
M_b &= S_x p_b \leq p_y Z \\
&= 1624 \times 10^3 \times 333 \times 10^{-6} \leq 345 \times 1408 \times 10^3 \times 10^{-6} \\
&= 540 \text{ kNm} \leq 486 \text{ kNm}
\end{aligned}$$

$\bar{M} = m_t M_A = 0.6 \times 313.1 = 187.9$ kNm

$\dfrac{\bar{M}}{M_b} = \dfrac{187.9}{486} = 0.39 < 1$ (Satisfied)

The beam selected is stable under all conditions of construction loading.

Composite stage: stability of bottom flange in compression
The worst case of stability of span 2 arises when spans 1 and 3 are fully loaded but spans 2 and 4 are loaded with dead load only. The analytical model allows for fully cracked section properties for the whole of span 2 and half of span 4. This produces the maximum bending moment in span 1 and all of span 2 has hogging moment, creating the most critical stability problem of the compressive bottom flange. By computer analysis the results shown in SK 8/21 are obtained.

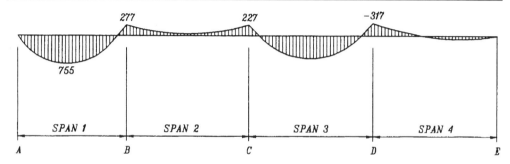

SK 8/21 Bending moment diagram when spans 1 and 3 only are fully loaded.

It should be noted that the maximum span moment in span 1 is not governed by the plastic hinge formation at B because the percentage of redistribution of negative moment is very low (see Step 4: 627 kNm vs 755 kNm). The loading condition shown in SK 8/21 produces higher bending moment in span 1 than when the plastic hinge is allowed to form at B. The support bending moments due to this loading condition do not exceed the plastic negative moment capacity of the steel beam. The section moment capacity of the composite beam for positive moments is checked against this maximum moment of 755 kNm

Determination of location of additional torsional restraints in span
Assume $m_t = 1.0$ and $\bar{M} = M_{max} = 277$ kNm.
The minimum bending strength p_b required to satisfy $\bar{M} = M_b$ is given by:

$$p_b = \frac{277 \times 10^6}{1624 \times 10^3} = 171 \, \text{N/mm}^2$$

From Table 11 of BS 5950: Part 1 it is observed that λ_{TB} should not be greater than 85 to obtain $p_b = 173 \, \text{N/mm}^2$. Taking $v_t = 0.5$, $n_t = 1.0$

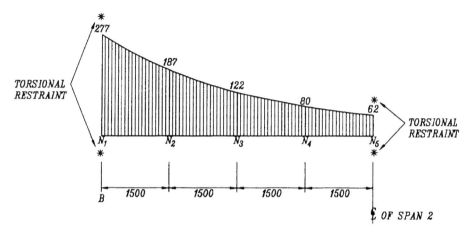

SK 8/22 Bending moment diagram of span 2 between torsional restraints.

and $u = 0.87$ as before, the slenderness ratio of the beam λ should not exceed:

$$\lambda = \frac{\lambda_{TB}}{n_t u v_t c} = \frac{85}{0.87 \times 0.5} = 195$$

Therefore, the maximum distance between the torsional restraints in the beam is given by:

$$L = \lambda r_y = 195 \times 32.6 = 6357 \, \text{mm}$$

Use a torsional stiffener at midspan of spans AB, BC, CD and DE when $L = 6000$.

Distance between adjacent torsional restraints $= L = 6000 \, \text{mm}$

Slenderness ratio $= \lambda = \dfrac{L}{r_y} = 184$

$$a = \frac{D}{2} + \frac{D_s}{2} = 293 \, \text{mm}$$

$h_s = 444 \, \text{mm}$; $x = 30$; $u = 0.87$

$$Y = \left[\frac{1 + \left(\dfrac{2 \times 293}{444}\right)^2}{1 + \left(\dfrac{2 \times 293}{444}\right)^2 + \dfrac{1}{20}\left(\dfrac{184}{30}\right)^2} \right]^{\frac{1}{2}} = 0.77$$

$$\beta_t = \frac{62}{277} = 0.22$$

Using $\beta_t = 0.22$ and $Y = 0.77$, determine the equivalent moment factor m_t from Table 39 of BS 5950: Part 1:

$m_t = 0.63$

From the bending moment diagram in SK 8/22, the moments at the quarter points are as follows:

$N_1 = 277 \, \text{kNm}$
$N_2 = 187 \, \text{kNm}$
$N_3 = 122 \, \text{kNm}$
$N_4 = 80 \, \text{kNm}$
$N_5 = 62 \, \text{kNm}$

The plastic moment capacity for negative moments of the steel beam is constant over the whole length of the beam and is equal to 560 kNm. Therefore:

$$M_1 = M_2 = M_3 = M_4 = M_5 = 560 \, \text{kNm}$$

$$\frac{N_s}{M_s} = \text{the greatest of } \frac{N_2}{M_2}, \frac{N_3}{M_3} \text{ and } \frac{N_4}{M_4} = \frac{N_2}{M_2}$$

$$\frac{N_E}{M_E} = \text{the greatest of } \frac{N_1}{M_1} \text{ and } \frac{N_5}{M_5} = \frac{N_1}{M_1}$$

$$n_t = \left[\frac{1}{12}\left\{\frac{N_1}{M_1} + \frac{3N_2}{M_2} + \frac{4N_3}{M_3} + \frac{3N_4}{M_4} + \frac{N_5}{M_5} + 2\left(\frac{N_s}{M_s} - \frac{N_E}{M_E}\right)\right\}\right]^{\frac{1}{2}}$$

= 0.492 (only positive values of the expressions are included in the computation)

$$v_t = \left[\frac{\frac{4a}{h_s}}{1 + \left(\frac{2a}{h_s}\right)^2 + \frac{1}{20}\left(\frac{\lambda}{x}\right)^2}\right]^{\frac{1}{2}}$$

$$= \left[\frac{\frac{4 \times 293}{444}}{1 + \left(\frac{2 \times 293}{444}\right)^2 + \frac{1}{20}\left(\frac{184}{30}\right)^2}\right]^{\frac{1}{2}} = 0.76$$

$$\therefore \quad \lambda_{TB} = n_t u v_t c \lambda = 0.492 \times 0.87 \times 0.76 \times 1 \times 184 = 60$$

From Table 11 of BS 5950: Part 1 the design strength $p_b = 250\,\text{N/mm}^2$, corresponding to $\lambda = 60$ and $p_y = 345\,\text{N/mm}^2$

$$M_b = S_x p_b = 1624 \times 10^3 \times 250 \times 10^{-6} = 406\,\text{kNm}$$

$$\bar{M} = m_t M = 0.63 \times 277 = 175\,\text{kNm}$$

$$\frac{\bar{M}}{M_b} = \frac{175}{406} = 0.43 < 1$$

Step 8 **Shear connection**

Select headed studs as shear connectors with a nominal shank diameter of 22 mm and nominal height of 100 mm. From Table 8.7, the characteristic resistance Q_k in concrete of grade $40\,\text{N/mm}^2$ is found to be 139 kN per connector.

For positive moment region the capacity of shear connector Q_p is given by:

$$Q_p = 0.8 Q_k = 0.8 \times 139 = 111.2\,\text{kN}$$

Shear connectors are not required in the negative moment region because tensile reinforcement has not been used.

Reduction of stud shear connectors in profiled sheets
Use one stud per rib.

$$k = 0.85\left(\frac{b_r}{D_p}\right)\left(\frac{h}{D_p} - 1\right) \leq 1$$

where b_r = breadth of concrete rib = b_a for open trough profile
   = 150 mm
   D_p = depth of profiled sheet = 50 mm
   h = height of stud = 100 mm

∴ $k = 2.55$, so k is taken as 1.0

Use partial shear connection as determined in Step 6.

$R_q = 1944$ kN (see Step 6)

The number of shear connectors N_p required between the point of maximum positive moment and the support is given by:

$$N_p = \frac{R_q}{Q_p} = \frac{1944}{111.2} = 18$$

Assume that the maximum positive moment occurs at about 0.45L from the end support in span 1. The distance to nearest support = $0.45 \times 12\,000 = 5400$ mm. The ribs in the profiled sheet are spaced at 300 mm. Using one stud per trough of the profiled sheet will give 19 studs in 5400 mm. This number is adequate for effective shear connection.

Use uniform spacing of one stud per trough for the whole four-span length of the continuous composite beam.
Maximum allowable spacing of studs is 600 mm or
   $4D_s = 4 \times 125 = 500$ mm (Satisfied)
Minimum allowable spacing of studs = $5d = 5 \times 22 = 110$ (Satisfied)
Maximum allowable diameter of stud shear connectors
   = $2.5T = 2.5 \times 17 = 42.5$ mm (Satisfied)

Step 9 Check longitudinal shear

Applied longitudinal shear on the concrete flange per unit length is given by:

$$v = \frac{NQ}{S}$$

where N = number of shear connectors at spacing $S = 1$
   $Q = Q_P$ or $Q_N = 111.2$ kN

∴ $$v = \frac{111.2 \times 10^3}{300} = 371 \text{ N/m}$$

The resistance to longitudinal shear of the beam per unit length is given by:

$$v_r = 0.7 A_{sv} f_y + 0.03 \eta A_{cv} f_{cu} + v_p \leq 0.8 \eta A_{cv} \sqrt{f_{cu}} + v_p$$

where A_{sv} = is the area of steel reinforcement crossing the potential shear failure plane
   $= \dfrac{78.5}{150}$ mm²/mm assumed (10 mm diameter bars at 150 centre to centre)

Design of Composite Beams and Columns

A_{cv} = the mean cross-sectional area of the concrete shear surface per unit length of the beam
= 100 mm²/mm (average depth of concrete flange is 100 mm)
$\eta = 1.0$ for normal weight concrete
$f_{cu} = 40 \text{ N/mm}^2$

Assume that the profiled steel sheeting is discontinuous across the top flange of the beam and the stud shear connectors are welded to the steel beam directly through the profiled sheet.

$$\therefore v_p = \left(\frac{N}{S}\right) n \, d t_p p_{yp} \leq t_p p_{yp}$$

where N = number of shear connectors in spacing S
= 1 number in 300 mm
$n = 4$
d = diameter of shank of shear connector = 22 mm
t_p = thickness of profiled steel sheet = 1.2 mm
p_{yp} = design strength of profiled steel sheet = 280 N/mm²

$$v_p = \frac{1}{300} \times 4 \times 22 \times 1.2 \times 280 = 99 \text{ N/mm}$$

$$v_r = \left(0.7 \times \frac{78.5}{150} \times 460\right) + (0.03 \times 1 \times 100 \times 40) + 99$$

$$= 387 \text{ N/mm} > 371 \text{ N/mm}$$

The concrete flange has adequate longitudinal shear resistance.

Step 10   Check serviceability limit state

Check stress
The midspan region of span 1 is checked for stress. At the serviceability limit state the steel stress should not exceed the design strength p_y and the stress in the concrete should not be higher than $0.5 f_{cu}$.

Construction stage
All spans are loaded with wet concrete (8.32 kN/m)
Bending moment at $0.45L$ in span $1 = 92.3$ kNm
Locked-in stresses in the steel beam = $\pm 92.3 \times 10^6 / Z_x$
= ± 65.6 N/mm² at the outermost tension and compression fibres

Composite stage
Spans 1 and 3 are loaded with full dead load and live load and spans 2 and 4 are loaded with dead load only.
Loading on spans 1 and 3 = 23.85 kN/m
Loading on spans 2 and 4 = 2.25 kN/m
Maximum positive bending moment = 377 kNm = M
(The analysis allows for full cracked section properties in span 2.)
The elastic neutral axis of the composite beam is 147.4 mm from the top of the slab.

I_g = gross moment of inertia of the composite section

$= 9.631 \times 10^8 \, \text{mm}^4$ (steel)

$= 9.631 \times 10^8 \times \alpha_e = 9.631 \times 10^9$ (concrete)

Stress in concrete $= \dfrac{377 \times 10^6 \times 147.4}{9.631 \times 10^9} = 5.77 \, \text{N/mm}^2 < 0.5 f_{cu}$

Section modulus (steel beam bottom fibre)

$= \dfrac{I_g}{(D + D_s - n)} = Z_b$

$= \dfrac{9.631 \times 10^8}{(44.3 + 125 - 147.4)} = 2.194 \times 10^6$

Maximum tensile stress in steel beam

$= \text{locked-in stress} + M/Z_b$

$= 65.6 + \dfrac{377 \times 10^6}{2.194 \times 10^6} = 237 \, \text{N/mm}^2 < p_y$

Check deflection of span 1 for live load only
Load spans 1 and 3 with live load.
Live loading $= 21.6 \, \text{kN/m}$
From Table 8.8 in Step 4, determine deflections:

$\delta = (1.082 \times 21.6) + (0.053 \times 21.6) = 24.52 \, \text{mm}$

$= \text{span}/489 < \text{span}/360$

This calculation of deflection allows for patterned loading and has been calculated by the rigorous method using the cracked section moment of inertia of the composite beam. The shake-down effects need not be considered because the percentage of moment redistribution is very small.

Step 11 **Check vibration**
Continuous beams have higher stiffness than simply supported beams with the same length of span and the same section properties. This results in higher frequency of vibration of the continuous beam.

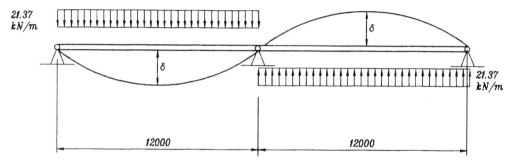

SK 8/23 Estimation of natural frequency by deflections due to gravity loads.

The lowest frequency of a composite beam is exhibited when the gravity load is applied in opposite directions on adjacent spans, as shown in SK 8/23. This mode shape is asymmetrical, and the deflection at the midspan of span 1 may be used to compute the natural frequency f_n, which is given by:

$$f_n = \frac{18}{\sqrt{\delta}}$$

where δ is the deflection at the midspan due to gravity loads in opposite directions.

The gravity load for the computation of this deflection may be taken as equal to all of the permanent dead load and half of the imposed load.

$$\therefore \quad \text{load} = 8.32 + 2.25 + 3 \times 3.6 = 21.37 \, \text{kN/m}$$

Calculate the deflection by using Table 8.8 in Step 4.

$$\delta = (1.082 \times 21.37) + (0.287 \times 21.37) = 29.25 \, \text{mm}$$

$$f_n = \frac{18}{\sqrt{29.25}} = 3.33 \, \text{Hz} < 4 \, \text{Hz}$$

This means that by this approximate method of analysis, the beam is not adequately stiff, and vibration could be a problem. The method is very conservative and the inclusion of half the imposed load for the computation of deflection may be regarded as excessive. It is not necessary to change the beam section to a stiffer alternative.

Step 12 **Check pre-cambering requirement**
The deflection at the midspan due to dead load is small, and the false ceiling under the floor can be adjusted to give a level ceiling. Pre-cambering may be unnecessary.

8.3 COMPOSITE COLUMNS

Composite columns are either concrete encased structural steel sections or hollow structural sections filled with concrete. They offer the following advantages:

(1) Increased fire resistance of steel column encased
(2) Increased load-carrying capacity of the steel section
(3) Reduced slenderness of the steel member
(4) Increased durability of the steel section
(5) Increased load-carrying capacity of the confined concrete inside a concrete filled structural hollow section

384 Structural Steelwork

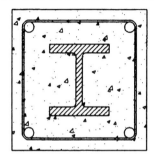

SK 8/24 Some examples of composite columns.

8.3.1 Step-by-step design of an encased composite column

Step 1 Determine ultimate axial section capacity in direct compression
The column is assumed to be 'stocky', which is not slender and is not subject to reduction of stresses due to elastic instability.

$$P_u = 0.45 f_{cu} A_c + A_s p_y + 0.87 A_r f_y$$

where P_u = ultimate maximum compressive load in a 'stocky' column
A_c = the net area of concrete in the encasement after deduction of the area of the steel section and any steel reinforcement used
A_r = area of steel reinforcement in the concrete encasement
A_s = area of steel section encased
f_{cu} = characteristic 28-day cube strength of the concrete
p_y = design strength of the steel section
f_y = yield strength of the steel reinforcement

The formula for P_u is the addition of ultimate resistances of all the components of the composite column. The expression includes a material factor γ_m equal to 1.5 for concrete and 1.15 for steel reinforcement.

Step 2 Determine axial capacity in direct tension
The ultimate maximum tensile load in the encased column P_t is given by:

$$P_t = A_s p_y + 0.87 A_r f_y$$

Step 3 Determine position of plastic neutral axis (PNA) about the major axis
Assume the position of the neutral axis is parallel to the X–X axis and then equate the tensile and compressive internal forces in the

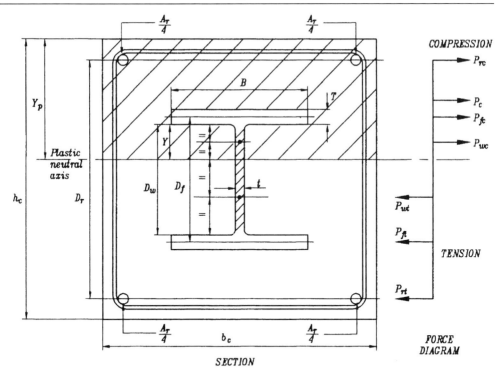

SK 8/25 Plastic neutral axis for moment about the major axis of a typical encased column.

components of the encased column on either side of the plastic neutral axis. This equation will give the position of the neutral axis Y_P.

P_{rc} = force in reinforcement in compression
P_{rt} = force in reinforcement in tension
P_{fc} = force in steel flange in compression
P_{ft} = force in steel flange in tension
P_{wc} = force in web in compression
P_{wt} = force in web in tension
P_{c} = force in concrete in compression

Equating tension and compression:

$$P_{rc} + P_{fc} + P_c + P_{wc} = P_{wt} + P_{ft} + P_{rt}$$

In most cases, with a symmetrical cross-section, this equation can be simplified:

$$P_c + P_{wc} = P_{wt}$$

or $P_c = P_w - 2P_{wc}$

where P_w = force in the total web of depth D_w at ultimate stress level

$$P_c = 0.45 f_{cu} \left(b_c Y_P - \frac{A_r}{2} - BT - Yt \right)$$

$$P_w - 2P_{wc} = (D_w - 2Y)tp_y$$

b_c = breadth of the concrete encasement
$D_w = D - 2T$
Y = depth of web in compression from the compression flange of the steel section
T = thickness of the flanges of the symmetrical steel section
t = thickness of web of the steel section
D = overall depth of the steel section

Step 4 *Determine plastic moment of resistance about the major axis (X–X)*
Take moments of the ultimate internal forces in the components about the plastic neutral axis to get the plastic moment of resistance. For a symmetrical section the plastic moment of resistance about the major axis M_{px} is given by:

$$M_{px} = P_f D_f + P_r D_r + P_c \frac{Y_P}{2} + \frac{Y^2}{2} tp_y + \frac{(D_w - Y)^2}{2} tp_y$$

where D_f = distance between the centre of the flanges
D_r = distance between the centre of tensile and compressive reinforcement

Step 5 *Determine plastic neutral axis for moment about the minor axis (Y–Y)*

X = distance of plastic neutral axis from the edge of web
X_P = distance of plastic neutral axis from edge of compressive face of concrete

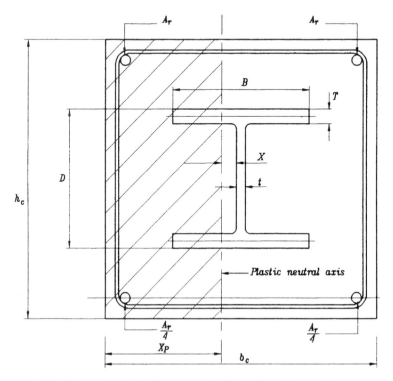

SK 8/26 Plastic neutral axis for moment about the minor axis.

b_c = breadth of concrete encasement
h_c = depth of concrete encasement
D = overall depth of steel section
B = width of flange of steel section
t = thickness of web of steel section
T = thickness of flange of steel section
$D_w = D - 2T$

Taking the symmetrical cross-section in SK 8/26 as an example, the plastic neutral axis may be determined by equating the tensile and compressive internal forces as follows:

$$\text{Tensile force} = 2\left(X + t + \frac{B}{2} - \frac{t}{2}\right)Tp_y + D_w tp_y + \frac{A_r}{2}f_y$$

$$\text{Compressive force} = 2\left(\frac{B}{2} - \frac{t}{2} - X\right)Tp_y + \frac{A_r}{2}f_y$$

$$+ 0.45 f_{cu}\left[\left(\frac{b_c}{2} - \frac{t}{2} - X\right)h_c - \frac{A_r}{2} - \left(\frac{B}{2} - \frac{t}{2} - X\right)T\right]$$

Equating the forces:

$$2(2X + t)Tp_y + D_w tp_y$$

$$= 0.45 f_{cu}\left[\left(\frac{b_c}{2} - \frac{t}{2} - X\right)h_c - \frac{A_r}{2} - \left(\frac{B}{2} - \frac{t}{2} - X\right)T\right]$$

The only unknown in the above equation is X. It can be determined easily.

$$X_P = \frac{b_c}{2} - \frac{t}{2} - X$$

Step 6 *Determine plastic moment of resistance about the minor axis*
Take moments of all internal forces about the plastic neutral axis.

$$M_{py} = 0.87 f_y \frac{A_r}{2} D_r + tD_w p_y \left(X + \frac{t}{2}\right)$$

$$+ \left[\left(\frac{B}{2} - X - \frac{t}{2}\right)^2 + \left(\frac{B}{2} + X + \frac{t}{2}\right)^2\right] Tp_y$$

$$+ 0.45 f_{cu}\left[\left(\frac{b_c}{2} - \frac{t}{2} - X\right)h_c - \frac{A_r}{2} - \left(\frac{B}{2} - \frac{t}{2} - X\right)T\right]\frac{X_P}{2}$$

Step 7 *Determine effective slenderness of column and compressive strength* p_c
A column is classed as slender if L/b_c is greater than 12, where L is the length of the column. The slenderness factor is given by:

$$\bar{\lambda} = \frac{L}{\pi}\sqrt{\frac{P_u}{E_s \sum I}}$$

where $\sum I$ = second moment of area of all the components in a composite column about the minor axis expressed in steel units
E_s = modulus of elasticity of steel
P_u = composite column section capacity in axial compression (see Step 1)

If the slenderness factor is less than 0.2 then the column may be classed as stocky.

Find effective slenderness ratio λ, which is given by:

$$\lambda_{\text{eff}} = \bar{\lambda}\pi\sqrt{\frac{E_s}{p_y}}$$

Using this slenderness ratio, find the compressive strength p_c from Table 27(c) of BS 5950: Part 1 corresponding to design strength p_y.

Determine $K_1 = \dfrac{p_c}{p_y}$

Reduced axial compressive strength of slender composite column
$= P_{cx} = K_1 P_u$

Step 8 ***Determine additional moment due to potential eccentricity of vertical load on slender column***
Take nominal minimum eccentricity $= 0.03 b_c$ about both the X–X and Y–Y axes. Add these additional bending moments to the bending moments from structural analyses and determine final moments M_x and M_y.

Step 9 ***Check column capacity for biaxial bending and direct load by interaction formula***

Check $\dfrac{P}{P_{cx}} + \dfrac{M_x}{M_{px}} + \dfrac{M_y}{M_{py}} \leq 1.0$

Alternatively check $\dfrac{M_x}{\mu_x M_{px}} + \dfrac{M_y}{\mu_y M_{py}} \leq 1.0$

where $\mu_x = \dfrac{P_{ux} - P}{P_{ux} - P_0}$ and $\mu_y = \dfrac{P_{uy} - P}{P_{uy} - P_0}$

$P_0 = 0.45 f_{cu}(h_c b_c - A_r - A_s)$

$P_{ux} = P_{cx}$ and $P_{uy} = P_u$

Step 10 ***Check overall geometry and detailing***
Minimum recommended cover to steel section and reinforcement $= 40$ mm
The area of steel reinforcement A_r should not be greater than 3% of the net area of concrete A_c.
Minimum shear links should be 5 mm in diameter at 150 mm centres.

8.3.2 Step-by-step design of concrete filled hollow circular section

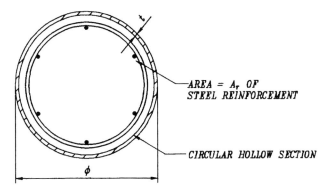

SK 8/27 Circular hollow section filled with concrete.

Step 1 **Determine axial section capacity for direct compression**

The concrete in the hollow section is confined, and the strength of the concrete is enhanced due to the Poisson ratio effect. The steel tube is subjected to hoop tension which has little effect on the axial capacity of the steel section. The enhancement of strength of the concrete largely depends on the proportions of the cross-section.

The axial compressive load capacity of a stocky circular hollow steel section filled with concrete is given by (see *Steel Designer's Manual*, Fifth Edition, published by Blackwell Science):

$$P_u = C_1 p_y A_s + 0.87 f_y A_r + \left(\frac{0.83 f_{cu}}{\gamma_{mc}}\right) A_c \left[1 + C_2 \left(\frac{t}{\phi}\right)\left(\frac{p_y}{0.83 f_{cu}}\right)\right]$$

Factor C_1 is less than one to allow for a reduction of the axial load capacity of the steel section due to the presence of hoop tension. Values of C_1 and C_2 are given in Table 8.9.

Table 8.9 Values of C_1 and C_2 for concrete filled hollow circular sections.

$\bar{\lambda}$	0	0.1	0.2	0.3	0.4	≥ 0.5
C_1	0.75	0.80	0.85	0.90	0.95	1.00
C_2	4.90	3.22	1.88	0.88	0.22	0.00

t = thickness of the steel tube
ϕ = diameter of the steel tube
A_s = area of steel in the steel tube cross-section
A_r = area of any steel reinforcement in the concrete
A_c = area of concrete in the steel tube
γ_{mc} = material factor for concrete = 1.5
$\sum I$ = second moment of area of all the components in a composite column about the minor axis expressed in steel units
E_s = modulus of elasticity of steel

P_u = composite column section capacity in axial compression (see Step 1)

$$\bar{\lambda} = \text{slenderness factor} = \frac{L}{\pi}\sqrt{\frac{P_u}{E_s \sum I}}$$

Finding the slenderness factor is an iterative process. First, find P_u, assuming $C_1 = 1.0$ and $C_2 = 0$. Using P_u, find $\bar{\lambda}$. Then, using $\bar{\lambda}$, find C_1 and C_2. Carry on with this iterative procedure until convergence.

Step 2 ***Check limits of application of the load capacity formula***
Check 1: $0.2P_u < A_s p_y < 0.9P_u$
Check 2: $\phi \leq 85t\varepsilon$ to avoid local buckling

$$\varepsilon = \sqrt{\frac{275}{p_y}}$$

Step 3 ***Determine column slenderness and reduction factor K_1***
The effective slenderness ratio λ is given by:

$$\lambda = \bar{\lambda}\pi\sqrt{\frac{E_s}{p_y}}$$

E_s is the modulus of elasticity of steel.
Using this value of λ, find the compressive strength p_c from Table 27(a) of BS 5950: Part 1 corresponding to design strength p_y.

Find $K_1 = \dfrac{p_c}{p_y}$ and $P_c = K_1 P_u$

Step 4 ***Determine ultimate moment capacity M_u of the concrete filled CHS***
See *SHS Design Manual*, published by British Steel, 1986 for concrete filled columns.

Step 5 ***Check interaction of direct load and moment***
The interaction formula is:

$$\frac{P}{P_c} + \frac{M}{M_u} \leq 1.0$$

where M = resultant moment on the circular section
$= \sqrt{(M_x^2 + M_y^2)}$

8.4 EXAMPLE 8.2: COMPOSITE COLUMN

This example relates to a check on an encased building column with biaxial moments.

Size of concrete encasement = 450×450
Size of steel column = $305 \times 305 \times 118$ kg/m
Grade of steel = 50

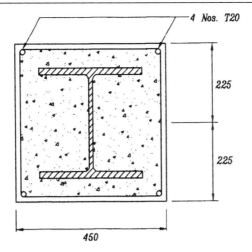

SK 8/28 Details of an encased column.

Design strength $= p_y = 355\,\text{N/mm}^2$
Grade of concrete in encasement $= f_{cu} = 40\,\text{N/mm}^2$
Steel reinforcement in encasement $= 4$ no. 20 mm diameter bars
Area of steel reinforcement $= A_r = 1256\,\text{mm}^2$
Grade of steel reinforcement $= f_y = 460\,\text{N/mm}^2$
Cover to steel reinforcement $= 35\,\text{mm}$ minimum
Length of column between floors $= 6000\,\text{mm}$
Loading condition: $P = 4200\,\text{kN}$, $M_x = 220\,\text{kNm}$

Step 1 Determine axial section capacity for direct compression

$$P_u = 0.45 f_{cu} A_c + A_s p_y + 0.87 A_r f_y$$

where $A_s =$ area of steel section $= 15\,000\,\text{mm}^2$
$A_c =$ net area of concrete
$= 450 \times 450 - A_s - A_r = 202\,500 - 15\,000 - 1256$
$= 186\,244\,\text{mm}^2$

$\therefore \quad P_u = ([0.45 \times 40 \times 186\,244] + [15\,000 \times 355]$
$\qquad\qquad + [0.87 \times 460 \times 1256]) \times 10^{-3}$

$\qquad = 9180\,\text{kN}$

Capacity of steel column section only $= 15 \times 355 = 5325\,\text{kN}$
Contribution of concrete only
$= P_0 = 0.45 \times 40 \times 186.244 = 3352\,\text{kN} = 0.365 P_u$

Step 2 Determine tensile load capacity of the encased column
This is necessary if an interaction diagram is required to be drawn.

$$P_t = \text{ultimate tensile capacity} = A_s p_y + 0.87 A_r f_y$$

$\qquad = ([15\,000 \times 355] + [0.87 \times 1256 \times 460]) \times 10^{-3} = 5828\,\text{kN}$

Step 3 *Determine position of plastic neutral axis about the major axis (X–X)*

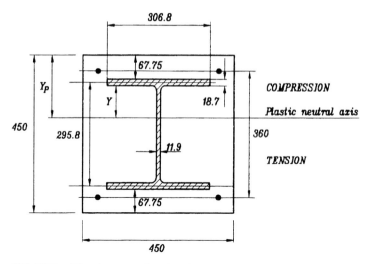

SK 8/29 Plastic neutral axis of the composite column about the major axis.

Y = distance of the plastic neutral axis from the compression flange
Y_P = distance of the plastic neutral axis from the compressive face of the concrete
D = overall depth of steel section = 314.5 mm
T = thickness of flange of the steel section = 18.7 mm
$D_w = D - 2T = 277.1$ mm
t = thickness of web = 11.9 mm
B = width of flange = 306.8 mm

Equating tensile and compressive internal forces:

$$(D_w - 2Y)tp_y = 0.45f_{cu}\left[b_c(Y + T + 67.75) - \frac{A_r}{2} - BT - Yt\right]$$

$$(277.1 - 2Y) \times 11.9 \times 355 = 0.45 \times 40 \times \left[450 \times (Y + 18.7 + 67.75)\right.$$

$$\left. - \frac{1256}{2} - (306.8 \times 18.7) - 11.9Y\right]$$

Simplifying gives:

$Y = 35.8$ mm

$Y_P = Y + T + 67.75 = 122.25$ mm

Step 4 *Determine plastic moment of resistance of composite column about major axis*

P_f = flange force = $BTp_y = 306.8 \times 18.7 \times 345 \times 10^{-3} = 1979$ kN

P_r = force in reinforcement = $0.87f_y\left(\dfrac{A_r}{2}\right)$

$= 0.87 \times 460 \times 628 \times 10^{-3} = 251$ kN

Design of Composite Beams and Columns 393

P_c = force in concrete in compression

$$= 0.45 f_{cu} \left(b_c Y_P - \frac{A_r}{2} - BT - Yt \right)$$

$$= 0.45 \times 40 \times ([450 \times 122.25] - 628 - [306.8 \times 18.7]$$
$$- [35.8 \times 11.9]) \times 10^{-3}$$

$$= 868 \, \text{kN}$$

D_f = distance between centroid of flanges = 295.8 mm
D_r = distance between centroid of steel reinforcement in compression and tension = 360 mm

Taking moments of these forces about the PNA, M_{px} is given by:

$$M_{px} = P_f D_f + P_r D_r + P_c \frac{Y_P}{2} + t p_y \frac{Y^2}{2} + t p_y \frac{(D_w - Y)^2}{2}$$

$$= (1979 \times 0.2958) + (251 \times 0.36) + \left(868 \times \frac{0.122}{2} \right)$$

$$+ 11.9 \times 355 \times 10^{-6} \left[\frac{35.8^2}{2} + \frac{(277.1 - 35.8)^2}{2} \right]$$

$$= 854 \, \text{kNm}$$

Step 5 *Determine plastic neutral axis for moment about minor axis (Y–Y)*
  X = distance of plastic neutral axis from edge of web on the compression side
  X_P = distance of plastic neutral axis from the concrete face in compression

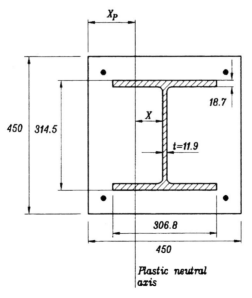

SK 8/30 Plastic neutral axis of the composite column about minor axis.

Equating tensile and compressive internal forces:

$$2Tp_y(2X+t) + D_w t p_y$$

$$= 0.45f_{cu}\left[\left(\frac{b_c}{2} - \frac{t}{2} - X\right)h_c - \frac{A_r}{2} - \left(\frac{B}{2} - \frac{t}{2} - X\right)T\right]$$

$$(2 \times 18.7 \times 345 \times [2X + 11.9]) + (277.1 \times 11.9 \times 355)$$

$$= 0.45 \times 40 \times \left[\left(\frac{450}{2} - \frac{11.9}{2} - X\right) \times 450 - \frac{1256}{2}\right.$$

$$\left. - \left(\frac{306.8}{2} - \frac{11.9}{2} - X\right) \times 18.7\right]$$

Simplifying gives:

$$X = 11.6 \text{ mm}$$

$$X_P = \frac{h_c}{2} - \frac{t}{2} - X = 207.45 \text{ mm}$$

Step 6 *Determine plastic moment of resistance of the encased column about the minor axis (Y–Y)*

Taking moments of the internal forces about the plastic neutral axis parallel to Y–Y, the plastic moment of resistance M_{py} is given by:

$$M_{py} = 0.87f_y \frac{A_r}{2} D_r + tD_w p_y\left(X + \frac{t}{2}\right)$$

$$+ \left[\left(\frac{B}{2} - X - \frac{t}{2}\right)^2 + \left(\frac{B}{2} + X + \frac{t}{2}\right)^2\right]Tp_y$$

$$+ 0.45f_{cu}\left[\left(\frac{b_c}{2} - \frac{t}{2} - X\right)h_c - \frac{A_r}{2} - \left(\frac{B}{2} - \frac{t}{2} - X\right)T\right]\frac{X_P}{2}$$

$$= 587 \times 10^6 \text{ Nmm}$$

Step 7 *Determine column slenderness and axial strength modification factor K_1*

$$\frac{L}{b_c} = \frac{6000}{450} = 13.33 > 12$$

$\therefore$ Design column as slender

$$\text{Slenderness factor} = \bar{\lambda} = \frac{L}{\pi}\sqrt{\frac{P_u}{E_s \sum I}}$$

Determine second moment of area $= \sum I$

Second moments of area of the composite column about the X–X and Y–Y axes in terms of steel using modular ratio α_e are determined as follows.

$$\alpha_e = \frac{E_s}{450 f_{cu}} = \frac{205 \times 10^3}{450 \times 40} = 11.4$$

$$I_{XX} = \left[I_{xs} + A_r \left(\frac{D_{rx}}{2}\right)^2\right]\left(1 - \frac{1}{\alpha_e}\right) + \frac{1}{12} b_c h_c^3 \frac{1}{\alpha_e}$$

$$= \left(27\,610 \times 10^4 + \frac{1256}{4} \times 360^2\right)\left(1 - \frac{1}{11.4}\right)$$

$$+ \left(\frac{1}{12} \times 450^4 \times \frac{1}{11.4}\right)$$

$$= 589 \times 10^6 \text{ mm}^4$$

$$\sum I = I_{YY} = \left[I_{ys} + A_r \left(\frac{D_{ry}}{2}\right)^2\right]\left(1 - \frac{1}{\alpha_e}\right) + \frac{1}{12} b_c^3 h_c \frac{1}{\alpha_e}$$

$$= \left(9006 \times 10^4 + \frac{1256}{4} \times 360^2\right)\left(1 - \frac{1}{11.4}\right)$$

$$+ \left(\frac{1}{12} \times 450^4 \times \frac{1}{11.4}\right)$$

$$= 419 \times 10^6 \text{ mm}^4$$

Using the second moment of area about the minor axis, the slenderness factor is given by:

$$\bar{\lambda} = \frac{6000}{\pi} \sqrt{\frac{9180 \times 10^3}{205 \times 10^3 \times 419 \times 10^6}} = 0.624$$

$$\lambda_{\text{eff}} = \bar{\lambda}\pi \sqrt{\frac{E_s}{p_y}} = 0.624 \times \pi \times \sqrt{\frac{205 \times 10^3}{355}} = 47$$

Using $\lambda_{\text{eff}} = 47$ and design strength $p_y = 355 \text{ N/mm}^2$, the compressive strength p_c is found to be 283 N/mm^2 from Table 27(c) of BS 5950: Part 1.

The reduction factor K_1 is given by:

$$K_1 = \frac{p_c}{p_y} = \frac{283}{355} = 0.80$$

$$\therefore\ P_{cx} = K_1 P_u = 0.80 \times 9180 = 7344 \text{ kN}$$

Step 8 *Determine additional moment due to potential eccentricity of vertical load in slender column*

Take nominal minimum eccentricity as $0.03b_c$ about both axes.
$M_{addx} = M_{addy} = P \times 0.03 \times 450 = 56.7$ kNm when applied load $P = 4200$ kN

Total design moments are given by:

$M_x = 220 + 56.7 = 276.7$ kNm
$M_y = 56.7$ kNm

Step 9 *Check encased column by interaction formula*

$P = 4200$ kN
$P_{cx} = 7344$ kN
$M_{px} = 854$ kNm
$M_{py} = 587$ kNm

Check $\begin{cases} \dfrac{P}{P_{cx}} + \dfrac{M_x}{M_{px}} + \dfrac{M_y}{M_{py}} \leq 1.0 \\ \dfrac{4200}{7344} + \dfrac{276.7}{854} + \dfrac{56.7}{587} = 0.99 < 1.0 \end{cases}$

Alternatively:

$P_{ux} = P_{cx} = K_1 P_u = 7344$ kN
$P_{uy} = P_u = 9180$ kN
$P_{0,x} = P_{0,y} = 0.45 f_{cu} A_c = 3352$ kN (see Step 1)

$\mu_x = \dfrac{P_{ux} - P}{P_{ux} - P_0} = \dfrac{7344 - 4200}{7344 - 3352} = 0.787$

$\mu_y = \dfrac{P_{uy} - P}{P_{uy} - P_0} = \dfrac{9180 - 4200}{9180 - 3352} = 0.854$

Check $\begin{cases} \dfrac{M_x}{\mu_x M_{px}} + \dfrac{M_y}{\mu_y M_{py}} \leq 1.0 \\ \dfrac{276.7}{0.787 \times 854} + \dfrac{56.7}{0.854 \times 587} = 0.52 \end{cases}$

The simple interaction formula may be considered as conservative and the alternative approach may be used.

Chapter 9
Connections in Steelwork

9.1 BOLTED CONNECTIONS

9.1.1 Types of bolts

Grade 4.6 and 8.8 bolts: Reference BS 4190 and BS 3692
The most frequently used bolts are Grade 4.6 and 8.8. Grade 4.6 bolts have a minimum ultimate tensile strength of 40 kg/mm² (392 N/mm²) and a minimum stress at yield of $0.6 \times 40 = 24$ kg/mm² (235 N/mm²). Similarly, Grade 8.8 bolts have a minimum ultimate strength of 80 kg/mm² (785 N/mm²) and a minimum stress at yield of $0.8 \times 80 = 64$ kg/mm² (627 N/mm²).

High strength friction grip bolts: Reference BS 4395
Generally Grade 8.8 bolts are used for this purpose up to a diameter of 24 mm.

Table 9.1 Geometry of bolted connections.

Size of hole (D)	Spacing of bolts	Edge distance	End distance
$d+2$ for $d \leq 24$ mm	$2.5d$ centres minimum	$1.25D$ to rolled, sawn or machine flame-cut edge minimum	$1.4D$ minimum recommended subject to providing adequate bearing capacity
$d+3$ for $d > 24$ mm	$14t$ maximum in the direction of stress	$1.4D$ to sheared or hand flame-cut edge minimum	(The end distance will not govern the bearing capacity if it is at least $2d$ for ordinary bolting and $3d$ for high-strength friction grip bolts)
$d+6$ short-slotted for $d \leq 22$ mm	$16t$ or 200 mm maximum in corrosive environment	$4t+40$ for corrosive environment maximum	
$d+8$ short-slotted for $d = 24$ mm	$t =$ thickness of thinner ply	$11t\varepsilon$ maximum for normal environment	
$d+10$ short-slotted for $d \geq 27$ mm			
$2.5d$ long-slotted			

$D =$ diameter of hole; $d =$ diameter of bolt; $t =$ thickness of connected ply.

Note: The edge distance is the distance from the centre of a hole for a fastener to the adjacent edge of ply at right angles to the direction of applied load.

9.1.2 Capacity of bolts

Effective area of bolts
A_t = effective area of bolt in tension (reference BS 4190 and BS 3692)
  = area of the bolt at the bottom of the threads
A_s = effective area of bolt in shear = A_t = generally, or
  = area of shank where threads do not appear in the shear plane

Shear capacity of bolts
P_s = shear capacity of bolt = $p_s A_s$
p_s = design strength of bolts in shear
  = 160 N/mm² for Grade 4.6 bolts
  = 375 N/mm² for Grade 8.8 bolts

Bearing capacity of bolts
P_{bb} = bearing capacity of bolt = $dt p_{bb}$
d = nominal diameter of bolt
t = thickness of the connected ply
p_{bb} = design strength of bolt in bearing
  = 460 N/mm² for Grade 4.6 bolts
  = 1035 N/mm² for Grade 8.8 bolts

Bearing capacity of connected ply
P_{bs} = bearing capacity of connected ply = $dt p_{bs} \leq \frac{1}{2} e t p_{bs}$
e = end distance in the direction of load
p_{bs} = design strength of connected ply in bearing
  = 460 N/mm² for Grade 43 material of the ply
  = 550 N/mm² for Grade 50 material of the ply
  = 650 N/mm² for Grade 55 material of the ply

Bolt shear capacity reduction
Long joints

P_s = reduced bolt shear capacity = $p_s A_s [(5500 - L_j)/5000]$

L_j = distance between the first row of bolts and the last row of bolts in the direction of loading in a long joint > 500 mm.
Large grip joints
When T_g = the total thickness of all the plies joined together by a bolt of nominal diameter d exceeds $5d$, then the bolt shear capacity is reduced as follows:

$$P_s = p_s A_s [8d/(3d + T_g)]$$

Tensile capacity of bolt
P_t = tensile capacity = $p_t A_t$
p_t = design strength of bolt in tension = 195 N/mm² for Grade 4.6 bolts
  = 450 N/mm² for Grade 8.8 bolts

Note: These tension capacities already include an allowance of 20 to 30% for prying effects and prying forces may not be included separately in the analysis of bolt tension.

Derivation of tension capacities of bolts

$$p_t = UTS \times \frac{1}{\gamma_m} \times \gamma_t \times \gamma_p$$

UTS = ultimate tensile strength
γ_m = material factor = 1.25
γ_t = factor to allow for thread stripping effects = 0.9 generally
γ_p = factor to allow for prying effects = 0.8 for Grade 8.8 bolts and 0.7 for Grade 4.6 bolts

Combined shear and tension in bolts

The interaction formula is:

$$(F_s/P_s) + (F_t/P_t) \leq 1.4$$

where F_s is applied shear force and F_t is applied direct tension

Friction grip bolts

Note: These bolts are much more expensive to purchase, install and inspect. The contact surfaces have to be masked off and then touched up after erection, which involves additional labour costs. They should be used with care, and should only be specified where joint slippage is unacceptable, as in beam splices.

Parallel shank bolts: slip resistance

P_{sl} = slip resistance = $1.1 K_s \mu P_0$
K_s = 1.0 for bolts in clearance holes
     = 0.85 for bolts in short-slotted holes
     = 0.60 for bolts in long-slotted holes
μ = slip factor ≤ 0.55 (generally taken as 0.45)
P_0 = minimum shank tension as per BS 4604

Parallel shank bolts: bearing resistance

P_{bg} = bearing capacity of friction grip bolt = $P_{bg} = dt p_{bg} \leq \frac{1}{3} et p_{bg}$
d = nominal diameter of bolt
e = end distance
t = thickness of ply
p_{bg} = 825 N/mm² for Grade 43 material of connected ply
      = 1065 N/mm² for Grade 50 material of connected ply
      = 1210 N/mm² for Grade 55 material of connected ply

Modification of slip resistance for long joints

$$P_{slr} = 0.6 P_0 [(5500 - L_j)/5000] \leq P_{sl}$$

L_j = distance between the first row of bolts and the last row of bolts in the direction of loading > 500 mm

Waisted shank friction grip bolts: slip resistance

P_{sl} = slip resistance of waisted shank friction grip bolt = $0.9 K_s \mu P_0$

Tension capacity of friction grip bolts

Permitted types are as per BS 4395: Parts 1 and 3.

P_t = tension capacity of friction grip bolt = $0.9 P_0$

Combined shear and tension for friction grip bolts
The interaction formula is:

$(F_s/P_{sl}) + 0.8\,(F_t/P_t) \leq 1.0$

9.1.3 In-plane loading of a group of bolts

A group of bolts is shown in SK 9/1. A load P is applied to the group at an angle θ_x to the axis X–X. The load acts at an eccentricity e from the centroid of the group of bolts. The coordinates of the centroid of a group of bolts is found as follows:
Choose any orthogonal set of axes X–X and Y–Y.

$\bar{x} = (\sum x)/n; \quad \bar{y} = (\sum y)/n$

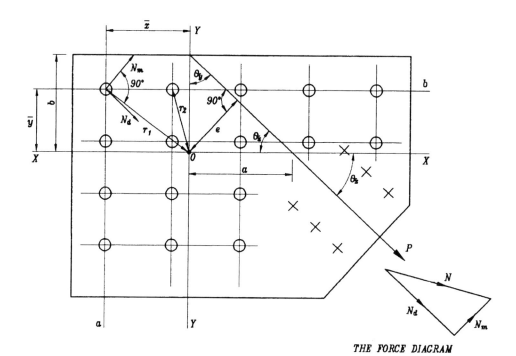

SK 9/1 In-plane loading of a group of bolts.

where $\sum x$ is the sum of the x co-ordinates and $\sum y$ is the sum of the y co-ordinates of n number of bolts in the group. Shift the origin of the orthogonal system of axes to O, where O has the co-ordinates $\bar{x}$ and $\bar{y}$.

Determine polar moment of inertia of the group of bolts about the origin at centroid O
Determine co-ordinates x_i and y_i of each bolt in the group relative to the origin at O.

I_P = polar moment of inertia of a group of bolts = $\sum x_i^2 + \sum y_i^2$

Determine eccentricity of load P about the centroid of the group of bolts
The eccentricity e is measured as the distance from the origin O to the line of action of the load P along a line perpendicular to the direction of P. If the line of action of load P makes an intercept a on the X–X axis through the origin O, then the eccentricity is given by:

$$e = a \sin \theta_x$$

where θ_x is the angle of inclination of P to the X–X axis. If the line of action of load P makes an intercept b on the Y–Y axis through the origin O, then the eccentricity is given by:

$$e = b \sin \theta_y$$

where θ_y is the angle of inclination of P to the Y–Y axis.

Determine torsional moment M_T on the group of bolts due to eccentric load P

$$M_T = Pe$$

M_T = in-plane moment on the group of bolts
P = in-plane load on the group of bolts
e = eccentricity of in-plane loading on the group of bolts

Determine bolt shear N_d due to load P

$$N_d = P/n \text{ (in the direction of action of load } P\text{)}$$

n = the number of bolts in the group

Determine bolt shear N_m due to torsional moment M_T on the group

$$N_{mi} = M_T r_i / I_P \text{ (in the direction perpendicular to the line joining the bolt } i \text{ to the origin O)}$$

r_i = distance from bolt i to origin O
I_P = polar moment of inertia of the group of bolts

The shear N_m produces the same directional moment at the origin O as the torsional moment M_T. Alternatively:

$$N_{mi,x} = M_T y_i / I_P \text{ (in the direction } X\text{–}X\text{)}$$
$$N_{mi,y} = M_T x_i / I_P \text{ (in the direction } Y\text{–}Y\text{)}$$

x_i and y_i are co-ordinates of both i with respect to the origin at O.

Determine the resultant shear in the bolt
Draw the triangle of forces with N_d and N_{mi} to find the resultant shear N_i in bolt i.
Check against the capacity of the bolt in shear and bearing.

9.1.4 Out-of-plane loading of a group of bolts

The bolt capacities in tension, obtained by the use of allowable stresses from BS 5950: Part 1, include an allowance for prying forces

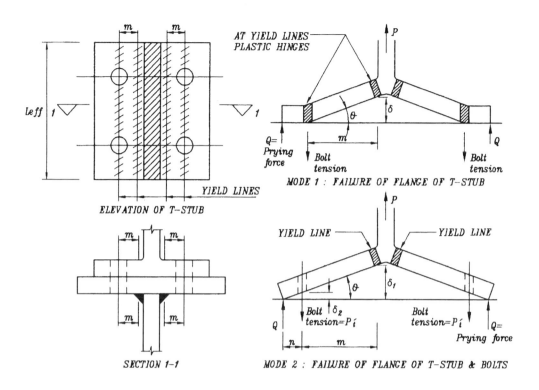
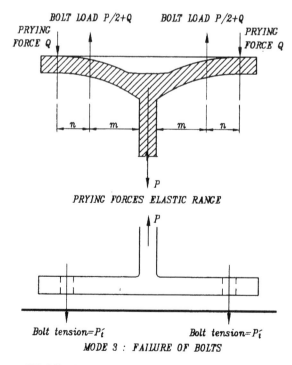

SK 9/2 Modes of failure of bolts in tension.

Connections in Steelwork 403

(20% for Grade 8.8 and 30% for Grade 4.6). The enhanced bolt tensions at failure are obtained after removing these margins because in the modes of failure illustrated in SK 9/2 the prying forces do not come into consideration.

Mode 1: Failure of flange or end plate of beam by yielding

The work done by rotation of plastic hinges in the flange or end plate of the beam is equated to the work done by the tensile load P.

$$P\delta = 4M_P\theta$$

$$\delta = m\theta$$

$$\therefore\ P = 4M_P/m$$

$$M_P = L_{\text{eff}} \frac{t^2}{4} p_y$$

L_{eff} = effective length of flange or end plate gone beyond yield
p_y = design strength of the material of flange or end plate
t = thickness of flange or end plate
m = distance from the centre of the bolt to 20% of the distance into the root of the rolled section or the fillet weld
M_P = the plastic moment of resistance of flange or end plate per unit length

δ and θ are as shown in SK 9/2.

Mode 2: Failure of flange or end plate and bolts by yielding

Apply the principle of work done as in mode 1: i.e. work done by load P = work done in rotation of plastic hinges in the flange + work done in plastic extension of bolts (see SK 9/2):

$$P\delta_1 = 2M_P\theta + (\sum P'_t)\delta_2$$

P'_t = enhanced bolt tensile capacity without allowance for prying
$\delta_1 = (m+n)\theta$
$\delta_2 = n\theta$

$$\therefore\ P = \frac{2M_P + (\sum P'_t)n}{(m+n)}$$

Here, n is the effective edge distance, which is the least of the following:

For end plate:
- end distance e for the column flange
- end distance e for the end plate
- $1.25m$ for the end plate

For column flange:
- end distance e for the column flange
- end distance e for the end plate
- $1.25m$ for the column flange

Mode 3: Failure of bolts by yielding

The ultimate load in bolts with enhanced capacities, excluding prying effects, equates to the load P applied.

$$P = \sum P'_t$$

404 Structural Steelwork

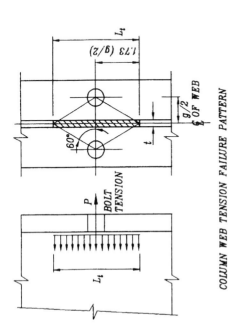
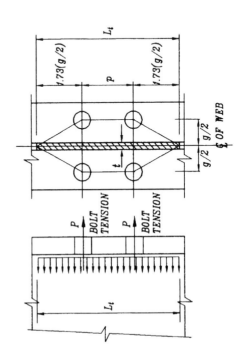

SK 9/3 Web tension failure pattern.

Table 9.2 Typical yield-line failure of plate around bolts.

Yield-line pattern		L_{eff}
Pattern 1	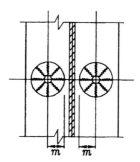	$L_{\text{eff}} = 2\pi m$
Pattern 2	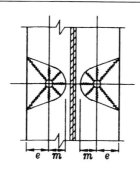	$L_{\text{eff}} = 4m + 1.25e$
Pattern 3	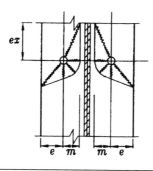	$L_{\text{eff}} = 2m + 0.625e + e_x$
Pattern 4	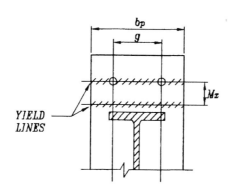	$L_{\text{eff}} = b_p/2$

Table 9.2 (contd)

Yield-line pattern		L_{eff}
Pattern 5	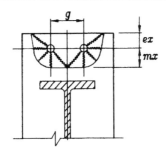	$L_{eff} = 2m_x + 0.625e_x + g/2$
Pattern 6	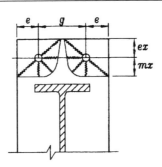	$L_{eff} = 2m_x + 0.625e_x + e$
Pattern 7	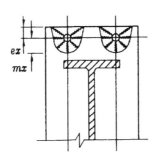	$L_{eff} = 4m_x + 1.25e_x$
Pattern 8	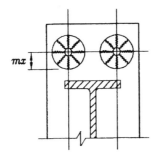	$L_{eff} = 2\pi m_x$

Table 9.2 (contd)

Yield-line pattern		L_{eff}
Pattern 9	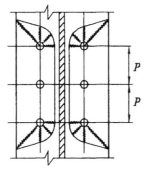	$L_{eff} = 4m + 1.25e + 2p$

Note: Detailed analysis of the yield-line patterns may be obtained by reference to *Joints in Steel Construction: Moment Connections*, The Steel Construction Institute, 1995.

Mode 4: Failure of web of beam or column in tension
See SK 9/3.

P_t = resistance of web in tension = $L_t t_w p_y$
L_t = effective length of web resisting tension assuming 60° spread of load
t_w = thickness of web
p_y = design strength of web

9.1.5 Out-of-plane bending moment, direct load and shear on a group of bolts

If the beam end plate and the column flange are thin and flexible, then it can be assumed that all bolts in the tension zone of the moment connection will reach yield or their maximum rated capacity according to mode of failure (modes 1 to 4). On the other hand, if the column flange or the beam end plate is very stiff and cannot bend enough under load to cause yield in all the bolts in the tension zone, then a triangular stress distribution in the bolt group has to be assumed, as shown in SK 9/4. The triangular distribution of loads should be adopted unless:

either $\quad t_p < \dfrac{d}{1.9}\sqrt{\dfrac{U_f}{p_{yp}}}$

or $\quad T_c < \dfrac{d}{1.9}\sqrt{\dfrac{U_f}{p_{yc}}}$

where t_p = thickness of end plate
d = diameter of bolt
U_f = ultimate tensile strength of bolt
p_{yp} = design strength of end plate
p_{yc} = design strength of column flange
T_c = thickness of column flange

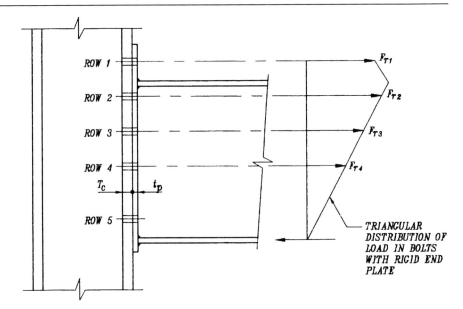

SK 9/4 Bolt load distribution in beam–column moment connection.

Modification of applied moment due to axial loads
M_m = equivalent bending moment:
 $= M - Nh$ (for axial compression)
 $= M + Nh$ (for axial tension)
M = applied bending moment

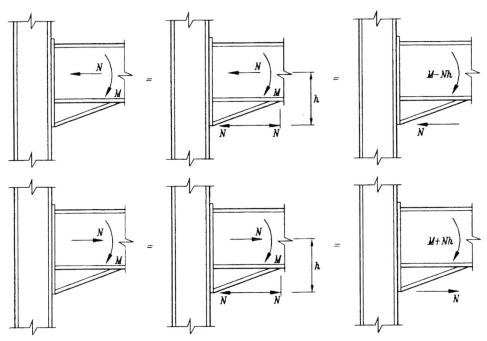

SK 9/5 Equivalent bending moment on the bolt group due to axial load.

N = applied direct load (+ve for axial compression on the connection)

h = distance from line of action of direct load N to the centre of compression

Moment capacity of connection

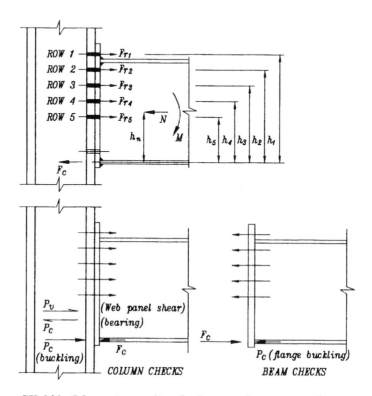

SK 9/6 **Moment capacity of a beam–column connection.**

Distribution of loads in the bolts

P_{ri} = maximum allowable tensile load in the bolt at the ith row depending on the modes of failure 1 to 4 and also limited by triangular stress distribution of bolt load with respect to the centre of rotation

F_{ri} = actual tensile load in the bolt at the ith row

h_i = the distance of the bolt in the ith row from the centre of compression

F_c = total compressive force acting at the centre of compression

M_c = moment capacity of the connection

The following conditions should be satisfied:

$F_{ri} \leq P_{ri}$

$F_c \leq P_c$ (web bearing capacity)

$\leq P_c$ (web buckling capacity)

$\leq P_c$ (beam flange bearing capacity)

$\leq P_v$ (web panel shear capacity)

$\leq \sum P_{ri} + N$

$F_c = \sum F_{ri} + N$

Determination of F_c from equivalent applied moment M_m
Determine P_{ri}
Determine $M_c = \sum (P_{ri} h_i)$. If M_c is greater than M_m then assume that the bolts in the top row have a tension F_{r1} equal to P_{r1} and progressively reduce the tensions F_{ri} below P_{ri} in the lower rows of bolts until $\sum (F_{ri} h_i)$ equates to M_m
Determine $F_c = \sum F_{ri} + N$
Use this F_c to check against P_c and P_v

Shear capacity of connection
The interaction formula in BS 5950: Part 1 for combined tension and shear on bolts, apart from friction grip bolts, requires the factor 1.4 not to be exceeded. If it is assumed that the applied tension F_t is equal to the tension capacity P_t of the bolt, then the applied shear F_s could have a maximum value equal to $0.4 P_s$, where P_s is the shear capacity of the bolt.

The design shear force F_v should not exceed the shear capacity of bolts in the tension and shear zones combined. The bolt layout in the end plate may be divided into tension zone bolts and shear zone bolts. The shear zone bolts are designed for shear force only and any tension due to moment in these bolts is neglected.

$F_v \leq n_t P_t'' + n_s P_s'$

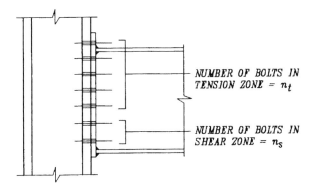

SK 9/7 Shear capacity of a connection by bolts.

$P_t'' =$ the smaller of $0.4P_s$, P_{bb} and P_{bs}
$P_s' =$ the smaller of P_s, P_{bb} and P_{bs}
$n_t =$ number of bolts in the tension zone
$n_s =$ number of bolts in the shear zone
$P_s =$ shear capacity of bolt
$P_{bb} =$ bearing capacity of bolt
$P_{bs} =$ bearing capacity of connected ply

9.1.6 Local capacity check of connected elements in a moment connection

Local compression check of column web
Web bearing
Assume the distribution of compressive load from the flange of the beam is at 45° through the stiff end plate and at 1:2.5 through the flange thickness and the root of the flange of the column as shown in SK 9/8.
The bearing resistance P_c of the column web is given by:

$$P_c = (b + n_2)t_c p_{yc}$$

where $b_1 =$ stiff bearing length determined by assuming 45° dispersion of load
$n_2 =$ dispersion length assuming 1:2.5 dispersion through the flange and the root
$t_c =$ thickness of web of the column
$p_{yc} =$ design strength of the column

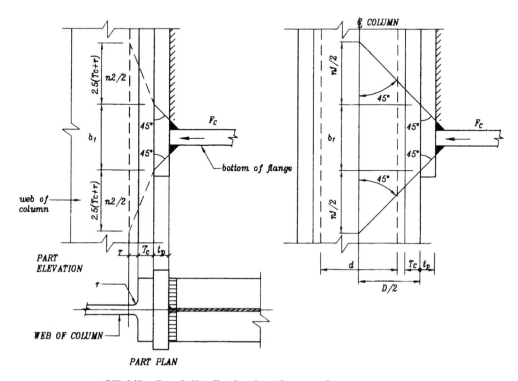

SK 9/8 Load distribution in column web.

Web buckling
Assume a dispersion of load at 45° through the stiff bearing length and the flange root and web up to the centre of the column web as shown in SK 9/8.

The buckling resistance P_c of column web is given by:

$$P_c = (b_1 + n_1)t_c p_c$$

where n_1 = depth of column D_c
p_c = compressive strength of the column web as per Table 27(c) of BS 5950: Part 1 with slenderness ratio λ taken as $2.5d/t_c$
d = depth of web between fillets

Local bearing check of compression flange of beam
The bearing resistance P_c of the beam flange in compression is given by:

$$P_c = p_y T B$$

where p_y = design strength of beam
T = thickness of flange of beam
B = width of flange of beam

The bearing stress in flange may be allowed to exceed p_y by up to 40% to allow for the effects of local strain hardening and dispersion of the load partly in the web of the beam.

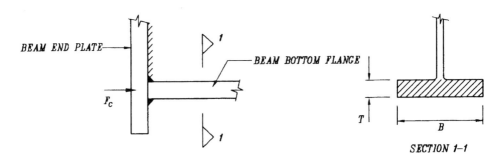

SK 9/9 Bearing by compression of flange of beam.

Local shear in column web
The column web panel shear at the connection as shown in SK 9/10 is the algebraic sum of the flange forces in the beams on either side of the column, assuming that the moment is fully resisted by the flanges in the beam and that the direct axial load in the beam is equally divided between the flanges.

$F_{1,t}$ = flange force in the top flange of beam B_1

$$= \frac{M_1}{D_1 - T_1} + \frac{N_1}{2}$$

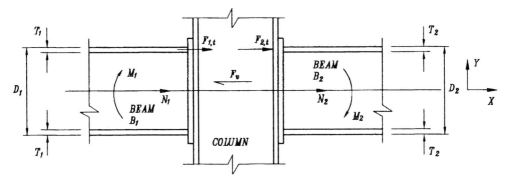

SK 9/10 Column web panel shear at the connection.

$F_{1,b}$ = flange force in the bottom flange of beam B_1

$$= -\frac{M_1}{D_1 - T_1} + \frac{N_1}{2}$$

$F_{2,t}$ = flange force in the top flange of beam B_2

$$= \frac{M_2}{D_2 - T_2} + \frac{N_2}{2}$$

$F_{2,b}$ = flange force in the bottom flange of beam B_2

$$= -\frac{M_2}{D_2 - T_2} + \frac{N_2}{2}$$

where F_v = column web panel shear = $F_{1,t} + F_{2,t}$ or $F_{1,b} + F_{2,b}$ (whichever is the greater)
M_1 = bending moment in beam B_1 (clockwise positive)
M_2 = bending moment in beam B_2 (clockwise positive)
N_1 = direct tension or compression in beam B_1 (positive in the direction of positive x)
N_2 = direct tension or compression in beam B_2 (positive in the direction of positive x)
D_1 = overall depth of beam B_1
D_2 = overall depth of beam B_2
T_1 = thickness of flange of beam B_1
T_2 = thickness of flange of beam B_2

The web shear capacity P_v of the column is given by:

$$P_v = 0.6 p_y A_v$$

where A_v = shear area of column section as per Clause 4.2.3 of BS 5950: Part 1
= tD for rolled I-, H- and channel sections
= td for built-up sections and boxes
= $0.9A$ for solid bars and plates
= $\left(\dfrac{D}{D+B}\right) A$ for rectangular hollow sections
= $0.6A$ for circular hollow sections

9.1.7 Design of stiffeners and haunched ends of beams

- If the column web fails by bearing or buckling due to the load F_c at the connection, then use column compression stiffeners or supplementary web plates.
- If the column flange fails due to yielding at the connection, use column flange backing plates.
- If the column web or the beam web fails due to local tension (mode 4) at the bolt, use tension stiffeners or supplementary web plates.
- If the column web fails due to web panel shear, use supplementary web plates or diagonal shear stiffeners.

Note: A haunched beam end helps to reduce the internal forces in the connection geometry. Use haunched ends at connections with high bending moment, axial load and shear force.

Column compression stiffener

Outstand of the column compression stiffener from the face of the web should not exceed $19 t_s \varepsilon$.

t_s = thickness of stiffener

$$\varepsilon = \sqrt{\frac{275}{p_{ys}}}$$

p_{ys} = design strength of stiffener

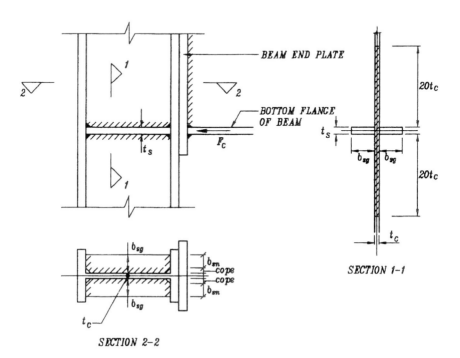

SK 9/11 Loading on column compression stiffener.

When the outstand is between $13t_s\varepsilon$ and $19t_s\varepsilon$, the structural strength computation should be done on the basis of an outstand equal to $13t_s\varepsilon$. Use the actual outstand when it is less than $13t_s\varepsilon$.

b_{sg} = gross outstand of the stiffener (limited to $13t_s\varepsilon$ for calculation of strength)

b_{sn} = net outstand in contact with the flange of the column allowing for corner snipe

A_{sg} = gross area of stiffener, assuming a pair of plates symmetrically placed about the web
$= 2b_{sg}t_s$

A = area of stiffener in contact with the flange of the column
$= 2b_{sn}t_s$

A_{eff} = effective area of stiffener considering 20 times thickness of the web on either side of the stiffener
$= 2b_{sg}t_s + 40t_c^2$

t_c = thickness of web of column

I_{eff} = effective moment of inertia of the section with area A_{eff} about the vertical axis through the web of the column

$$= \frac{1}{12} t_s(2b_{sg} + t_c)^3$$

r_{eff} = effective radius of gyration

$$= \sqrt{(I_{eff}/A_{eff})}$$

L_{eff} = effective length of the stiffener
$= 0.7(D_c - 2T_c)$ assuming that the column flanges are restrained against rotation relative to each other

D_c = overall depth of the column

T_c = thickness of flange of column

λ_{eff} = slenderness ratio = L_{eff}/r_{eff}

Using this slenderness ratio and the design strength p_y find the compressive strength p_c from Table 27(c) of BS 5950: Part 1. The design strength p_y should be taken as the lesser of the design strengths of the column and the stiffener.

The buckling resistance P_c of the stiffener assembly is given by:

$$P_c = A_{eff} p_c \geq F_c$$

The bearing strength P_c of the stiffener assembly is given by:

$$P_c = Ap_y + \text{bearing resistance of web alone}$$

$$= Ap_y + (b_1 + n_2)t_c p_{yc} \geq F_c \text{ (see bearing resistance of column web)}$$

Assuming that 80% of the load passes through the contact surface between the stiffener and the flange, the net bearing resistance P_c is given by:

$$P_c = \frac{Ap_y}{0.8} \geq F_c$$

Column flange backing plates

This method of stiffening may only be used where column flanges are very thin and mode 1 type of failure dictates the connection geometry.

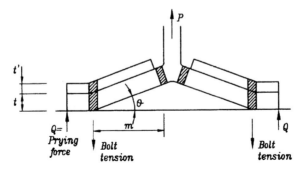

SK 9/12 Mode of failure with column flange backing plates.

As in mode 1 type of failure, the work done by the load P is equated to the work done in the rotation of plastic hinges in the flange plate and the backing plate. This gives:

$$P = \frac{4M_P + 2M'_P}{m}$$

$$M_P = L_{\text{eff}} \frac{t^2}{4} p_y$$

$$M'_P = L_{\text{eff}} \frac{(t')^2}{4} p'_y$$

where t' = thickness of backing plate
p'_y = design strength of backing plate

Supplementary column web plate

Design philosophy
(1) t_s = thickness of web plate $\geq t_c$ = thickness of column web.
(2) p_{ys} = design strength of web plate = p_{yc} = design strength of column.
(3) Leg length of fillet weld all round = thickness of web plate t_s.
(4) b_s = width of web plate $\geq d - 2t_s$.
(5) Use fill-in weld between web plate and the flange where resistance to web tension is required.
(6) $b_s = d$ where fill-in weld is used.
(7) L_s = length of web plate $\geq g + L_c + D/2$.
(8) Use plug welds where $b_s > 37t_s$ (Grade 43) or $33t_s$ (Grade 50).
(9) Diameter of plug weld when used should be greater than or equal to t_s, and the spacing of plug welds should be less than or equal to $37t_s$.

g = horizontal spacing of bolts at the connection
L_c = length of end plate
D = overall depth of column section
d = depth between fillets of column section

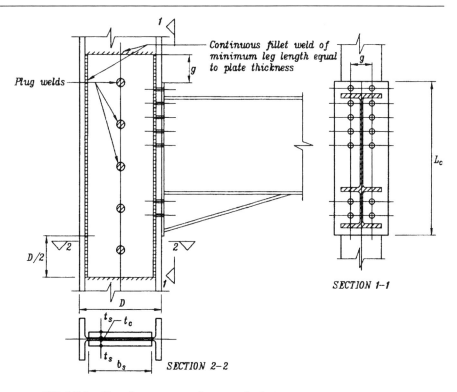

SK 9/13 Supplementary column web plates.

Web tension capacity with supplementary web plate
t_{eff} = effective thickness of web
   $= 1.5t_c$ for web plates on one side of web
   $= 2.0t_c$ for web plates on both sides of the web

Use the effective thickness t_{eff} of the web to determine web tension capacity.

Web bearing and buckling capacity with supplementary web plate
$t_{\text{eff}} = 1.5t_c$ for web plates on one side of web
   $= 2.0t_c$ for web plates on both sides of the web

Use the effective thickness t_{eff} of the web to determine web bearing and buckling capacity.

Web panel shear capacity
A_v = shear area of the web with web plates
   = shear area of column web + $t_c b_s$

The shear area remains unchanged if supplementary web plates are used on one side of web or both sides of web.
   Use the modified shear area A_v to determine web panel shear capacity.

Tension stiffeners
P_t = web tension capacity
   $= L_t t_c p_y$ (see mode 4 type of failure)

P_{st} = tension capacity of the tension stiffener
 = Ap_y
A = area of stiffener in contact with the flange
 = $2b_{sn}t_s$
b_{sn} = net outstand of stiffener in contact with the flange
t_s = thickness of stiffener

Check $P_t + P_{st} \geq F_{ri} + F_{rj}$

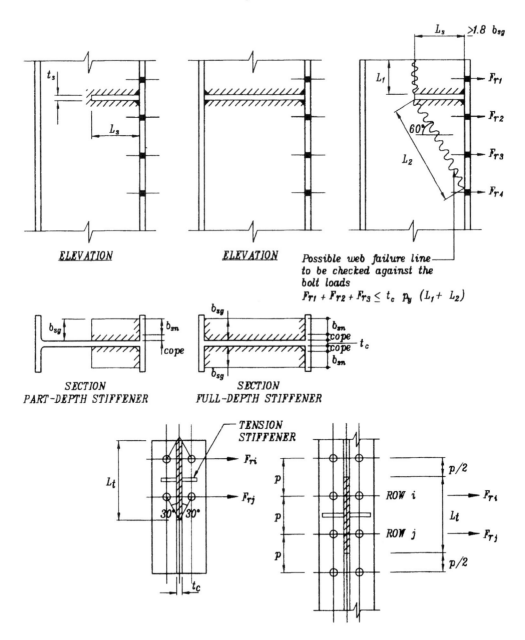

SK 9/14 Distribution of loads in tension stiffeners.

The tension stiffener is used between bolt rows i and j

F_{ri} = actual load in the bolts in the row i
F_{rj} = actual load in the bolts in the row j

For a part-depth tension stiffener, the length L_s should be more than or equal to $1.8b_{sg}$ and full-strength fillet welds equivalent to the thickness of the stiffener should be used. (See SK 9/14.)

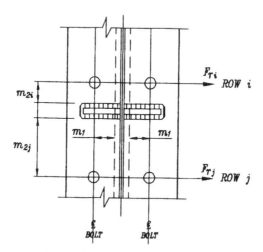

SK 9/15 Distribution of loads from the bolts on the tension stiffener.

The bolt tensions F_{ri} and F_{rj} may be assumed to be resisted partly by the web and partly by the stiffener. Assuming that the load carried by the stiffener is inversely proportional to its distance from the row of bolts, the load F_s in the stiffener is given by:

$$F_s = \frac{m_1 F_{ri}}{(m_1 + m_{2i})} + \frac{m_1 F_{rj}}{(m_1 + m_{2j})}$$

Check $P_{st} \geq F_s$

m_1, m_{2i} and m_{2j} are the distances from the bolts as shown in SK 9/15.

Diagonal shear stiffeners
These stiffeners are used when the capacity of the web in panel shear has been exceeded due to applied shear force F_v.

P_s = resistance of horizontal shear of an inclined diagonal stiffener
     = $A_{sg} p_y \cos \theta$
A_{sg} = area of the stiffener assembly
     = $2b_{sg} t_s$
b_{sg} = width of stiffener on each side of the web
t_s = thickness of stiffener
θ = angle the diagonal stiffener makes to the horizontal
P_v = resistance of the web in panel shear as found earlier

Check $P_s + P_v \geq F_v$

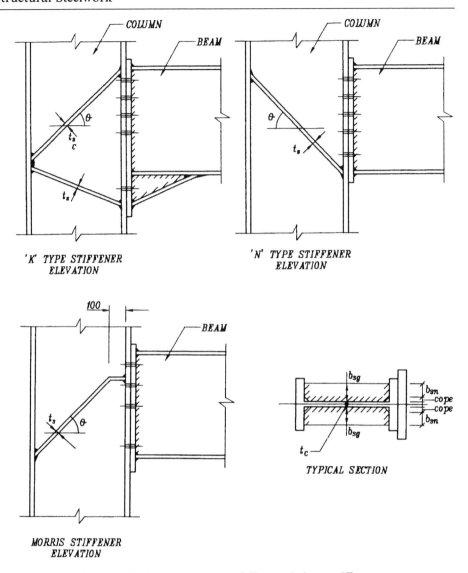

SK 9/16 Typical arrangements of diagonal shear stiffeners.

K-stiffener: Use where the depth of connection is large compared to the depth of the column. The bottom part of the stiffener is in compression and must be checked as a web compression stiffener.

N-stiffener: This compression stiffener is used to avoid problems of access to bolts in the tension zone. Design as a compression web stiffener.

Morris stiffener: This is a diagonal stiffener which also acts as a tension stiffener against failure of the web in tension (mode 4). This stiffener provides unimpeded access to the bolts in the tension zone of the connection.

Connections in Steelwork 421

Haunched end of beam
Geometrical requirements
(1) ϕ, the haunch angle, should be greater than or equal to 45° (see SK 9/17).
(2) The thickness of the web and flange of the haunch should not be less than those of the beam. Use the same beam section cut to suit.

Compressive flange force in the haunch $= \dfrac{F_c}{\sin \phi}$

F_c = horizontal compressive force for connection equilibrium (see moment capacity of connection)

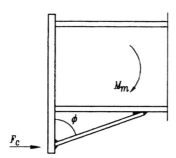

SK 9/17 Typical haunched end geometry of beams.

9.1.8 Bolted splices

Design philosophy
(1) Continuity of the member about both axes has to be maintained at the splice.
(2) The applied moment at the splice is resisted by the flange cover plates.
(3) The applied shear force at the splice is resisted by the web cover plates.
(4) The applied axial direct load in compression or tension is resisted equally by the flange cover plates in the case of beams. For columns, the direct load is shared between the flange and the web cover plates in proportion to their areas.
(5) Additional minor axis bending moment due to lateral torsional buckling (strut action) should be considered for the design of splices (reference Appendix C3 of BS 5950: Part 1).

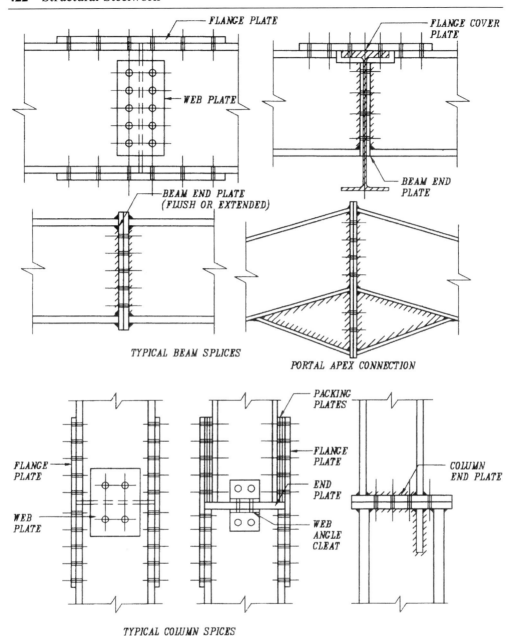

SK 9/18 Typical details of splices in columns and beams.

Flange cover plates
The flange forces are given by (see SK 9/19):

$$F_f = \pm \frac{M_x}{h} + \frac{N}{2}$$

Compressive forces are positive.

M_x = applied moment about the major axis at the splice
h = distance between centroids of flanges

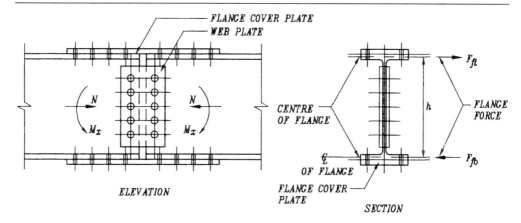

SK 9/19 Distribution of forces at a splice.

N = applied direct axial load (compression positive)
A = area of beam or column

Design philosophy
(1) Design flange cover plates as members subject to direct compression or tension, with due allowance for holes for the bolts at the splice. (See SK 9/19.)
(2) Check flange plates for combined direct load and moment due to bending about the minor axis (applied + strut action). It may be assumed that the total minor axis bending moment is carried equally by the two flange plates.
(3) Also check the existing flanges of the beam or column for tension or compression resistance, making due allowance for bolt holes.
(4) Bolts are in direct shear or bearing (see section on capacity of bolts). Preferably, use high-strength friction grip (HSFG) bolts at all splices to prevent slippage at the connection (see section on HSFG bolts for slip resistance and bearing resistance).
(5) The minor axis bending moment in the flange creates in-plane shear on the group of bolts in the splice. This shear should be combined with shear due to direct tension or compression in the flange to find the resultant shear in the bolt.

Strut action at the splice (see SK 9/20)
Minor axis bending due to strut action should be considered only when the splice is away from a point of lateral restraint.
The minor axis bending moment M_{max} due to strut action is given by:

$$M_{max} = \frac{\eta f_c S}{\left(1 - \dfrac{f_c}{p_E}\right)}$$

where η = Perry factor = $0.001a(\lambda - \lambda_0) \leq 0$
a = Robertson constant = 5.5 (use Table 27(c) of BS 5950: Part 1)

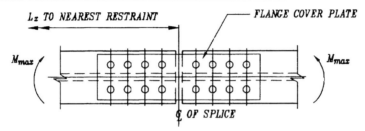

PLAN VIEW OF FLANGE AT SPLICE

SK 9/20 Minor axis bending due to strut action.

The slenderness ratio λ is given by:

$$\lambda = \frac{L_E}{r_y}$$

where L_E = effective length of the beam or column between restraints (see Chapter 7)
  r_y = radius of gyration of the section about the minor axis

The limiting slenderness λ_0 is given by:

$$\lambda_0 = 0.2\sqrt{\frac{\pi^2 E}{p_y}}$$

S is the minor axis plastic modulus of the smaller section at the splice. A beam has equal sections on either side of a splice, whereas a column may have different sections on either side.

f_c = stress in compression due to axial load N
$$= \frac{N}{A}$$

Euler's buckling strength P_E is given by:

$$P_E = \frac{\pi^2 E}{\lambda^2}$$

where E = modulus of elasticity of steel

The maximum strut action bending moment occurs at the midpoint between lateral restraints along the length of the beam or column. The splice may not be at the point of maximum bending moment due to applied loading.

The minor axis bending moment due to strut action at any other point in the beam, assuming sinusoidal distribution, is given by:

$$M_{\max,x} = M_{\max} \sin(180 L_x / L)$$

where L_x = length along the beam or column to a point of interest from the nearest lateral restraint
  L = length along the beam or column between adjacent points of lateral restraint

Interaction formula for flange plates

$$\frac{F_f}{A_e p_y} + \frac{M_f}{M_c} \leq 1$$

A_e = effective area of flange plates, allowing for bolt holes
p_y = design strength of flange plates
M_f = flange plate bending moment due to minor axis bending moment M_{yy} in beam or column
    $= 0.5 M_{yy}$
M_{yy} = minor axis bending moment due to applied minor axis moment M_y and additional minor axis bending moment due to strut action $M_{max,x}$

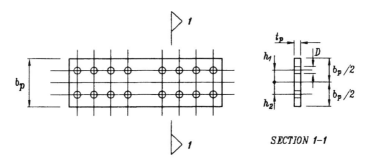

SK 9/21 Elastic section modulus of flange plates.

M_c = elastic moment of resistance of flange plates about the plate centre line parallel to the axis of the beam or column, allowing for bolt holes (see SK 9/21)

$$= \left[\frac{1}{6} t_p b_p^2 - \frac{2}{b_p} \left(\sum t_p D h_i^2 \right) \right] p_y$$

D = diameter of the holes for bolts
t_p = thickness of flange plates
h_i = distance of the ith bolt hole in the plate from the centre-line parallel to the axis of the beam or column

The group of bolts in the flange plates on each side of the splice is subjected to an in-plane torsional moment equal to M_f and an in-plane shear equal to F_f. The bolt shear should be found by the principle of in-plane loading on bolt groups explained earlier in this chapter. Assume that the flange force F_f acts at the centre of the bolt group in the direction along the flange.

See SK 9/22.

N_{dx} = direct load shear in each bolt along the X–X axis

$= F_f / n$

n = number of bolts on each side of the splice

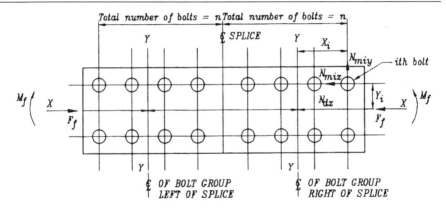

SK 9/22 Distribution of loads in bolts in flange plates.

N_{mix} = shear in the ith bolt in the X–X direction due to moment M_f
   = $M_f/Z_{x,i}$
N_{miy} = shear in the ith bolt in the Y–Y direction due to moment M_f
   = $M_f/Z_{y,i}$

Note: The directions of N_{mix} and N_{miy} depend on the direction of the moment M_f and should be considered to find the resultant shear in the bolt.

$Z_{x,i} = I_P/Y_i; \qquad Z_{y,i} = I_P/X_i$

I_P = polar moment of inertia of the group of bolts
   = $\sum_i X_i^2 + \sum_i Y_i^2$
X_i = distance along the X–X axis of the ith bolt from the centroid of the bolt group
Y_i = distance along the Y–Y axis of the ith bolt from the centroid of the bolt group

The resultant shear in the ith bolt is given by:

$$N_i = \sqrt{(N_{dx} + N_{mix})^2 + (N_{miy})^2}$$

Check $N_i \leq P_s$ = shear capacity of the bolt

Web plates (see SK 9/23)
Shear capacity P_v of each web plate on either side of the web is given by:

$P_v = 0.6(0.9A)p_y$ (as per Clause 4.2.3 of BS 5950: Part 1)

where A = net area of plate allowing for holes = $t_p L - nDt_p$
   t_p = thickness of web plate
   L = length of web plate
   n = number of holes in a vertical line
   D = diameter of holes

Check $F_v \leq n_p P_v$

n_p = number of web plates = 2, generally

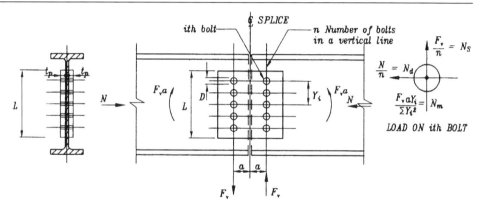

SK 9/23 Distribution of loads in bolts in web plates.

Bolts in web plates (see SK 9/23)
Check end and edge distances. For a friction grip connection the end distances should be greater than three times the bolt diameter.

In-plane moment on the bolt group on each side of the splice is $F_v a$, where F_v is the applied shear and a is the distance from the centre of splice to the centroid of the bolt group.

Design the group of bolts to resist the combined action of in-plane moment $F_v a$ and shear F_v.

N_{mi} = shear in ith bolt due to the moment $F_v a = F_v a r_i / \sum r_i^2$

The direction of bolt shear N_{mi} is at right angles to the line joining the ith bolt to the centroid of the group. The length of this line is r_i.

N_s = shear in the bolt due to direct shear $F_v = F_v/n$

N_d = shear in the bolt due to axial load

Find the resultant of N_{mi}, N_s and N_d by vectorial combination and check against the capacity of the bolt in shear.

Note: The splice plates and bolts may be designed assuming that these elements carry the actual stresses in the components spliced. In deep plate girders the flange splice plates and bolts should be designed to resist the actual flange forces, depending on the bending moment at the splice and the section modulus of the plate girder at the splice allowing for the moment capacity of the web. At the same time, the top row of bolts in the web cover plates should be checked against the combined action of vertical shear and horizontal shear due to average longitudinal bending stress in the web.

9.2 WELDS AND WELDING

Electric arc welding is the most common method of connecting metal to metal by fusion. A low-voltage (10–50 V), high-current (10–2000 A) generator is used to create an electric arc between an electrode and the elements to be welded. The heat generated melts the metal, which then fuses to form a weld. The molten metal is protected either by flux or by inert gas to prevent oxidation and embrittlement.

Shielded metal arc welding (SMAW)
The consumable electrode has a core of filler metal with an outer covering of flux. The flux melts during the welding operation to protect the molten metal.

Gas metal arc welding (GMAW)
The electrode is a continuous spool of wire acting as the filler metal. Inert gas supplied continuously during the welding process protects the molten metal from oxidation.

Submerged arc welding (SAW)
The electrode is a continuous spool of wire and the flux is a granular material fed from a hopper directly in front of the arc. The welding takes place submerged under a flow of granular flux.

9.2.1 Types of weld

Fillet weld
Two steel faces may be joined by a fillet weld which normally has a triangular cross-section. The strength of the weld depends on the throat thickness which is a multiple of the leg length. The effective throat thickness varies with the angle between the surfaces of steel joined together. When the angle between the surfaces to be joined is less than 60° or more than 120° then fillet welds are not permitted.

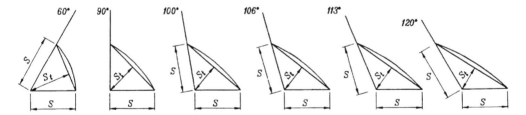

SK 9/24 Effective throat thickness of fillet welds.

Angle between the faces to be joined	Effective throat thickness as a multiple of leg length
60°–90°	$S_t/S = 0.70$
91°–100°	$S_t/S = 0.65$
101°–106°	$S_t/S = 0.60$
107°–113°	$S_t/S = 0.55$
114°–120°	$S_t/S = 0.50$

S_t = throat thickness of fillet weld; S = leg length of fillet weld.

Design rules as per BS 5950: Part 1: Clause 6.6

(1) *End returns*

Fillet welds terminating at ends of connected parts should be returned round the corners for a distance of at least twice the leg length.

(2) *Lap joints*

The minimum lap should not be less than $4t$, where t is the thickness of the thinner ply joined.

(3) *Single fillet welds*

A single line of fillet weld must not be subject to bending moment about its longitudinal axis.

(4) *Intermittent fillet welds*

Do not use intermittent weld where moisture ingress and rust formation cannot be avoided. Intermittent fillet welds should not be subject to frequent stress changes (e.g. as in a gantry girder or a wind girder). The longitudinal spacing between intermittent runs of fillet weld should not exceed $16t$ or 300 mm for compression elements and $24t$ for tension elements, where t is the minimum thickness of the connected ply.

(5) *Effective length of fillet weld*

The effective length of a fillet weld is its actual length less twice the length of the leg. It should not be less than four times the leg length.

Butt welds

Partial penetration butt welds

The throat thickness of a partial penetration butt weld should be taken as the minimum depth of penetration. In the case of a V- or bevel weld,

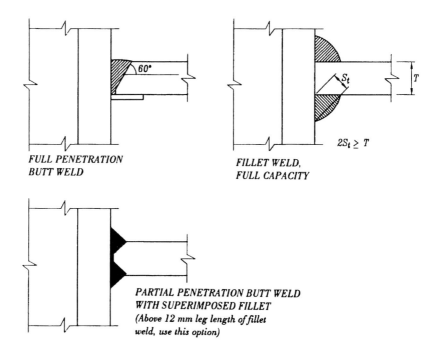

FULL PENETRATION BUTT WELD

FILLET WELD, FULL CAPACITY

PARTIAL PENETRATION BUTT WELD WITH SUPERIMPOSED FILLET
(Above 12 mm leg length of fillet weld, use this option)

SK 9/25 Full penetration and partial penetration butt welds.

the depth of penetration should be taken as the depth of preparation minus 3 mm. In the case of J- or U-preparation, the depth of penetration should be taken as the depth of preparation. The specified penetration of a partial penetration butt weld should not be less than $2\sqrt{t}$, where t is the thickness of the thinner part joined.

Partial penetration butt weld with superimposed fillet weld
The capacity of a partial penetration butt weld with superimposed fillet weld should be calculated by reference to its throat thickness as illustrated in SK 9/26.

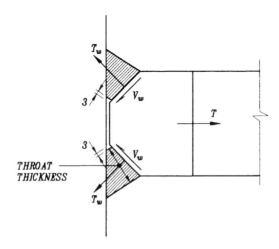

SK 9/26 Partial penetration butt weld with superimposed fillet weld.

The vector sum of all loads should be used to determine the stress in the weld using the throat thickness. This stress should not exceed the design strength p_w of the weld, as given in Table 9.4. Assuming that the angle of bevel preparation is 45°, the force equilibrium at the connection is given by:

$$T = 2(T_w \cos 45° + V_w \cos 45°)$$

where T = applied tensile force at the connection
T_w = tension on the fusion face
V_w = shear on the fusion face

From this analogy check that the shear and tensile stresses on the fusion face do not exceed $0.7p_y$ and $1.0p_y$ respectively.

Full penetration butt weld
The strength of a full penetration butt weld is equal to the strength of the parent material, provided a suitable welding consumable is used. The preparation of a butt weld is very important. The preferred shapes of butt weld preparation are given in Table 9.3. The shape of weld preparation depends on accessibility and the thickness of the metal parts being joined. When joined by butt welds, plates of differing thickness may give rise to additional bending moments if their

centroidal axes are not properly aligned. Sometimes the thicker element is tapered to meet the thinner element. The slope of the taper should not be more than 1 in 4.

9.2.2 Defects in welds

Undercut
This defect occurs when insufficient weld metal is incorporated. The specification should lay down the limits of undercut acceptable.

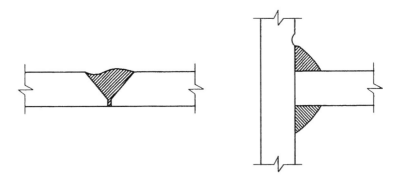

SK 9/27   Example of undercut weld.

Incomplete penetration
This happens when the electrode is too large for the geometry of the joint or the wrong welding angle is used, or if there is insufficient current or inadequate backgouging of a root run back to sound weld metal.

Slag inclusions
This defect is the result of non-metallic particles from the flux getting trapped in the weld in multipass welding operations. The flux must be removed carefully to avoid this problem.

Porosity
The trapping of gas in the weld metal gives rise to this defect.

Lack of fusion
Runs of weld may not fully fuse with the parent metal because of incorrect welding practice.

9.2.3 Weld inspection methods: non-destructive testing (NDT)

Visual checks
- accuracy of fit-up
- cleanliness of parts to be welded
- preparation
- examination of visible cracks
- interpass cleanliness
- final weld geometry

Dye penetrant inspection (DPI)
- The size and extent of cracks can be found by applying a liquid dye on the surface, which then penetrates the cracks. The cracks are then made visible by the use of a developer.
- Look for signs of leakage through the weld on the opposite side.

Magnetic particle inspection (MPI)
- This involves the detection of surface and subsurface cracks in ferro-magnetic materials by magnetising the specimen and applying fine iron particles. These particles are drawn into the cracks because of the formation of secondary magnetic poles at the crack faces.

Ultrasonic flaw detection (UFD)
- With this technique, flaws are detected by applying ultrasonic pulses from a transmitter probe (frequency range 1–5 MHz) which are then picked up by a receiving probe. The pulses are deflected at points where discontinuities exist.

Radiography
- This method of detection of flaws, cracks and cavities involves projecting X-rays through the specimen on to film.

Table 9.3 Butt weld preparations.

t	g	R	r	L	α	t	g	R	r	L	α
3–6	3	–	–	–	–	3–5	6	–	–	–	–
						5–8	8	–	–	–	–
						8–16	10	–	–	–	–

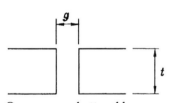

Open square butt weld

Open square butt weld with plate

t	g	R	r	L	α	t	g	R	r	L	α
5–12	2	1	–	–	60°	>10	6	0	–	–	45°
>12	2	2	–	–	60°	>10	10	0	–	–	20°

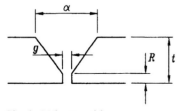
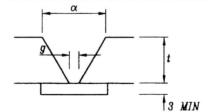

Single V-butt weld

Single V-butt weld with plate

Table 9.3 (contd)

t	g	R	r	L	α		t	g	R	r	L	α
>12	3	2	–	–	60°		>12	3	2	–	–	60°

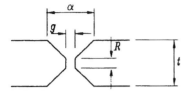

Double V-butt weld: Type 1   Double V-butt weld: Type 2

t	g	R	r	L	α		t	g	R	r	L	α
>20	0	5	5	–	20°		>40	0	5	5	–	20°

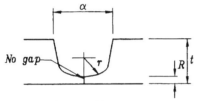

Single U-butt weld   Double U-butt weld

t	g	R	r	L	α		t	g	R	r	L	α
>20	0	5	5	5	20°		>40	0	5	5	5	20°

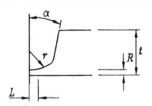

Single J-butt weld   Double J-butt weld

t	g	R	r	L	α		t	g	R	r	L	α
5–12	3	1	–	–	45°		>12	3	2	–	–	45°
>12	3	2	–	–	45°							

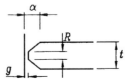

Single bevel butt weld   Double bevel butt weld

All of these NDT methods are fully described in *National Specification of Structural Steelwork for Building Construction*, BCSA & SCI Publication No. 203/91, available from the British Constructional Steelwork Association.

9.2.4 Design of fillet welds

The design strength p_w of fillet welds made with electrodes complying with BS 639 should be as shown in Table 9.4.

Table 9.4 Design strength p_w of fillet welds.

Design grade of steel	Electrode strength as per BS 639	
	E43	E51
43	215 N/mm²	215 N/mm²
50	215 N/mm²	255 N/mm²
55	–	255 N/mm²

The strength of fillet welds may be taken as equal to the strength of the parent metal if the following conditions are satisfied.

- The welds should be symmetrical about the plate connected.
- Suitable electrodes must be used.
- The sum of the throat sizes should be greater than or equal to the thickness of the plate connected.
- The weld must be subjected to tension or compression only.

The stress in the fillet weld should be checked using the following principles:

(1) Find the stress in the weld from the vector sum of all loads using the throat thickness and check that this stress does not exceed the design strength p_w.
(2) For fillet welds with unequal size legs, check the stress on the fusion line. The shear stress on the fusion line should not exceed $0.7p_y$ and the tensile stress should not exceed $1.0p_y$, where p_y is the design strength of the parent metal.

9.2.5 Design of butt welds

The strength of full penetration butt welds may be taken as equal to that of the parent metal provided appropriate electrodes have been used, as per BS 639.

Partial penetration butt welds with a superimposed fillet weld should be designed as a fillet weld using the throat thickness shown in SK 9/26.

Connections in Steelwork 435

9.2.6 Analysis of weld group

F_v = the applied shear on the weld group
M = the applied in-plane moment on the weld group
a = length of run of weld in the X–X direction
b = length of run of weld in the Y–Y direction

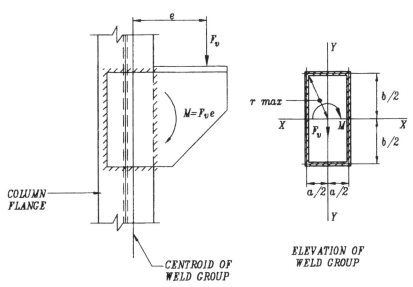

SK 9/28 Analysis of shear stress in weld due to in-plane loading of weld group.

For a four-sided weld group as shown in SK 9/28, the analysis is carried out by the polar inertia method. The moment about the centroid of the weld group is considered as torsion acting on the group and causing shear stress proportional to the distance of the point of weld from the centroid of the group.

I_{xx} = moment of inertia of the weld group about the X–X axis
$$= 2\left[\frac{b^3}{12} + a\left(\frac{b}{2}\right)^2\right] \text{ mm}^3$$

I_{yy} = moment of inertia of the weld group about the Y–Y axis
$$= 2\left[\frac{a^3}{12} + b\left(\frac{a}{2}\right)^2\right] \text{ mm}^3$$

I_P = polar moment of inertia of the weld group
$= I_{xx} + I_{yy}$

Z_P = polar modulus of the weld group
$$= \frac{I_P}{r_{max}} \text{ mm}^2$$

r_{max} = maximum distance to a point on the weld from the centroid of the weld group
$$= \frac{1}{2}\sqrt{a^2 + b^2} \text{ mm}$$

f_m = maximum shear stress in the weld group due to the applied moment M

$$= \frac{M}{Z_\text{P}} \text{ kN/mm}$$

The direction of this shear vector is perpendicular to the direction of the line r_max.

Alternatively:

$$Z_{\text{P},x} = \frac{I_\text{P}}{\left(\dfrac{b}{2}\right)}$$

$$Z_{\text{P},y} = \frac{I_\text{P}}{\left(\dfrac{a}{2}\right)}$$

$f_{\text{m},x}$ = shear in the weld in the X–X direction due to moment on the weld group

$$= \frac{M}{Z_{\text{P},x}} \text{ kN/mm}$$

$f_{\text{m},y}$ = shear in the weld in the Y–Y direction due to moment on the weld group

$$= \frac{M}{Z_{\text{P},y}} \text{ kN/mm}$$

f_v = shear stress due to direct shear F_v in the Y–Y direction

$$= \frac{F_\text{v}}{\text{total length of weld}} = \frac{F_\text{v}}{2(a+b)} \text{ kN/mm}$$

f_r = resultant shear stress in weld due to combined moment and shear

$$= \sqrt{(f_{\text{m},x})^2 + (f_{\text{m},y} + f_\text{v})^2} \text{ kN/mm}$$

The resultant shear should be checked against the strength of the weld per unit length for the size of weld and grade of steel.

9.2.7 Welded beam-to-column connection

The welds connecting the flanges of the beam to the column may be designed to withstand tensile and compressive flange forces only. The beam-end shear may be designed to be resisted fully by the welds connecting the web of the beam to the column.

The beam-to-column welded connection may be designed using the following principles:

- If the flanges of the beam are to be connected to the flange of the column using a full penetration butt weld, the choice of electrode is governed by the grade of steel in the beam and column to be connected.

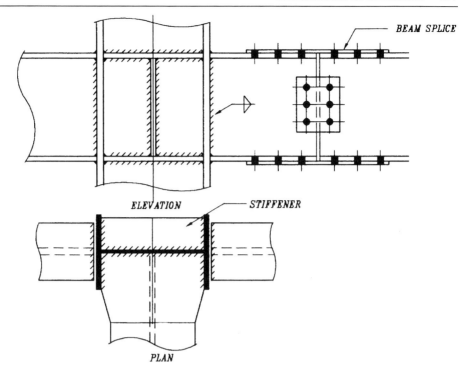

SK 9/29 Typical shop welded beam-to-column connection details.

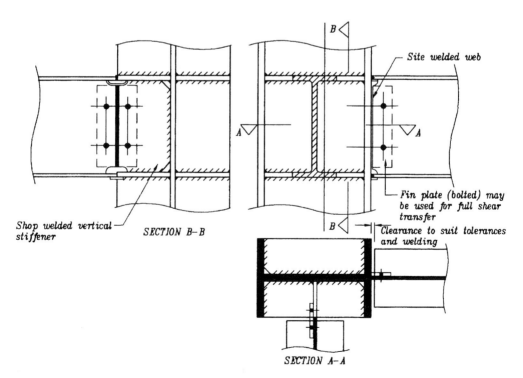

SK 9/30 Typical site welded beam-to-column connection details.

- If the flanges of the beam are to be connected to the flange of the column using fillet welds, the total throat thickness of the fillet welds at the connection should be more than or equal to the thickness of the flange of the beam. Generally do not exceed a leg size of 12 mm for fillet welds.
- If the flanges of the beam are to be connected to the flange of the column using a partial penetration butt weld, with or without superimposed fillet weld, see SK 9/26 for the method of design of the weld.
- If the web of the beam is to be connected to the column using two runs of fillet welds, the total throat thickness of the fillet welds should be more than or equal to the thickness of the web of the beam. For very thick webs partial penetration butt welds may be used.

Check column flange at connection
The flange force tension T and compression C from the beam are transferred to the flange of the column. (See SK 9/31.)

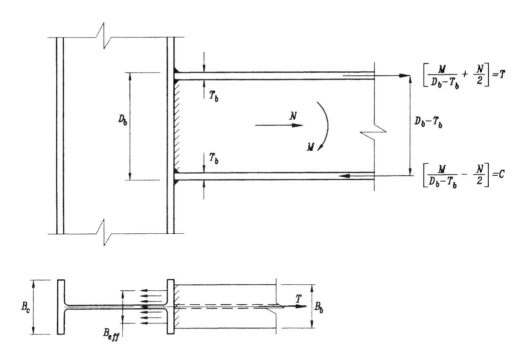

SK 9/31 Column–beam welded connection showing load transfer.

Because the flange of the column is flexible, the load is transferred over an effective width of flange B_{eff} which is a function of the thickness T_c of the flange of column. This is a local capacity check where the load path is from the flange of the beam to the web of the column. If the design strength and flange thickness of the column and the beam are different, then further adjustment of the effective width is necessary as per the following formulae:

$$B_{\text{eff}} = t_c + 2r_c + 7T_c$$

$$\leq t_c + 2r_c + 7\left[\frac{T_c}{T_b}\frac{p_{yc}}{p_{yb}}\right]T_c$$

$$\leq B_b \text{ and } B_c$$

where t_c = thickness of column web
r_c = root radius of column section
T_c = thickness of column flange
T_b = thickness of beam flange
p_{yc} = design strength of the column
p_{yb} = design strength of the beam
B_b = width of beam flange
B_c = width of column flange

The tension capacity of the flange at the connection P_t and the flange tension T are given by:

$$P_t = B_{\text{eff}} T_b p_{yb}$$

$$T = M/(D_b - T_b) + N/2$$

where M = applied moment at the connection
N = applied direct tension
D_b = overall depth of beam

Check $P_t \geq T$ and $B_{\text{eff}} \geq 0.7 B_b$

Use column flange stiffeners if these conditions are not satisfied.

Note: The tension stiffener in the column flange in line with the beam flange should be of the same width and thickness as the beam flange and should be of the same grade of material.

Check column web at the connection
P_t = local tension capacity of column web
   $= p_{yc} t_c [T_b + 2s + 5(T_c + r_c)]$
s = leg length of fillet weld

assuming dispersion of load to the web at 1:2.5 vertically along the web (see SK 9/32).

Check $P_t \geq T$

Use tension stiffener if this condition is not satisfied.

Check column for compression at the beam compression flange
This check in principle is exactly the same as that for bolted connections. Also check the column web panel shear as in bolted connection.

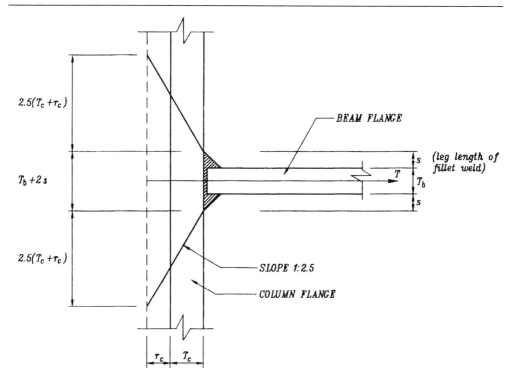

SK 9/32 Effective width of column web resisting tension from beam flange.

9.3 NOTCHED BEAMS

BS 5950: Part 1 has no clear guidelines regarding shear capacity of webs with holes for fasteners. It is considered advisable to check the shear capacity of webs of beams where one or both flanges have been notched at the connection. This check should also make due allowance for holes required for fasteners. Two checks are usually necessary at the notched end of a beam, viz. plain shear check and block shear check.

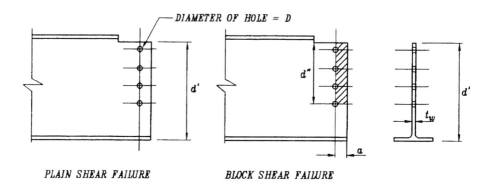

SK 9/33 Failure pattern of notched beam.

9.3.1 Plain shear check at notched end of beam

There are two prescribed methods of carrying out this check.

Method 1 (As per *Joints in Simple Construction: Volumes 1 and 2*, published by BCSA/SCI)

P_v = plain shear capacity of notched beam
 $= 0.6 p_y A_v$ or $0.5 U_s A_{v,net}$ (whichever is the smaller)
$A_v = k t_w d'$
t_w = thickness of web
d' = total depth of web remaining after notches, without deduction of any holes for fasteners
$k = 0.9$ for single-notched or double-notched beam
$A_{v,net}$ = net shear area at the notch after deduction of holes
 $= A_v - nDt_w$
n = number of fastener holes in any single vertical line
D = diameter of holes for fasteners
U_s = ultimate tensile strength of material of the beam

Method 2 (As per *Steel Designer's Manual*, 5th edn, published by Blackwell Science)

$P_v = 0.6 p_y (0.9 A_{ve})$
A_{ve} = effective shear area at the notch
 $= K_e A_{vn} \leq A_{vg}$
A_{vn} = net shear area at the notch after deduction of holes
 $= t_w d' - nDt_w$
A_{vg} = gross shear area at the notch
 $= t_w d'$
$K_e = 1.2$ for Grade 43
 $= 1.1$ for Grade 50 or WR50
 $= 1.0$ for Grade 55

(See Clause 3.3.3 of BS 5950: Part 1)

9.3.2 Block shear check at notched end of beam

A block tear-out failure can also occur if the connection is towards the top of the beam. Again there are two different methods suggested in the two references cited above. Block shear capacity is the combined shear and tensile capacity of the corner of the beam at the notch. The failure mode could be by shear along the vertical fastener line at the notch and then by a tension failure along the horizontal line from the bottom fastener hole. A triangular stress distribution along the horizontal line may be assumed.

Method 1 (As per *Joints in Simple Construction: Volumes 1 and 2*, published by BCSA/SCI)

P_v = block shear capacity
 $= 0.6 p_y A'_v + 0.5 U_s A_h$

A'_v = vertical shear area of web from the notch to the bottom fastener hole
$= t_w d''$
d'' = depth of web from the notch to the bottom fastener hole
A_h = tensile area of web from the innermost vertical line of fastener holes to the edge of the web along a horizontal line
$= (a - 2.5D)t_w$

The factor 2.5 comes from experimental observations.

a = distance from the innermost vertical line of fastener holes to the edge of the web
D = diameter of holes for fasteners

Method 2 (As per *Steel Designer's Manual*, 5th edn, published by Blackwell Science)

P_v = block shear capacity
$= 0.6p_y(0.9A'_{ve}) + 0.5p_y A_{he}$
A'_{ve} = effective vertical shear area of web from the notch to the bottom fastener hole
$= K_e A'_{vn} \leq A'_{vg}$
A'_{vn} = net vertical shear area of web from the notch to the bottom fastener hole
$= t_w d'' - nDt_w$
A'_{vg} = gross vertical shear area of web from the notch to the bottom fastener hole
$= t_w d''$
A_{he} = effective tensile area of web from the innermost vertical line of fastener holes to the edge of the web along a horizontal line
$= (a - n'D)t_w$
n' = number of fastener holes along a horizontal line from the innermost line of fasteners to the edge of the web.

9.4 BEAM-TO-BEAM CONNECTION: SHEAR CAPACITY

The shear capacity of a beam-to-beam connection expressed as a percentage of the beam (UB) shear capacity is shown in Table 9.5 for different types of simple end connection.

Table 9.5 Maximum shear capacity of connection expressed as a percentage of UB shear capacity.

Type of UB (BS range)	Type of connection			
	Fin plate	Double angle web cleat with single row of bolts	Double angle web cleat with double row of bolts	Flexible end plate
203–356	25–40%	25–40%	40–60%	70–100%
406–610	50%	60%	75–90%	75–100%
686–914	N/A	60%	80%	80–100%

9.5 COLUMN BASES

9.5.1 General rules

(1) The bearing pressure on concrete should be assumed to be an average plastic rectangular stress block in compression with a maximum stress level of $0.4 f_{cu}$, where f_{cu} is the characteristic 28-day cube strength of the concrete. The factor 0.4 corresponds to the reinforced concrete theory of design of sections in direct compression or bending, where a material factor γ_m is taken as equal to 1.5 for concrete. (See *Reinforced Concrete: Analysis and Design* by S. S. Ray, published by Blackwell Science, 1995.)

(2) The thickness limitation to prevent brittle fracture of base plates in Grade 43A steel does not apply to base plates subject to compression only.

(3) Base plates in Grade 43A steel should not exceed 50 mm in thickness if it is required to transmit moment to the foundation.

(4) The pressure on the effective area of the base plate should not exceed the allowable bearing pressure on the concrete (i.e. $0.4 f_{cu}$). The effective area depends on the plastic moment capacity of the base plate, which is determined from its thickness. The effective area in an oversized base plate is found by limiting beyond the column the cantilever overhang, which is dictated by the plastic capacity of the base plate.

(5) If all contact surfaces between the base plate and the column, including any stiffeners used, are machined to have full and continuous contact, then the compressive forces may be assumed to be transmitted by direct bearing from the column to the base plate. Otherwise, the welds or the bolts connecting these elements should be designed to transmit all compressive forces. The welds and the fasteners should be designed to transmit all shear forces and tensile forces at the connection.

(6) The thickness of the base plate should not be less than the thickness of the column flange.

(7) The fundamental assumption in the design of the base plate is that it is rigid enough to distribute the load uniformly over the assumed contact surface. Adequate thickness of the base plate determines this rigidity.

9.5.2 Empirical design of base plates for concentric forces

If a rectangular base plate supported by concrete is loaded concentrically by I-, H-, channel, box or rolled hollow section, then the minimum thickness t of the base plate is given by:

$$t \geq \sqrt{\frac{2.5}{p_{yp}} w(a^2 - 0.3b^2)}$$

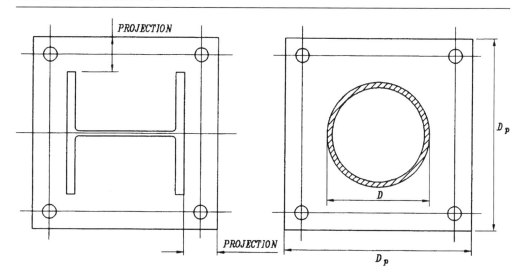

SK 9/34 Empirical design of base plates.

where t = thickness of the base plate
a = greatest projection of the plate beyond the column
b = smallest projection of the plate beyond the column
w = pressure on the underside of the plate, assuming uniform distribution
p_{yp} = design strength of the base plate

If a square base plate supported by concrete is loaded concentrically by a solid round or a hollow circular section, then the minimum thickness t of the base plate is given by:

$$t \geq \sqrt{\frac{w}{2.4 p_{yp}} D_p(D_p - 0.9D)}$$

where D_p = length of a side of the square base plate $\geq 1.5(D + 75)$
D = diameter of the column

9.5.3 Analysis of column-to-foundation connection

Fixed bases transfer moment, shear and direct loads to the foundation. Pinned bases transfer only direct load and shear force. It is difficult to achieve a pinned connection at the base with full rotational freedom. Most bases exhibit a certain degree of fixity. The fixed bases are used in industrial structures with cantilever columns and portal frames with large horizontal loads. Again, full fixity is very difficult to achieve because the foundation is not infinitely rigid rotationally. For most structures the concept of a fixed and pinned base in the analysis still remains valid provided that the designer takes adequate care in the design of the connection of the structure to the foundation and reproduces as closely as possible the assumptions made in the analyses.

Connections in Steelwork 445

Assume the X–X and Y–Y axes lie along the centre-lines of the column section (see SK 9/36).

$X_1 =$ the larger side of the assumed trapezoidal contact pressure diagram under the base plate.
$X_2 =$ the smaller side of the assumed trapezoidal contact pressure diagram under the base plate
$D =$ length of base plate along the X–X axis
$B =$ width of base plate along the Y–Y axis

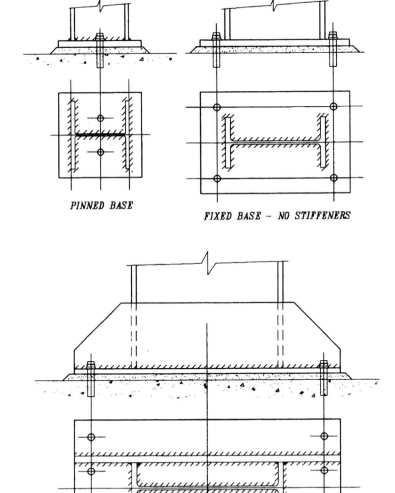

SK 9/35 Typical column-to-foundation connection details.

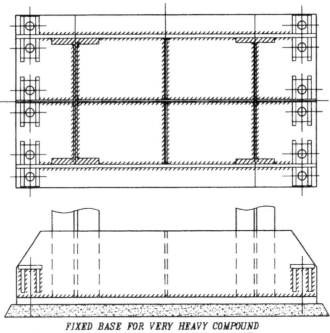

FIXED BASE FOR VERY HEAVY COMPOUND
COLUMN WITH LARGE MOMENT AT BASE

SK 9/35 (contd)

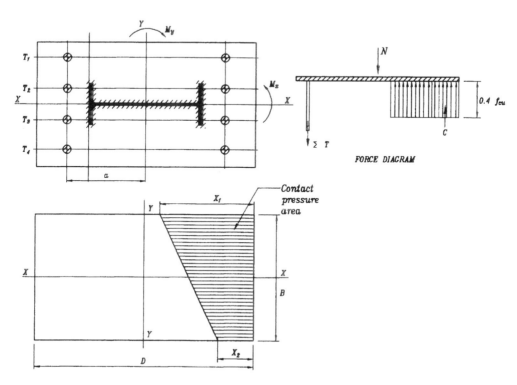

SK 9/36 Biaxial bending of column: pressure distribution under the base plate.

Assume uniform ultimate contact pressure equal to $0.4 f_{cu}$ over the trapezoidal contact pressure diagram.

C = total upward contact reaction of the concrete foundation on the base plate

$$= \left(\frac{X_1 + X_2}{2}\right)(0.4 f_{cu} B) \qquad (9.1)$$

T_1 = tension in holding-down bolt 1
T_2 = tension in holding-down bolt 2
T_i = tension in ith holding-down bolt
N = direct axial compression in the column applied at the intersection of the X–X and the Y–Y axis

$$\sum T + N = C \qquad (9.2)$$

$T_1 = T_2 = T_i = (\sum T)/n$ because all bolts are assumed to have gone past yield
$\sum T$ = sum of all the tensile forces in the bolts in the tensile zone
n = number of holding-down bolts in the tension zone
M_x = applied moment about the X–X axis
M_y = applied moment about the Y–Y axis
a = distance from the centroid of bolts in tension to the Y–Y axis

Taking moments about the Y–Y axis:

$$M_y = (\sum T)a + 0.4 f_{cu} B X_2 \left(\frac{D - X_2}{2}\right)$$

$$+ 0.4 f_{cu} B \left(\frac{X_1 - X_2}{2}\right)\left(\frac{D}{2} - X_2 - \frac{(X_1 - X_2)}{3}\right) \qquad (9.3)$$

Taking moments about the X–X axis:

$$M_x = 0.4 f_{cu} B \left(\frac{X_1 - X_2}{2}\right)\frac{B}{6} \qquad (9.4)$$

There are four unknown parameters (i.e. C, $\sum T$, X_1 and X_2) and there are four equations to solve for them.

Note: For base plates subject to uniaxial moment M_y about the major axis Y–Y and direct load N, the same equations apply but substitute X for X_1 and X_2 (i.e. $X = X_1 = X_2$).

9.5.4 Analysis of base plate (see SK 9/37)

Compression-side analysis of base plate
M_c = bending moment in the cantilever base plate overhang on the compression side

$$= 0.4 f_{cu} B \frac{m_c^2}{2} \text{ Nmm}$$

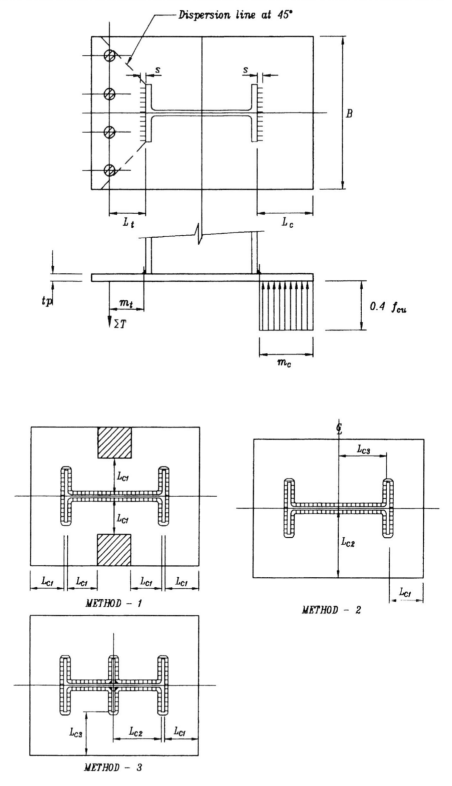

SK 9/37 Analysis of base plate.

m_c = effective overhang of base plate on the compression side
$= L_c - 0.8s$
L_c = actual overhang of base plate on the compression side (see SK 9/37)
s = leg length of weld
M_{rc} = maximum moment of resistance of the base plate: should be as per Clause 4.13.2.3 of BS 5950: Part 1
$= 1.2 p_{yp} Z$
p_{yp} = design strength of base plate (limited to 270 N/mm² for calculation of strength)
Z = elastic modulus of base plate along the axis of moment M_c
$= \dfrac{1}{6} B t_p^2$
t_p = thickness of base plate

Check $M_c \leq M_{rc}$

The base plate is designed on the basis of cantilever overhang L_c. There are three different methods to check if the overhang L_c produces the maximum stress in the base plate.

Method 1 (See SK 9/37): The hatched area of the total contact pressure surface may be ignored in the force equilibrium calculations. The design of the base plate is based on the overhang L_{c1}.

Method 2 (See SK 9/37): The overhang L_{c2} may be used to determine the bending moment in the base plate if $L_{c2} > L_{c1}$ and also $L_{c3} > L_{c1}$.

Method 3 (See SK 9/37): A central stiffener may be used where $L_{c2} < 2L_{c1}$. The analysis of the base plate may be based on a cantilever overhang of L_{c3} or L_{c1}, whichever is the greater.

Tension-side analysis of base plate
M_t = bending moment in the base plate overhang on the tension side
$= (\sum T) m_t$
m_t = effective overhang of the base plate from the centre of tension of the holding-down bolts
$= L_t - 0.8s$
L_t = actual overhang of base plate (see SK 9/37)
s = leg length of weld

Check $M_t \leq M_{rc} = 1.2 p_{yp} Z$

Note: To avoid bending along the corner of the base plate, it is advisable to have the bolts within the 45° load dispersion line from the flanges of the column, as shown in SK 9/37.

9.5.5 Base plate with stiffeners

The maximum moment of resistance M_{rc} of a base plate with stiffeners should be as per Clause 4.13.2.4 of BS 5950: Part 1.

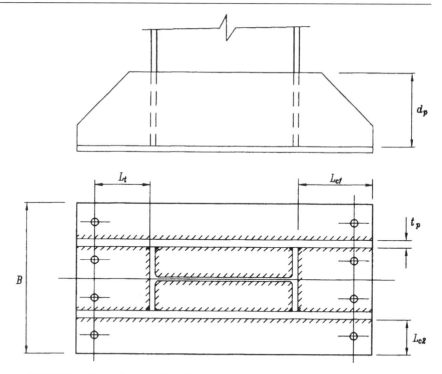

SK 9/38 Base plate with stiffeners.

$M_{rc} = p_{yg}Z$ per stiffener
p_{yg} = design strength of the plate stiffener (limited to 270 N/mm^2 for calculation of strength)
Z = elastic modulus of the plate stiffener
$= \frac{1}{6} t_p d_p^2$ per stiffener
t_p = thickness of the plate stiffener
d_p = depth of the plate stiffener at the point of maximum moment in the base plate

Check $nM_{rc} \geq M_c$ and M_t based on L_{c1} and L_t respectively (see SK 9/38)

n = number of plate stiffeners resisting the compression side moment M_c and the tension side moment M_t in the base plate

Check F_v = shear in the stiffeners
$= BL_{c1} 0.4 f_{cu} < P_v$ = shear capacity of the stiffener assembly
$= n(0.9 t_p d_p)(0.6 p_{yg})$

The base plate itself should be designed to resist the contact pressure on the concrete. The panels of the base plate between stiffeners and the column may be analysed using the yield-line theory as described in *Reinforced Concrete: Analysis and Design* by S. S. Ray, published by Blackwell Science 1995.

Note: When bolt boxes are used with stiffeners, a gap should be maintained between the underside of the bolt boxes and the base plate so as to transmit the bolt tension directly to the stiffeners.

Connections in Steelwork 451

9.5.6 Holding-down bolts

Note: The bolt capacities in Table 9.6 do not include the prying effect, as in Table 32 of BS 5950: Part 1. These are enhanced capacities, including a material factor of 1.25 and a reduction factor of 0.9, to allow for the effects of thread stripping.

The punching shear perimeter is at 1.5 times the depth of effective embedment from the sides of the anchor plate, as shown in SK 9/39. This perimeter is adjusted to suit the location of the bolt in relation to the edge of the concrete foundation, as also shown in SK 9/39. If the shear perimeter zones of two adjacent bolts overlap, then consider the two bolts together as a group.

The punching shear stress f_v at the effective perimeter is given by:

$$f_v = \frac{T}{Ud}$$

where T = tension in a single bolt or in a group of bolts as relevant
 U = effective perimeter for punching shear stress
 d = effective depth of embedment = $D - c$
 D = total depth of embedment of the holding-down bolt
 c = average cover to the two orthogonal layers of reinforcement at the top of the foundation

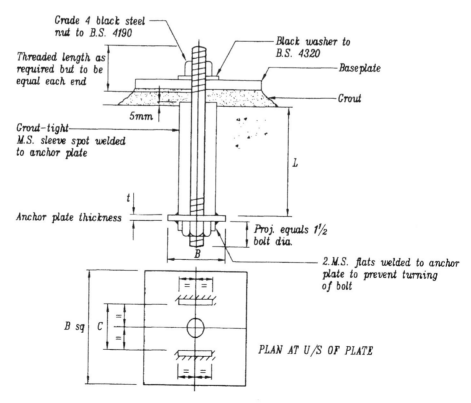

SK 9/39 Details of typical holding-down bolts and anchor plates showing punching shear perimeters in the concrete embedment.

452 Structural Steelwork

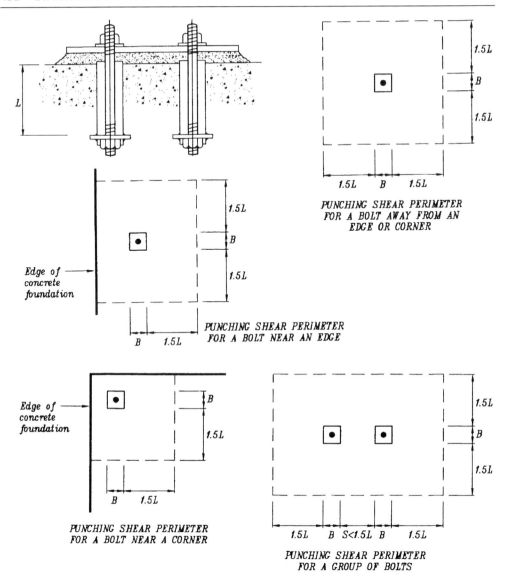

SK 9/39 (contd)

Table 9.6 Capacity and spacing of holding-down bolts.

Diameter of bolts	Tension capacity		Minimum spacing of bolts in sleeves
	Grade 8.8–560 N/mm²	Grade 4.6–280 N/mm²	
M20	137 kN	68 kN	120 mm
M24	198 kN	98 kN	150 mm
M30	314 kN	157 kN	180 mm
M36	445 kN	228 kN	220 mm
M42	615 kN	308 kN	260 mm

The concrete design shear stress v_c can be obtained from BS 8110: Part 1, corresponding to the average percentage of area of steel in the two orthogonal layers of reinforcement at the top of foundation.

Check $f_v \leq v_c$

For a single holding-down bolt the effective perimeter U for punching shear check is given by:

$$U = 4(3d + B)$$

where B = length of side of square anchor plate

Note: The anchor plate for a single holding-down bolt is generally square, of a size and thickness as given in Table 9.7. The design of this plate is based on an average bearing stress of 12 N/mm^2 on the concrete. A minimum grade of concrete with $f_{cu} = 30 \text{ N/mm}^2$ is assumed for this purpose.

Table 9.7 Anchor plates for single holding-down (HD) bolts.

Diameter of HD bolts Grades 8.8 and 4.6	Size of square anchor plate Grade 43	Thickness of square anchor plate
M20	110 × 110 mm	12 mm
M24	130 × 130 mm	15 mm
M30	165 × 165 mm	20 mm
M36	200 × 200 mm	25 mm
M42	230 × 230 mm	25 mm

9.5.7 Shear transfer from column to concrete

General rules
(1) Do not use holding-down bolts to transfer shear force to the foundation if they are cast in sleeves.
(2) Do not use friction to transfer shear force to the foundation in cases of high shear forces and alternating dynamic loads. Transfer of shear forces by friction must only be allowed in the cases where the coefficient of friction used can be guaranteed by tests and the contact pressure is also guaranteed, as in the case of prestressed anchor bolts.
(3) It is always prudent to transfer shear forces to the foundation in important structures by using shear keys. The shear keys are welded to the bottom of the column base plate and cast in a pocket in the foundation.
(4) Fully cast-in bolts with no tolerance for erection of columns should normally be avoided but, if used, may be relied on to resist shear forces by dowel action. These bolts are designed for combined tension and shear. The bearing stress on the concrete should be checked assuming the shear is transferred to the top of the concrete in a depth equal to three times the diameter of the holding-down bolt.

9.5.8 Rules to determine trial size of base plate

(1) Allow, generally, 150 mm overhang from the column to arrive at a base plate size rounded off to the nearest 50 mm.
(2) Choose a trial thickness of base plate without stiffeners using the following rules:
- thickness of 30–35 mm for 100–125 mm outstand beyond column;
- thickness of 40–50 mm for 150–200 mm outstand beyond column.

(3) The bolt spacing should be more than or equal to six times the diameter of the bolt.
(4) For stiffened base plates the thickness may be reduced by about 10 mm from the recommended thickness.
(5) The edge distance of the base plate from the centre of the holding-down bolt should be between 2.5 and 3 times the diameter of the bolt.
(6) Grade 8.8 bolts should be used where large bending moments are transmitted to the foundation.
(7) The stiffeners, when used, should extend from the column to 20 mm short of the edge of the base plate.
(8) The depth (vertical height) of the stiffener plate should be two times the outstand from the face of the column.
(9) The corner of the stiffener plate may be trimmed at a ratio of 2 vertical : 1 horizontal.
(10) The thickness of the stiffener plate should not be less than 10 mm or depth $\div$ 16 (whichever is the greater).
(11) The stiffener should be designed to withstand compression from the compressive stress block, allowing for buckling effects.
(12) The minimum area of base plate should be the direct axial column load divided by $0.4 f_{cu}$.
(13) There will be no tension in the holding-down bolts if the following condition is satisfied:

$$e = \frac{M}{N} \leq \frac{D}{2} - \frac{N}{0.8 f_{cu} B}$$

where D = depth of the base plate
B = width of the base plate
M = applied bending moment about an axis perpendicular to D
N = axial direct load from column
f_{cu} = 28-day cube strength of concrete in the foundation

9.6 WORKED EXAMPLES

9.6.1 Example 9.1: Rigid bolted connection of column-to-roof truss

This example relates to a rigid bolted connection of column-to-roof truss of a 50 m span portal frame structure, as shown in SK 9/40.

Connections in Steelwork

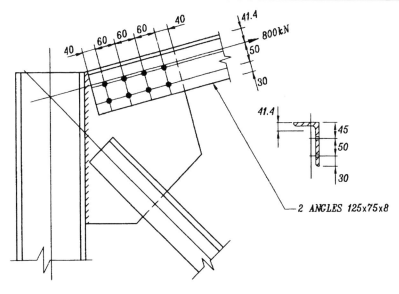

SK 9/40 Rigid connection of column-to-roof truss.

Steel is Grade 50 for the roof truss and the gusset plate.
Bolts are Grade 8.8.
Diameter of bolt is M20 in clearance holes.
Thickness of gusset plate is 12 mm between the pair of angles.
The tensile load of 800 kN is assumed to act along the centroidal axis of the two angles placed back to back.

Step 1 *Determine capacity of bolt*
Effective area of M20 bolts = tensile area = 245 mm², assuming that threads can appear in the shear plane.

P_s = shear capacity of bolt in double shear = $2p_s A_s$
$\quad = 2 \times 375 \times 245 \times 10^{-3} = 183.7$ kN
P_{bb} = bearing capacity of bolt = dtp_{bb}
$\quad = 20 \times 12 \times 1035 \times 10^{-3} = 248.4$ kN
P_{bs} = bearing capacity of connected ply = $dtp_{ss} \leq \frac{1}{2} e t p_{bs}$
e = end distance = 40 mm = $2d$
$P_{bs} = 20 \times 12 \times 550 \times 10^{-3} = 132$ kN

$\therefore$ capacity of bolt = 132 kN

Step 2 *Determine resultant maximum shear in bolt*
Assume eight M20 bolts in a group as shown in SK 9/41.
Eccentricity of load on the bolt group $e = 45 + 25 - 41.4 = 28.6$ mm
In-plane moment on the group of bolts = $800 \times 28.6 = 22\,880$ kNmm

I_P = polar moment of inertia of the group of bolts
$\quad = \sum(x^2 + y^2)$
$\quad = (4 \times 90^2) + (4 \times 30^2) + (8 \times 25^2)$
$\quad = 41\,000$ mm²

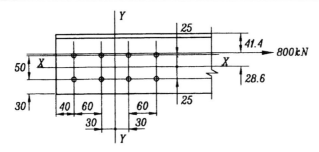

SK 9/41 Distribution of shear in the group of bolts.

$$Z_x = \frac{I_P}{Y} = \frac{41\,000}{25} = 1640\,\text{mm}$$

$$Z_y = \frac{I_P}{X} = \frac{41\,000}{90} = 455.5\,\text{mm}$$

f_{mx} = shear in the X-direction due to in-plane moment

$$= \frac{22\,880}{1640} = 14.0\,\text{kN}$$

f_{my} = shear in the Y-direction due to in-plane moment

$$= \frac{22\,880}{455.5} = 50.2\,\text{kN}$$

f_v = shear in the X-direction due to direct axial load

$$= \frac{800}{8} = 100\,\text{kN}$$

The resultant maximum shear in the bolts is given by:

$$[(f_v + f_{mx})^2 = f_{my}^2]^{\frac{1}{2}} = \sqrt{114^2 + 50.2^2} = 125\,\text{kN} < 132\,\text{kN}$$

Step 3 *Determine strength of gusset plate (see SK 9/42)*

Assuming a 30° dispersion of load from the first row of bolts to the last, the width w of the gusset plate at section a–a is given by:

$$w = 125 - 30 + 180\tan 30° = 199\,\text{mm}$$

Assuming a failure line at a–a due to direct axial tension, the capacity of the gusset plate P_{gt} in tension is given by:

$$P_{gt} = wt_p p_{yp} = 199 \times 12 \times 355 \times 10^{-3} = 848\,\text{kN} < 800\,\text{kN}$$

The number of bolts and the gusset plate are adequate for the connection.

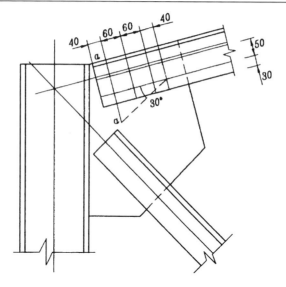

SK 9/42 Load distribution in the gusset plate.

9.6.2 Example 9.2: Welded connections of members in a roof truss

The results of the analysis of a roof truss in Example 3.1 have been used to demonstrate the procedure. The calculations shown are for one joint in the roof truss (see SK 9/43).
Steel is Grade 43.
Strength of electrode to BS 639 is E43.
Design strength of fillet weld is 215 N/mm².

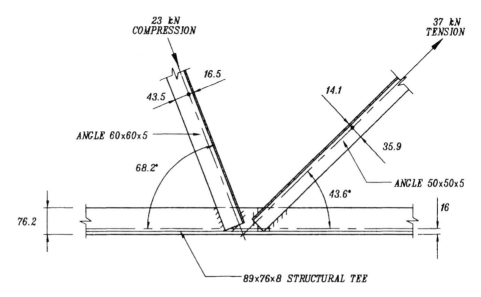

SK 9/43 Geometry of a joint in a roof truss and the internal forces.

Maximum size of fillet weld to toe of angles is limited by the radius on the toe of the angle section. For angle with 5 mm thickness of leg, the maximum size of fillet weld is 3 mm (5–2.4), because the radius at the toe is 2.4 mm.

Capacity of 3 mm fillet weld $= 0.7 \times 3 \times 215 \times 10^{-3} = 0.451$ kN/mm.

Member 16

Two runs of 3 mm fillet weld are used, one each at the toe and heel of the angle $60 \times 60 \times 5$. The internal force (23 kN) may be assumed to act at the centroid of the angle. The force is assumed to be distributed in proportion to the distance of its line of action from the weld runs. The nearer the line of action to a weld run, the more load it carries.

The force on the run of fillet weld at the heel of the angle

$$= 23 \times \frac{43.5}{60} = 16.7 \text{ kN}$$

Length of 3 mm run of weld required at the heel of the angle

$$= \frac{16.7}{0.451} = 37 \text{ mm}$$

Available length of weld at the heel is greater than 37 mm.

Member 19

The force on the run of fillet weld at the heel of the angle

$$= 31 \times \frac{35.9}{50} = 22.3 \text{ kN}$$

Length of 3 mm run of weld required at the heel of the angle

$$= \frac{22.3}{0.451} = 49 \text{ mm}$$

Available length of weld at the heel is greater than 49 mm.

9.6.3 Example 9.3: Beam-to-beam connection using fin plates

See SK 9/44
Use M24 bolts of Grade 8.8.
Size of main beam $= $ UB $914 \times 305 \times 201$ kg/m
Size of secondary beams
 $= $ UB $610 \times 229 \times 140$ kg/m: end shear $= 825$ kN (right)
 and $= $ UB $457 \times 152 \times 82$ kg/m: end shear $= 400$ kN (left)
Assumed size of fin plates $= 12 \times 160 \times 420$ (right)
 and $= 12 \times 100 \times 340$ (left)
Assume steel is Grade 50 for all materials including fin plates.

Step 1 **Check detailing requirements**

Thickness of fin plate t_p should be less than or equal to half the diameter d of bolt:

$t_p = 12 \text{ mm} \leq 0.5d = 12$

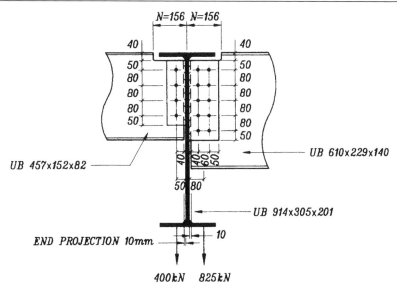

SK 9/44 Beam-to-beam connection using fin plate.

The length of fin plate l should be more than or equal to 0.6 times the overall depth of the secondary beams:

$l = 420\,\text{mm} \geq 0.6 \times 617 = 370$

and $= 340\,\text{mm} \geq 0.6 \times 465.1 = 279$

Minimum spacing of M24 bolts = 60 mm
Diameter of clearance holes = 26 mm

Step 2 *Determine capacity of bolt*

Effective area of bolt is taken as the tensile area, assuming that threads may appear in the shear plane:

$A_t = 353\,\text{mm}^2$ for M24

Single shear capacity of bolt $= P_s = A_t p_s = 353 \times 375 \times 10^{-3} = 132\,\text{kN}$
Bearing capacity of bolt $= P_{bb} = dt p_{bb}$
Bearing capacity of connected ply $= P_{bs} = dt p_{bs} \leq \frac{1}{2} e t p_{bs}$

Thickness of finplate	12 mm	$P_{bb} = 298\,\text{kN}$	$P_{bs} = 158\,\text{kN}$
Web of UB 457×152	10.7 mm	$P_{bb} = 266\,\text{kN}$	$P_{bs} = 141\,\text{kN}$
Web of UB 610×229	13.1 mm	$P_{bb} = 325\,\text{kN}$	$P_{bs} = 173\,\text{kN}$

e = end distance = 50 mm > $2d$
$p_{bs} = 550\,\text{N/mm}^2$ for Grade 50 material
$p_{bb} = 1035\,\text{N/mm}^2$ for Grade 8.8 bolts

∴ capacity of bolt = 132 kN in single shear

Step 3 *Determine maximum shear in bolt*

UB $457 \times 152 \times 82\,kg/m$
Size of fin plate = $12 \times 100 \times 340$
Use four M24 bolts (see SK 9/44)
Eccentricity of the line of bolts from the edge of the main beam web = 50 mm

End shear of the secondary beam = 400 kN
In-plane moment on the line of four bolts = 400 × 50 = 20 000 kNmm

I_P = polar moment of inertia of the group of bolts
$= \sum(X^2 + Y^2) = 2 \times (40^2 + 120^2)$
$= 32\,000\,\text{mm}^2$ (where $X = 0$ for all bolts)

Z_{px} = modulus of the group of bolts
$= \dfrac{I_P}{Y} = \dfrac{32\,000}{120} = 266.7\,\text{mm}$

f_v = vertical shear in bolt due to direct shear
$= \dfrac{400}{4} = 100\,\text{kN}$ per bolt

f_{mx} = horizontal shear in bolt due to in-plane moment
$= \dfrac{M}{Z_{px}} = \dfrac{20\,000}{266.7} = 75\,\text{kN}$

Resultant maximum shear in bolt
$= \sqrt{f_v^2 + f_{mx}^2} = \sqrt{100^2 + 75^2} = 125\,\text{kN} < 132\,\text{kN}$

The bolt arrangement is adequate.

UB 610 × 229 × 140 kg/m
Size of fin plate = 12 × 160 × 420
Use ten M24 bolts in two lines of five bolts (see SK 9/44)
Eccentricity of the group of bolts from the web of the main beam = 80 mm
End shear of the secondary beam = 825 kN
In-plane moment on the group of ten bolts = 825 × 80 = 66 000 kNmm

I_P = polar moment of inertia of the group of bolts
$= \sum(X^2 + Y^2) = (10 \times 30^2) + 4 \times (160^2 + 80^2) = 137\,000\,\text{mm}^2$

$Z_{px} = \dfrac{I_P}{Y} = \dfrac{137\,000}{160} = 856.3\,\text{mm}$

$Z_{py} = \dfrac{I_P}{X} = \dfrac{137\,000}{30} = 4566.7\,\text{mm}$

f_v = vertical shear in bolt due to direct shear $= \dfrac{825}{10} = 82.5\,\text{kN}$

f_{mx} = horizontal shear in bolt due to in-plane moment
$= \dfrac{M}{Z_{px}} = \dfrac{66\,000}{856.3} = 77\,\text{kN}$

f_{my} = vertical shear in bolt due to in-plane moment
$= \dfrac{M}{Z_{py}} = \dfrac{66\,000}{4566.7} = 14.5\,\text{kN}$

Resultant maximum shear in bolts at the extreme corner of the group
$= \sqrt{(f_v + f_{my})^2 + f_{mx}^2} = \sqrt{(82.5 + 14.5)^2 + 77^2}$
$= 124\,\text{kN} < 132\,\text{kN}$

The arrangement of bolts has adequate capacity.

Connections in Steelwork 461

Step 4 *Determine shear capacity of notched beams*
UB $457 \times 152 \times 82\,kg/m$

Method 1
$P_v = 0.6 p_y A_v$ or $0.5 U_s A_{v,net}$ (whichever is the smaller)
$A_v = k t_w d'$
$t_w =$ thickness of web $= 10.7\,$mm
$k = 0.9$
$d' =$ total depth after notches $= 465.1 - 40 = 425.1\,$mm
$A_v = 0.9 \times 10.7 \times 425.1 = 4093.7\,$mm^2
$P_v = 0.6 \times 4093.7 \times 355 \times 10^{-3} = 872\,$kN
$A_{v,net} = A_v - nDt_w$
  $n =$ number of bolts in one vertical line $= 4$
  $D =$ diameter of clearance hole $= 26\,$mm
$A_{v,net} = 4093.7 - (4 \times 26 \times 10.7) = 2980.9\,$mm^2
$$\therefore P_v = 0.5 U_s A_{v,net} = 0.5 \times 490 \times 2980.9 \times 10^{-3} = 730\,\text{kN}$$

The shear capacity of the notched beam by method $1 = 730\,$kN

Method 2
$P_v = 0.6 p_y (0.9 A_{ve})$
$A_{ve} = K_e A_{vn} \leq A_{vg}$
$A_{vn} = t_w d' - nDt_w = (10.7 \times 425.1) - (4 \times 26 \times 10.7) = 3435.8\,$mm^2
$A_{vg} = 10.7 \times 425.1 = 4548.6\,$mm^2
$K_e = 1.1$ (as per Clause 3.3.3 of BS 5950: Part 1)
$K_e A_{vn} = 1.1 \times 3435.8 = 3779.4 < 4548.6$
$$\therefore P_v = 0.6 \times 355 \times 0.9 \times 3779.4 \times 10^{-3} = 724.5\,\text{kN}$$

The shear capacity of the notched beam by method 2
$= 724.5\,$kN $> 400\,$kN
The notched beam has adequate shear capacity.
Use the same procedure to carry out the shear capacity check of the notched UB $610 \times 229 \times 140$.

Step 5 *Determine block shear capacity of notched beam*
UB $610 \times 229 \times 140$

Method 1
$P_v =$ block shear capacity $= 0.6 p_y A'_v + 0.5 U_s A_h$
$A'_v = t_w d'' = 13.1 \times 370 = 4847\,$mm^2
$A_h = (a - 2.5D) t_w = (100 - [2.5 \times 26]) \times 13.1 = 458.5\,$mm^2
$$\therefore P_v = (0.6 \times 355 \times 4847 \times 10^{-3}) + (0.5 \times 490 \times 458.5 \times 10^{-3})$$
$$= 1145\,\text{kN} > 825\,\text{kN}$$

Method 2
$P_v = 0.6 p_y (0.9 A'_{ve}) + 0.5 p_y A_{he}$
$A'_{ve} = K_e A'_{vn} \leq A'_{vg}$
$A'_{vn} = t_w d'' - nDt_w = (13.1 \times 370) - (5 \times 26 \times 13.1) = 3144\,$mm^2
$A'_{vg} = t_w d'' = 4847\,$mm^2
$A'_{ve} = 1.1 \times 3144 = 3458.4\,$mm^2 $< 4847\,$mm^2
$A_{he} = K_e (a - n'D) t_w = 1.1 \times (100 - [2 \times 26]) \times 13.1 = 692\,$mm^2
$$\therefore P_v = (0.6 \times 355 \times 0.9 \times 3458.4 \times 10^{-3}) + (0.5 \times 355 \times 692 \times 10^{-3})$$
$$= 786\,\text{kN} < 825\,\text{kN}$$

462 Structural Steelwork

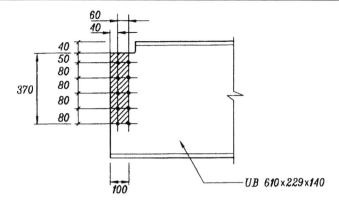

SK 9/45 Block shear capacity of UB $610 \times 229 \times 140$.

This slight overstress may be allowed because this method is very conservative.

Use the same procedure to carry out the block shear capacity check for UB $457 \times 152 \times 82$ kg/m.

Note See Step 7 where the bolt spacing has been increased to 90 mm. This gives the notched beam additional block shear capacity.

Step 6 ***Check moment capacity of beams at the notch***
UB $457 \times 152 \times 82$
$\bar{y} =$ depth of neutral axis from the flange of beam

$$= \frac{\left(153.5 \times 18.9 \times \frac{18.9}{2}\right) + \left\{(425.1 - 18.9) \times 10.7 \times \left[\left(\frac{425.1 - 18.9}{2}\right) + 18.9\right]\right\}}{(153.5 \times 18.9) + [(425.1 - 18.9) \times 10.7]}$$

$= 136.9$ mm

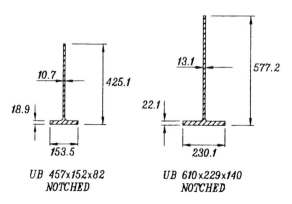

SK 9/46 Geometry of section at the notch.

Connections in Steelwork

$$I_{yy} = \left(\frac{1}{12} \times 10.7 \times 406.2^3\right) + (10.7 \times 406.2 \times 85.1^2)$$

$$+ \left(\frac{1}{12} \times 18.9^3 \times 153.5\right) + (18.9 \times 153.5 \times 127.45^2)$$

$$= 138.45 \times 10^6 \, \text{mm}^4 \text{ (moment of inertia about neutral axis)}$$

$$Z_{min} = \frac{138.45 \times 10^6}{425.1 - 136.9} = 480\,396 \, \text{mm}^3$$

Moment capacity of the section at the notch
$= p_y Z_{min} = 355 \times 480\,396 \times 10^{-6} = 170.5 \, \text{kNm}$
Applied moment at the section
$= 400 \times 156 \times 10^{-3} = 62.4 \, \text{kNm} < 170.5 \, \text{kNm}$
Length of the notch is 156 mm (see SK 9/44)
Use the same procedure to check the moment capacity at the notch of UB $610 \times 229 \times 140$.

Step 7 **Check shear capacity of fin plates**
UB $610 \times 229 \times 140$
Size of fin plate $= 12 \times 160 \times 420$

$$A_v = 0.9td = 0.9 \times 12 \times 420 = 4536 \, \text{mm}^2$$
$$A_{v,net} = 4536 - (5 \times 26 \times 12) = 2976 \, \text{mm}^2$$

P_v is the lesser of $0.6 p_y A_v$ and $0.5 U_s A_{v,net}$

$$\therefore \quad P_v = 0.6 \times 355 \times 4536 \times 10^{-3} = 966 \, \text{kN}$$
$$\text{or} = 0.5 \times 490 \times 2976 \times 10^{-3} = 729 \, \text{kN} < 825 \, \text{kN}$$

The capacity of the fin plate in shear is not adequate. Increase the spacing of the bolts to 90 mm from 80 mm. The depth of the fin plate becomes 460 mm.

$$A_{v,net} = (0.9 \times 12 \times 460) - (5 \times 26 \times 12) = 3408 \, \text{mm}^2$$
$$P_v = 0.5 \times 490 \times 3408 \times 10^{-3} = 835 \, \text{kN} > 825 \, \text{kN}$$

This increase in the bolt spacing decreases the resultant shear on the bolts and improves the block shear capacity of the beam.

Use the same procedure to check the shear capacity of the fin plate of UB $457 \times 152 \times 82$.

Step 8 **Check moment capacity of fin plate (see SK 9/47)**
UB $610 \times 229 \times 140$
Revised fin plate size $= 12 \times 160 \times 460$
Number of holes $= 5$
Diameter of holes $= 26 \, \text{mm}$
Gross moment of inertia of fin plate

$$= \frac{1}{12} \times 12 \times 460^3 = 97.34 \times 10^6 \, \text{mm}^4$$

Moment of inertia of holes
$= 2 \times 12 \times 26 \times (180^2 + 90^2) = 25.27 \times 10^6 \, \text{mm}^4$

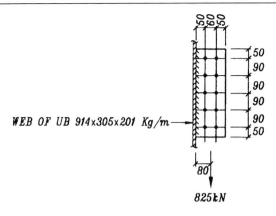

SK 9/47 Fin plate for UB 610 × 229 × 140.

Net moment of inertia $= (97.34 - 25.27) \times 10^6 = 72.07 \times 10^6$ mm^4

Net section modulus $= \dfrac{72.07 \times 10^6}{230} = 31.33 \times 10^4$ mm^3

Net moment capacity $= p_y Z = 355 \times 31.33 \times 10^4 \times 10^{-6} = 111$ kNm

Gross moment capacity $= \dfrac{355 \times 97.34}{230} = 150$ kNm

At the second line of bolts, which is the farthest from the web of the main beam, the eccentricity is 110 mm and the applied moment on the fin plate is $825 \times 0.11 = 90.75$ kNm (which is less than the capacity $= 111$ kNm). The fin plate has adequate moment capacity.

Use the same procedure to check the moment capacity of the fin plate for UB 457 × 152 × 82.

Note: A lateral torsional buckling check of long fin plates is required only where the thickness of the fin plate is less than 0.15 times the distance from the support to the first line of bolts in the fin plate.

Step 9 **Check weld of fin plate to supporting beam**
Provide continuous fillet weld on both sides of the fin plate to the main beam connection. The leg length of the weld run should not be less than 0.8 times the thickness of the fin plate. Use 10 mm continuous fillet weld.

Step 10 **Check local shear capacity of supporting main beam**
Main beam: UB 914 × 305 × 201
Average local shear capacity of the web of the main beam per unit depth of web
$= 0.6(0.9 t_w) p_y = 0.6 \times 0.9 \times 15.2 \times 355 \times 10^{-3} = 2.914$ kN/mm
Revised depth of fin plate for UB 457 × 152 × 82 $= 370$ mm
Average local shear force on the main beam per unit depth from UB 457 × 152 × 82
$= \dfrac{400}{2} \times \dfrac{1}{370} = 0.541$ kN/mm

Revised depth of fin plate for UB $610 \times 229 \times 140 = 460$ mm
Average local shear force on the main beam per unit depth from UB $610 \times 229 \times 140$

$$= \frac{825}{2} \times \frac{1}{460} = 0.897 \, \text{kN/mm}$$

Total local shear force per unit depth of main beam
$= 0.541 + 0.897 = 1.438 \, \text{kN/mm} < 2.914 \, \text{kN/mm}$
The main beam has adequate local shear capacity.

9.6.4 Example 9.4: Beam-to-beam connection using double angle web cleats

Use M20 bolts Grade 8.8.
Main beam = UB $914 \times 305 \times 201$: Grade 50
Secondary beams
 = UB $610 \times 229 \times 140$: Grade 50: end shear = 1100 kN
 = UB $457 \times 152 \times 82$: Grade 50: end shear = 650 kN
Angle cleats = $150 \times 90 \times 10$: Grade 43

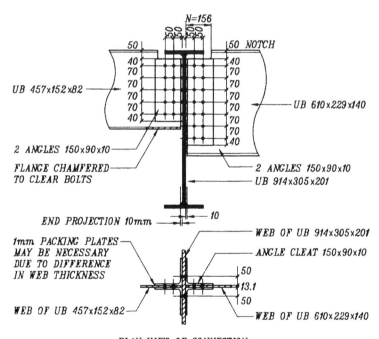

SK 9/48 Beam-to-beam connection using double angle web cleats.

Step 1 **Check detailing requirements**
(1) Place the cleats as close to the top of the beam as possible to provide directional restraints to the top compressive flange of the beam.
(2) The cleats should be at least 0.6 times the depth of beam to provide torsional restraint to the beam at the support.

(3) The cleats should be thin enough to allow the necessary flexibility at the connection. This is required to prevent attracting significant moment at the simple connection. Generally, an 8 or 10 mm thickness of angle cleat should be chosen.

(4) The spacing of bolts should be such that the bolt heads at right angles should not interfere with each other during the tightening operation.

Cleats used have 10 mm thickness of legs.
Length of cleat for UB $457 \times 152 \times 82 = 360$ mm $> 0.6 \times 465.1$
Length of cleat for UB $610 \times 229 \times 140 = 500$ mm $> 0.6 \times 617$

Step 2 Check capacity of bolt
Effective area of bolt = tensile stress area = 245 mm^2

P_s = single shear capacity of bolt
$= A_t p_s = 245 \times 375 \times 10^{-3} = 91.9$ kN
P_{bb} = bearing capacity of bolt = $dt p_{bb}$; $p_{bb} = 1035$ N/mm^2
P_{bs} = bearing capacity of connected ply = $dt p_{bs} \leq \frac{1}{2} et p_{bs}$
$= 460$ N/mm^2 for Grade 43 material
$= 550$ N/mm^2 for Grade 50 material
e = end distance = $2d = 2 \times 20 = 40$ mm

Web cleat
  Thickness = 10.0 mm   Grade = 43 $P_{bb} = 207$ kN $P_{bs} = 92$ kN
UB $457 \times 152 \times 82$
  Thickness = 10.7 mm   Grade = 50 $P_{bb} = 221$ kN $P_{bs} = 118$ kN
UB $610 \times 229 \times 140$
  Thickness = 13.1 mm   Grade = 50 $P_{bb} = 271$ kN $P_{bs} = 144$ kN
UB $914 \times 305 \times 201$
  Thickness = 15.2 mm   Grade = 50 $P_{bb} = 315$ kN $P_{bs} = 167$ kN

Therefore, the capacity of the bolt is 91.9 kN in single shear and $2 \times 91.9 = 183.8$ kN in double shear.

Step 3 Check resultant shear in bolt (see SK 9/49)
In this double web cleat connection, two angle cleats are used to connect each secondary beam to the main beam. The beam-end shear is divided equally between the two web cleats.

UB $610 \times 229 \times 140$
I_P = polar moment of inertia of the group of bolts
$= \sum (X^2 + Y^2) = [4 \times (210^2 + 140^2 + 70^2)] + [14 \times 25^2]$
$= 283\,150$ mm^2

$Z_x = \dfrac{I_P}{Y} = \dfrac{283\,150}{210} = 1348.3$ mm

$Z_y = \dfrac{I_P}{X} = \dfrac{283\,150}{25} = 11\,326$ mm

Applied in-plane moment on the group of bolts
$= \dfrac{1100}{2} \times 75 = 41\,250$ kNmm

Connections in Steelwork

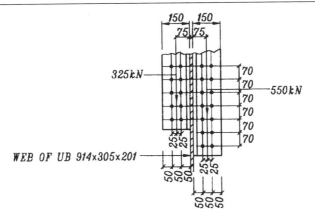

SK 9/49 Bolt arrangement in web cleat connection of secondary beams to main beam.

f_v = vertical shear in the bolt due to beam-end shear
$$= \frac{550}{14} = 39.3 \text{ kN}$$

f_{mx} = horizontal shear in the bolt due to in-plane moment
$$= \frac{M}{Z_x} = \frac{41\,250}{1348.3} = 30.6 \text{ kN}$$

f_{my} = vertical shear in the bolt due to in-plane moment
$$= \frac{M}{Z_y} = \frac{41\,250}{11\,326} = 3.6 \text{ kN}$$

Resultant shear in bolt

$$= \sqrt{(f_v + f_{my})^2 + f_{mx}^2} = \sqrt{(39.3 + 3.6)^2 + 30.6^2} = 52.7 \text{ kN}$$

Use the same procedure to check the resultant shear in bolt for UB $457 \times 152 \times 82$.

Step 4 *Check shear capacity of web cleat*
UB $457 \times 152 \times 82$
l = length of angle cleat = 360 mm
This cleat is carrying $650/2 = 325$ kN vertical shear in each leg.
Shear capacity P_v of each leg of the cleat is the smaller of $0.6 p_y A_v$ and $0.5 U_s A_{v,net}$.

A_v = shear area = $0.9 tl = 0.9 \times 10 \times 360 = 3240 \text{ mm}^2$
$A_{v,net}$ = shear area after deduction of holes
$= A_v - nDt = 3240 - (5 \times 22 \times 10) = 2140 \text{ mm}^2$

$\therefore \quad P_v = 0.6 \times 3240 \times 275 \times 10^{-3} = 534.6 \text{ kN}$
or $= 0.5 \times 410 \times 2140 \times 10^{-3} = 438.7 \text{ kN} > 325 \text{ kN}$

The angle cleat has adequate shear capacity.
Use the same procedure to check the shear capacity of the cleat for the connection of UB $610 \times 229 \times 140$.

468 Structural Steelwork

Step 5 **Check shear capacity of notched beam**
See Example 9.3 for the method of analysis.

Step 6 **Check block shear capacity of notched beam**
See Example 9.3 for the method of analysis.

Note: To increase the block shear capacity at the notch of the secondary beams, increase the bolt spacing or the number of bolts.

Step 7 **Check moment capacity of notched beam**
See Example 9.3 for the method of analysis.

Note: The moment capacity of UB $457 \times 152 \times 82$ should be checked with reduced bottom flange width due to chamfer at the connection. (See SK 9/48.)

Step 8 **Check shear capacity of group of bolts connecting angle cleats to the main beam**
Capacity of each bolt in single shear $= 91.9$ kN
Capacity of connection
$= 14 \times 91.9 = 1286.6$ kN > 1100 kN applied shear
Number of bolts is adequate.
Use the same procedure to check the connection of UB $457 \times 152 \times 82$.

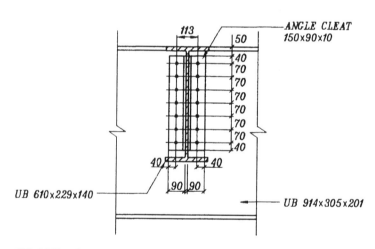

SK 9/50 Connection of UB $610 \times 229 \times 140$ to UB $914 \times 305 \times 201$ by angle cleats $150 \times 90 \times 10$.

Step 9 **Check shear capacity of the leg of the angle cleat connecting to the main beam**
This check is similar to the check in Step 4.

Step 10 **Check local shear capacity of main beam**
Pitch of bolt $= 70$ mm
Diameter of clearance hole $= 22$ mm
Thickness of web of main beam $= 15.2$ mm
Gross shear area per pitch of bolt
$= 0.9 \times 70 t_w = 0.9 \times 70 \times 15.2 = 957.6$ mm^2

Shear capacity of web per pitch of bolt
$= 0.6 \times 957.6 \times 355 \times 10^{-3} = 204\,\text{kN}$
Net shear area per pitch of bolt $= 957.6 - (22 \times 15.2) = 623.2\,\text{mm}^2$
Shear capacity per pitch of bolt $= 0.5 A_{v,\text{net}} U_s$
$= 0.5 \times 623.2 \times 490 \times 10^{-3} = 152.7\,\text{kN}$
Applied shear per pitch of bolt from UB $610 \times 229 \times 140$
$= 550 \div 7 = 78.6\,\text{kN}$
Applied shear per pitch of bolt from UB $457 \times 152 \times 82$
$= 325 \div 5 = 65\,\text{kN}$
Total applied local shear per pitch of bolt
$= 78.6 + 65 = 143.6\,\text{kN} < 152.7\,\text{kN}$
The local shear capacity of the main beam is adequate.

9.6.5 Example 9.5: Beam-to-beam connection using end plates

Use M20 bolts of Grade 8.8.
Main beam $=$ UB $914 \times 305 \times 201$ Grade 50
Secondary beams
$=$ UB $610 \times 229 \times 140$ Grade 50: end shear $= 1200\,\text{kN}$
and $=$ UB $457 \times 152 \times 82$ Grade 50: end shear $= 720\,\text{kN}$
Flexible end plate $= 200 \times 10$ Grade 50

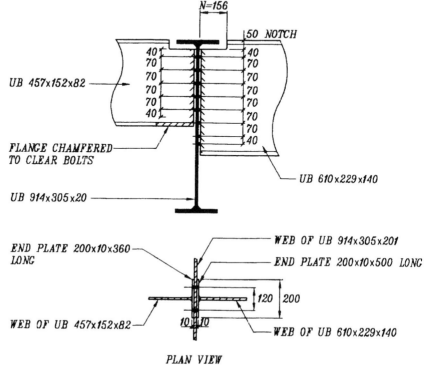

SK 9/51 Beam-to-beam connection using flexible end plates.

Step 1 Check detailing requirements

(1) The end plate should be relatively thin to allow flexibility at the connection. The assumption in the design is that the joint is theoretically pinned.
(2) The bolt centres should be positioned so as not to cause too much rigidity about the vertical axis.
(3) The end plate should be as close to the top flange as possible to give directional restraint to the top flange.
(4) The length of the end plate should be at least 0.6 times the depth of the beam to give adequate torsional restraint at the connection.
(5) The partial-depth end plate should not be too thin because the welding distortion may give rise to significant bowing. The thickness should be between 8 and 10 mm.
(6) The end plate in a simple pinned connection should not be full depth with welds to flanges, as this may invalidate the design assumptions. Sometimes in small beams a full-depth plate may be necessary but the flexibility can be maintained by the choice of thinner plates.

UB $457 \times 152 \times 82$:
  l = length of end plate used = 360 mm > 0.6×460.3
UB $610 \times 229 \times 140$:
  l = length of end plate used = 500 mm > 0.6×616.8
Thickness of partial-depth end plate = 10 mm
Bolt gauge in the main beam
  = 120 mm for an end plate width of 200 mm
The end and edge distances are 40 mm = $2 \times$ diameter of bolt

Step 2 Check capacity of bolt

Diameter of bolt = 20 mm
Grade of bolt used = 8.8
Single shear capacity = $P_s = A_t p_s = 245 \times 375 \times 10^{-3} = 91.9$ kN
Bearing capacity of connected ply = $P_{bs} = dt p_{bs} \leq \frac{1}{2} e t p_{bs}$
$p_{bs} = 550$ N/mm^2 for Grade 50 material of connected ply and $e = 2d$
Minimum thickness of connected ply = 10 mm
$P_{bs} = 20 \times 10 \times 550 \times 10^{-3} = 110$ kN

$\therefore$ capacity of bolt = 91.9 kN

Step 3 Check shear capacity of bolt group at end plate

UB $457 \times 152 \times 82$ (see SK 9/52)
Applied end shear = 720 kN
Capacity of ten M20 bolts = $10 \times 91.9 = 919$ kN > 720 kN

UB $610 \times 229 \times 140$ (see SK 9/53)
Applied end shear = 1200 kN
Capacity of 14 M20 bolts = $14 \times 91.9 = 1287$ kN > 1200 kN

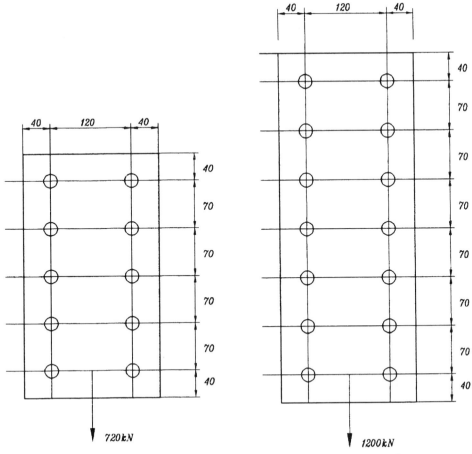

SK 9/52 Arrangement of bolts at end plate of UB 457 × 152 × 82.

SK 9/53 Arrangement of bolts at end plate of UB 610 × 229 × 140.

Step 4 **Check shear capacity of end plate**
  UB 610 × 229 × 140
  Applied shear on the end plate = 1200 ÷ 2 = 600 kN
  Shear capacity of the end plate is the smaller of $0.6p_y A_v$ and $0.5 U_s A_{v,net}$

$$A_v = 0.9tl = 0.9 \times 10 \times 500 = 4500 \text{ mm}^2$$
$$A_{v,net} = A_v - nDt = 4500 - (7 \times 22 \times 10) = 2960 \text{ mm}^2$$
$$n = \text{number of bolts in one vertical line} = 7$$
$$D = \text{diameter of clearance hole} = 22 \text{ mm}$$
$$0.6p_y A_v = 0.6 \times 355 \times 4500 \times 10^{-3} = 958.5 \text{ kN}$$
$$0.5 U_s A_{v,net} = 0.5 \times 490 \times 2960 \times 10^{-3} = 725.2 \text{ kN}$$

Shear capacity of end plate = 725.2 kN > 600 kN
Use the same procedure to check the shear capacity of end plate for UB 457 × 152 × 82.

472 Structural Steelwork

Step 5 **Check shear capacity of the web of secondary beam at the connection with the end plate**
UB 457 × 152 × 82
Thickness of web $= t_w = 10.7$ mm
Applied shear force $= 720$ kN
Length of end plate $= l = 360$ mm
Shear capacity of web of beam $= 0.9 \times 0.6 l t_w p_y$
$= 0.9 \times 0.6 \times 360 \times 10.7 \times 355 \times 10^{-3} = 738$ kN > 720 kN
The web of the beam has adequate shear capacity.

UB 610 × 229 × 140
Thickness of web $= t_w = 13.1$ mm
Applied shear force $= 1200$ kN
Length of end plate $= l = 500$ mm
Shear capacity of web of beam $= 0.9 \times 0.6 l t_w p_y$
$= 0.9 \times 0.6 \times 500 \times 13.1 \times 355 \times 10^{-3} = 1256$ kN > 1200 kN
The web of the beam has adequate shear capacity.

Step 6 **Check capacity of fillet weld connecting end plate to web**
Size of weld $= 8$ mm
Throat size $= 0.7 \times 8 = 5.6$ mm
Design grade of steel $=$ Grade 50
Electrode strength to be used $=$ E51
The design strength p_w of weld $= 255$ N/mm^2

UB 610 × 229 × 140
Length of end plate $= 500$ mm
Use two 500 mm runs of 8 mm fillet weld between the end plate and the web of the beam.
Effective length of weld $= 2 \times (500 - 2s) = 968$ mm
Capacity of weld $= 968 \times 5.6 \times 255 \times 10^{-3} = 1382$ kN > 1200 kN

UB 457 × 152 × 82
Length of end plate $= 360$ mm
Use two 360 mm runs of 8 mm fillet weld between end plate and web of beam.
Effective length of weld $= 2 \times (360 - 2s) = 688$ mm
Capacity of weld $= 688 \times 5.6 \times 255 \times 10^{-3} = 982$ kN > 720 kN

Step 7 **Check reduced moment capacity of notched beam**
See Example 9.3 for method of analysis.

Note: The moment capacity of UB 457 × 152 × 82 should be checked using a reduced width of bottom flange to account for the chamfer at the connection.

Step 9 **Check local shear capacity of main beam**
The method is similar to Step 10 of Example 9.4.

Connections in Steelwork 473

9.6.6 Example 9.6: Portal frame eaves haunch connection

Use M24 Grade 8.8 bolts in clearance holes of 26 mm diameter.
Check connection for the following loading condition:
   Clockwise bending moment from the rafter on the connection
   = 1000 kNm
   Axial tension from the rafter on the connection = 100 kN
   Shear force from the rafter on the connection = 350 kN
Rafter 457 × 191 × 82 Grade 50 steel
Column 610 × 305 × 149 Grade 50 steel
End plate 200 × 20 × 1100 long Grade 50 steel

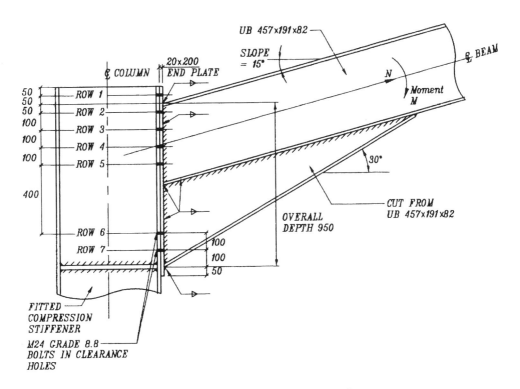

SK 9/54 Portal frame eaves haunch connection.

Step 1 **Check detailing requirement**
Check limits of thickness of the end plate and the flange of the column to make sure that the bolts in the tension zone reach yield or their maximum rated capacity.

$$t_p < \frac{d}{1.9}\sqrt{\frac{U_f}{p_{yp}}} < \frac{24}{1.9}\sqrt{\frac{785}{345}} = 19.1 \text{ mm}$$

t_p = thickness of end plate = 20 mm (may be allowed)
d = diameter of bolt = 24 mm

U_f = ultimate strength of bolt material = 785 N/mm²
p_{yp} = design strength of end plate = 345 N/mm²

$$T_c < \frac{d}{1.9}\sqrt{\frac{U_f}{p_{yp}}} < \frac{24}{1.9}\sqrt{\frac{785}{345}} = 19.1\,\text{mm}$$

T_c = thickness of flange of column = 19.7 mm

This slight excess thickness of flange of column may be allowed.

Step 2 *Determine equivalent bending moment considering the direct axial load*
The horizontal component N_e of the direct tension N and shear V is given by:

$$F_n = N\cos 15° + V\sin 15° = 100\cos 15° + 350\sin 15° = 187.2\,\text{kN}$$

$$h_n = 950 - \frac{1}{2}\frac{460.2}{\cos 15°} = 711.8\,\text{mm}$$

The equivalent moment M_m is given by:

$$M_m = M + F_n h_n = 1000 + (187.2 \times 0.7118) = 1133\,\text{kNm}$$

The equivalent vertical shear force at the connection F_v is given by:

$$F_v = V\cos 15° - N\sin 15° = 350\cos 15° - 100\sin 15° = 312.2\,\text{kN}$$

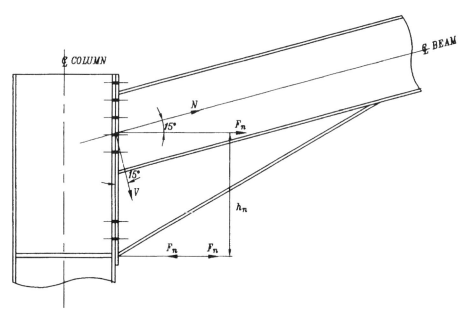

SK 9/55 Equivalent bending moment in the presence of axial tension.

Step 3 *Determine potential resistance of the bolts*
Diameter of bolt = 24 mm

P'_t = enhanced tensile capacity without prying allowance
     = $353 \times 560 \times 10^{-3} = 198\,\text{kN}$

Check column flange for potential resistance of bolts
T_c = thickness of column flange = 19.7 mm
t_w = thickness of column web = 11.9 mm
B = width of column flange = 304.8 mm
r = radius of fillet at flange-to-web connection = 16.5 mm
g = gauge of bolt in the column flange = 120 mm
p = pitch of bolts in the flange = 100 mm

$$m = \frac{g}{2} - \frac{t_w}{2} - 0.8r = 60 - \frac{11.9}{2} - (0.8 \times 16.5) = 40.8 \text{ mm (see SK 9/56)}$$

e_x = end distance of bolt in flange = 50 mm

$$e = \frac{B}{2} - \frac{g}{2} = \frac{304.8}{2} - 60 = 92.4 \text{ mm}$$

n = the lowest of e = 92.4, the end distance of flange plate
= 40 or 1.25m = 51

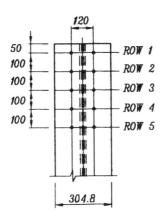

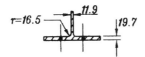

SK 9/56 Geometry of column flange.

Mode 1 failure
Refer to Table 9.2 to find L_{eff} for different types of possible yield lines in the flange. The following L_{eff} calculations are for a pair of bolts in each row.

(1) Circular yielding, pattern 1 failure:
  $L_{eff} = 2\pi m = 2 \times \pi \times 40.8 = 256.4$ mm
(2) Side yielding, pattern 2 failure:
  $L_{eff} = 4m + 1.25e = 278.7$ mm
(3) Corner yielding, pattern 3 failure:
  $L_{eff} = 2m + 0.625e + e_x = 189.4$ mm
(4) Combined five rows, pattern 9 failure:
  $L_{eff} = 2m + e_x + 0.625e + 4p = 589$ mm

By inspection, $L_{eff} = 589$ mm for 2×5 bolts gives rise to minimum potential resistance P_r for each bolt in rows 1 to 5. For rows 1 to 5 combined, $\sum P_r$ is given by:

$$\frac{4}{m} M_P = \frac{4}{m} L_{eff} \frac{t^2}{4} p_y = \frac{4}{40.8} \times 589 \times \frac{19.7^2}{4} \times 345 \times 10^{-3} = 1933 \text{ kN}$$

$$P_{r1} = P_{r2} = P_{r3} = P_{r4} = P_{r5} = \frac{1933}{10} = 193.3 \text{ kN per bolt}$$

Mode 2 failure
Combined five rows, pattern 9 yielding of ten bolts in mode 2 type failure gives the potential resistance of the group as:

$$\sum P_r = \frac{2M_P + (\sum P_t')n}{m+n}$$

$$= \frac{\left(2 \times 589 \times \frac{19.7^2}{4} \times 345\right) + (10 \times 198 \times 40 \times 10^3)}{40.8 + 40} \times 10^{-3}$$

$$= 1468 \text{ kN}$$

Potential resistance per bolt $= 146.8$ kN

Mode 3 failure
Potential resistance per bolt $= P_t' = 198$ kN

Mode 4 failure (failure of column web in tension)
Assume that a pair of bolts at the top of the column is trying to tear off the flange from the web.

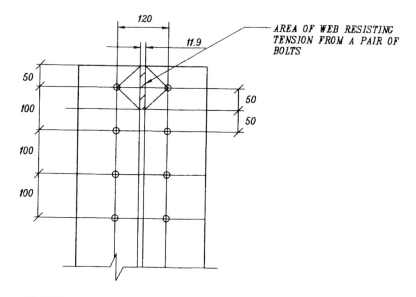

SK 9/57 Web tension failure.

Length of web resisting tension from bolts $L = 100$ mm (see SK 9/57)
Resistance of web $= Lt_w p_y = 100 \times 11.9 \times 355 \times 10^{-3} = 422$ kN
Potential tension in the pair of bolts to cause web tension failure
$= 422$ kN

$\therefore \quad P_{r1} = 422 \div 2 = 211$ kN

$\therefore$ *Considering all modes of failure in the column flange and column web, the lowest potential bolt resistance for rows 1 to 5 $= 146.8$ kN.*

Check end plate of beam for potential resistance of bolts
Steel in end plate = Grade 50

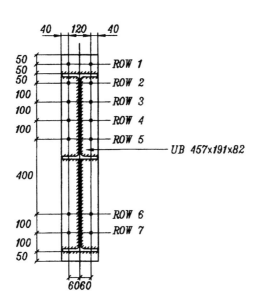

SK 9/58 **Geometry of end plate of beam.**

Potential resistance of bolt in row 1
The pair of bolts in row 1 may have failure patterns 4, 5, 6, 7 and 8, as described in Table 9.2.
T_b = thickness of beam flange = 16 mm
t_p = thickness of end plate = 20 mm
b_p = width of end plate = 200 mm
p = pitch of bolts in end plate = 100 mm
g = gauge of bolt in end plate = 120 mm
e_x = end distance of bolt in end plate = 50 mm
e = edge distance of bolt in end plate = 40 mm
s = leg length of fillet weld between flange of beam and end plate
 = 8 mm

$m_x = \dfrac{p}{2} - \dfrac{T_b}{2} - 0.8s = 50 - 8 - 6.4 = 35.6$ mm

n_x = the smaller of e_x and $1.25 m_x = 44.5$ mm

By inspection the minimum L_{eff} for pattern 4, 5, 6, 7 and 8 yield lines is given by:

$$L_{eff} = \frac{b_p}{2} = 100 \text{ mm}$$

$$M_P = L_{eff} \frac{t_p^2}{4} p_y = 100 \times \frac{20^2}{4} \times 345 \times 10^{-3} = 3450 \text{ kNmm}$$

Mode 1 failure

$$\sum P_{r1} = \frac{4M_P}{m} = \frac{4 \times 3450}{35.6} = 387.6 \text{ kN}$$

$$P_{r1} = 0.5 \times 387.6 = 193.8 \text{ kN per bolt on row 1}$$

Mode 2 failure

$$\sum P_{r1} = \frac{2M_P + n(\sum P'_t)}{m + n}$$

$$= \frac{(2 \times 3450) + (44.5 \times 2 \times 198)}{35.6 + 44.5} = 306 \text{ kN}$$

$$P_{r1} = 0.5 \times 306 = 153 \text{ kN per bolt on row 1}$$

∴ *Potential maximum resistance of bolt in row 1 = 153.0 kN*

Potential resistance of bolts in rows 2, 3, 4 and 5
t_w = thickness of web of beam = 9.9 mm
s = length of leg of fillet weld between web of beam and end plate = 8 mm

$$m = \frac{g}{2} - \frac{t_w}{2} - 0.8s = 60 - \frac{9.9}{2} - 0.8 \times 8 = 48.65 \text{ mm}$$

By inspection, the minimum L_{eff} is for a combined failure of 4 rows of bolts in the end plate, which is given by:

$$L_{eff} = 4m + 1.25e + 3p = (4 \times 48.65) + (1.25 \times 40) + (3 \times 100)$$
$$= 544.6 \text{ mm}$$

$$M_P = L_{eff} \frac{t_p^2}{4} p_y = 544.6 \times \frac{20^2}{4} \times 345 \times 10^{-3} = 18\,789 \text{ kNmm}$$

Mode 1 failure
Combined rows 2, 3, 4 and 5 pattern 9 failure:

$$\sum P_{ri} = \frac{4}{m} M_P = \frac{4}{48.65} \times 18\,789 = 1544.8 \text{ kN (for eight bolts)}$$

Potential resistance of bolts in rows 2 to 5 = 1544.8 ÷ 8 = 193.1 kN per bolt

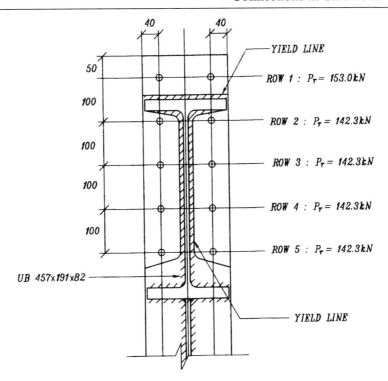

SK 9/59 Modes of failure of end plate for rows 1, 2, 3, 4 and 5.

Mode 2 failure

$$\sum P_{\text{ri}} = \frac{2M_P + n(\sum P'_t)}{m+n} = \frac{(2 \times 18\,789) + (40 \times 8 \times 198)}{48.65 + 40}$$

$$= 1138.6\,\text{kN (for eight bolts)}$$

Potential resistance of each bolt in rows 2 to 5 = $1138.6 \div 8 = 142.3\,\text{kN}$

Mode 3 failure
Potential resistance per bolt = $P'_t = 198\,\text{kN}$

Mode 4 failure (beam web tension failure)
Spacing of bolts = $p = 100\,\text{mm}$
Beam web tension resistance per pair of bolts
$= 100 \times 9.9 \times 355 \times 10^{-3} = 351.4\,\text{kN}$
Potential resistance per bolt = $351.4 \div 2 = 175.7\,\text{kN}$

∴ *Potential maximum resistance of bolt in rows 2 to 5 = 142.3 kN*

Step 4 **Determine moment of resistance of connection**
Bolt row 1:
  Potential resistance = 146.8 kN; Column flange failure Mode 2
Bolt row 2:
  Potential resistance = 142.3 kN; Beam end plate failure Mode 2
Bolt row 3:
  Potential resistance = 142.3 kN; Beam end plate failure Mode 2

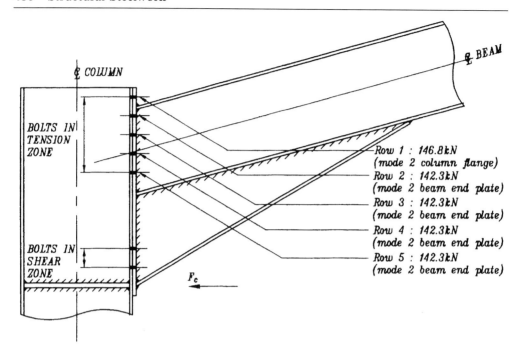

SK 9/60 Potential maximum resistance of bolts in each row.

Bolt row 4:
  Potential resistance = 142.3 kN; Beam end plate failure Mode 2
Bolt row 5:
  Potential resistance = 142.3 kN; Beam end plate failure Mode 2

Assume that the bolts in rows 6 and 7 are in the shear zone and do not contribute towards the moment of resistance of the connection.

The point of rotation is assumed to lie at the connection of the end plate to the bottom flange of the haunch.

Assuming that all bolts reach their potential resistance, the moment of resistance of the connection M_c is given by:

$$M_c = \sum P_{ri} h_i = (2 \times 146.8 \times 1) + [2 \times 142.3 \times (0.9 + 0.8 + 0.7 + 0.6)]$$

$$= 1147.4 \, \text{kNm}$$

$M_c > M_m = 1133 \, \text{kNm}$

∴ Some adjustment of the bolt tensions is necessary for equilibrium.

Assume that bolts in rows 1 to 4 reach their potential capacities but the tension in bolts in row 5 only is adjusted to achieve equilibrium.

M_c for bolts in rows 1 to 4 $= (2 \times 146.8 \times 1)$

$$+ [2 \times 142.3 \times (0.9 + 0.8 + 0.7)]$$

$$= 976.6$$

$M_m - 976.6 = 156.4 \, \text{kNm}$

Bolts in row 5 have to provide this 156.4 kNm moment of resistance for equilibrium.

F_{r5} = actual tension in row $5 = 156.4 \div 0.6 = 260.6$ kN
  actual tension in bolts in row $5 = 260.6 \div 2 = 130.3$ kN
$F_{r1} = 146.8$ kN per bolt
$F_{r2} = F_{r3} = F_{r4} = 142.3$ kN per bolt
$F_{r5} = 130.3$ kN per bolt

The horizontal compression F_c at the connection of the bottom flange of the haunch to the end plate is given by:

$$F_c = \sum F_{ri} + F_n \quad \text{(see Step 2)}$$
$$= (2 \times 146.8) + (2 \times 3 \times 142.3) + (2 \times 130.3) - 187.2$$
$$= 1220.8 \text{ kN}$$

Step 5 Check shear capacity of connection

A_t = tensile stress area of M24 bolt = 353 mm²
P_s = shear capacity of bolt = $353 \times 375 \times 10^{-3} = 132$ kN
P_{bb} = bearing capacity of bolt = dtp_{bb}
  $= 24 \times 19.7 \times 1035 \times 10^{-3} = 489$ kN
P_{bs} = bearing capacity of connected ply = $dtp_{ss} \leq \frac{1}{2} etp_{bs}$
  $= 24 \times 19.7 \times 550 \times 10^{-3} = 260$ kN
e = end distance in the direction of load = 50 mm > $2d = 48$ mm

The capacity of the bolt in shear is 132 kN. The shear capacity of the connection P_v is given by:

$$P_v = n_t P_t'' + n_c P_s'$$

where n_t = number of bolts in the tension zone = 10
  n_c = number of bolts in the shear zone = 4
  $P_t'' = 0.4 P_s = 0.4 \times 132 = 52.8$ kN
  $P_s' = P_s = 132$ kN

$$\therefore \quad P_v = (10 \times 52.8) + (4 \times 132) = 1056 \text{ kN} > F_v = 312.2 \text{ kN}$$

The shear capacity of the connection is adequate.

Step 6 Local compression check of column web

Check web bearing (see SK 9/8)
D_c = overall depth of column = 609.6 mm
d = depth between fillets = 537.2 mm
r = column root radius = 16.5 mm
T_c = thickness of flange of column = 19.7 mm
t_p = thickness of end plate = 20 mm
t_w = thickness of web of column = 11.9 mm
b_1 = stiff bearing length by dispersion through end plate at 45°
  = thickness of flange of the haunch + leg lengths of fillet weld of the flange of haunch + two times thickness of end plate

$$= \frac{16}{\cos 30°} + (2 \times 8) + (2 \times 20) = 74 \text{ mm}$$

n_2 = dispersion length through flange of column and the root of flange at 1:2.5
$= 2 \times 2.5 \times (19.7 + 16.5) = 181$ mm

P_c = web bearing resistance of column
$= (b_1 + n_2)t_w p_y$
$= (74 + 181) \times 11.9 \times 355 \times 10^{-3}$
$= 1077$ kN $< F_c = 1220.8$ kN (see Step 4)

A fitted compression stiffener is required in the column.

Check column web buckling resistance

n_1 = dispersion length of flange compression force at 45° through the web of column up to the centre-line of the column section
$= 2 \times \dfrac{D_c}{2} = 609.6$ mm

The slenderness ratio of the web of the column is given by:

$$2.5 \dfrac{d}{t_w} = 2.5 \times \dfrac{537.2}{11.9} = 112.9$$

(See Table 27(c) of BS 5950: Part 1)
$p_c = 117$ N/mm^2 for $p_y = 355$ N/mm^2
P_c = web buckling capacity of column
$= (b_1 + n_1)t_w p_c$
$= (74 + 609.6) \times 11.9 \times 117 \times 10^{-3}$
$= 951.8$ kN $< F_c = 1220.8$ kN

A compression stiffener is required.

Step 7 **Check column web shear capacity**

h = distance between the top flange of the beam and the bottom flange of the haunch
$= 950$ mm

F_v = column web panel shear = flange force at the top of the beam
$= \dfrac{M_m}{h} = \dfrac{1133}{0.95} = 1193$ kN

P_v = shear capacity of web of column $= 0.6 p_y D_c t_w$
$= 0.6 \times 355 \times 609.6 \times 11.9 \times 10^{-3} = 1545$ kN > 1193 kN

Web shear capacity is adequate and a stiffener is not required.

Step 8 **Check local bearing of compression flange of haunch**

F_c = horizontal component of compressive force at the bottom flange of haunch
$= 1220.8$ kN (see Step 4)

Compressive force in the direction of the flange of the haunch
$= \dfrac{F_c}{\cos 30°} = \dfrac{1220.8}{0.866} = 1410$ kN

Bearing resistance of flange
$= 1.4 p_y TB = 1.4 \times 355 \times 16 \times 191.3 \times 10^{-3}$
$= 1521.2$ kN > 1410 kN

Connections in Steelwork 483

Step 9 *Design of column compression stiffener*

Selected steel for compression stiffener = Grade 43

Generally the net bearing resistance P_c governs the design of the compression stiffener: it is assumed that 80% of the compression force passes through the contact surface between the column flange and the stiffener.

$$\therefore \quad P_c = 0.8F_c = Ap_y$$

A = area of the contact surface

$$Ap_y = 0.8 \times 1220.8 = 977 \,\text{kN}$$

$$\therefore \quad A = \frac{977 \times 10^3}{265} = 3698 \,\text{mm}^2$$

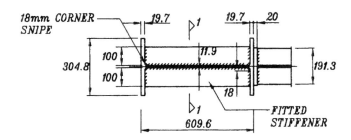

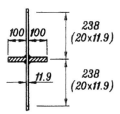

SK 9/61 Compression stiffener of column.

Assuming a thickness of plate stiffener t_p equal to 25 mm, the minimum contact length b_{sn} is given by:

$$2b_{sn}t_p = 3698 \,\text{mm}^2$$

$$\therefore \quad b_{sn} = \frac{3698}{2 \times 25} = 74 \,\text{mm (minimum required)}$$

Use 100×25 Grade 43 stiffeners with 18 mm corner snipes to allow for a column root radius of 16.5 mm.

$$\varepsilon = \sqrt{\frac{275}{p_{ys}}}$$

$= 1$ (because p_{ys} = design strength of stiffener = 275 N/mm²)

b_{sg} = outstand of the stiffener = $100 < 13t_s\varepsilon = 260$
b_{sn} = net outstand in contact with flange of column
$\quad = 100 - 18 = 82$ mm
t_s = thickness of stiffener = 25 mm
t_w = thickness of web of column
A_{sg} = gross area of stiffener only
$\quad = 2b_{sg}t_s = 2 \times 100 \times 25 = 5000$ mm^2
A = area of stiffener in contact with the flange
$\quad = 2b_{sn}t_s = 2 \times 82 \times 25 = 4100$ mm^2
A_{eff} = area of effective section of stiffener, including 20 times the thickness of column web on either side of the stiffeners as part of the assembly
$\quad = 2b_{sg}t_s + 40t_w^2 = (2 \times 100 \times 25) + (40 \times 11.9^2)$
$\quad = 10\,664.4$ mm^2
I_{eff} = effective moment of inertia of the effective section about the centroidal axis of the effective section parallel to the web of the column

$$= \frac{1}{12} t_s (2b_{sg} + t_w)^3 = \frac{1}{12} \times 25 \times [(2 \times 100) + 11.9]^3$$

$\quad = 19.82 \times 10^6$ mm^2
r_{eff} = radius of gyration of effective section

$$= \sqrt{\frac{I_{eff}}{A_{eff}}} = \sqrt{\frac{19.82 \times 10^6}{10\,664.4}} = 43.1 \text{ mm}$$

L_{eff} = effective length of the effective section of the stiffener
$\quad = 0.7(D_c - 2T_c) = 0.7 \times [609.6 - (2 \times 19.7)] = 399.1$ mm
λ_{eff} = effective slenderness ratio

$$= \frac{L_{eff}}{r_{eff}} = \frac{399.1}{43.1} = 9.3$$

Using this slenderness ratio and design strength $p_y = 265$ N/mm^2, obtain p_c from Table 27(c) of BS 5950: Part 1.

$p_c = 265$ N/mm^2

Buckling resistance $P_c = A_{eff} p_c = 10\,961.9 \times 265 \times 10^{-3}$
$\quad = 2905$ kN $> F_c$
Bearing strength $P_c = A p_y + (b_1 + n_2) t_w p_{yc}$
$\quad = (4100 \times 265 \times 10^{-3}) + 1077$ (see Step 6)
$\quad = 2163.5$ kN $> F_c$

The column compression stiffener fitted as $2 \times 100 \times 25$ in Grade 43 steel has adequate capacity.

Step 10 **Design tension stiffeners**
Tension stiffeners are not required because both the beam and the column webs have adequate tension capacity.

Step 11 **Design column web stiffener or diagonal stiffener**
Column web stiffeners are not required because the web has got adequate web panel shear capacity.

Step 12 *Design welded connections*

Flange of beam to end plate connection
Maximum allowable flange force = area of flange × design strength of flange $= 191.3 \times 16 \times 355 \times 10^{-3} = 1086.6$ kN

For a full-strength weld, the combined throat thickness of the two runs of fillet weld connecting the flange with the end plate should be equal to the thickness of the flange, provided the electrode strength is comparable. The end plate is Grade 50 and the beam is Grade 50 so, therefore, the chosen electrode is E51.

The thickness of the flange is 16 mm: hence, the equivalent leg length for two runs of fillet weld $= 16 \div 1.4 = 11.43$ mm. Use two runs of 12 mm fillet weld using an E51 electrode.

For the web to end plate connection, use a weld size of equivalent thickness to the web of the beam.

The minimum size of weld to be used is given by:

$$\frac{t_w}{1.4} = \frac{9.9}{1.4} = 7.07$$

Use two runs of 8 mm fillet weld to connect the web of the beam to the end plate.

Assume that the compression in the bottom flange of the haunch is transferred to the end plate by the weld. Use the same size and type of weld as for the top flange.

Connection of column compression stiffeners to column flange
Assume that $0.8 F_c = 977$ kN (see Step 9) is transferred through the welded connection between the stiffeners and the flange of the column. The contact length is 2×82 mm on each side of the flange. The stiffeners are 25 mm thick.
Full-strength weld requires $25 \div 1.4 = 18$ mm fillet welds
To transfer the load of 977 kN using E43 electrodes and four rows of fillet welds of 82 mm length each, the throat thickness required is given by:

$$\frac{977 \times 10^3}{4 \times 82 \times 265} = 11.24 \text{ mm}$$

Leg length of fillet weld required $= 11.24 \div 0.7 = 16$ mm
Use 16 mm fillet weld to connect the stiffeners to the column flange.
Use 8 mm fillet weld everywhere else.

9.6.7 Example 9.7: Bolted splice of a beam

Span of beam between columns = 8000 mm
Location of splice from the centre of column = 1200 mm
Size of beam = UB 533 × 210 × 122 (Grade 43)
Selected loading condition:
  Bending moment at the splice = +200 kNm (bottom tension)
  Axial compression at the splice = 100 kN
  Vertical shear at the splice = 300 kN

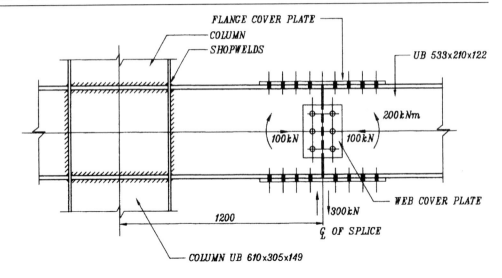

SK 9/62 Beam splice to shop welded beam stub on column.

D = overall depth of beam = 544.5 mm
B = width of flange of beam = 211.9 mm
T = thickness of flange = 21.3 mm
t_w = thickness of web = 12.8 mm
r = root radius = 12.7 mm
h = distance between centre of flanges = $D - T$ = 523.2 mm
A = area of the beam = 155.8 cm^2

Use high-strength friction grip (HSFG) bolts to prevent slippage at the splice.

Step 1 Find flange forces
F_f = flange force at splice

$$= \pm \frac{M}{h} + \frac{N}{2} = \pm \frac{200}{0.5232} + \frac{100}{2} = +432.3 \text{ and } -332.3 \text{ kN}$$

Step 2 Check capacity of flange in tension
Assume M20 HSFG bolts in 22 mm diameter clearance holes.
Assume two holes in a row perpendicular to the axis of the beam.
Effective area of flange = $A_e = K_e A_n$

K_e = 1.2 for Grade 43 steel
A_n = net area of flange after deduction of holes
   = [211.9 − (2 × 22)] × 21.3 = 3576 mm^2
A_g = gross area of flange = BT = 211.9 × 21.3 = 4513 mm^2
A_e = 1.2 × 3576 = 4292 < A_g = 4513 mm^2

The tension capacity of flange at the splice is given by:

$$A_e p_y = 4292 \times 265 \times 10^{-3} = 1137 \text{ kN} > 332.3 \text{ kN}$$

Step 3 *Determine size of flange cover plate*
Assume Grade 43 cover plate for the flange.
The effective area required for the cover plate is given by

$$\frac{F_f}{p_y} = \frac{432.3 \times 10^3}{265} = 1631 \text{ mm}^2$$

Use 200×12 cover plate for the flange with 2×22 holes.

A_g = gross area of cover plate = 2400 mm^2
A_e = effective area of cover plate = $K_e A_n$
$= 1.2 \times [200 - (2 \times 22)] \times 12 = 2246 \text{ mm}^2$

Capacity of cover plate based on effective area
$= 2246 \times 275 \times 10^{-3} = 617.8 \text{ kN} > 432.3 \text{ kN}$
200×12 cover plate is adequate for the flange.

Step 4 *Check capacity of flange cover plate including strut action*

Strut action at the splice
M_{max} = maximum minor axis bending moment due to strut action

$$= \frac{\eta f_c S_y}{1 - \dfrac{f_c}{p_E}}$$

η = Perry factor = $0.001 a(\lambda - \lambda_0) \leq 0$
a = Robertson constant = 5.5

$$\lambda = \text{slenderness ratio} = \frac{L_E}{r_y}$$

L_E = effective length of beam between restraints
$= 0.7 \times 8000 = 5600 \text{ mm}$
$r_y = 46.7 \text{ mm}$

$$\lambda = \frac{5600}{46.7} = 120$$

$$\lambda_0 = 0.2\sqrt{\frac{\pi^2 E}{p_y}} = 0.2 \times \sqrt{\frac{\pi^2 \times 205 \times 10^3}{275}} = 17$$

$\eta = 0.001 \times 5.5 \times (120 - 17) = 0.5665$
f_c = compressive stress due to axial compressive force

$$= \frac{100 \times 10^3}{155.8 \times 10^2} = 6.4 \text{ N/mm}^2$$

S_y = minor axis plastic modulus of beam section = 500.6 cm^3
p_E = Euler buckling strength

$$= \frac{\pi^2 E}{\lambda^2} = \frac{\pi^2 \times 205 \times 10^3}{120^2} = 140.5 \text{ N/mm}^2$$

$$\therefore M_{max} = \frac{0.5665 \times 6.4 \times 500.6 \times 10^3}{1 - \dfrac{6.4}{140.5}} = 1901.6 \text{ kNmm}$$

$M_{max,x}$ = minor axis strut action bending moment at the splice

$$= M_{max} \sin\left(\frac{180 L_x}{L}\right) = 1901.6 \times \sin\left(\frac{180 \times 1.2}{8}\right)$$

$$= 863.3 \text{ kNmm}$$

M_f = bending moment in each flange cover plate (top and bottom)
= $M_{max,x} \div 2 = 432$ kNmm about the vertical plane

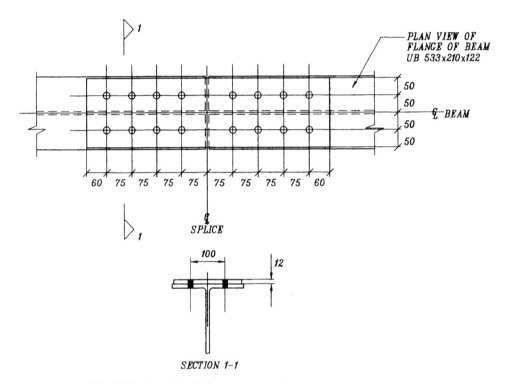

SK 9/63 Details of flange cover plate.

Allow end distance in the direction of load = $3d = 3 \times 20 = 60$ mm for HSFG bolts.

M_c = elastic moment of resistance of flange cover plate

$$= \left[\frac{1}{6} t_p b_p^2 - \frac{2}{b_p} \left(\sum t_p D h_i^2\right)\right] p_y$$

$$= \left[\left(\frac{1}{6} \times 12 \times 200^2\right) - \frac{2}{200} \times (2 \times 12 \times 22 \times 50^2)\right] \times 275 \times 10^{-3}$$

$$= 18\,370 \text{ kNmm}$$

Check interaction formula for flange cover plate:

$$\frac{F_f}{A_e p_y} + \frac{M_f}{M_c} = \frac{432.3 \times 10^3}{2246 \times 275} + \frac{432}{18\,370} = 0.72 < 1$$

The flange cover plate has adequate strength.

Connections in Steelwork 489

Step 5 Check resultant shear in bolts for flange cover plate

The polar moment of inertia of the group of eight bolts on either side of the splice is given by:

$$I_P = \sum X^2 + \sum Y^2 = (4 \times 112.5^2) + (4 \times 37.5^2) + (8 \times 50^2)$$
$$= 76250 \, \text{mm}^2$$

$$Z_x = \frac{I_P}{Y} = \frac{76250}{50} = 1525 \, \text{mm}$$

$$Z_y = \frac{I_P}{X} = \frac{76250}{112.5} = 677.8 \, \text{mm}$$

$$f_{m,x} = \frac{M_f}{Z_x} = \frac{432}{1525} = 0.3 \, \text{kN in the } X\text{-direction}$$

$$f_{m,y} = \frac{M_f}{Z_y} = \frac{432}{677.8} = 0.6 \, \text{kN in the } Y\text{-direction}$$

$$f_v = \frac{F_f}{n} = \frac{432.3}{8} = 54 \, \text{kN in the } X\text{-direction}$$

$$f_r = \text{resultant shear in bolt} = \sqrt{(f_v + f_{m,x})^2 + f_{m,y}^2} = 54.3 \, \text{kN}$$

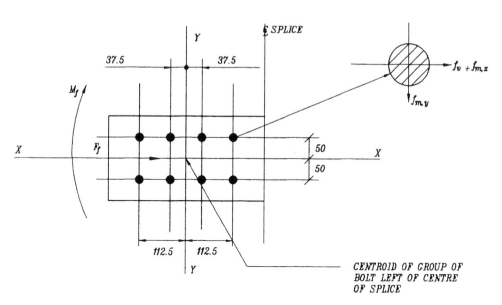

SK 9/64   Forces in bolts in flange cover plate.

Step 6 Check capacity of bolt

Preloaded M20 HSFG bolts:
Slip capacity $= 1.1 K_s \mu P_0$
$K_s = 1.0$ for parallel shank fasteners
$\mu =$ slip factor $= 0.45$
$P_0 =$ Proof load of bolt $= 144 \, \text{kN}$ for M20

∴ Slip resistance of bolt for single shear
$= 1.1 \times 1.0 \times 0.45 \times 144 = 71.3 \, \text{kN} > 54.3 \, \text{kN}$

Bearing capacity of 12 mm cover plate is greater than 71.3 kN.
The bolt has adequate capacity.

Step 7 **Determine size of cover plates for web**
Assume two cover plates 240 × 10 × 300 long.
Shear capacity P_v of each cover plate is given by:

$$P_v = 0.6(0.9A)p_y$$

Assume two vertical rows of three holes in each plate.

A = net area of web plate allowing for holes
$= (10 \times 300) - (3 \times 22 \times 10) = 2340 \, \text{mm}^2$
$P_v = 0.6 \times 0.9 \times 2340 \times 275 \times 10^{-3} = 347.5 \, \text{kN}$

Shear force carried by each web plate $= F_v/2 = 300 \div 2 = 150 \, \text{kN}$
Web plates have adequate shear capacity.

Step 8 **Check resultant shear in bolts for web cover plates**
F_v = applied shear force at the splice = 300 kN
a = eccentricity of shear force on the group of bolts = 60 mm
f_m = shear in the bolt farthest from the centroid of the group of three bolts on either side of the splice due to the applied eccentric moment $F_v a$

$$= \frac{F_v a r}{\sum r^2} \text{ (in the horizontal direction)}$$

$$= \frac{300 \times 60 \times 90}{2 \times 90^2} = 100 \, \text{kN}$$

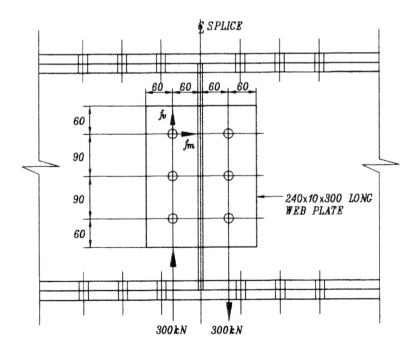

SK 9/65 Bolts in web cover plate.

f_v = shear in the bolt due to direct shear F_v in the vertical direction
$$= \frac{F_v}{n} = \frac{300}{3} = 100 \, \text{kN}$$
f_r = resultant shear in bolt $= \sqrt{f_m^2 + f_v^2} = \sqrt{100^2 + 100^2} = 141 \, \text{kN}$
Slip resistance of M20 HSFG bolt in double shear $2 \times 71.3 = 142.6 \, \text{kN}$
Bearing resistance of web plates $= 2 \times 165 = 330 \, \text{kN}$
Bearing resistance of web of beam $= 12.8 \times 20 \times 825 \times 10^{-3} = 211.2 \, \text{kN}$
The arrangement of bolts in the web cover plate is adequate.

9.6.8 Example 9.8: Bolted splice of a column

Upper level UC $254 \times 254 \times 89$
 $D = 260.4 \, \text{mm}$
 $B = 255.9 \, \text{mm}$
 $t = 10.5 \, \text{mm}$
 $T = 17.3 \, \text{mm}$
 $d = 200.3 \, \text{mm}$
 $r_y = 6.52 \, \text{cm}$
 $S_y = 575 \, \text{cm}^3$
 $h = D - T = 243.1 \, \text{mm}$

Lower level UC $254 \times 254 \times 132$
 $D = 276.4 \, \text{mm}$
 $B = 261.0 \, \text{mm}$
 $t = 15.6 \, \text{mm}$

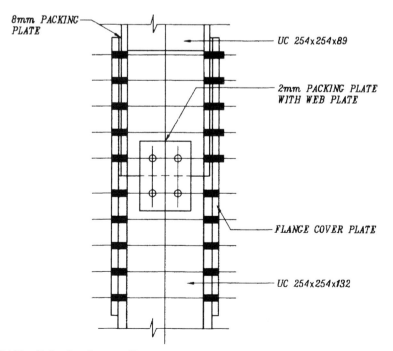

SK 9/66 Bolted column splice.

$T = 25.3$ mm
$d = 200.3$ mm
$r_y = 6.67$ cm
$S_y = 879$ cm^3
$h = D - T = 251.1$ mm

Steel = Grade 50 for all components
The column is part of a rigid jointed frame with unrestricted sidesway about the major axis. Refer to BS 5950: Part 1: Appendix E2.

$k_3 = 0$ (unrestricted sidesway)
$k_1 = k_2 = 0.5$

The height of columns between floors is 5 m.
From Figure 24 of BS 5950: Part 1 the effective length ratio $L_E/L = 1.45$ in the direction of the major axis.
The column is effectively braced against sidesway about the minor axis, giving $k_3 = \infty$.
The effective length ratio $L_E/L = 1.0$ in the direction of the minor axis.
The column splice is located 1.2 m from the centre of the joint between the beams and columns at the floor level.
All loads are carried by the splice plates and the fasteners because the column ends are not prepared for direct bearing.
Loads to be transferred at the splice are as follows:

N = axial compression = 1300 kN
M_x = bending moment about the major axis = 100 kNm
V_x = horizontal shear about the major axis = 100 kN

Step 1 *Determine flange forces*

The axial load in the column is assumed to be distributed in the web and the flanges of the column in proportion to their respective areas.

Area of column UC $254 \times 254 \times 89 = 114$ cm^2
Area of web only of the column
$= [260.4 - (2 \times 17.3)] \times 10.5 = 2370.9$ mm^2
Axial compression carried by the web of the column
$$= 1300 \times \frac{2370.9}{11\,400} = 270 \text{ kN}$$
Direct axial compression carried by each flange of column
$$N_f = 0.5 \times (1300 - 270) = 515 \text{ kN}$$

F_f = flange force at splice
$$= \pm \frac{M_x}{h} + N_f = \pm \frac{100}{0.2431} + 515 = 926 \text{ or } 104 \text{ kN}$$

It is assumed that the horizontal shear at the splice is carried by the web cover plates.

Step 2 *Check capacity of flange in tension*

This check is not required for the loading condition.

Connections in Steelwork

Step 3 *Determine size of flange cover plate*
Assume cover plates = Grade 50

$$\text{Net area required} = \frac{F_f}{p_y} = \frac{926 \times 10^3}{345} = 2684\,\text{mm}^2$$

Assuming 20 mm thick cover plates, the required width of flange cover plates is given by:

$$\frac{2684}{20} = 134\,\text{mm}$$

Use 250 × 20 flange cover plates to allow for bending of cover plates due to strut action.

Note: No deduction for holes is made in the cover plates because the plates are in compression.

Step 4 *Check capacity of flange cover plates including strut action*

Strut action at the splice

λ = slenderness ratio of the column

$$= \frac{L_{Ey}}{r_y} \text{ or } \frac{L_{Ex}}{r_x} \text{ (whichever is the greater)}$$

$$= \frac{5000}{65.2} \text{ or } \frac{1.45 \times 5000}{112} = 76.7 \text{ or } 64.7$$

$$\lambda_0 = 0.2\sqrt{\frac{\pi^2 E}{p_y}} = 0.2 \times \sqrt{\frac{\pi^2 \times 205 \times 10^3}{355}} = 15$$

$\eta = 0.001a(\lambda - \lambda_0) = 0.001 \times 5.5 \times (76.7 - 15) = 0.3394$
a = Robertson constant = 5.5
f_c = average axial compressive stress in column section

$$= \frac{N}{A} = \frac{1300 \times 10^3}{11\,400} = 114\,\text{N/mm}^2$$

p_E = Euler buckling strength

$$= \frac{\pi^2 E}{\lambda^2} = \frac{\pi^2 \times 205 \times 10^3}{76.7^2} = 343.9\,\text{N/mm}^2$$

S_y = minor axis plastic section modulus = $575 \times 10^3\,\text{mm}^3$

$$M_{\max} = \frac{\eta f_c S_y}{1 - \frac{f_c}{p_E}} = \frac{0.3394 \times 114 \times 575 \times 10^3}{\left(1 - \frac{114}{343.9}\right) \times 10^3} = 33\,279\,\text{kNmm}$$

The minor axis bending moment at the splice is given by:

$$M_f = M_{\max,x} = M_{\max} \sin\left(\frac{180 L_x}{L}\right) = 33\,279 \times \sin\left(\frac{180 \times 1.2}{5}\right)$$

$$= 22\,781\,\text{kNmm}$$

Minor axis bending moment carried by each cover plate = $0.5 \times 22\,781$
= 11 391 kNmm

t_p = thickness of flange cover plates = 20 mm
b_p = width of flange cover plates = 250 mm

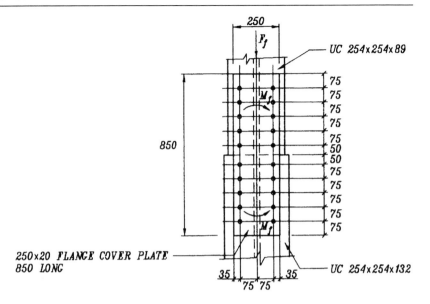

SK 9/67 Details of flange cover plates.

Use general grade M24 preloaded HSFG bolts.
Minimum end distance = $3d = 72$ mm
Minimum edge distance = $1.4d = 34$ mm

D = clearance hole size = 26 mm
M_c = elastic moment of resistance of flange cover plate resisting the column minor axis bending moment

$$= \left[\frac{1}{6} t_p b_p^2 - \frac{2}{b_p}\left(\sum t_p D h_i^2\right)\right] p_y$$

$$= \left[\left(\frac{1}{6} \times 20 \times 250^2\right) - \frac{2}{250} \times (2 \times 20 \times 26 \times 90^2)\right] \times 345 \times 10^{-3}$$

$$= 48\,625\,\text{kNmm}$$

Check interaction formula for flange plate:

$$\frac{F_f}{A_e p_y} + \frac{M_f}{M_c} = \frac{926 \times 10^3}{20 \times 250 \times 345} + \frac{11\,391}{48\,625} = 0.77 < 1.0$$

Note: No deduction for holes is made in the cover plates because the plates are in compression.

Step 5 *Check resultant shear of bolts in flange cover plates*

The polar moment of inertia of the group of ten bolts on either side of the splice is given by:

$$I_P = \sum X^2 + \sum Y^2 = (4 \times 150^2) + (4 \times 75^2) + (10 \times 90^2)$$

$$= 193\,500\,\text{mm}^2$$

$$Z_x = \frac{I_P}{Y} = \frac{193\,500}{90} = 2150\,\text{mm}$$

$$Z_y = \frac{I_P}{X} = \frac{193\,500}{150} = 1290\,\text{mm}$$

Bolt shear due to strut action:

$$f_{m,x} = \frac{M_f}{Z_x} = \frac{11\,391}{2150} = 5.3\,\text{kN (acting vertically)}$$

$$f_{m,y} = \frac{M_f}{Z_y} = \frac{11\,391}{1290} = 8.8\,\text{kN (acting horizontally)}$$

Bolt shear due to axial load acting vertically:

$$f_v = \frac{F_f}{n} = \frac{926}{10} = 92.6\,\text{kN (acting vertically)}$$

$f_r =$ resultant maximum shear in bolt

$$= \sqrt{(f_v + f_{m,x})^2 + f_{m,y}^2} = \sqrt{(92.6 + 5.3)^2 + 8.8^2} = 98.3\,\text{kN}$$

Step 6 *Check capacity of bolt*
Slip capacity $= 1.1 K_s \mu P_0$
For M24 HSFG: $K_s = 1.0$; $\mu = 0.45$; $P_0 = 207\,\text{kN}$
Slip resistance for single shear $= 1.1 \times 1.0 \times 0.45 \times 207 = 102\,\text{kN}$
Flange thickness $= 17.3\,\text{mm}$
Flange plate thickness $= 20\,\text{mm}$
The bearing capacity in 17.3 mm connected ply of Grade 50 material is given by:

$$P_{bg} = dt p_{bg} \leq \tfrac{1}{3} e t p_{bg}$$

End distance e is greater than $3d$.

$$P_{bg} = 24 \times 17.3 \times 1065 \times 10^{-3} = 442\,\text{kN}$$

The capacity of bolt $= 102\,\text{kN} > 98.3\,\text{kN}$ maximum applied shear
The arrangement of bolts in the flange cover plate is adequate.

Step 7 *Determine size of web cover plates*
The horizontal shear at the splice is 100 kN, resisted by two web cover plates measuring $190 \times 10 \times 200$ long.
The shear capacity P_v of each web cover plate is given by:

$$P_v = 0.6(0.9A)p_y$$

$A =$ net area of plate allowing for holes

$$= (10 \times 190) - (2 \times 26 \times 10) = 1380\,\text{mm}^2$$

$$P_v = 0.6 \times 0.9 \times 1380 \times 355 \times 10^{-3} = 265\,\text{kN} > 0.5 \times 100\,\text{kN}$$

Axial load capacity of the web plate $= 190 \times 10 \times 355 \times 10^{-3}$
$= 674\,\text{kN} > 0.5 \times 270\,\text{kN}$ (see Step 1)

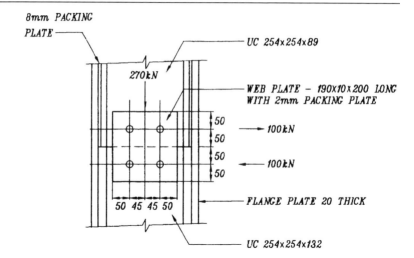

SK 9/68 Details of web cover plates.

Step 8 *Check resultant shear in bolts in web cover plates (see SK 9/68)*
(See SK 9/68) Use four M24 HSFG bolts.
Capacity of bolt = slip resistance of M24 in double shear = 204 kN
(see Step 6)
Bearing capacity in each 10 mm web plate = $\frac{1}{3} e t p_{bg}$
   = $(50 \times 10 \times 1065 \times 10^{-3})/3$
   = 177.5 kN > 102 kN bolt capacity in single shear
Moment due to eccentricity of horizontal shear on the group of bolts
   = $100 \times 50 = 5000$ kNmm
Vertical shear in each bolt due to this moment = $5000 \div 90 = 55.6$ kN
Vertical shear in each bolt due to axial load = $270 \div 2 = 135$ kN
Horizontal shear in each bolt due to shear force = $100 \div 2 = 50$ kN
Resultant shear in bolt = $\sqrt{(135 + 55.6)^2 + 50^2} = 197$ kN < 204 kN
The arrangement of bolts in the web cover plates is adequate.

9.6.9 Example 9.9: Design of a column base

Loading condition:
   Bending moment about the major axis = 750 kNm
   Axial direct compression
      = 2500 kN maximum and 450 kN minimum
   Size of column = UC $356 \times 368 \times 202$ kg/m Grade 50 steel
   Assume grade of concrete in foundation = 30 N/mm^2

Step 1 *Select trial size of base plate*
Assume M30 Grade 8.8 bolts
Minimum spacing of bolts = $6d = 6 \times 30 = 180$ mm
Use 200 mm spacing
Minimum size of base plate = size of column + 300 = 674.7 mm
Use 750×750 base plate
Minimum edge distance = 2.5 diameter of bolt = 75 mm
Use trial thickness of base plate = 50 mm

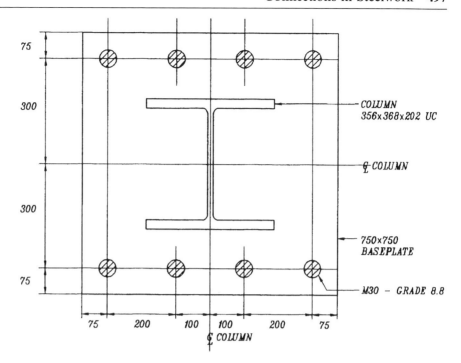

SK 9/69 Column base plate connection.

Step 2 *Determine the concrete stress block and the tension in holding-down (HD) bolts*

Load case 1
$N = 2500 \, \text{kN}$
$M_x = 750 \, \text{kNm}$

Check $e = \dfrac{M_x}{N} = \dfrac{750 \times 10^3}{2500} = 300 \, \text{mm}$

$> \dfrac{D}{2} - \dfrac{N}{0.8 f_{cu} B} = 375 - \dfrac{2500 \times 10^3}{0.8 \times 30 \times 750} = 236 \, \text{mm}$

∴ HD bolts will be in tension in load case 1.

C = ultimate contact reaction of concrete rectangular stress block of depth X
$= 0.4 f_{cu} BX = 0.4 \times 30 \times 750 X = 9000 X$
$\sum T$ = sum of tension in all HD bolts in tension
$\sum T + N = C$, or $\sum T + (2500 \times 10^3) = 9000 X$

Taking moments about the centre of the column:

$$M_x = 0.4 f_{cu} BX \left(\dfrac{D-X}{2}\right) + \left(\sum T\right) a$$

$$750 \times 10^6 = \left[0.4 \times 30 \times 750 X \left(\dfrac{750-X}{2}\right)\right] + [(\sum T) \times 300]$$

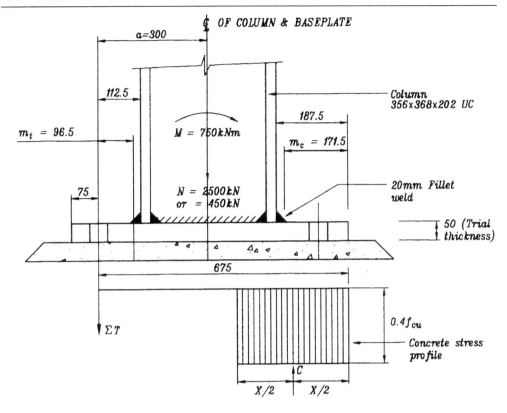

SK 9/70 Concrete stress profile under base plate.

Substituting $\sum T = 9000X - (2500 \times 10^3)$ in the above equation and simplifying gives:

$$15X^2 - 20\,250X + (5.0 \times 10^6) = 0$$

$$X = \frac{20\,250 \pm \sqrt{20\,250^2 - (4 \times 15 \times 5 \times 10^6)}}{2 \times 15} = 325.3 \text{ mm}$$

$$C = 9000X = 9000 \times 325.3 \times 10^{-3} = 2927.7 \text{ kN}$$

$$\sum T = C - N = 2927.7 - 2500 = 427.7 \text{ kN}$$

$$\therefore \text{ tension in each HD bolt} = \frac{427.7}{4} = 106.9 \text{ kN}$$

Capacity of Grade 8.8 M30 bolt in tension $= 314$ kN (see Table 9.5)

Load case 2
$N = 450$ kN
$M_x = 750$ kNm

$$\text{Check } e = \frac{750 \times 10^6}{450 \times 10^3} = 1667 > \frac{D}{2} - \frac{N}{0.8f_{cu}B}$$

Connections in Steelwork

The HD bolts will be in tension due to this combination of !
By similar analytical procedure as before it is determined th:

$X = 166.1$ mm
$C = 1495$ kN
$\sum T = 1045$ kN

Tension in each holding-down bolt $= 1045 \div 4 = 261.3$ kN < 314 kN
The HD bolts have adequate capacity.

Step 3   *Carry out compression-side analysis of base plate*
$m_c =$ effective overhang of base plate on the compression side in contact with concrete
$= L_c - 0.8s = 187.5 - (0.8 \times 20) = 171.5$ mm
$L_c =$ actual overhang of base plate $= 187.5$ mm
$s =$ leg length of fillet weld between base plate and column
$= 20$ mm
$M_c =$ bending moment in the cantilever overhang of base plate
$= 0.4 f_{cu} B \dfrac{m_c^2}{2} = 0.4 \times 30 \times 750 \times \dfrac{171.5^2}{2} = 132.36$ kNm

$M_{rc} =$ moment of resistance of base plate $= 1.2 p_{yp} Z$

$Z = \dfrac{1}{6} B t_p^2 = \dfrac{1}{6} \times 750 \times 50^2 = 312\,500$ mm^3

$p_{yp} = 255$ N/mm^2 for Grade 43 steel in base plate
$M_{rc} = 1.2 \times 255 \times 312\,500 \times 10^{-6}$ kNm $= 95.6$ kNm < 132.36 kNm

The base plate capacity is inadequate.

Select 60 mm thick base plate in Grade 50 steel.
$Z = 450\,000$ mm^3; $p_{yp} = 270$ N/mm^2 for base plates in Grade 50
$M_{rc} = 1.2 \times 270 \times 450\,000 \times 10^{-6} = 145.8$ kNm > 132.36 kNm

Step 4   *Carry out tension-side analysis of base plate*
$m_t =$ effective overhang of base plate on the tension side
$= L_t - 0.8s = 112.5 - 16 = 96.5$ mm
$L_t =$ actual overhang of base plate from centre of the bolt to the face of the column
$= 187.5 - 75 = 112.5$ mm
$M_t =$ bending moment in base plate on the tension side
$= (\sum T) m_t = 1045 \times 0.0965 = 100.84$ kNm < 145.8 kNm

The 60 mm thick base plate has adequate moment of resistance.

Step 5   *Determine size of weld*

Weld between base plate and flange

Minimum size of fillet weld $= \dfrac{T}{1.4} = \dfrac{27}{1.4} = 19.3$ (say, 20 mm)

Axial compression in flange is proportional to its area.
Area of column $= 25\,800$ mm^2
Area of flange $= 374.4 \times 27 = 10\,108.8$ mm^2
Direct axial load $= 2500$ kN

Compressive load carried by flange of column

$$= \frac{2500 \times 10\,108.8}{25\,800} = 980\,\text{kN}$$

Flange force due to moment only neglecting the web

$$= \frac{750}{(D-T)} = \frac{750}{(0.3747 - 0.027)} = 2157\,\text{kN}$$

Total maximum compression in flange $= 980 + 2157 = 3137\,\text{kN}$
Use electrode grade E51 as per BS 639.
Length of weld available around the flange
$= (2 \times 374.4) - 16.8 = 732\,\text{mm}$
Direct stress and shear stress on the fusion face due to compressive load of 3137 kN using 20 mm fillet weld

$$= \frac{3137 \times 10^3}{732 \times 20} = 214\,\text{N/mm}^2 < 0.7 p_y = 241.5\,\text{N/mm}^2$$

Use 20 mm fillet weld for flange to base plate connection and 12 mm fillet weld for web to base plate connection.

Chapter 10
Corrosion Protection

10.1 PROCESS OF CORROSION IN STEEL

Corrosion in steel is an electrochemical process in which ferrous ions go into solution. The electrons released from the anode go to the adjacent cathodic sites to combine with oxygen and water. The hydroxyl ions thus produced combine with the ferrous ions from the anode to give ferrous hydroxide. After further oxidation in contact with air this product becomes ferric oxide, or rust. Chemically speaking this reaction may be written as:

$$4Fe + 3O_2 + 2H_2O = 2Fe_2O_3H_2O$$

It is necessary to have both air and water present to start the corrosion process at the anode. The process may also be explained in electrochemical terms as follows:

Anodic reaction $\quad Fe \rightarrow Fe^{2+} + 2e^-$
Cathodic reaction $\quad O_2 + 2H_2O + 4e^- \rightarrow 4(OH)^-$
Anodic reaction $\quad Fe^{2+} + 2(OH)^- \rightarrow Fe(OH)_2$
Anodic reaction $\quad 2Fe(OH)_2 + O_2 \rightarrow Fe_2O_3H_2O$

The corrosion cells are microscopic, with anodic and cathodic sites very close to each other. The electrons can flow from anode to cathode if the transfer medium is conductive.

10.2 PROTECTIVE SYSTEM

The way the corrosion protection system is designed to work may be summarised in the following statement of principles:

- The coating should provide a high electrical resistance, making it difficult for the electrons to flow.
- Use of pigments in the coating system affects the anodic or the cathodic reaction. Pigments with zinc act as sacrificial metal at the anodes, whereas the zinc phosphates act as inhibitors at the cathodic sites.
- The coating should be designed to reduce the rate of diffusion of water and oxygen through the coating. The reduction of microporosity of the coating may be achieved by the addition of inert pigments such as micaceous iron oxide (MCO). These materials block the pores in the coating system.

10.3 TYPES OF COATING

(1) *Air drying*: The protective film is formed by the oxidation of oils in the medium.
(2) *Polymerisation*: The protective film is formed by the chemical reaction between a resin and a curing agent. Single pack materials require atmospheric moisture as the curing agent.
(3) *Solvent release*: The protective film is formed by the release of solvent by evaporation.
(4) *Water evaporation*: The protective film is formed by the evaporation of water.

10.4 COMPONENTS OF A CORROSION PROTECTION SYSTEM

- *Primers*: The first coat to be applied on bare steelwork is the primer. This coat is most important as it is responsible for the corrosion inhibition. Primers consist of a pigment in a binder. The pigments most commonly used are metallic zinc and zinc phosphate. The most commonly used binder is an epoxy resin.
- *Barrier coats*: These are applied on the primer coat to increase the film thickness. The barrier coats may have micaceous iron oxide as the pigment, which results in a decrease in the microporosity of the film. The pigment is used with a binder which is generally an alkyd or an epoxy. The barrier coats do not have any corrosion inhibitors or sacrificial metal such as zinc, and hence should not be applied on bare metal.
- *Finishes*: The purpose of the finishing coat is decorative and the pigments are suspended in a media which does not require two-part mixing. The finishing coat may also serve the purpose of promoting run off of surface water by being smooth. A finishing coat with chemical resistance may require two-part mixing.

10.5 SPECIFICATION OF A PROTECTIVE SYSTEM

The following should be specified:

(1) *Surface preparation*: to Swedish Standard SIS 055900 or BS 7079: Part A1.
(2) *Application*: conforming to the Environmental Protection Act, controlling the emission level of solvents by demonstrating that the volatile organic compounds have emission rates below a threshold level.
(3) *Coating thickness*: the coating thickness is specified as a target DFT (dry film thickness). No DFT reading should be less than 75% of the target value.
(4) *Galvanising*: this should generally be specified for all fasteners because they are difficult to paint.

10.6 SAMPLE CORROSION PROTECTION SYSTEMS

Table 10.1 Painting specification for controlled internal environment.

Preparation	Blast clean to Sa 2.5	BS 7079: Part A1
1 primer coat	Oil/resin zinc phosphate	40 μm DFT
2 undercoats	Oil/alkyd with normal pigment	40 μm DFT
1 finishing coat	Oil/alkyd gloss finish	40 μm DFT

Table 10.2 Painting specification for uncontrolled internal environment.

Preparation	Blast clean to Sa 2.5	BS 7079: Part A1
1 primer coat	Epoxy zinc phosphate	50 μm DFT
1 barrier coat	Epoxy micaceous iron oxide	75 μm DFT
2 undercoats	Oil/alkyd with normal pigment	40 μm DFT
1 finishing coat	Oil/alkyd gloss finish	40 μm DFT

Table 10.3 Painting specification for corrosive internal environment, such as in swimming pools.

Preparation	Blast clean to Sa 2.5	BS 7079: Part A1
1 primer coat	Two pack epoxy zinc phosphate	50 μm DFT
1 barrier coat	Two pack epoxy micaceous iron oxide	75 μm DFT
1 undercoat	Acrylated rubber	40 μm DFT
1 finishing coat	Acrylated rubber finishing	25 μm DFT

Table 10.4 Painting specification for external environment.

Preparation	Blast clean to Sa 2.5	BS 7079: Part A1
1 primer coat	Epoxy zinc rich	75 μm DFT
1 barrier coat	Two pack epoxy micaceous iron oxide	75 μm DFT
1 undercoat	Silicone alkyd enamel	35 μm DFT
1 finishing coat	Silicone alkyd enamel	35 μm DFT

Chapter 11
Material Properties

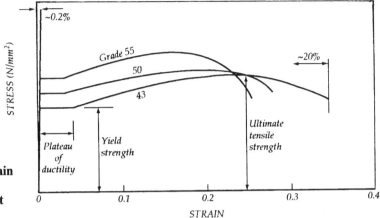

SK 11/1 Stress/strain curves of tensile specimens of different grades of steel.

Modulus of elasticity $E = 205 \, \text{kN/mm}^2$
Poisson's ratio $\nu = 0.30$
Coefficient of linear expansion 12×10^{-6} per °C

Fracture toughness

(1) Brittle fracture shall be considered when the material is under tensile stress.
(2) There is no requirement to check brittle fracture if the service temperature does not fall below −5°C for internal elements and −15°C for external elements.
(3) Limit the thickness in tension elements to control brittle fracture if the expected service temperature range falls outside the stated limits.
(4) Obtain the value of K from Table 3 of BS 5950 and, correspnding to this value of K, find the maximum thickness of tension elements using different grades of steel from Table 4 of BS 5950.
(5) Alternatively, the Charpy impact value C_v in Joules determined from tests at the appropriate service temperature should not be less than:

$$\frac{Y_s t}{710 K}$$

where Y_s = minimum yield strength of the material in N/mm²
t = thickness of material from which the specimen for the test has been taken
K = value obtained from Table 3 of BS 5950

Table 11.1 Influence coefficients for continuous beams.

Span configuration	Inertia ratio	Loaded span	Reaction influence coefficient			Support and	
			R_A	R_B	R_C	M_B	M_C
2-span Continuous	1	AB	+0.438	+0.625	−0.063	−0.0625	−
		BC	−0.063	+0.625	+0.438	−0.0625	−
		All	+0.375	+1.250	+0.375	−0.1250	−
	2	AB	+0.450	+0.600	−0.050	−0.0500	−
		BC	−0.050	+0.600	+0.450	−0.0500	−
		All	+0.400	+1.200	+0.400	−0.1001	−
	3	AB	+0.457	+0.586	−0.043	−0.0430	−
		BC	−0.043	+0.586	+0.457	−0.0430	−
		All	+0.414	+1.172	+0.414	−0.0860	−
3-Span Continuous	1	AB	+0.433	+0.650	−0.100	−0.0667	+0.0167
		BC	−0.050	+0.550	+0.550	−0.0500	−0.0500
		All	+0.400	+1.100	+1.100	−0.1000	−0.1000
	2	AB	+0.448	+0.615	−0.073	−0.0521	+0.0105
		BC	−0.042	+0.542	+0.542	−0.0420	+0.0420
		All	+0.416	+1.084	+1.084	−0.0836	−0.0836
	3	AB	+0.456	+0.596	−0.060	−0.0443	+0.0078
		BC	−0.037	+0.537	+0.537	−0.0372	−0.0372
		All	+0.426	+1.074	+1.074	−0.0738	−0.0738
4-Span Continuous	1	AB	+0.433	+0.652	−0.107	−0.0670	+0.0179
		BC	−0.049	+0.545	+0.571	−0.0491	−0.0536
		All	+0.393	+1.143	+0.929	−0.1071	−0.0714
	2	AB	+0.448	+0.615	−0.076	−0.0522	+0.0110
		BC	−0.042	+0.539	+0.555	−0.0417	−0.0438
		All	+0.413	+1.109	+0.957	−0.0872	−0.0657
	3	AB	+0.456	+0.597	−0.062	−0.0443	+0.0080
		BC	−0.037	+0.536	+0.546	−0.0370	−0.0384
		All	+0.424	+1.091	+0.969	−0.0760	−0.0608

Note: *signifies the maximum value in span due to patterned loading.

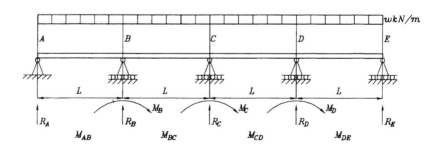

Material Properties

span bending moment influence coefficient					Span deflection influence coefficient			
M_D	M_{AB}	M_{BC}	M_{CD}	M_{DE}	δ_{AB}	δ_{BC}	δ_{CD}	δ_{DE}
–	+0.0956*	–0.0313	–	–	–9.1266*	+3.9063	–	–
–	–0.0313	+0.0956*	–	–	+3.9063	–9.1266*	–	–
–	+0.0675*	+0.0675*	–	–	–5.3883*	–5.3883*	–	–
–	+0.1012*	–0.0250	–	–	–9.8898	+3.3806	–	–
–	–0.0250	+0.1012*	–	–	+3.3806	–9.8898	–	–
–	+0.0787*	+0.0787*	–	–	–6.6479*	–6.6479*	–	–
–	+0.1044*	–0.0215	–	–	–10.3970	+3.1231	–	–
–	–0.0215	+0.1044*	–	–	+3.1231	–10.3970	–	–
–	+0.0850*	+0.0850*	–	–	–7.3860*	–7.3860*	–	–
–	+0.0938*	–0.0250	+0.0083	–	–8.8773*	+3.1250	–1.0417	–
–	–0.0250	+0.0750	–0.0250	–	+3.1250	–6.7710	+3.1250	–
–	+0.0788*	+0.0250	+0.0788*	–	–6.7708	–0.05208	–6.7708	–
–	+0.1003*	–0.0208	+0.0053	–	–9.7502	+2.8323	–0.7113	–
–	–0.0210	+0.0830	–0.0210	–	+2.8893	–7.7943	+2.8393	–
–	+0.0861*	+0.0414	+0.0861*	–	–7.6223	–2.1296	–7.6223	–
–	+0.1038*	–0.0183	+0.0039	–	–10.3044	+2.6938	–0.5629	–
–	–0.0186	+0.0878	–0.0186	–	+2.7049	–8.5257	+2.7049	–
–	+0.0906*	+0.0512	+0.0906*	–	–8.2295*	–3.1380	–8.2295*	–
–0.0045	+0.0936*	–0.0246	+0.0067	–0.0022	–8.8595*	+3.0692	–0.8371	+0.2790
+0.0134	–0.0246	+0.0737	–0.0210	+0.0067	+3.0692	–6.6034	+2.5112	–0.8371
–0.1071	+0.0755*	+0.0357	+0.0357	+0.0755*	–6.4564*	–1.8601	–1.8601	–6.4564*
–0.0022	+0.1003*	–0.0206	+0.0044	–0.0011	–9.7440	+2.8077	–0.5970	+0.1499
+0.0089	–0.0208	+0.0823	–0.0175	+0.0044	+2.8146	–7.6958	+2.3828	–0.5984
–0.0872	+0.0845*	+0.0485	+0.0485	+0.0845*	–7.4756*	–3.1023	–3.1023	–7.4756*
–0.0014	+0.1038*	–0.0182	+0.0033	–0.0007	–10.3014	+2.6794	–0.4860	+0.1016
+0.0067	–0.0185	+0.0873	–0.0158	+0.0034	+2.6904	–8.4565	+2.3352	–0.4880
–0.0760	+0.0895*	+0.0566	+0.0566	+0.0895*	–8.0728*	–3.9278	–3.9278	–8.0728*

$$\text{Inertia ratio} = \frac{\text{moment of inertia of composite beam at midspan}}{\text{moment of inertia of steel beam}}$$

w = uniformly distributed load on the beam (kN/m)
L = span of beam assumed equal (m)
E = modulus of elasticity (kN/mm^2)
I = moment of inertia of section of steel beam (cm^4)

$$\delta = \text{deflection} = \frac{\text{coefficient} \times L^4 \times w}{EI} \times 10^2 \text{ (mm)}$$

M = bending moment at midspan or support = coefficient $\times wL^2$ (kNm)
R = reaction at support = coefficient $\times wL$ (kN)

Table 11.2 Bolt data and capacities.

Diameter of bolt	Clearance hole diameter	Dimension of short slot	Dimension of long slot maximum	Minimum spacing	Tensile stress	Shank area
(mm)	(mm)	(mm)	(mm)	(mm)	(mm^2)	(mm^2)
M12	14	18	30	30	84	113
M16	18	22	40	40	157	201
M20	22	26	50	50	245	314
M22	24	28	55	55	303	380
M24	26	32	60	60	353	452
M27	30	37	67	68	459	573
M30	33	40	75	75	561	707

Table 11.3 Capacities of welds.

		Fillet welds		
Leg length	Throat thickness	E43 electrode for Grade 43 steel @ 215 N/mm^2	E51 electrode for Grade 50 steel @ 255 N/mm^2	E51[a] electrode for Grade 55 steel @ 275 N/mm^2
(mm)	(mm)	(kN/mm)	(kN/mm)	(kN/mm)
3	2.1	0.451	0.535	0.577
4	2.8	0.602	0.714	0.770
5	3.5	0.752	0.892	0.962
6	4.2	0.903	1.071	1.155
8	5.6	1.204	1.428	1.540
10	7.0	1.505	1.785	1.925
12	8.4	1.806	2.142	2.310
15	10.5	2.257	2.677	2.887
18	12.6	2.709	3.213	3.465
20	14.0	3.010	3.570	3.850
22	15.4	3.311	3.927	4.235
25	17.5	3.762	4.462	4.812

Note: The throat thickness in full penetration butt weld is the thickness of connected plies.
The throat thickness for partial penetration butt weld is the minimum depth of penetration, which is depth of penetration minus 3 mm for V- or bevel welds.
Partial penetration butt weld with superimposed fillet weld should be designed as described in Section 9.2.1.

Bolt data and capacities.

Grade 4.6 capacity (kN)			Grade 8.8 capacity (kN)			HSFG bolts preloaded		
Tensile capacity @ 195 N/mm^2	Enhanced tensile capacity @ 280 N/mm^2	Single shear capacity @ 160 N/mm^2	Tensile capacity @ 450 N/mm^2	Enhanced tensile capacity @ 560 N/mm^2	Single shear capacity @ 375 N/mm^2	Proof load (kN)	Tensile capacity (kN)	Slip resistance $K_s = 1$ $\mu = 0.45$
16.4	23.5	13.5	37.9	47	31.6	49.4	44.5	24.5
30.6	44.0	25.1	70.7	87.9	58.9	92.1	82.9	45.5
47.8	68.6	39.2	110.3	137.2	91.9	144	129.6	71.2
59.1	84.8	48.5	136.4	169.7	113.6	177	159.3	87.6
68.8	98.8	56.5	158.9	197.7	132.4	207	186.3	102.4
89.5	128.5	73.4	206.6	257	172.1	234	210.6	115.8
109.4	157.1	89.8	252.5	314.2	210.4	286	257.4	141.5

Capacities of welds.

Butt welds

Plate thickness t (mm)	E43 electrode @ 215 N/mm^2 Shear capacity (kN/mm)	Grade 43 steel Tension or compresssion capacity (kN/mm)	E51 electrode @ 255 N/mm^2 Shear capacity (kN/mm)	Grade 50 steel Tension or compression capacity (kN/mm)	E51^a electrode @ 275 N/mm^2 Shear capacity (kN/mm)	Grade 55 steel Tension or compression capacity (kN/mm)
3	0.495	0.825	0.639	1.065	0.810	1.350
4	0.660	1.100	0.852	1.420	1.080	1.800
5	0.825	1.375	1.065	1.775	1.350	2.250
6	0.990	1.650	1.278	2.130	1.620	2.700
8	1.320	2.200	1.704	2.840	2.160	3.600
10	1.650	2.750	2.130	3.550	2.700	4.500
12	1.980	3.300	2.556	4.260	3.240	5.400
15	2.475	4.125	3.195	5.325	4.050	6.750
18	2.862	4.770	3.726	6.210	4.644	7.740
20	3.180	5.300	4.140	6.900	5.160	8.600
22	3.498	5.830	4.554	7.590	5.676	9.460
25	3.975	6.625	5.175	8.625	6.450	10.750

Shear capacity $= 0.6 t p_y$
Tension capacity $= t p_y$

Structural Steelwork

Table 11.4 Dimensions and properties of steel sections.

UNIVERSAL BEAMS
To BS 4: Part 1: 1993

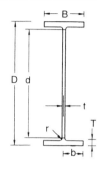

Designation	Mass per metre	Depth of Section D	Width of Section B	Thickness Web t	Thickness Flange T	Root Radius r	Depth between fillets d	Ratios for Local Buckling Flange b/T	Ratios for Local Buckling Web d/t	Second Moment of Area Axis x-x	Second Moment of Area Axis y-y
	kg/m	mm	mm	mm	mm	mm	mm			cm⁴	cm⁴
†1016 x 305 x 487	487·0	1036·1	308·5	30·0	54·1	30·0	867·8	2·85	29·0	1020400	26720
†1016 x 305 x 438	438·0	1025·9	305·4	26·9	49·0	30·0	868·0	3·12	32·3	908900	23440
†1016 x 305 x 393	393·0	1016·0	303·0	24·4	43·9	30·0	868·2	3·45	35·7	806600	20490
†1016 x 305 x 349	349·0	1008·1	302·0	21·1	40·0	30·0	868·0	3·78	41·2	722100	18460
†1016 x 305 x 314	314·0	1000·0	300·0	19·1	35·9	30·0	868·2	4·18	45·6	643200	16230
†1016 x 305 x 272	272·0	990·1	300·0	16·5	31·0	30·0	868·0	4·84	52·7	552900	14000
†1016 x 305 x 249	249·0	980·2	300·0	16·5	26·0	30·0	868·0	5·77	52·7	480300	11750
†1016 x 305 x 222	222·0	970·3	300·0	16·0	21·1	30·0	867·8	7·11	54·4	406900	9544
914 x 419 x 388	388·0	921·0	420·5	21·4	36·6	24·1	799·6	5·74	37·4	719600	45440
914 x 419 x 343	343·3	911·8	418·5	19·4	32·0	24·1	799·6	6·54	41·2	625800	39160
914 x 305 x 289	289·1	926·6	307·7	19·5	32·0	19·1	824·4	4·81	42·3	504200	15600
914 x 305 x 253	253·4	918·4	305·5	17·3	27·9	19·1	824·4	5·47	47·7	436300	13300
914 x 305 x 224	224·2	910·4	304·1	15·9	23·9	19·1	824·4	6·36	51·8	376400	11240
914 x 305 x 201	200·9	903·0	303·3	15·1	20·2	19·1	824·4	7·51	54·6	325300	9423
838 x 292 x 226	226·5	850·9	293·8	16·1	26·8	17·8	761·7	5·48	47·3	339700	11360
838 x 292 x 194	193·8	840·7	292·4	14·7	21·7	17·8	761·7	6·74	51·8	279200	9066
838 x 292 x 176	175·9	834·9	291·7	14·0	18·8	17·8	761·7	7·76	54·4	246000	7799
762 x 267 x 197	196·8	769·8	268·0	15·6	25·4	16·5	686·0	5·28	44·0	240000	8175
762 x 267 x 173	173·0	762·2	266·7	14·3	21·6	16·5	686·0	6·17	48·0	205300	6850
762 x 267 x 147	146·9	754·0	265·2	12·8	17·5	16·5	686·0	7·58	53·6	168500	5455
762 x 267 x 134	133·9	750·0	264·4	12·0	15·5	16·5	686·0	8·53	57·2	150700	4788
686 x 254 x 170	170·2	692·9	255·8	14·5	23·7	15·2	615·1	5·40	42·4	170300	6630
686 x 254 x 152	152·4	687·5	254·5	13·2	21·0	15·2	615·1	6·06	46·6	150400	5784
686 x 254 x 140	140·1	683·5	253·7	12·4	19·0	15·2	615·1	6·68	49·6	136300	5183
686 x 254 x 125	125·2	677·9	253·0	11·7	16·2	15·2	615·1	7·81	52·6	118000	4383
610 x 305 x 238	238·1	635·8	311·4	18·4	31·4	16·5	540·0	4·96	29·3	209500	15840
610 x 305 x 179	179·0	620·2	307·1	14·1	23·6	16·5	540·0	6·51	38·3	153000	11410
610 x 305 x 149	149·1	612·4	304·8	11·8	19·7	16·5	540·0	7·74	45·8	125900	9308
610 x 229 x 140	139·9	617·2	230·2	13·1	22·1	12·7	547·6	5·21	41·8	111800	4505
610 x 229 x 125	125·1	612·2	229·0	11·9	19·6	12·7	547·6	5·84	46·0	98610	3932
610 x 229 x 113	113·0	607·6	228·2	11·1	17·3	12·7	547·6	6·60	49·3	87320	3434
610 x 229 x 101	101·2	602·6	227·6	10·5	14·8	12·7	547·6	7·69	52·2	75780	2915
533 x 210 x 122	122·0	544·5	211·9	12·7	21·3	12·7	476·5	4·97	37·5	76040	3388
533 x 210 x 109	109·0	539·5	210·8	11·6	18·8	12·7	476·5	5·61	41·1	66820	2943
533 x 210 x 101	101·0	536·7	210·0	10·8	17·4	12·7	476·5	6·03	44·1	61520	2692
533 x 210 x 92	92·1	533·1	209·3	10·1	15·6	12·7	476·5	6·71	47·2	55230	2389
533 x 210 x 82	82·2	528·3	208·8	9·6	13·2	12·7	476·5	7·91	49·6	47540	2007
457 x 191 x 98	98·3	467·2	192·8	11·4	19·6	10·2	407·6	4·92	35·8	45730	2347
457 x 191 x 89	89·3	463·4	191·9	10·5	17·7	10·2	407·6	5·42	38·8	41020	2089
457 x 191 x 82	82·0	460·0	191·3	9·9	16·0	10·2	407·6	5·98	41·2	37050	1871
457 x 191 x 74	74·3	457·0	190·4	9·0	14·5	10·2	407·6	6·57	45·3	33320	1671
457 x 191 x 67	67·1	453·4	189·9	8·5	12·7	10·2	407·6	7·48	48·0	29380	1452

† These dimensions are in addition to the standard range to BS4 specifications.

Table 11.4 (contd)

UNIVERSAL BEAMS
To BS 4: Part 1: 1993

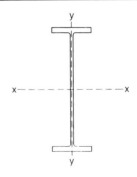

Radius of Gyration		Elastic Modulus		Plastic Modulus		Buckling Parameter u	Torsional Index x	Warping Constant H	Torsional Constant J	Area of Section	Mass per metre	Designation
Axis x-x	Axis y-y	Axis x-x	Axis y-y	Axis x-x	Axis y-y							
cm	cm	cm³	cm³	cm³	cm³			dm⁶	cm⁴	cm²	kg/m	
40·6	6·57	19700	1732	23180	2799	0·867	21·2	64·4	4276	619	487·0	1016 x 305 x 487
40·4	6·49	17720	1535	20740	2467	0·868	23·2	55·9	3166	556	438·0	1016 x 305 x 438
40·2	6·40	15880	1353	18520	2167	0·868	25·6	48·4	2314	500	393·0	1016 x 305 x 393
40·3	6·44	14330	1222	16570	1940	0·872	28·0	43·2	1706	445	349·0	1016 x 305 x 349
40·1	6·37	12860	1082	14830	1712	0·871	30·8	37·7	1253	400	314·0	1016 x 305 x 314
40·0	6·36	11170	934	12800	1469	0·872	35·1	32·2	826	346	272·0	1016 x 305 x 272
39·0	6·09	9799	784	11330	1244	0·861	40·1	26·8	575	316	249·0	1016 x 305 x 249
38·0	5·81	8387	636	9784	1019	0·849	46·0	21·5	384	282	222·0	1016 x 305 x 222
38·2	9·59	15630	2161	17670	3341	0·885	26·7	88·9	1734	494	388·0	914 x 419 x 388
37·8	9·46	13730	1871	15480	2890	0·883	30·1	75·8	1193	437	343·3	914 x 419 x 343
37·0	6·51	10880	1014	12570	1601	0·867	31·9	31·2	926	368	289·1	914 x 305 x 289
36·8	6·42	9501	871	10940	1371	0·866	36·2	26·4	626	323	253·4	914 x 305 x 253
36·3	6·27	8269	739	9535	1163	0·861	41·3	22·1	422	286	224·2	914 x 305 x 224
35·7	6·07	7204	621	8351	982	0·854	46·8	18·4	291	256	200·9	914 x 305 x 201
34·3	6·27	7985	773	9155	1212	0·870	35·0	19·3	514	289	226·5	838 x 292 x 226
33·6	6·06	6641	620	7640	974	0·862	41·6	15·2	306	247	193·8	838 x 292 x 194
33·1	5·90	5893	535	6808	842	0·856	46·5	13·0	221	224	175·9	838 x 292 x 176
30·9	5·71	6234	610	7176	959	0·869	33·2	11·3	404	251	196·8	762 x 267 x 197
30·5	5·58	5387	514	6198	807	0·864	38·1	9·39	267	220	173·0	762 x 267 x 173
30·0	5·40	4470	411	5156	647	0·858	45·2	7·40	159	187	146·9	762 x 267 x 147
29·7	5·30	4018	362	4644	570	0·854	49·8	6·46	119	171	133·9	762 x 267 x 134
28·0	5·53	4916	518	5631	811	0·872	31·8	7·42	308	217	170·2	686 x 254 x 170
27·8	5·46	4374	455	5000	710	0·871	35·5	6·42	220	194	152·4	686 x 254 x 152
27·6	5·39	3987	409	4558	638	0·868	38·7	5·72	169	178	140·1	686 x 254 x 140
27·2	5·24	3481	346	3994	542	0·862	43·9	4·80	116	159	125·2	686 x 254 x 125
26·3	7·23	6589	1017	7486	1574	0·886	21·3	14·5	875	303	238·1	610 x 305 x 238
25·9	7·07	4935	743	5547	1144	0·886	27·7	10·2	340	228	179·0	610 x 305 x 179
25·7	7·00	4111	611	4594	937	0·886	32·7	8·17	200	190	149·1	610 x 305 x 149
25·0	5·03	3622	391	4142	611	0·875	30·6	3·99	216	178	139·9	610 x 229 x 140
24·9	4·97	3221	343	3676	535	0·873	34·1	3·45	154	159	125·1	610 x 229 x 125
24·6	4·88	2874	301	3281	469	0·870	38·0	2·99	111	144	113·0	610 x 229 x 113
24·2	4·75	2515	256	2881	400	0·864	43·1	2·52	77·0	129	101·2	610 x 229 x 101
22·1	4·67	2793	320	3196	500	0·877	27·6	2·32	178	155	122·0	533 x 210 x 122
21·9	4·60	2477	279	2828	436	0·875	30·9	1·99	126	139	109·0	533 x 210 x 109
21·9	4·57	2292	256	2612	399	0·874	33·2	1·81	101	129	101·0	533 x 210 x 101
21·7	4·51	2072	228	2360	356	0·872	36·5	1·60	75·7	117	92·1	533 x 210 x 92
21·3	4·38	1800	192	2059	300	0·864	41·6	1·33	51·5	105	82·2	533 x 210 x 82
19·1	4·33	1957	243	2232	379	0·881	25·7	1·18	121	125	98·3	457 x 191 x 98
19·0	4·29	1770	218	2014	338	0·880	28·3	1·04	90·7	114	89·3	457 x 191 x 89
18·8	4·23	1611	196	1831	304	0·877	30·9	0·922	69·2	104	82·0	457 x 191 x 82
18·8	4·20	1458	176	1653	272	0·877	33·9	0·818	51·8	94·6	74·3	457 x 191 x 14
18·5	4·12	1296	153	1471	237	0·872	37·9	0·705	37·1	85·5	67·1	457 x 191 x 67

Table 11.4 (contd)

UNIVERSAL BEAMS
To BS 4: Part 1: 1993

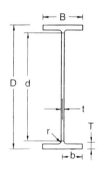

Designation	Mass per metre	Depth of Section D	Width of Section B	Thickness Web t	Thickness Flange T	Root Radius r	Depth between fillets d	Ratios for Local Buckling		Second Moment of Area	
								Flange b/T	Web d/t	Axis x-x	Axis y-y
	kg/m	mm	mm	mm	mm	mm	mm			cm⁴	cm⁴
457 x 152 x 82	82.1	465.8	155.3	10.5	18.9	10.2	407.6	4.11	38.8	36590	1185
457 x 152 x 74	74.2	462.0	154.4	9.6	17.0	10.2	407.6	4.54	42.5	32670	1047
457 x 152 x 67	67.2	458.0	153.8	9.0	15.0	10.2	407.6	5.13	45.3	28930	913
457 x 152 x 60	59.8	454.6	152.9	8.1	13.3	10.2	407.6	5.75	50.3	25500	795
457 x 152 x 52	52.3	449.8	152.4	7.6	10.9	10.2	407.6	6.99	53.6	21370	645
406 x 178 x 74	74.2	412.8	179.5	9.5	16.0	10.2	360.4	5.61	37.9	27310	1545
406 x 178 x 67	67.1	409.4	178.8	8.8	14.3	10.2	360.4	6.25	41.0	24330	1365
406 x 178 x 60	60.1	406.4	177.9	7.9	12.8	10.2	360.4	6.95	45.6	21600	1203
406 x 178 x 54	54.1	402.6	177.7	7.7	10.9	10.2	360.4	8.15	46.8	18720	1021
406 x 140 x 46	46.0	403.2	142.2	6.8	11.2	10.2	360.4	6.35	53.0	15690	538
406 x 140 x 39	39.0	398.0	141.8	6.4	8.6	10.2	360.4	8.24	56.3	12510	410
356 x 171 x 67	67.1	363.4	173.2	9.1	15.7	10.2	311.6	5.52	34.2	19460	1362
356 x 171 x 57	57.0	358.0	172.2	8.1	13.0	10.2	311.6	6.62	38.5	16040	1108
356 x 171 x 51	51.0	355.0	171.5	7.4	11.5	10.2	311.6	7.46	42.1	14140	968
356 x 171 x 45	45.0	351.4	171.1	7.0	9.7	10.2	311.6	8.82	44.5	12070	811
356 x 127 x 39	39.1	353.4	126.0	6.6	10.7	10.2	311.6	5.89	47.2	10170	358
356 x 127 x 33	33.1	349.0	125.4	6.0	8.5	10.2	311.6	7.38	51.9	8249	280
305 x 165 x 54	54.0	310.4	166.9	7.9	13.7	8.9	265.2	6.09	33.6	11700	1063
305 x 165 x 46	46.1	306.6	165.7	6.7	11.8	8.9	265.2	7.02	39.6	9899	896
305 x 165 x 40	40.3	303.4	165.0	6.0	10.2	8.9	265.2	8.09	44.2	8503	764
305 x 127 x 48	48.1	311.0	125.3	9.0	14.0	8.9	265.2	4.47	29.5	9575	461
305 x 127 x 42	41.9	307.2	124.3	8.0	12.1	8.9	265.2	5.14	33.2	8196	389
305 x 127 x 37	37.0	304.4	123.3	7.1	10.7	8.9	265.2	5.77	37.4	7171	336
305 x 102 x 33	32.8	312.7	102.4	6.6	10.0	7.6	275.9	1.74	41.8	6501	194
305 x 102 x 28	28.2	308.7	101.8	6.0	8.8	7.6	275.9	5.78	46.0	5366	155
305 x 102 x 25	24.8	305.1	101.6	5.8	7.0	7.6	275.9	7.26	47.6	4455	123
254 x 146 x 43	43.0	259.6	147.3	7.2	12.7	7.6	219.0	5.80	30.4	6544	677
254 x 146 x 37	37.0	256.0	146.4	6.3	10.9	7.6	219.0	6.72	34.8	5537	571
254 x 146 x 31	31.1	251.4	146.1	6.0	8.6	7.6	219.0	8.49	36.5	4413	448
254 x 102 x 28	28.3	260.4	102.2	6.3	10.0	7.6	225.2	5.11	35.7	4005	179
254 x 102 x 25	25.2	257.2	101.9	6.0	8.4	7.6	225.2	6.07	37.5	3415	149
254 x 102 x 22	22.0	254.0	101.6	5.7	6.8	7.6	225.2	7.47	39.5	2841	119
203 x 133 x 30	30.0	206.8	133.9	6.4	9.6	7.6	172.4	6.97	26.9	2896	385
203 x 133 x 25	25.1	203.2	133.2	5.7	7.8	7.6	172.4	8.54	30.2	2340	308
203 x 102 x 23	23.1	203.2	101.8	5.4	9.3	7.6	169.4	5.47	31.4	2105	164
178 x 102 x 19	19.0	177.8	101.2	4.8	7.9	7.6	146.8	6.41	30.6	1356	137
152 x 89 x 16	16.0	152.4	88.7	4.5	7.7	7.6	121.8	5.76	27.1	834	89.8
127 x 76 x 13	13.0	127.0	76.0	4.0	7.6	7.6	96.6	5.00	24.1	473	55.7

Table 11.4 (contd)

UNIVERSAL BEAMS
To BS 4: Part 1: 1993

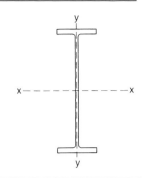

Radius of Gyration		Elastic Modulus		Plastic Modulus		Buckling Parameter u	Torsional Index x	Warping Constant H	Torsional Constant J	Area of Section	Mass per metre	Designation
Axis x-x	Axis y-y	Axis x-x	Axis y-y	Axis x-x	Axis y-y							
cm	cm	cm³	cm³	cm³	cm³			dm⁶	cm⁴	cm²	kg/m	
18·7	3·37	1571	153	1811	240	0·873	27·4	0·591	89·2	105	82·1	457 x 152 x 82
18·6	3·33	1414	136	1627	213	0·873	30·1	0·518	65·9	94·5	74·2	457 x 152 x 74
18·4	3·27	1263	119	1453	187	0·869	33·6	0·448	47·7	85·6	67·2	457 x 152 x 67
18·3	3·23	1122	104	1287	163	0·868	37·5	0·387	33·8	76·2	59·8	457 x 152 x 60
17·9	3·11	950	84·6	1096	133	0·859	43·9	0·311	21·4	66·6	52·3	457 x 152 x 52
17·0	4·04	1323	172	1501	267	0·882	27·6	0·608	62·8	94·5	74·2	406 x 178 x 74
16·9	3·99	1189	153	1346	237	0·880	30·5	0·533	46·1	85·5	67·1	406 x 178 x 67
16·8	3·97	1063	135	1199	209	0·880	33·8	0·466	33·3	76·5	60·1	406 x 178 x 60
16·5	3·85	930	115	1055	178	0·871	38·3	0·392	23·1	69·0	54·1	406 x 178 x 54
16·4	3·03	778	75·7	888	118	0·871	38·9	0·207	19·0	58·6	46·0	406 x 140 x 46
15·9	2·87	629	57·8	724	90·8	0·858	47·5	0·155	10·7	49·7	39·0	406 x 140 x 39
15·1	3·99	1071	157	1211	243	0·886	24·4	0·412	55·7	85·5	67·1	356 x 171 x 67
14·9	3·91	896	129	1010	199	0·882	28·8	0·330	33·4	72·6	57·0	356 x 171 x 57
14·8	3·86	796	113	896	174	0·881	32·1	0·286	23·8	64·9	51·0	356 x 171 x 51
14·5	3·76	687	94·8	775	147	0·874	36·8	0·237	15·8	57·3	45·0	356 x 171 x 45
14·3	2·68	576	56·8	659	89·1	0·871	35·2	0·105	15·1	49·8	39·1	356 x 127 x 39
14·0	2·58	473	44·7	543	70·3	0·863	42·2	0·0812	8·79	42·1	33·1	356 x 127 x 33
13·0	3·93	754	127	846	196	0·889	23·6	0·234	34·8	68·8	54·0	305 x 165 x 54
13·0	3·90	646	108	720	166	0·891	27·1	0·195	22·2	58·7	46·1	305 x 165 x 46
12·9	3·86	560	92·6	623	142	0·889	31·0	0·164	14·7	51·3	40·3	305 x 165 x 40
12·5	2·74	616	73·6	711	116	0·873	23·3	0·102	31·8	61·2	48·1	305 x 127 x 48
12·4	2·70	534	62·6	614	98·4	0·872	26·5	0·0846	21·1	53·4	41·9	305 x 127 x 42
12·3	2·67	471	54·5	539	85·4	0·872	29·7	0·0725	14·8	47·2	37·0	305 x 127 x 37
12·5	2·15	416	37·9	481	60·0	0·866	31·6	0·0442	12·2	41·8	32·8	305 x 102 x 33
12·2	2·08	348	30·5	403	48·5	0·859	37·4	0·0349	7·40	35·9	28·2	305 x 102 x 28
11·9	1·97	292	24·2	342	38·8	0·846	43·4	0·0273	4·77	31·6	24·8	305 x 102 x 25
10·9	3·52	504	92·0	566	141	0·891	21·2	0·103	23·9	54·8	43·0	254 x 146 x 43
10·8	3·48	433	78·0	483	119	0·890	24·3	0·0857	15·3	47·2	37·0	254 x 146 x 37
10·5	3·36	351	61·3	393	94·1	0·880	29·6	0·0660	8·55	39·7	31·1	254 x 146 x 31
10·5	2·22	308	34·9	353	54·8	0·874	27·5	0·0280	9·57	36·1	28·3	254 x 102 x 28
10·3	2·15	266	29·2	306	46·0	0·866	31·5	0·0230	6·42	32·0	25·2	254 x 102 x 25
10·1	2·06	224	23·5	259	37·3	0·856	36·4	0·0182	4·15	28·0	22·0	254 x 102 x 22
8·71	3·17	280	57·5	314	88·2	0·881	21·5	0·0374	10·3	38·2	30·0	203 x 133 x 30
8·56	3·10	230	46·2	258	70·9	0·877	25·6	0·0294	5·96	32·0	25·1	203 x 133 x 25
8·46	2·36	207	32·2	234	49·8	0·888	22·5	0·0154	7·02	29·4	23·1	203 x 102 x 23
7·48	2·37	153	27·0	171	41·6	0·888	22·6	0·00987	4·41	24·3	19·0	178 x 102 x 19
6·41	2·10	109	20·2	123	31·2	0·890	19·6	0·00470	3·56	20·3	16·0	152 x 89 x 16
5·35	1·84	74·6	14·7	84·2	22·6	0·895	16·3	0·00199	2·85	16·5	13·0	127 x 76 x 13

514 Structural Steelwork

Table 11.4 (contd)

UNIVERSAL COLUMNS
To BS 4: Part 1: 1993

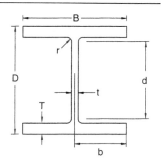

Designation	Mass per metre	Depth of Section D	Width of Section B	Thickness of Web t	Thickness of Flange T	Root Radius r	Depth between Fillets d	Ratios for Local Buckling		Second Moment of Area	
								Flange b/T	Web d/t	Axis x-x	Axis y-y
	kg/m	mm	mm	mm	mm	mm	mm			cm^4	cm^4
*356 x 406 x 634	633.9	474.6	424.0	47.6	77.0	15.2	290.2	2.75	6.10	274800	98130
*356 x 406 x 551	551.0	455.6	418.5	42.1	67.5	15.2	290.2	3.10	6.89	226900	82670
356 x 406 x 467	467.0	436.6	412.2	35.8	58.0	15.2	290.2	3.55	8.11	183000	67830
356 x 406 x 393	393.0	419.0	407.0	30.6	49.2	15.2	290.2	4.14	9.48	146600	55370
356 x 406 x 340	339.9	406.4	403.0	26.6	42.9	15.2	290.2	4.70	10.9	122500	46850
356 x 406 x 287	287.1	393.6	399.0	22.6	36.5	15.2	290.2	5.47	12.8	99880	38680
356 x 406 x 235	235.1	381.0	394.8	18.4	30.2	15.2	290.2	6.54	15.8	79080	30990
356 x 368 x 202	201.9	374.6	374.7	16.5	27.0	15.2	290.2	6.94	17.6	66260	23690
356 x 368 x 177	177.0	368.2	372.6	14.4	23.8	15.2	290.2	7.83	20.2	57120	20530
356 x 368 x 153	152.9	362.0	370.5	12.3	20.7	15.2	290.2	8.95	23.6	48590	17550
356 x 368 x 129	129.0	355.6	368.6	10.4	17.5	15.2	290.2	10.5	27.9	40250	14610
305 x 305 x 283	282.9	365.3	322.2	26.8	44.1	15.2	246.7	3.65	9.21	78870	24630
305 x 305 x 240	240.0	352.5	318.4	23.0	37.7	15.2	246.7	4.22	10.7	64200	20310
305 x 305 x 198	198.1	339.9	314.5	19.1	31.4	15.2	246.7	5.01	12.9	50900	16300
305 x 305 x 158	158.1	327.1	311.2	15.8	25.0	15.2	246.7	6.22	15.6	38750	12570
305 x 305 x 137	136.9	320.5	309.2	13.8	21.7	15.2	246.7	7.12	17.9	32810	10700
305 x 305 x 118	117.9	314.5	307.4	12.0	18.7	15.2	246.7	8.22	20.6	27670	9059
305 x 305 x 97	96.9	307.9	305.3	9.9	15.4	15.2	246.7	9.91	24.9	22250	7308
254 x 254 x 167	167.1	289.1	265.2	19.2	31.7	12.7	200.3	4.18	10.4	30000	9870
254 x 254 x 132	132.0	276.3	261.3	15.3	25.3	12.7	200.3	5.16	13.1	22530	7531
254 x 254 x 107	107.1	266.7	258.8	12.8	20.5	12.7	200.3	6.31	15.6	17510	5928
254 x 254 x 89	88.9	260.3	256.3	10.3	17.3	12.7	200.3	7.41	19.4	14270	4857
254 x 254 x 73	73.1	254.1	254.6	8.6	14.2	12.7	200.3	8.96	23.3	11410	3908
203 x 203 x 86	86.1	222.2	209.1	12.7	20.5	10.2	160.8	5.10	12.7	9449	3127
203 x 203 x 71	71.0	215.8	206.4	10.0	17.3	10.2	160.8	5.97	16.1	7618	2537
203 x 203 x 60	60.0	209.6	205.8	9.4	14.2	10.2	160.8	7.25	17.1	6125	2065
203 x 203 x 52	52.0	206.2	204.3	7.9	12.5	10.2	160.8	8.17	20.4	5259	1778
203 x 203 x 46	46.1	203.2	203.6	7.2	11.0	10.2	160.8	9.25	22.3	4568	1548
152 x 152 x 37	37.0	161.8	154.4	8.0	11.5	7.6	123.6	6.71	15.4	2210	706
152 x 152 x 30	30.0	157.6	152.9	6.5	9.4	7.6	123.6	8.13	19.0	1748	560
152 x 152 x 23	23.0	152.4	152.2	5.8	6.8	7.6	123.6	11.2	21.3	1250	400

*Discuss with supplier.

Table 11.4 (contd)

UNIVERSAL COLUMNS
To BS 4: Part 1: 1993

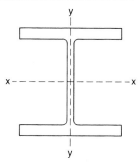

Radius of Gyration		Elastic Modulus		Plastic Modulus		Buckling Parameter	Torsional Index	Warping Constant	Torsional Constant	Area of Section	Mass per metre	Designation
Axis x-x	Axis y-y	Axis x-x	Axis y-y	Axis x-x	Axis y-y	u	x	H	J			
cm	cm	cm^3	cm^3	cm^3	cm^3			dm^6	cm^4	cm^2	kg/m	
18.4	11.0	11580	4629	14240	7108	0.843	5.46	38.8	13720	808	633.9	356 x 406 x 634
18.0	10.9	9962	3951	12080	6058	0.841	6.05	31.1	9240	702	551.0	356 x 406 x 551
17.5	10.7	8383	3291	10000	5034	0.839	6.86	24.3	5809	595	467.0	356 x 406 x 467
17.1	10.5	6998	2721	8222	4154	0.837	7.86	18.9	3545	501	393.0	356 x 406 x 393
16.8	10.4	6031	2325	6999	3544	0.836	8.85	15.5	2343	433	339.9	356 x 406 x 340
16.5	10.3	5075	1939	5812	2949	0.835	10.2	12.3	1441	366	287.1	356 x 406 x 287
16.3	10.2	4151	1570	4687	2383	0.834	12.1	9.54	812	299	235.1	356 x 406 x 235
16.1	9.60	3538	1264	3972	1920	0.844	13.4	7.16	558	257	201.9	356 x 368 x 202
15.9	9.54	3103	1102	3455	1671	0.844	15.0	6.09	381	226	177.0	356 x 368 x 177
15.8	9.49	2684	948	2965	1435	0.844	17.0	5.11	251	195	152.9	356 x 368 x 153
15.6	9.43	2264	793	2479	1199	0.844	19.9	4.18	153	164	129.0	356 x 368 x 129
14.8	8.27	4318	1529	5105	2342	0.855	7.65	6.35	2034	360	282.9	305 x 305 x 283
14.5	8.15	3643	1276	4247	1951	0.854	8.74	5.03	1271	306	240.0	305 x 305 x 240
14.2	8.04	2995	1037	3440	1581	0.854	10.2	3.88	734	252	198.1	305 x 305 x 198
13.9	7.90	2369	808	2680	1230	0.851	12.5	2.87	378	201	158.1	305 x 305 x 158
13.7	7.83	2048	692	2297	1053	0.851	14.2	2.39	249	174	136.9	305 x 305 x 137
13.6	7.77	1760	589	1958	895	0.850	16.2	1.98	161	150	117.9	305 x 305 x 118
13.4	7.69	1445	479	1592	726	0.850	19.3	1.56	91.2	123	96.9	305 x 305 x 97
11.9	6.81	2075	744	2424	1137	0.851	8.49	1.63	626	213	167.1	254 x 254 x 167
11.6	6.69	1631	576	1869	878	0.850	10.3	1.19	319	168	132.0	254 x 254 x 132
11.3	6.59	1313	458	1484	697	0.848	12.4	0.898	172	136	107.1	254 x 254 x 107
11.2	6.55	1096	379	1224	575	0.850	14.5	0.717	102	113	88.9	254 x 254 x 89
11.1	6.48	898	307	992	465	0.849	17.3	0.562	57.6	93.1	73.1	254 x 254 x 73
9.28	5.34	850	299	977	456	0.850	10.2	0.318	137	110	86.1	203 x 203 x 86
9.18	5.30	706	246	799	374	0.853	11.9	0.250	80.2	90.4	71.0	203 x 203 x 71
8.96	5.20	584	201	656	305	0.846	14.1	0.197	47.2	76.4	60.0	203 x 203 x 60
8.91	5.18	510	174	567	264	0.848	15.8	0.167	31.8	66.3	52.0	203 x 203 x 52
8.82	5.13	450	152	497	231	0.847	17.7	0.143	22.2	58.7	46.1	203 x 203 x 46
6.85	3.87	273	91.5	309	140	0.848	13.3	0.0399	19.2	47.1	37.0	152 x 152 x 37
6.76	3.83	222	73.3	248	112	0.849	16.0	0.0308	10.5	38.3	30.0	152 x 152 x 30
6.54	3.70	164	52.6	182	80.2	0.840	20.7	0.0212	4.63	29.2	23.0	152 x 152 x 23

Table 11.4 (contd)

UNIVERSAL BEARING PILES
To BS 4: Part 1: 1993

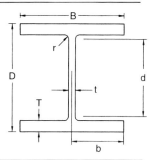

Designation	Mass per metre	Depth of Section D	Width of Section B	Thickness of Web t	Thickness of Flange T	Root Radius r	Depth between Fillets d	Ratios for Local Buckling		Second Moment of Area	
								Flange b/T	Web d/t	Axis x-x	Axis y-y
	kg/m	mm	mm	mm	mm	mm	mm			cm^4	cm^4
356 x 368 x 174	173.9	361.4	378.5	20.3	20.4	15.2	290.2	9.28	14.3	51010	18460
356 x 368 x 152	152.0	356.4	376.0	17.8	17.9	15.2	290.2	10.5	16.3	43970	15880
356 x 368 x 133	133.0	352.0	373.8	15.6	15.7	15.2	290.2	11.9	18.6	37980	13680
356 x 368 x 109	108.9	346.4	371.0	12.8	12.9	15.2	290.2	14.4	22.7	30630	10990
305 x 305 x 223	222.9	337.9	325.7	30.3	30.4	15.2	246.7	5.36	8.14	52700	17580
305 x 305 x 186	186.0	328.3	320.9	25.5	25.6	15.2	246.7	6.27	9.67	42610	14140
305 x 305 x 149	149.1	318.5	316.0	20.6	20.7	15.2	246.7	7.63	12.0	33070	10910
305 x 305 x 126	126.1	312.3	312.9	17.5	17.6	15.2	246.7	8.89	14.1	27410	9002
305 x 305 x 110	110.0	307.9	310.7	15.3	15.4	15.2	246.7	10.1	16.1	23560	7709
305 x 305 x 95	94.9	303.7	308.7	13.3	13.3	15.2	246.7	11.6	18.5	20040	6529
305 x 305 x 88	88.0	301.7	307.8	12.4	12.3	15.2	246.7	12.5	19.9	18420	5984
305 x 305 x 79	78.9	299.3	306.4	11.0	11.1	15.2	246.7	13.8	22.4	16440	5326
254 x 254 x 85	85.1	254.3	260.4	14.4	14.3	12.7	200.3	9.10	13.9	12280	4215
254 x 254 x 71	71.0	249.7	258.0	12.0	12.0	12.7	200.3	10.8	16.7	10070	3439
254 x 254 x 63	63.0	247.1	256.6	10.6	10.7	12.7	200.3	12.0	18.9	8860	3016
203 x 203 x 54	53.9	204.0	207.7	11.3	11.4	10.2	160.8	9.11	14.2	5027	1705
203 x 203 x 45	44.9	200.2	205.9	9.5	9.5	10.2	160.8	10.8	16.9	4100	1384

Table 11.4 (contd)

UNIVERSAL BEARING PILES
To BS 4: Part 1: 1993

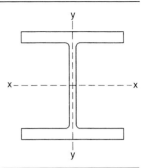

Radius of Gyration		Elastic Modulus		Plastic Modulus		Buckling Parameter	Torsional Index	Warping Constant	Torsional Constant	Area of Section	Mass per metre	Designation
Axis x-x	Axis y-y	Axis x-x	Axis y-y	Axis x-x	Axis y-y	u	x	H	J			
cm	cm	cm^3	cm^3	cm^3	cm^3			dm^6	cm^4	cm^2	kg/m	
15.2	9.13	2823	976	3186	1497	0.821	15.8	5.37	330	221	173.9	**356 x 368 x 174**
15.1	9.05	2468	845	2767	1293	0.821	17.8	4.55	223	194	152.0	**356 x 368 x 152**
15.0	8.99	2158	732	2406	1119	0.822	20.1	3.87	151	169	133.0	**356 x 368 x 133**
14.9	8.90	1769	592	1956	903	0.823	24.2	3.05	84.6	139	108.9	**356 x 368 x 109**
13.6	7.87	3119	1079	3653	1680	0.826	9.55	4.15	943	284	222.9	**305 x 305 x 223**
13.4	7.73	2596	881	3003	1366	0.827	11.1	3.24	560	237	186.0	**305 x 305 x 186**
13.2	7.58	2076	691	2370	1066	0.828	13.5	2.42	295	190	149.1	**305 x 305 x 149**
13.1	7.49	1755	575	1986	885	0.829	15.7	1.95	182	161	126.1	**305 x 305 x 126**
13.0	7.42	1531	496	1720	762	0.830	17.7	1.65	122	140	110.0	**305 x 305 x 110**
12.9	7.35	1320	423	1474	648	0.830	20.2	1.38	80.0	121	94.9	**305 x 305 x 95**
12.8	7.31	1221	389	1360	595	0.830	21.6	1.25	64.2	112	88.0	**305 x 305 x 88**
12.8	7.28	1099	348	1218	531	0.832	23.9	1.11	46.9	100	78.9	**305 x 305 x 79**
10.6	6.24	966	324	1092	498	0.825	15.6	0.607	81.8	108	85.1	**254 x 254 x 85**
10.6	6.17	807	267	904	409	0.826	18.4	0.486	48.4	90.4	71.0	**254 x 254 x 71**
10.5	6.13	717	235	799	360	0.827	20.5	0.421	34.3	80.2	63.0	**254 x 254 x 63**
8.55	4.98	493	164	557	252	0.827	15.8	0.158	32.7	68.7	53.9	**203 x 203 x 54**
8.46	4.92	410	134	459	206	0.827	18.6	0.126	19.2	57.2	44.9	**203 x 203 x 45**

Table 11.4 (contd)

JOISTS
To BS 4: Part 1: 1993

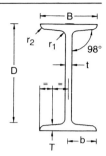

Inside slope 8°

Designation	Mass per metre	Depth of Section D	Width of Section B	Thickness		Radius		Depth between fillets d	Ratios for Local Buckling		Second Moment of Area		Radius of gyration	
				Web t	Flange T	Root r_1	Toe r_2		Flange b/T	Web d/t	Axis x-x	Axis y-y	Axis x-x	Axis y-y
	kg/m	mm	mm	mm	mm	mm	mm	mm			cm⁴	cm⁴	cm	cm
254 x 203 x 82	82.0	254.0	203.2	10.2	19.9	19.6	9.7	166.6	5.11	16.3	12020	2280	10.7	4.67
203 x 152 x 52	52.3	203.2	152.4	8.9	16.5	15.5	7.6	133.2	4.62	15.0	4798	816	8.49	3.50
152 x 127 x 37	37.3	152.4	127.0	10.4	13.2	13.5	6.6	94.3	4.81	9.07	1818	378	6.19	2.82
127 x 114 x 29	29.3	127.0	114.3	10.2	11.5	9.9	4.8	79.5	4.97	7.79	979	242	5.12	2.54
127 x 114 x 27	26.9	127.0	114.3	7.4	11.4	9.9	5.0	79.5	5.01	10.7	946	236	5.26	2.63
102 x 102 x 23	23.0	101.6	101.6	9.5	10.3	11.1	3.2	55.2	4.93	5.81	486	154	4.07	2.29
102 x 44 x 7	7.5	101.6	44.5	4.3	6.1	6.9	3.3	74.6	3.65	17.3	153	7.82	4.01	0.907
89 x 89 x 19	19.5	88.9	88.9	9.5	9.9	11.1	3.2	44.2	4.49	4.65	307	101	3.51	2.02
76 x 76 x 13	12.8	76.2	76.2	5.1	8.4	9.4	4.6	38.1	4.54	7.47	158	51.8	3.12	1.79

PARALLEL FLANGE CHANNELS

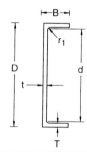

Designation	Mass per metre	Depth of Section D	Width of Section B	Thickness		Distance of Cy	Radius Root r_1	Depth between Fillets d	Ratios for local buckling		Second Moment of Area	
				Web t	Flange T				Flange b/T	Web d/t	Axis x-x	Axis y-y
	kg/m	mm	mm	mm	mm	cm	mm	mm			cm⁴	cm⁴
430 x 100 x 64	64.4	430	100	11.0	19.0	2.62	15	362	5.26	32.9	21940	722
380 x 100 x 54	54.0	380	100	9.5	17.5	2.79	15	315	5.71	33.2	15030	643
300 x 100 x 46	45.5	300	100	9.0	16.5	3.05	15	237	6.06	26.3	8229	568
300 x 90 x 41	41.4	300	90	9.0	15.5	2.60	12	245	5.81	27.2	7218	404
260 x 90 x 35	34.8	260	90	8.0	14.0	2.74	12	208	6.43	26.0	4728	353
260 x 75 x 28	27.6	260	75	7.0	12.0	2.10	12	212	6.25	30.3	3619	185
230 x 90 x 32	32.2	230	90	7.5	14.0	2.92	12	178	6.43	23.7	3518	334
230 x 75 x 26	25.7	230	75	6.5	12.5	2.30	12	181	6.00	27.8	2748	181
200 x 90 x 30	29.7	200	90	7.0	14.0	3.12	12	148	6.43	21.1	2523	314
200 x 75 x 23	23.4	200	75	6.0	12.5	2.48	12	151	6.00	25.2	1963	170
180 x 90 x 26	26.1	180	90	6.5	12.5	3.17	12	131	7.20	20.2	1817	277
180 x 75 x 20	20.3	180	75	6.0	10.5	2.41	12	135	7.14	22.5	1370	146
150 x 90 x 24	23.9	150	90	6.5	12.0	3.30	12	102	7.50	15.7	1162	253
150 x 75 x 18	17.9	150	75	5.5	10.0	2.58	12	106	7.50	19.3	861	131
125 x 65 x 15	14.8	125	65	5.5	9.5	2.25	12	82	6.84	14.9	483	80.0
100 x 50 x 10	10.2	100	50	5.0	8.5	1.73	9	65	5.88	13.0	208	32.3

Table 11.4 (contd)

JOISTS
To BS 4: Part 1: 1993

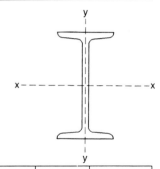

Elastic modulus		Plastic modulus		Buckling Parameter	Torsional Index	Warping Constant	Torsional Constant	Area of Section	Mass per metre	Designation
Axis x-x	Axis y-y	Axis x-x	Axis y-y	u	x	H	J	A		
cm^3	cm^3	cm^3	cm^3			dm^6	cm^4	cm^2	kg/m	
947	224	1077	371	0.890	11.0	0.312	152	105	82.0	**254 x 203 x 82**
472	107	541	176	0.891	10.7	0.0711	64.8	66.6	52.3	**203 x 152 x 52**
239	59.6	279	99.8	0.866	9.33	0.0183	33.9	47.5	37.3	**152 x 127 x 37**
154	42.3	181	70.8	0.853	8.76	0.00807	20.8	37.4	29.3	**127 x 114 x 29**
149	41.3	172	68.2	0.868	9.32	0.00788	16.9	34.2	27.1	**127 x 114 x 27**
95.6	30.3	113	50.6	0.836	7.43	0.00321	14.2	29.3	23.0	**102 x 102 x 23**
30.1	3.51	35.4	6.03	0.872	14.9	0.000178	1.25	9.50	7.5	**102 x 44 x 7**
69.0	22.8	82.7	38.0	0.830	6.57	0.00158	11.5	24.9	19.5	**89 x 89 x 19**
41.5	13.6	48.7	22.4	0.852	7.22	0.000595	4.59	16.2	12.8	**76 x 76 x 13**

PARALLEL FLANGE CHANNELS

Radius of Gyration		Elastic modulus		Plastic modulus		Buckling Parameter	Torsional Index	Warping constant	Torsional constant	Area of Section	Mass per metre	Designation
Axis x-x	Axis y-y	Axis x-x	Axis y-y	Axis x-x	Axis y-y	u	x	H	J	A		
cm	cm	cm^3	cm^3	cm^3	cm^3			dm^6	cm^4	cm^2	kg/m	
16.3	2.97	1020	97.9	1222	176	0.917	22.5	0.219	63.0	82.1	64.4	**430 x 100 x 64**
14.8	3.06	791	89.2	933	161	0.932	21.2	0.150	45.7	68.7	54.0	**380 x 100 x 54**
11.9	3.13	549	81.7	641	148	0.944	17.0	0.0813	36.8	58.0	45.5	**300 x 100 x 46**
11.7	2.77	481	63.1	568	114	0.934	18.4	0.0581	28.8	52.7	41.4	**300 x 90 x 41**
10.3	2.82	364	56.3	425	102	0.942	17.2	0.0379	20.6	44.4	34.8	**260 x 90 x 35**
10.1	2.30	278	34.4	328	62.0	0.932	20.5	0.0203	11.7	35.1	27.6	**260 x 75 x 28**
9.27	2.86	306	55.0	355	98.9	0.950	15.1	0.0279	19.3	41.0	32.2	**230 x 90 x 32**
9.17	2.35	239	34.8	278	63.2	0.947	17.3	0.0153	11.8	32.7	25.7	**230 x 75 x 26**
8.16	2.88	252	53.4	291	94.5	0.954	12.9	0.0197	18.3	37.9	29.7	**200 x 90 x 30**
8.11	2.39	196	33.8	227	60.6	0.956	14.8	0.0107	11.1	29.9	23.4	**200 x 75 x 23**
7.40	2.89	202	47.4	232	83.5	0.949	12.8	0.0141	13.3	33.2	26.1	**180 x 90 x 26**
7.27	2.38	152	28.8	176	51.8	0.946	15.3	0.00754	7.34	25.9	20.3	**180 x 75 x 20**
6.18	2.89	155	44.4	179	76.9	0.936	10.8	0.00890	11.8	30.4	23.9	**150 x 90 x 24**
6.15	2.40	115	26.6	132	47.2	0.946	13.1	0.00467	6.10	22.8	17.9	**150 x 75 x 18**
5.07	2.06	77.3	18.8	89.9	33.2	0.942	11.1	0.00194	4.72	18.8	14.8	**125 x 65 x 15**
4.00	1.58	41.5	9.89	48.9	17.5	0.942	10.0	0.000491	2.53	13.0	10.2	**100 x 50 x 10**

Table 11.4 (contd)

EQUAL ANGLES
To BS 4848: Part 4: 1972

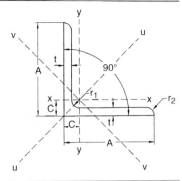

Designation	Mass per metre	Radius		Area of Section	Distance of centre of gravity C_x and C_y
		Root r_1	Toe r_2		
	kg/m	mm	mm	cm²	cm
200 x 200 x 24	71.1	18.0	4.8	90.8	5.85
200 x 200 x 20	59.9	18.0	4.8	76.6	5.70
200 x 200 x 18	54.2	18.0	4.8	69.4	5.62
200 x 200 x 16	48.5	18.0	4.8	62.0	5.54
150 x 150 x 18	40.1	16.0	4.8	51.2	4.38
150 x 150 x 15	33.8	16.0	4.8	43.2	4.26
150 x 150 x 12	27.3	16.0	4.8	35.0	4.14
150 x 150 x 10	23.0	16.0	4.8	29.5	4.06
120 x 120 x 15	26.6	13.0	4.8	34.0	3.52
120 x 120 x 12	21.6	13.0	4.8	27.6	3.41
120 x 120 x 10	18.2	13.0	4.8	23.3	3.32
120 x 120 x 8	14.7	13.0	4.8	18.8	3.24
100 x 100 x 15	21.9	12.0	4.8	28.0	3.02
100 x 100 x 12	17.8	12.0	4.8	22.8	2.91
100 x 100 x 10	15.0	12.0	4.8	19.2	2.83
100 x 100 x 8	12.2	12.0	4.8	15.6	2.75
90 x 90 x 12	15.9	11.0	4.8	20.3	2.66
90 x 90 x 10	13.4	11.0	4.8	17.2	2.58
90 x 90 x 8	10.9	11.0	4.8	13.9	2.50
90 x 90 x 7	9.61	11.0	4.8	12.3	2.46
90 x 90 x 6	8.30	11.0	4.8	10.6	2.41

Material Properties

Table 11.4 (contd)

EQUAL ANGLES
To BS 4848: Part 4: 1972

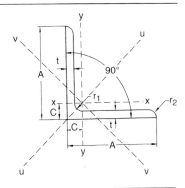

Second Moment of Area			Radiation of gyration			Elastic modulus	
Axis x-x, y-y	Axis u-u Max.	Axis v-v Min.	Axis x-x, y-y	Axis u-u Max.	Axis v-v Min.	Axis x-x, y-y	Designation
cm^4	cm^4	cm^4	cm	cm	cm	cm^3	
3356	5332	1391	6.08	7.65	3.91	237	200 x 200 x 24
2877	4569	1185	6.13	7.72	3.93	201	200 x 200 x 20
2627	4174	1080	6.15	7.76	3.95	183	200 x 200 x 18
2369	3765	973	6.18	7.79	3.96	164	200 x 200 x 16
1060	1680	440	4.55	5.73	2.93	99.8	150 x 150 x 18
909	1442	375	4.59	5.78	2.95	84.6	150 x 150 x 15
748	1187	308	4.62	5.82	2.97	68.9	150 x 150 x 12
635	1008	262	4.64	5.85	2.99	58.0	150 x 150 x 10
448	710	186	3.63	4.57	2.34	52.8	120 x 120 x 15
371	589	153	3.66	4.62	2.35	43.1	120 x 120 x 12
316	502	130	3.69	4.65	2.37	36.4	120 x 120 x 10
259	411	107	3.71	4.67	2.38	29.5	120 x 120 x 8
250	395	105	2.99	3.76	1.94	35.8	100 x 100 x 15
208	330	86.5	3.02	3.81	1.95	29.4	100 x 100 x 12
178	283	73.7	3.05	3.84	1.96	24.8	100 x 100 x 10
146	232	60.5	3.07	3.86	1.97	20.2	100 x 100 x 8
149	235	62.0	2.71	3.40	1.75	23.5	90 x 90 x 12
128	202	52.9	2.73	3.43	1.76	19.9	90 x 90 x 10
105	167	43.4	2.75	3.46	1.77	16.2	90 x 90 x 8
93.2	148	38.6	2.76	3.47	1.77	14.3	90 x 90 x 7
81.0	128	33.6	2.76	3.48	1.78	12.3	90 x 90 x 6

Table 11.4 (contd)

UNEQUAL ANGLES
To BS 4848: Part 4: 1972

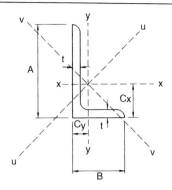

Designation	Mass per Metre	Radius		Area of Section	Distance of Centre of Gravity	
		Root r_1	Toe r_2		Cx	Cy
	kg/m	mm	mm	cm²	cm	cm
200 x 150 x 18	47.1	15.0	4.8	60.1	6.34	3.86
200 x 150 x 15	39.6	15.0	4.8	50.6	6.22	3.75
200 x 150 x 12	32.0	15.0	4.8	40.9	6.10	3.36
200 x 100 x 15	33.7	15.0	4.8	43.1	7.17	2.23
200 x 100 x 12	27.3	15.0	4.8	34.9	7.04	2.11
200 x 100 x 10	23.0	15.0	4.8	29.4	6.95	2.03
150 x 90 x 15	26.6	12.0	4.8	34.0	5.21	2.24
150 x 90 x 12	21.6	12.0	4.8	27.6	5.09	2.12
150 x 90 x 10	18.2	12.0	4.8	23.2	5.00	2.04
150 x 75 x 15	24.8	11.0	4.8	31.7	5.53	1.81
150 x 75 x 12	20.2	11.0	4.8	25.7	5.41	1.70
150 x 75 x 10	17.0	11.0	4.8	21.7	5.32	1.62
125 x 75 x 12	17.8	11.0	4.8	22.7	4.31	1.84
125 x 75 x 10	15.0	11.0	4.8	19.2	4.23	1.76
125 x 75 x 8	12.2	11.0	4.8	15.5	4.14	1.68
100 x 75 x 12	15.4	10.0	4.8	19.7	3.27	2.03
100 x 75 x 10	13.0	10.0	4.8	16.6	3.19	1.96
100 x 75 x 8	10.6	10.0	4.8	13.5	3.11	1.88
100 x 65 x 10	12.3	10.0	4.8	15.6	3.36	1.64
100 x 65 x 8	9.94	10.0	4.8	12.7	3.28	1.56
100 x 65 x 7	8.77	10.0	4.8	11.2	3.23	1.51

Table 11.4 (contd)

UNEQUAL ANGLES
To BS 4848: Part 4: 1972

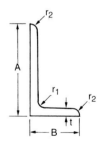

Second Moment of Area				Radiation of Gyration				Elastic Modules		
Axis x-x	Axis y-y	Axis u-u Max.	Axis v-v Min.	Axis x-x	Axis y-y	Axis x-x Max.	Axis y-y Min.	Axis x-x	Axis y-y	Designation
cm⁴	cm⁴	cm⁴	cm⁴	cm	cm	cm	cm	cm³	cm³	
2390	1155	2922	623	6.30	4.38	6.97	3.22	175	104	**200 x 150 x 18**
2037	989	2495	531	6.34	4.42	7.02	3.24	148	87.8	**200 x 150 x 15**
1667	812	2045	435	6.38	4.46	7.07	3.26	120	71.4	**200 x 150 x 12**
1772	303	1879	197	6.41	2.65	6.60	2.13	138	39.0	**200 x 100 x 15**
1454	252	1544	162	6.45	2.68	6.65	2.15	112	31.9	**200 x 100 x 12**
1233	215	1310	138	6.48	2.70	6.68	2.17	94.5	26.9	**200 x 100 x 10**
764	207	844	127	4.74	2.47	4.99	1.93	78.0	30.6	**150 x 90 x 15**
630	172	698	105	4.78	2.50	5.03	1.95	63.6	25.0	**150 x 90 x 12**
536	147	595	89.1	4.81	2.52	5.06	1.96	53.6	21.2	**150 x 90 x 10**
715	120	756	79.2	4.75	1.95	4.89	1.58	75.5	21.1	**150 x 75 x 15**
591	100	626	65.2	4.79	1.98	4.93	1.59	61.6	17.3	**150 x 75 x 12**
503	86.3	534	55.7	4.82	2.00	4.96	1.60	52.0	14.7	**150 x 75 x 10**
355	96.0	392	58.8	3.95	2.06	4.15	1.61	43.4	17.0	**125 x 75 x 12**
303	82.5	336	50.2	3.98	2.08	4.19	1.62	36.7	14.4	**125 x 75 x 10**
249	68.1	275	41.2	4.00	2.10	4.21	1.63	29.7	11.7	**125 x 75 x 8**
189	90.3	230	49.5	3.10	2.14	3.42	1.59	28.1	16.5	**100 x 75 x 12**
162	77.7	198	42.2	3.12	2.16	3.45	1.60	23.8	14.0	**100 x 75 x 10**
133	64.2	163	34.7	3.15	2.18	3.48	1.61	19.3	11.4	**100 x 75 x 8**
154	51.1	175	30.2	3.14	1.81	3.35	1.39	23.2	10.5	**100 x 65 x 10**
127	42.3	145	24.9	3.17	1.83	3.38	1.40	18.9	8.6	**100 x 65 x 8**
113	37.7	128	22.1	3.18	1.84	3.39	1.41	16.6	7.6	**100 x 65 x 7**

Table 11.4 (contd)

STRUCTURAL TEES SPLIT FROM UNIVERSAL BEAMS
To BS 4: Part 1: 1993

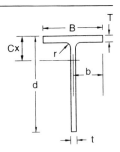

Properties have been calculated assuming that there is no loss of material due to splitting.

Designation	Mass per metre	Cut from universal beam	Section		Thickness		Radius	Ratios For Local Buckling		Dimension
			Width B	Depth d	Web t	Flange T	root r	d/t	b/T	Cx
	kg/m		mm	mm	mm	mm	mm			cm
254 x 343 x 63	62.6	686 x 254 x 125	253.0	338.9	11.7	16.2	15.2	29.0	7.81	8.8
305 x 305 x 90	89.5	610 x 305 x 179	307.1	310.0	14.1	23.6	16.5	22.0	6.51	6.7
305 x 305 x 75	74.6	610 x 305 x 149	304.8	306.1	11.8	19.7	16.5	25.9	7.74	6.4
229 x 305 x 70	69.9	610 x 229 x 140	230.2	308.5	13.1	22.1	12.7	23.5	5.21	7.6
229 x 305 x 63	62.5	610 x 229 x 125	229.0	306.0	11.9	19.6	12.7	25.7	5.84	7.5
229 x 305 x 57	56.5	610 x 229 x 113	228.2	303.7	11.1	17.3	12.7	27.4	6.60	7.6
229 x 305 x 51	50.6	610 x 229 x 101	227.6	301.2	10.5	14.8	12.7	28.7	7.69	7.8
210 x 267 x 61	61.0	533 x 210 x 122	211.9	272.2	12.7	21.3	12.7	21.4	4.97	6.7
210 x 267 x 55	54.5	533 x 210 x 109	210.8	269.7	11.6	18.8	12.7	23.2	5.61	6.6
210 x 267 x 51	50.5	533 x 210 x 101	210.0	268.3	10.8	17.4	12.7	24.8	6.03	6.5
210 x 267 x 46	46.1	533 x 210 x 92	209.3	266.5	10.1	15.6	12.7	26.4	6.71	6.5
210 x 267 x 41	41.1	533 x 210 x 82	208.8	264.1	9.6	13.2	12.7	27.5	7.91	6.7
191 x 229 x 49	49.2	457 x 191 x 98	192.8	233.5	11.4	19.6	10.2	20.5	4.92	5.5
191 x 229 x 45	44.6	457 x 191 x 89	191.9	231.6	10.5	17.7	10.2	22.1	5.42	5.5
191 x 229 x 41	41.0	457 x 191 x 82	191.3	229.9	9.9	16.0	10.2	23.2	5.98	5.5
191 x 229 x 37	37.1	457 x 191 x 74	190.4	228.4	9.0	14.5	10.2	25.4	6.57	5.4
191 x 229 x 34	33.6	457 x 191 x 67	189.9	226.6	8.5	12.7	10.2	26.7	7.48	5.5
152 x 229 x 41	41.0	457 x 152 x 82	155.3	232.8	10.5	18.9	10.2	22.2	4.11	6.0
152 x 229 x 37	37.1	457 x 152 x 74	154.4	230.9	9.6	17.0	10.2	24.1	4.54	5.9
152 x 229 x 34	33.6	457 x 152 x 67	153.8	228.9	9.0	15.0	10.2	25.4	5.13	5.9
152 x 229 x 30	29.9	457 x 152 x 60	152.9	227.2	8.1	13.3	10.2	28.0	5.75	5.8
152 x 229 x 26	26.2	457 x 152 x 52	152.4	224.8	7.6	10.9	10.2	29.6	6.99	6.0

SIZES. These pages list the structural tees produced by rotary splitter from British Steel plc's range of universal beams. A number of the heavier sections cannot be split in this way but certain of these may be supplied by profile burning. Please verify with British Steel.

QUALITIES. Tees provided from beams 610 x 229 and above can only be supplied to Grade S275. All other sizes listed can be supplied in Grades S355 J2 G3 and S355 J2 G4 (Grades as published in BS EN 10 025:1993).

LENGTHS. Lengths above 18.3 metres may be supplied by negotiation.

SUPPLY. For prices and delivery times please refer to British Steel.

Table 11.4 (contd)

STRUCTURAL TEES SPLIT FROM UNIVERSAL BEAMS
To BS 4: Part 1: 1993

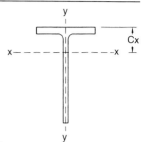

Properties have been calculated assuming that there is no loss of material due to splitting.

Second Moment of Area		Radius of Gyration		Elastic Modulus			Plastic Modulus		Buckling Parameter u	Torsional Index x	Torsional Constant J	Area A	Designation
Axis x-x	Axis y-y	Axis x-x	Axis y-y	Axis x-x		Axis y-y	Axis x-x	Axis y-y					
				Flange	Toe								
cm^4	cm^4	cm	cm	cm^3	cm^3	cm^3	cm^3	cm^3			cm^4	cm^2	
8976	2191	10.6	5.24	1014	358	173	643	271	0.651	22.0	57.9	79.7	**254 x 343 x 63**
9043	5704	8.91	7.07	1353	372	371	656	572	0.484	13.8	170	114	**305 x 305 x 90**
7415	4654	8.83	7.00	1150	307	305	538	469	0.483	16.4	99.8	95.0	**305 x 305 x 75**
7741	2253	9.32	5.03	1017	333	196	592	306	0.613	15.3	108	89.1	**229 x 305 x 70**
6898	1966	9.31	4.97	915	299	172	531	268	0.617	17.1	76.9	79.7	**229 x 305 x 63**
6265	1717	9.33	4.88	826	275	150	489	235	0.627	19.0	55.5	72.0	**229 x 305 x 57**
5691	1457	9.40	4.76	732	255	128	456	200	0.644	21.6	38.3	64.4	**229 x 305 x 51**
5161	1694	8.15	4.67	775	251	160	446	250	0.600	13.8	88.9	77.7	**210 x 267 x 61**
4604	1471	8.14	4.60	697	226	140	401	218	0.605	15.5	63.0	69.4	**210 x 267 x 55**
4245	1346	8.12	4.57	650	209	128	371	200	0.606	16.6	50.3	64.3	**210 x 267 x 51**
3885	1195	8.14	4.51	593	193	114	343	178	0.613	18.3	37.7	58.7	**210 x 267 x 46**
3527	1004	8.21	4.38	523	179	96.1	320	150	0.634	20.8	25.7	52.3	**210 x 267 x 41**
2967	1174	6.88	4.33	536	167	122	296	189	0.573	12.9	60.5	62.6	**191 x 229 x 49**
2684	1045	6.87	4.29	491	152	109	269	169	0.576	14.1	45.2	56.9	**191 x 229 x 45**
2474	935	6.88	4.23	452	141	97.8	250	152	0.583	15.5	34.5	52.2	**191 x 229 x 41**
2224	836	6.86	4.20	413	127	87.8	225	136	0.583	16.9	25.8	47.3	**191 x 229 x 37**
2034	726	6.90	4.12	372	118	76.5	209	119	0.597	18.9	18.5	42.7	**191 x 229 x 34**
2596	592	7.05	3.37	436	150	76.3	267	120	0.634	13.7	44.5	52.3	**152 x 229 x 41**
2332	523	7.03	3.33	397	135	67.8	242	107	0.636	15.1	32.9	47.2	**152 x 229 x 37**
2121	456	7.04	3.27	359	125	59.3	223	93.3	0.646	16.8	23.8	42.8	**152 x 229 x 34**
1879	397	7.02	3.23	322	111	52.0	199	81.5	0.648	18.8	16.9	38.1	**152 x 229 x 30**
1670	322	7.08	3.11	276	102	42.3	183	66.7	0.671	22.0	10.7	33.3	**152 x 229 x 26**

Table 11.4 (contd)

STRUCTURAL TEES SPLIT FROM UNIVERSAL BEAMS
To BS 4: Part 1: 1993

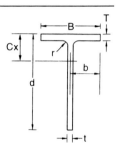

Properties have been calculated assuming that there is no loss of material due to splitting.

Designation	Mass per metre	Cut from universal beam	Section		Thickness		Radius	Ratios For Local Buckling		Dimension
			Width B	Depth d	Web t	Flange T	root r	d/t	b/T	C_x
	kg/m		mm	mm	mm	mm	mm			cm
178 x 203 x 37	37.1	406 x 178 x 74	179.5	206.3	9.5	16.0	10.2	21.7	5.61	4.8
178 x 203 x 34	33.6	406 x 178 x 67	178.8	204.6	8.8	14.3	10.2	23.2	6.25	4.7
178 x 203 x 30	30.0	406 x 178 x 60	177.9	203.1	7.9	12.8	10.2	25.7	6.95	4.6
178 x 203 x 27	27.1	406 x 178 x 54	177.7	201.2	7.7	10.9	10.2	26.1	8.15	4.8
140 x 203 x 23	23.0	406 x 140 x 46	142.2	201.5	6.8	11.2	10.2	29.6	6.35	5.0
140 x 203 x 20	19.5	406 x 140 x 39	141.8	198.9	6.4	8.6	10.2	31.1	8.24	5.3
171 x 178 x 34	33.5	356 x 171 x 67	173.2	181.6	9.1	15.7	10.2	20.0	5.52	4.0
171 x 178 x 29	28.5	356 x 171 x 57	172.2	178.9	8.1	13.0	10.2	22.1	6.62	4.0
171 x 178 x 26	25.5	356 x 171 x 51	171.5	177.4	7.4	11.5	10.2	24.0	7.46	3.9
171 x 178 x 23	22.5	356 x 171 x 45	171.1	175.6	7.0	9.7	10.2	25.1	8.82	4.1
127 x 178 x 20	19.5	356 x 127 x 39	126.0	176.6	6.6	10.7	10.2	26.8	5.89	4.4
127 x 178 x 17	16.5	356 x 127 x 33	125.4	174.4	6.0	8.5	10.2	29.1	7.38	4.6
165 x 152 x 27	27.0	305 x 165 x 54	166.9	155.1	7.9	13.7	8.9	19.6	6.09	3.2
165 x 152 x 23	23.1	305 x 165 x 46	165.7	153.2	6.7	11.8	8.9	22.9	7.02	3.1
165 x 152 x 20	20.1	305 x 165 x 40	165.0	151.6	6.0	10.2	8.9	25.3	8.09	3.0
127 x 152 x 24	24.0	305 x 127 x 48	125.3	155.4	9.0	14.0	8.9	17.3	4.48	3.9
127 x 152 x 21	21.0	305 x 127 x 42	124.3	153.5	8.0	12.1	8.9	19.2	5.14	3.9
127 x 152 x 19	18.5	305 x 127 x 37	123.4	152.1	7.1	10.7	8.9	21.4	5.77	3.8
102 x 152 x 17	16.4	305 x 102 x 33	102.4	156.3	6.6	10.8	7.6	23.7	4.74	4.1
102 x 152 x 14	14.1	305 x 102 x 28	101.8	154.3	6.0	8.8	7.6	25.7	5.78	4.2
102 x 152 x 13	12.4	305 x 102 x 25	101.6	152.5	5.8	7.0	7.6	26.3	7.26	4.4
146 x 127 x 22	21.5	254 x 146 x 43	147.3	129.7	7.2	12.7	7.6	18.0	5.80	2.6
146 x 127 x 19	18.5	254 x 146 x 37	146.4	127.9	6.3	10.9	7.6	20.3	6.72	2.6
146 x 127 x 16	15.6	254 x 146 x 31	146.1	125.6	6.0	8.6	7.6	20.9	8.49	2.7
102 x 127 x 14	14.2	254 x 102 x 28	102.2	130.1	6.3	10.0	7.6	20.7	5.11	3.2
102 x 127 x 13	12.6	254 x 102 x 25	101.9	128.5	6.0	8.4	7.6	21.4	6.07	3.3
102 x 127 x 11	11.0	254 x 102 x 22	101.6	126.9	5.7	6.8	7.6	22.3	7.47	3.5
133 x 102 x 15	15.0	203 x 133 x 30	133.9	103.3	6.4	9.6	7.6	16.1	6.97	2.1
133 x 102 x 13	12.5	203 x 133 x 25	133.2	101.5	5.7	7.8	7.6	17.8	8.54	2.1

SIZES. These pages list the structural tees produced by rotary splitter from British Steel plc's range of universal beams. A number of the heavier sections cannot be split in this way but certain of these may be supplied by profile burning. Please verify with British Steel.

QUALITIES. Tees provided from beams 610 x 229 and above can only be supplied to Grade S275. All other sizes listed can be supplied in Grades S355 J2 G3 and S355 J2 G4 (Grades as published in BS EN 10 025:1993).

LENGTHS. Lengths above 18.3 metres may be supplied by negotiation.

SUPPLY. For prices and delivery times please refer to British Steel.

Table 11.4 (contd)

STRUCTURAL TEES SPLIT FROM UNIVERSAL BEAMS
To BS 4: Part 1: 1993

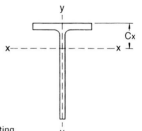

Properties have been calculated assuming that there is no loss of material due to splitting.

Second Moment of Area		Radius of Gyration		Elastic Modulus			Plastic Modulus		Buckling Parameter u	Torsional Index x	Torsional Constant J	Area A	Designation
Axis x-x	Axis y-y	Axis x-x	Axis y-y	Axis x-x		Axis y-y	Axis x-x	Axis y-y					
				Flange	Toe								
cm^4	cm^4	cm	cm	cm^3	cm^3	cm^3	cm^3	cm^3			cm^4	cm^2	
1736	773	6.06	4.04	365	109	86.1	194	133	0.555	13.8	31.3	47.2	178 x 203 x 37
1573	682	6.07	3.99	332	100	76.3	177	118	0.561	15.2	23.0	42.8	178 x 203 x 34
1395	602	6.04	3.97	301	89.0	67.6	157	105	0.561	16.9	16.6	38.3	178 x 203 x 30
1294	511	6.13	3.85	268	84.6	57.5	150	89.1	0.588	19.2	11.5	34.5	178 x 203 x 27
1123	269	6.19	3.03	224	74.2	37.8	132	59.1	0.633	19.5	9.49	29.3	140 x 203 x 23
979	205	6.28	2.87	184	67.2	28.9	121	45.4	0.668	23.8	5.33	24.8	140 x 203 x 20
1154	681	5.20	3.99	288	81.5	78.6	145	121	0.500	12.2	27.8	42.7	171 x 178 x 34
986	554	5.21	3.91	248	70.9	64.4	125	99.4	0.515	14.4	16.6	36.3	171 x 178 x 29
882	484	5.21	3.86	224	63.9	56.5	113	87.1	0.521	16.1	11.9	32.4	171 x 178 x 26
798	406	5.28	3.76	197	59.1	47.4	104	73.3	0.546	18.4	7.90	28.7	171 x 178 x 23
728	179	5.41	2.68	164	55.0	28.4	98.1	44.5	0.632	17.6	7.53	24.9	127 x 178 x 20
626	140	5.45	2.58	137	48.6	22.3	87.2	35.1	0.655	21.1	4.38	21.1	127 x 178 x 17
642	531	4.32	3.93	200	52.2	63.7	92.8	97.8	0.389	11.8	17.3	34.4	165 x 152 x 27
536	448	4.27	3.91	174	43.7	54.1	77.1	82.8	0.380	13.6	11.1	29.4	165 x 152 x 23
468	382	4.27	3.86	155	38.6	46.3	67.7	70.9	0.393	15.5	7.35	25.7	165 x 152 x 20
662	231	4.65	2.74	168	57.1	36.8	102	58.0	0.602	11.7	15.8	30.6	127 x 152 x 24
573	194	4.63	2.70	148	49.9	31.3	88.9	49.2	0.606	13.3	10.5	26.7	127 x 152 x 21
501	168	4.61	2.67	132	43.8	27.2	77.9	42.7	0.606	14.9	7.36	23.6	127 x 152 x 19
487	97.1	4.82	2.15	118	42.3	19.0	75.8	30.0	0.656	15.8	6.08	20.9	102 x 152 x 17
420	77.7	4.84	2.08	100	37.4	15.3	67.5	24.2	0.673	18.7	3.69	17.9	102 x 152 x 14
377	61.5	4.88	1.97	85.0	34.8	12.1	63.3	19.4	0.702	21.8	2.37	15.8	102 x 152 x 13
343	339	3.54	3.52	130	33.2	46.0	59.5	70.5	0.202	10.6	11.9	27.4	146 x 127 x 22
292	285	3.52	3.48	115	28.5	39.0	50.7	59.7	0.233	12.2	7.65	23.6	146 x 127 x 19
259	224	3.61	3.36	97.4	26.2	30.6	46.0	47.1	0.376	14.8	4.26	19.8	146 x 127 x 16
277	89.3	3.92	2.22	85.5	28.3	17.5	50.4	27.4	0.607	13.8	4.77	18.0	102 x 127 x 14
250	74.3	3.95	2.15	75.3	26.2	14.6	46.9	23.0	0.628	15.8	3.20	16.0	102 x 127 x 13
223	59.7	3.99	2.06	64.5	24.1	11.7	43.5	18.6	0.656	18.2	2.06	14.0	102 x 127 x 11
154	192	2.84	3.17	73.1	18.8	28.7	33.5	44.1	–	–	5.13	19.1	133 x 102 x 15
131	154	2.86	3.10	62.4	16.2	23.1	28.7	35.5	–	–	2.97	16.0	133 x 102 x 13

– indicates that no u and x are given as there is no possibility of lateral torsional buckling due to bending about the x-x axis because the second moment of area about y-y axis exceeds the second moment of area about x-x axis.

Table 11.4 (contd)

STRUCTURAL TEES SPLIT FROM UNIVERSAL COLUMNS
To BS 4: Part 1: 1993

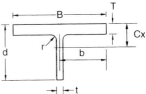

Properties have been calculated assuming that there is no loss of material due to splitting.

Designation	Mass per metre	Cut from universal column	Section		Thickness		Radius	Ratios For Local Buckling		Dimension
			Width B	Depth d	Web t	Flange T	root r	d/t	b/T	C_x
	kg/m		mm	mm	mm	mm	mm			cm
305 x 152 x 79	79.0	305 x 305 x 158	311.2	163.5	15.8	25.0	15.2	10.3	6.22	3.0
305 x 152 x 69	68.5	305 x 305 x 137	309.2	160.2	13.8	21.7	15.2	11.6	7.12	2.9
305 x 152 x 59	58.9	305 x 305 x 118	307.4	157.2	12.0	18.7	15.2	13.1	8.22	2.7
305 x 152 x 49	48.4	305 x 305 x 97	305.3	153.9	9.9	15.4	15.2	15.5	9.91	2.5
254 x 127 x 66	66.0	254 x 254 x 132	261.3	138.1	15.3	25.3	12.7	9.03	5.16	2.7
254 x 127 x 54	53.5	254 x 254 x 107	258.8	133.3	12.8	20.5	12.7	10.4	6.31	2.4
254 x 127 x 45	44.5	254 x 254 x 89	256.3	130.1	10.3	17.3	12.7	12.6	7.41	2.2
254 x 127 x 37	36.5	254 x 254 x 73	254.6	127.0	8.6	14.2	12.7	14.8	8.96	2.0
203 x 102 x 43	43.0	203 x 203 x 86	209.1	111.0	12.7	20.5	10.2	8.74	5.10	2.2
203 x 102 x 36	35.5	203 x 203 x 71	206.4	107.8	10.0	17.3	10.2	10.8	5.97	2.0
203 x 102 x 30	30.0	203 x 203 x 60	205.8	104.7	9.4	14.2	10.2	11.1	7.25	1.9
203 x 102 x 26	26.0	203 x 203 x 52	204.3	103.0	7.9	12.5	10.2	13.0	8.17	1.7
203 x 102 x 23	23.0	203 x 203 x 46	203.6	101.5	7.2	11.0	10.2	14.1	9.25	1.7
152 x 76 x 19	18.5	152 x 152 x 37	154.4	80.8	8.0	11.5	7.6	10.1	6.71	1.5
152 x 76 x 15	15.0	152 x 152 x 30	152.9	78.7	6.5	9.4	7.6	12.1	8.13	1.4
152 x 76 x 12	11.5	152 x 152 x 23	152.2	76.1	5.8	6.8	7.6	13.1	11.2	1.4

SIZES. These pages list the structural tees produced by rotary splitter from British Steel plc's range of universal beams. A number of the heavier sections cannot be split in this way but certain of these may be supplied by profile burning. Please verify with British Steel.

QUALITIES. Tees provided from beams 610 x 229 and above can only be supplied to Grade S275. All other sizes listed can be supplied in Grades S355 J2 G3 and S355 J2 G4 (Grades as published in BS EN 10 025:1993).

LENGTHS. Lengths above 18.3 metres may be supplied by negotiation.

SUPPLY. For prices and delivery times please refer to British Steel.

Table 11.4 (contd)

STRUCTURAL TEES SPLIT FROM UNIVERSAL COLUMNS
To BS 4: Part 1: 1993

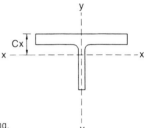

Properties have been calculated assuming that there is no loss of material due to splitting.

Second Moment of Area		Radius of Gyration		Elastic Modulus			Plastic Modulus		Torsional Constant J	Area A	Designation
Axis x-x	Axis y-y	Axis x-x	Axis y-y	Axis x-x		Axis y-y	Axis x-x	Axis y-y			
				Flange	Toe						
cm^4	cm^4	cm	cm	cm^3	cm^3	cm^3	cm^3	cm^3	cm^4	cm^2	
1532	6285	3.90	7.90	503	115	404	225	615	188	101	**305 x 152 x 79**
1286	5350	3.84	7.83	450	97.7	346	188	526	124	87.2	**305 x 152 x 69**
1079	4530	3.79	7.77	401	82.8	295	156	448	80.3	75.1	**305 x 152 x 59**
858	3654	3.73	7.69	343	66.5	239	123	363	45.5	61.7	**305 x 152 x 49**
871	3766	3.22	6.69	323	78.3	288	159	439	159	84.1	**254 x 127 x 66**
676	2964	3.15	6.59	276	62.1	229	122	349	85.9	68.2	**254 x 127 x 54**
524	2429	3.04	6.55	237	48.5	190	94.1	288	51.1	56.7	**254 x 127 x 45**
417	1954	2.99	6.48	204	39.2	153	74.1	233	28.8	46.5	**254 x 127 x 37**
373	1564	2.61	5.34	169	41.9	150	84.6	228	68.1	54.8	**203 x 102 x 43**
280	1269	2.49	5.30	143	31.8	123	63.6	187	40.0	45.2	**203 x 102 x 36**
244	1032	2.53	5.20	129	28.4	100	54.4	153	23.5	38.2	**203 x 102 x 30**
200	889	2.46	5.18	115	23.4	87.0	44.5	132	15.8	33.1	**203 x 102 x 26**
177	774	2.45	5.13	105	20.9	76.0	39.0	115	11.0	29.4	**203 x 102 x 23**
93.1	353	1.99	3.87	60.7	14.2	45.7	27.1	69.8	9.54	23.5	**152 x 76 x 19**
72.2	280	1.94	3.83	51.4	11.2	36.7	20.9	55.8	5.24	19.1	**152 x 76 x 15**
58.5	200	2.00	3.70	41.9	9.41	26.3	16.9	40.1	2.30	14.6	**152 x 76 x 12**

Table 11.5 Dimensions for detailing and surface areas.

UNIVERSAL BEAMS
To BS 4: Part 1: 1993

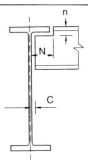

Designation	End Clearance C mm	Notch N mm	Notch n mm	Surface Area per metre m²	Surface Area per tonne m²	Surface Area two end faces m²
1016 x 305 x 487	17	150	84	3.20	6.56	0.124
1016 x 305 x 438	15	150	78	3.17	7.24	0.111
1016 x 305 x 393	14	150	74	3.15	8.00	0.100
1016 x 305 x 349	13	152	70	3.13	8.97	0.089
1016 x 305 x 314	12	152	66	3.11	9.91	0.080
1016 x 305 x 272	10	152	60	3.10	11.4	0.069
1016 x 305 x 249	10	152	56	3.08	12.4	0.064
1016 x 305 x 222	10	152	52	3.06	13.8	0.056
914 x 419 x 388	13	210	62	3.44	8.87	0.0988
914 x 419 x 343	12	210	58	3.42	9.95	0.0875
914 x 305 x 289	12	156	52	3.01	10.4	0.0737
914 x 305 x 253	11	156	48	2.99	11.8	0.0646
914 x 305 x 224	10	156	44	2.97	13.3	0.0571
914 x 305 x 201	10	156	40	2.96	14.7	0.0512
838 x 292 x 226	10	150	46	2.81	12.4	0.0577
838 x 292 x 194	9	150	40	2.79	14.4	0.0494
838 x 292 x 176	9	150	38	2.78	15.8	0.0448
762 x 267 x 197	10	138	42	2.55	13.0	0.0501
762 x 267 x 173	9	138	40	2.53	14.6	0.0441
762 x 267 x 147	8	138	34	2.51	17.1	0.0374
762 x 267 x 134	8	138	32	2.51	18.7	0.0341
686 x 254 x 170	9	132	40	2.35	13.8	0.0434
686 x 254 x 152	9	132	38	2.34	15.4	0.0388
686 x 254 x 140	8	132	36	2.33	16.6	0.0357
686 x 254 x 125	8	132	32	2.32	18.5	0.0319
610 x 305 x 238	11	158	48	2.45	10.3	0.0607
610 x 305 x 179	9	158	42	2.41	13.5	0.0456
610 x 305 x 149	8	158	38	2.39	16.0	0.0380
610 x 229 x 140	9	120	36	2.11	15.1	0.0356
610 x 229 x 125	8	120	34	2.09	16.7	0.0319
610 x 229 x 113	8	120	30	2.08	18.4	0.0288
610 x 229 x 101	7	120	28	2.07	20.5	0.0258
533 x 210 x 122	8	110	34	1.89	15.5	0.0311
533 x 210 x 109	8	110	32	1.88	17.2	0.0278
533 x 210 x 101	7	110	32	1.87	18.5	0.0257
533 x 210 x 92	7	110	30	1.86	20.2	0.0235
533 x 210 x 82	7	110	26	1.85	22.5	0.0209
457 x 191 x 98	8	102	30	1.67	16.9	0.0251
457 x 191 x 89	7	102	28	1.66	18.5	0.0228
457 x 191 x 82	7	102	28	1.65	20.1	0.0209
457 x 191 x 74	7	102	26	1.64	22.1	0.0189
457 x 191 x 67	6	102	24	1.63	24.3	0.0171

The dimension $n = (D-d)/2$ to the nearest 2 mm above. The dimension $C = t/2 + 2$ mm to the nearest 1 mm.

The dimension N is based upon the outstand from web face to flange edge + 10 mm, to the nearest 2 mm above and makes due allowance for rolling tolerance.

Table 11.5 (contd)

UNIVERSAL BEAMS
To BS 4: Part 1: 1993

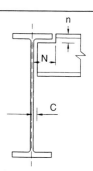

Designation	End Clearance C	Notch N	Notch n	Surface Area per metre	Surface Area per tonne	Surface Area two end faces
	mm	mm	mm	m²	m²	m²
457 x 152 x 82	7	84	30	1.51	18.4	0.0209
457 x 152 x 74	7	84	28	1.50	20.3	0.0189
457 x 152 x 67	7	84	26	1.50	22.3	0.0171
457 x 152 x 60	6	84	24	1.50	24.9	0.0152
457 x 152 x 52	6	84	22	1.48	28.2	0.0133
406 x 178 x 74	7	96	28	1.51	20.3	0.0189
406 x 178 x 67	6	96	26	1.50	22.3	0.0171
406 x 178 x 60	6	96	24	1.49	24.8	0.0153
406 x 178 x 54	6	96	22	1.48	27.4	0.0138
406 x 140 x 46	5	78	22	1.34	29.2	0.0117
406 x 140 x 39	5	78	20	1.33	34.2	0.0099
356 x 171 x 67	7	94	26	1.38	20.6	0.0171
356 x 171 x 57	6	94	24	1.37	24.1	0.0145
356 x 171 x 51	6	94	22	1.36	26.7	0.0130
356 x 171 x 45	6	94	20	1.36	30.1	0.0115
356 x 127 x 39	5	70	22	1.18	30.2	0.0100
356 x 127 x 33	5	70	20	1.17	35.4	0.0084
305 x 165 x 54	6	90	24	1.26	23.3	0.0138
305 x 165 x 46	5	90	22	1.25	27.1	0.0117
305 x 165 x 40	5	90	20	1.24	30.8	0.0103
305 x 127 x 48	7	70	24	1.09	22.7	0.0122
305 x 127 x 42	6	70	22	1.08	25.8	0.0107
305 x 127 x 37	6	70	20	1.07	29.0	0.0094
305 x 102 x 33	5	58	20	1.01	30.8	0.0084
305 x 102 x 28	5	58	18	1.00	35.4	0.0072
305 x 102 x 25	5	58	16	0.992	40.0	0.0063
254 x 146 x 43	6	82	22	1.08	25.1	0.0110
254 x 146 x 37	5	82	20	1.07	29.0	0.0094
254 x 146 x 31	5	82	18	1.06	34.2	0.0079
254 x 102 x 28	5	58	18	0.904	31.9	0.0072
254 x 102 x 25	5	58	16	0.897	35.6	0.0064
254 x 102 x 22	5	58	16	0.890	40.5	0.0056
203 x 133 x 30	5	74	18	0.923	30.8	0.0076
203 x 133 x 25	5	74	16	0.915	36.4	0.0064
203 x 102 x 23	5	60	18	0.790	34.2	0.0059
178 x 102 x 19	4	60	16	0.738	38.8	0.0049
152 x 89 x 16	4	54	16	0.638	39.8	0.0041
127 x 76 x 13	4	46	16	0.537	41.3	0.0033

The dimension $n = (D-d)/2$ to the nearest 2 mm above. The dimension $C = t/2 + 2$ mm to the nearest 1 mm.

The dimension N is based upon the outstand from web face to flange edge + 10 mm, to the nearest 2 mm above and makes due allowance for rolling tolerance.

Table 11.5 (contd)

UNIVERSAL COLUMNS
To BS 4: Part 1: 1993

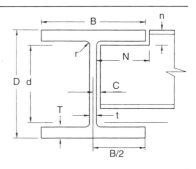

Designation	End Clearance C	Notch N	Notch n	Surface Area per metre	Surface Area per tonne	two end faces
	mm	mm	mm	m²	m²	m²
356 x 406 x 634	26	200	94	2.52	3.98	0.1615
356 x 406 x 551	23	200	84	2.47	4.49	0.1404
356 x 406 x 467	20	200	74	2.42	5.19	0.1190
356 x 406 x 393	17	200	66	2.38	6.05	0.1001
356 x 406 x 340	15	200	60	2.35	6.90	0.0866
356 x 406 x 287	13	200	52	2.31	8.05	0.0731
356 x 406 x 235	11	200	46	2.28	9.69	0.0599
356 x 368 x 202	10	190	44	2.19	10.8	0.0514
356 x 368 x 177	9	190	40	2.17	12.3	0.0451
356 x 368 x 153	8	190	36	2.16	14.1	0.0390
356 x 368 x 129	7	190	34	2.14	16.6	0.0329
305 x 305 x 283	15	158	60	1.94	6.86	0.0721
305 x 305 x 240	14	158	54	1.91	7.94	0.0612
305 x 305 x 198	12	158	48	1.87	9.46	0.0505
305 x 305 x 158	10	158	42	1.84	11.6	0.0403
305 x 305 x 137	9	158	38	1.82	13.3	0.0349
305 x 305 x 118	8	158	34	1.81	15.3	0.0300
305 x 305 x 97	7	158	32	1.79	18.5	0.0247
254 x 254 x 167	12	134	46	1.58	9.45	0.0426
254 x 254 x 132	10	134	38	1.55	11.7	0.0336
254 x 254 x 107	8	134	34	1.52	14.2	0.0273
254 x 254 x 89	7	134	30	1.50	16.9	0.0227
254 x 254 x 73	6	134	28	1.49	20.4	0.0186
203 x 203 x 86	8	110	32	1.24	14.4	0.0219
203 x 203 x 71	7	110	28	1.22	17.2	0.0181
203 x 203 x 60	7	110	26	1.21	20.1	0.0153
203 x 203 x 52	6	110	24	1.20	23.0	0.0133
203 x 203 x 46	6	110	22	1.19	25.8	0.0117
152 x 152 x 37	6	84	20	0.912	24.7	0.0094
152 x 152 x 30	5	84	18	0.901	30.0	0.0077
152 x 152 x 23	5	84	16	0.889	38.7	0.0058

The dimension $n = (D-d)/2$ to the nearest 2 mm above. The dimension $C = t/2 + 2$ mm to the nearest 1 mm.

The dimension N is based upon the outstand from web face to flange edge + 10 mm, to the nearest 2 mm above and makes due allowance for rolling tolerance.

Table 11.5 (contd)

JOISTS
To BS 4: Part 1: 1993

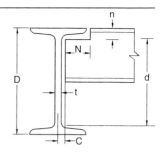

Designation	End Clearance	Notch		Surface Area		
	C	N	n	per metre	per tonne	two end faces
	mm	mm	mm	m²	m²	m²
254 x 203 x 82	7	104	44	1.21	14.8	0.021
203 x 152 x 52	6	78	36	0.932	17.9	0.013
157 x 127 x 37	7	66	30	0.737	19.8	0.009
127 x 114 x 29	7	60	24	0.646	21.7	0.009
127 x 114 x 27	6	60	24	0.650	24.3	0.007
102 x 102 x 23	7	54	24	0.549	23.8	0.006
102 x 44 x 7	4	28	14	0.350	46.6	0.002
89 x 89 x 19	7	46	24	0.476	24.6	0.005
76 x 76 x 13	5	42	20	0.411	32.5	0.003

The dimension N is equal to the outstand from web face edge + 6 mm to nearest 2 mm above.
The dimension $n = (D-d)/2$ to the nearest 2 mm above.

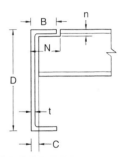

PARALLEL FLANGE CHANNELS

Designation	End Clearance	Notch		Surface Area	
	C	N	n	per metre	per tonne
	mm	mm	mm	m²	m²
430 x 100 x 64	13	96	36	1.23	19.0
380 x 100 x 54	12	98	34	1.13	20.9
300 x 100 x 46	11	98	32	0.969	21.3
300 x 90 x 41	11	88	28	0.932	22.5
260 x 90 x 35	10	88	28	0.854	24.5
260 x 75 x 28	9	74	26	0.796	28.8
230 x 90 x 32	10	90	28	0.795	24.7
230 x 75 x 26	9	76	26	0.737	28.7
200 x 90 x 30	9	90	28	0.736	24.8
200 x 75 x 23	8	76	26	0.678	28.9
180 x 90 x 26	9	90	26	0.697	26.7
180 x 75 x 20	8	76	24	0.638	31.4
150 x 90 x 24	9	90	26	0.637	26.7
150 x 75 x 18	8	76	24	0.579	32.4
125 x 65 x 15	8	66	22	0.489	33.1
100 x 50 x 10	7	52	18	0.382	37.5

The dimension N is equal to $(B-t) + 6$ mm (rounded up to a multiple of 2 mm).
The dimension C is equal to $t + 2$ mm (rounded up to the nearest mm).
The dimension n is equal to $(D-d)/2$ (taken to the next higher multiple of 2 mm)

Table 11.5 (contd)

EQUAL AND UNEQUAL ANGLES
To BS 4848: Part 4: 1972

EQUAL ANGLES

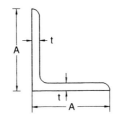

UNEQUAL ANGLES

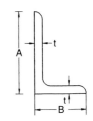

Designation	Surface Area	
	per metre	per tonne
A x A x t	m²	m²
200 x 200 x 24	0.79	11.05
200 x 200 x 20	0.79	13.11
200 x 200 x 18	0.79	14.48
200 x 200 x 16	0.79	16.18
150 x 150 x 18	0.59	14.65
150 x 150 x 15	0.59	17.37
150 x 150 x 12	0.59	21.43
150 x 150 x 10	0.59	25.48
120 x 120 x 15	0.47	17.61
120 x 120 x 12	0.47	21.69
120 x 120 x 10	0.47	25.75
120 x 120 x 8	0.47	31.83
100 x 100 x 15	0.39	17.80
100 x 100 x 12	0.39	21.86
100 x 100 x 10	0.39	25.91
100 x 100 x 8	0.39	31.97
90 x 90 x 12	0.35	22.01
90 x 90 x 10	0.35	26.07
90 x 90 x 8	0.35	32.13
90 x 90 x 7	0.35	36.54
90 x 90 x 6	0.35	42.20

Designation	Surface Area	
	per metre	per tonne
A x A x t	m²	m²
200 x 150 x 18	0.69	14.60
200 x 150 x 15	0.69	17.35
200 x 150 x 12	0.69	21.45
200 x 100 x 15	0.59	17.41
200 x 100 x 12	0.59	21.49
200 x 100 x 10	0.59	25.55
150 x 90 x 15	0.47	17.66
150 x 90 x 12	0.47	21.75
150 x 90 x 10	0.47	25.84
150 x 75 x 15	0.44	17.75
150 x 75 x 12	0.44	21.85
150 x 75 x 10	0.44	25.94
125 x 75 x 12	0.40	21.93
125 x 75 x 10	0.40	26.01
125 x 75 x 8	0.40	32.10
100 x 75 x 12	0.34	22.12
100 x 75 x 10	0.34	26.19
100 x 75 x 8	0.34	32.29
100 x 65 x 10	0.32	26.23
100 x 65 x 8	0.32	32.32
100 x 65 x 7	0.32	36.66

Table 11.6 Hp/A values.

UNIVERSAL BEAMS
To BS 4: Part 1: 1993

Designation	Hp/A ratio (m^{-1})			
	Profile 3 sides	Profile 4 sides	Box 3 sides	Box 4 sides
1016 x 305 x 487	45	50	40	45
1016 x 305 x 438	50	55	40	50
1016 x 305 x 393	55	65	45	55
1016 x 305 x 349	65	70	50	60
1016 x 305 x 314	70	80	55	65
1016 x 305 x 272	80	90	65	75
1016 x 305 x 249	90	95	70	80
1016 x 305 x 222	95	110	80	90
914 x 419 x 388	60	70	45	55
914 x 419 x 343	70	80	50	60
914 x 305 x 289	75	85	60	65
914 x 305 x 253	85	95	65	75
914 x 305 x 224	95	105	75	85
914 x 305 x 201	105	115	80	95
838 x 292 x 226	90	100	70	80
838 x 292 x 194	100	115	80	90
838 x 292 x 176	110	125	90	100
762 x 267 x 197	90	100	70	85
762 x 267 x 173	105	115	80	95
762 x 267 x 147	120	135	95	110
762 x 267 x 134	135	150	105	120
686 x 254 x 170	100	110	75	90
686 x 254 x 152	110	120	85	95
686 x 254 x 140	120	130	90	105
686 x 254 x 125	130	150	100	115
610 x 305 x 238	70	80	50	65
610 x 305 x 179	95	105	70	80
610 x 305 x 149	110	125	80	95
610 x 229 x 140	105	120	80	95
610 x 229 x 125	120	135	90	105
610 x 229 x 113	130	145	100	115
610 x 229 x 101	145	160	110	130
533 x 210 x 122	110	125	85	95
533 x 210 x 109	120	135	95	110
533 x 210 x 101	130	145	100	115
533 x 210 x 92	145	160	110	125
533 x 210 x 82	160	180	120	140
457 x 191 x 98	120	135	90	105
457 x 191 x 89	130	145	100	115
457 x 191 x 82	140	160	105	125
457 x 191 x 74	155	175	115	135
457 x 191 x 67	170	195	130	150

Table 11.6 (contd)

UNIVERSAL BEAMS
To BS 4: Part 1: 1993

Designation	Hp/A ratio (m^{-1})			
	Profile 3 sides	Profile 4 sides	Box 3 sides	Box 4 sides
457 x 152 x 82	130	145	105	120
457 x 152 x 74	145	160	115	130
457 x 152 x 67	160	175	125	145
457 x 152 x 60	175	195	140	160
457 x 152 x 52	200	225	160	180
406 x 178 x 74	140	160	105	125
406 x 178 x 67	155	175	115	140
406 x 178 x 60	175	195	130	155
406 x 178 x 54	190	220	145	170
406 x 140 x 46	210	230	160	185
406 x 140 x 39	245	270	190	215
356 x 171 x 67	145	165	105	125
356 x 171 x 57	170	190	120	145
356 x 171 x 51	185	215	135	160
356 x 171 x 45	210	240	150	180
356 x 127 x 39	215	240	165	195
356 x 127 x 33	250	280	195	225
305 x 165 x 54	160	185	115	140
305 x 165 x 46	185	215	135	160
305 x 165 x 40	210	245	150	185
305 x 127 x 48	160	180	120	145
305 x 127 x 42	180	205	140	160
305 x 127 x 37	205	230	155	180
305 x 102 x 33	220	245	175	200
305 x 102 x 28	255	280	200	230
305 x 102 x 25	285	320	225	255
254 x 146 x 43	175	200	120	150
254 x 146 x 37	200	230	140	170
254 x 146 x 31	235	270	165	200
254 x 102 x 28	225	255	175	200
254 x 102 x 25	250	285	190	225
254 x 102 x 22	285	320	220	255
203 x 133 x 30	210	245	145	180
203 x 133 x 25	250	290	170	210
203 x 102 x 23	240	275	175	205
178 x 102 x 19	270	310	190	230
152 x 89 x 16	275	320	195	235
127 x 76 x 13	285	335	200	245

Table 11.6 (contd)

UNIVERSAL COLUMNS
To BS 4: Part 1: 1993

Designation	Hp/A ratio (m⁻¹)			
	Profile 3 sides	Profile 4 sides	Box 3 sides	Box 4 sides
356 x 406 x 634	25	30	15	20
356 x 406 x 551	30	35	20	25
356 x 406 x 467	35	40	20	30
356 x 406 x 393	40	50	25	35
356 x 406 x 340	45	55	30	35
356 x 406 x 287	55	65	30	45
356 x 406 x 235	65	75	40	50
356 x 368 x 202	70	85	45	60
356 x 368 x 177	80	100	50	65
356 x 368 x 153	95	110	55	75
368 x 368 x 129	110	130	65	90
305 x 305 x 283	45	55	30	40
305 x 305 x 240	55	65	35	45
305 x 305 x 198	65	75	40	50
305 x 305 x 158	75	95	50	65
305 x 305 x 137	90	105	55	70
305 x 305 x 118	100	120	60	85
305 x 305 x 97	120	145	75	100
254 x 254 x 167	65	75	40	50
254 x 254 x 132	80	95	50	65
254 x 254 x 107	95	115	60	75
254 x 254 x 89	110	135	70	90
254 x 254 x 73	135	160	80	110
203 x 203 x 86	95	115	60	80
203 x 203 x 71	115	135	70	95
203 x 203 x 60	135	160	80	110
203 x 203 x 52	150	185	95	125
203 x 203 x 46	170	205	105	140
152 x 152 x 37	165	190	100	135
152 x 152 x 30	200	240	120	160
152 x 152 x 23	255	310	155	210

Table 11.6 (contd)

JOISTS
To BS 4: Part 1: 1993

Designation	Hp/A ratio (m⁻¹)			
	Profile 3 sides	Profile 4 sides	Box 3 sides	Box 4 sides
254 x 203 x 82	95	115	70	90
203 x 152 x 52	115	140	85	105
152 x 127 x 37	130	155	90	120
127 x 114 x 29	140	175	100	130
127 x 114 x 27	155	190	110	140
102 x 102 x 23	150	185	105	140
102 x 44 x 7	320	365	260	305
89 x 89 x 19	155	190	105	145
76 x 76 x 13	205	250	140	185

PARALLEL FLANGE CHANNELS

Designation	Hp/A ratio (m⁻¹)*							
	Profile 3 sides	Profile 3 sides	Profile 3 sides	Profile 4 sides	Box 3 sides	Box 3 sides	Box 3 sides	Box 4 sides
430 x 100 x 64	135	95	75	150	115	75	75	130
380 x 100 x 54	150	110	85	165	125	85	85	140
300 x 100 x 46	150	115	85	165	120	85	85	140
300 x 90 x 41	160	120	90	175	130	90	90	150
260 x 90 x 35	170	135	100	190	135	100	100	160
260 x 75 x 28	205	150	115	225	170	115	115	190
230 x 90 x 32	170	140	100	195	135	100	100	155
230 x 75 x 26	200	155	115	225	165	115	115	185
200 x 90 x 30	170	140	100	195	130	100	100	155
200 x 75 x 23	200	160	115	225	160	115	115	185
180 x 90 x 26	185	155	110	210	135	110	110	165
180 x 75 x 20	215	175	125	245	170	125	125	195
150 x 90 x 24	180	160	110	210	130	110	110	160
150 x 75 x 18	220	190	130	255	165	130	130	200
125 x 65 x 15	225	195	135	260	170	135	135	200
100 x 50 x 10	255	215	155	295	190	155	155	230

* Root Radius included in calculations.

Table 11.6 (contd)

EQUAL ANGLES
To BS 4848: Part 4: 1972

Designation	Hp/A ratio (m⁻¹)				
	Profile 3 sides	Profile 3 sides	Profile 4 sides	Box 3 sides	Box 3 sides
200 x 200 x 24	65	85	85	65	90
200 x 200 x 20	75	100	105	80	105
200 x 200 x 18	85	110	115	85	115
200 x 200 x 16	95	125	125	95	130
150 x 150 x 18	185	110	115	190	115
150 x 150 x 15	100	135	135	105	140
150 x 150 x 12	125	165	170	130	170
150 x 150 x 10	150	200	200	155	205
120 x 120 x 15	105	135	140	105	140
120 x 120 x 12	125	170	170	130	175
120 x 120 x 10	150	200	200	155	205
120 x 120 x 8	185	250	250	190	255
100 x 100 x 15	105	135	140	105	145
100 x 100 x 12	130	170	170	130	175
100 x 100 x 10	150	200	205	155	210
100 x 100 x 8	185	250	250	195	255
90 x 90 x 12	130	170	175	135	175
90 x 90 x 10	150	200	205	155	210
90 x 90 x 8	190	250	250	195	260
90 x 90 x 7	215	285	290	220	295
90 x 90 x 6	245	330	330	255	340

Table 11.6 (contd)

UNEQUAL ANGLES
To BS 4848: Part 4: 1972

Designation	Hp/A ratio (m⁻¹)									
	Profile 3 sides	Profile 3 sides	Profile 3 sides	Profile 4 sides	Profile 3 sides	Box 3 sides	Box 3 sides	Box 3 sides	Box 3 sides	Box 4 sides
200 x 150 x 18	110	110	90	80	115	90	85	90	85	115
200 x 150 x 15	135	135	105	95	135	110	100	110	100	140
200 x 150 x 12	165	165	130	120	170	135	120	135	120	170
200 x 100 x 15	135	135	115	90	135	115	95	115	95	140
200 x 100 x 12	165	165	140	110	170	145	115	145	115	170
200 x 100 x 10	195	195	165	135	200	170	135	170	135	205
150 x 90 x 15	135	135	110	95	115	115	95	115	95	140
150 x 90 x 12	165	165	140	115	170	140	120	140	120	175
150 x 90 x 10	200	200	165	140	205	170	140	170	140	205
150 x 75 x 15	135	135	115	90	140	120	95	120	95	140
150 x 75 x 12	165	165	140	115	170	145	115	145	115	175
150 x 75 x 10	200	200	170	135	205	175	140	175	140	210
125 x 75 x 12	165	165	140	115	170	145	120	145	120	175
125 x 75 x 10	200	200	165	140	205	170	145	170	145	210
125 x 75 x 8	245	245	205	170	250	210	175	210	175	260
100 x 75 x 12	170	170	135	125	175	140	125	140	125	180
100 x 75 x 10	200	200	160	145	205	165	150	165	150	210
100 x 75 x 8	250	250	200	180	255	205	185	205	185	260
100 x 65 x 10	200	200	165	140	205	170	145	170	145	210
100 x 65 x 8	245	245	200	175	255	210	180	210	180	260
100 x 65 x 7	280	280	230	200	290	235	205	235	205	295

Table 11.6 (contd)

STRUCTURAL TEES SPLIT FROM UNIVERSAL BEAMS

Designation	Split from	Hp/A ratio (m⁻¹)					
		Profile 3 sides	Profile 3 sides	Profile 4 sides	Box 3 sides	Box 3 sides	Box 4 sides
254 x 343 x 63	686 x 254 x 125	115	145	150	115	115	150
305 x 305 x 90	610 x 305 x 179	80	105	110	80	80	110
305 x 305 x 75	610 x 305 x 149	95	125	130	95	95	130
229 x 305 x 70	610 x 229 x 140	95	120	120	95	95	120
229 x 305 x 63	610 x 229 x 125	105	135	135	105	105	135
229 x 305 x 57	610 x 229 x 113	115	145	150	115	115	150
229 x 305 x 51	610 x 229 x 101	130	160	165	130	130	165
210 x 267 x 61	533 x 210 x 122	95	125	125	95	95	125
210 x 267 x 55	533 x 210 x 109	110	135	140	110	110	140
210 x 267 x 51	533 x 210 x 101	115	145	150	115	115	150
210 x 267 x 46	533 x 210 x 92	125	160	160	125	125	160
210 x 267 x 41	533 x 210 x 82	140	180	180	140	140	180
191 x 229 x 49	457 x 191 x 98	105	135	135	105	105	135
191 x 229 x 45	457 x 191 x 89	115	145	150	115	115	150
191 x 229 x 41	457 x 191 x 82	125	160	160	125	125	160
191 x 229 x 37	457 x 191 x 74	135	175	175	135	135	175
191 x 229 x 34	457 x 191 x 67	150	195	195	150	150	195
152 x 229 x 41	457 x 152 x 82	120	145	150	120	120	150
152 x 229 x 37	457 x 152 x 74	130	160	165	130	130	165
152 x 229 x 34	457 x 152 x 67	145	175	180	145	145	180
152 x 229 x 30	457 x 152 x 60	160	195	200	160	160	200
152 x 229 x 26	457 x 152 x 52	180	225	225	180	180	225
178 x 203 x 37	406 x 178 x 74	125	160	165	125	125	165
178 x 203 x 34	406 x 178 x 67	140	175	180	140	140	180
178 x 203 x 30	406 x 178 x 60	155	195	200	155	155	200
178 x 203 x 27	406 x 178 x 54	170	220	220	170	170	220

Table 11.6 (contd)

STRUCTURAL TEES SPLIT FROM UNIVERSAL BEAMS

Designation	Split from	Hp/A ratio (m⁻¹)					
		Profile 3 sides	Profile 3 sides	Profile 4 sides	Box 3 sides	Box 3 sides	Box 4 sides
140 x 203 x 23	406 x 140 x 46	185	230	235	185	185	235
140 x 203 x 20	406 x 140 x 39	215	270	275	215	215	275
171 x 178 x 34	356 x 171 x 67	125	165	165	125	125	165
171 x 178 x 29	356 x 171 x 57	145	190	195	145	145	195
171 x 178 x 26	356 x 171 x 51	160	215	215	160	160	215
171 x 178 x 23	356 x 171 x 45	180	240	240	180	180	240
127 x 178 x 20	356 x 127 x 39	195	240	245	195	195	245
127 x 178 x 17	356 x 127 x 33	225	280	285	225	225	285
165 x 152 x 27	305 x 165 x 54	140	185	185	140	140	185
165 x 152 x 23	305 x 165 x 46	160	205	210	160	160	210
127 x 152 x 19	305 x 127 x 37	180	230	235	180	180	235
102 x 152 x 17	305 x 102 x 33	200	245	245	200	200	245
102 x 152 x 14	305 x 102 x 28	230	280	285	230	230	285
102 x 152 x 13	305 x 102 x 25	255	320	320	255	255	320
146 x 127 x 22	254 x 146 x 43	150	200	200	150	150	200
146 x 127 x 19	254 x 146 x 37	170	230	235	170	170	235
146 x 127 x 16	254 x 146 x 31	200	270	275	200	200	275
102 x 127 x 14	254 x 102 x 28	200	255	260	200	200	260
102 x 127 x 13	254 x 102 x 25	225	285	290	225	225	290
102 x 127 x 11	254 x 102 x 22	255	320	325	255	255	325
133 x 102 x 15	203 x 133 x 30	180	245	250	180	180	250
133 x 102 x 13	203 x 133 x 25	210	290	295	210	210	295

Table 11.6 (contd)

STRUCTURAL TEES SPLIT FROM UNIVERSAL COLUMNS

		Hp/A ratio (m^{-1})					
		Profile	Profile	Profile	Box	Box	Box
Designation	Split from	3 sides	3 sides	4 sides	3 sides	3 sides	4 sides
305 x 152 x 79	305 x 305 x 158	65	95	95	65	65	95
305 x 152 x 69	305 x 305 x 137	70	115	110	70	70	110
305 x 152 x 59	305 x 305 x 118	85	120	125	85	85	125
305 x 152 x 49	305 x 305 x 97	100	145	150	100	100	150
254 x 127 x 66	254 x 254 x 132	65	95	95	65	65	95
254 x 127 x 54	254 x 254 x 107	75	115	115	75	75	115
254 x 127 x 45	254 x 254 x 89	90	135	135	90	90	135
254 x 127 x 37	254 x 254 x 73	110	160	165	110	110	165
203 x 102 x 43	203 x 203 x 86	80	115	115	80	80	115
203 x 102 x 36	203 x 203 x 71	95	135	140	95	95	140
203 x 102 x 30	203 x 203 x 60	110	160	160	110	110	165
203 x 102 x 26	203 x 203 x 52	125	180	185	125	125	185
203 x 102 x 23	203 x 203 x 46	140	205	210	140	140	210
152 x 76 x 19	152 x 152 x 37	135	195	200	135	135	200
152 x 76 x 15	152 x 152 x 30	160	240	240	160	160	240
152 x 76 x 12	152 x 152 x 23	210	310	310	210	210	310

Table 11.7 Back marks in channel flanges and angles.

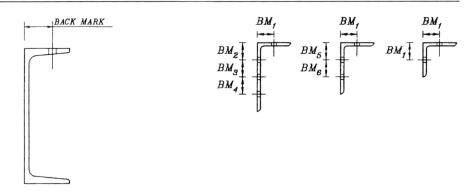

Back marks in channel flanges

Type of section	Width of flange	Back mark	Edge distance	Diameter of bolt
Channel	102	55	47	24
	89	55	34	20
	76	45	31	20
	64	35	29	16
	51	30	21	10
	38	22	–	–

Back marks in angles

Leg length	BM_1	Max. bolt dia.	BM_2	Max. bolt dia.	BM_3	Max. bolt dia.	BM_4	Max. bolt dia.	BM_5	Max. bolt dia.	BM_6	Max. bolt dia.
200	–	–	55	20	55	20	55	20	75	30	75	30
150	–	–	–	–	–	–	–	–	55	20	55	20
125	–	–	–	–	–	–	–	–	45	20	50	20
120	–	–	–	–	–	–	–	–	45	16	50	16
100	55	24	–	–	–	–	–	–	–	–	–	–
90	50	24	–	–	–	–	–	–	–	–	–	–
75	45	20	–	–	–	–	–	–	–	–	–	–
65	35	20	–	–	–	–	–	–	–	–	–	–

Material Properties 545

Table 11.8 Back marks in flanges of joists, UBs and UCs.

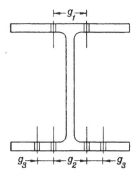

Type of section	Length of flange	g_1 min.	Max. bolt dia.	g_1 max	g_1	g_2	Max. bolt dia.	g_3	Max. bolt dia.	g_4	Max. bolt dia.
Joists	44	27	5	30	30	–	–	–	–	–	–
	64	38	10	39	40	–	–	–	–	–	–
	76	48	10	51	48	–	–	–	–	–	–
	89	54	12	59	56	–	–	–	–	–	–
	102	60	16	62	60	–	–	–	–	–	–
	114	66	16	74	70	–	–	–	–	–	–
	127	72	20	77	75	–	–	–	–	–	–
	152	75	20	102	90	–	–	–	–	–	–
	203	91	24	143	140	–	–	–	–	–	–
UC	152	65	24	92	90	–	–	–	–	–	–
	203	75	24	143	140	–	–	–	–	–	–
	254	87	24	194	140	–	–	–	–	–	–
	305	100	24	245	140	120	24	60	24	240	24
	368	88	24	308	140	140	24	75	24	290	24
	406	120	24	346	140	140	24	75	24	290	24
UB	102	50	16	62	54	–	–	–	–	–	–
	127	62	20	77	70	–	–	–	–	–	–
	133	57	20	83	70	–	–	–	–	–	–
	140	69	24	80	70	–	–	–	–	–	–
	146	64	24	86	70	–	–	–	–	–	–
	152	73	24	92	90	–	–	–	–	–	–
	165	67	24	105	90	–	–	–	–	–	–
	171	72	24	111	90	–	–	–	–	–	–
	178	72	24	118	90	–	–	–	–	–	–
	191	74	24	131	90	–	–	–	–	–	–
	210	80	24	150	140	–	–	–	–	–	–
	229	80	24	169	140	–	–	–	–	–	–
	254	87	24	194	140	–	–	–	–	–	–
	267	91	24	207	140	90	20	50	20	190	20
	292	94	24	232	140	100	24	60	24	220	24
	305	100	24	245	140	120	24	60	24	240	24
	419	112	24	359	140	140	24	75	24	290	24

546 Structural Steelwork

Table 11.9 Bending moments, shear forces and deflections.

Structure and loading	Bending moment at A	Bending moment at x from A
	$\dfrac{Wa}{2}$	$\dfrac{Wx^2}{2a}$
	$W\left(a+\dfrac{b}{2}\right)$	—
	$\dfrac{Wa}{3}$	$\dfrac{W(a-x)^3}{3a^2}$
	$\dfrac{2Wa}{3}$	$\dfrac{Wa}{3}\left[\left(\dfrac{x}{a}\right)^3 - \dfrac{3x}{a} + 2\right]$
	Pa	$P(a-x)$
	M	M

Cantilever beams

Shear force at A	Shear force at B	Maximum deflection
W	–	$\dfrac{Wa^3}{8EI}\left(1+\dfrac{4b}{3a}\right)$
W	–	$\dfrac{W}{24EI}(8a^3 + 18a^2b + 12ab^2 + 3b^3 + 12a^2c + 12abc + 4b^2c)$
W	–	$\dfrac{Wa^3}{15EI}\left(1+\dfrac{5b}{4a}\right)$
W	–	$\dfrac{11Wa^3}{60EI}\left(1+\dfrac{15b}{11a}\right)$
P	–	$\dfrac{Pa^3}{3EI}\left(1+\dfrac{3b}{2a}\right)$
–	–	$\dfrac{Ma^2}{2EI}\left(1+\dfrac{2b}{a}\right)$

548 Structural Steelwork

Table 11.9 (contd)

Structure and loading	Bending moment at midspan	Bending moment at x from A
Uniformly distributed load W over span L, supports A, B	$\dfrac{WL}{8}$	$\dfrac{Wx}{2}\left(1 - \dfrac{x}{L}\right)$
Triangular load increasing from A to B, total W	$\dfrac{WL}{8}$	$\dfrac{Wx}{3}\left(1 - \dfrac{x^2}{L^2}\right)$
Triangular load peaking at midspan, total W	$\dfrac{WL}{6}$	$\dfrac{Wx}{2}\left(1 - \dfrac{4x^2}{3L^2}\right)$
Two triangular loads $W/2$ each decreasing to midspan	$\dfrac{WL}{12}$	$Wx\left(\dfrac{1}{2} - \dfrac{x}{L} + \dfrac{2x^2}{3L^2}\right)$
Point load P at distance a from A, b from B	$M_{\max} = \dfrac{Pab}{L}$	—
Applied moment M at distance a from A	$\dfrac{M}{2}$	$\dfrac{Mx}{L}$ (when $x \leq a$)
Two triangular loads $W/2$ at each end, over length a	$\dfrac{Wa}{3}$	—
Uniform load W over central length $L-2a$	$\dfrac{W}{8}(L + 2a)$	—

Simply supported beams

Shear force at A	Shear force at B	Maximum deflection
$\dfrac{W}{2}$	$\dfrac{W}{2}$	$\dfrac{5}{384}\dfrac{WL^3}{EI}$
$\dfrac{W}{3}$	$2\dfrac{W}{3}$	$\dfrac{0.01304 WL^3}{EI}$
$\dfrac{W}{2}$	$\dfrac{W}{2}$	$\dfrac{WL^3}{60EI}$
$\dfrac{W}{2}$	$\dfrac{W}{2}$	$\dfrac{3}{320}\dfrac{WL^3}{EI}$
$\dfrac{Pb}{L}$	$\dfrac{Pa}{L}$	$\delta_{\text{centre}} = \dfrac{PL^3}{48EI}\left[\dfrac{3b}{L} - 4\left(\dfrac{b}{L}\right)^3\right]$ (when $a \geq b$)
$-\dfrac{M}{L}$	$\dfrac{M}{L}$	—
$\dfrac{W}{2}$	$\dfrac{W}{2}$	$\dfrac{Wa}{120EI}(16a^2 + 20ab + 5b^2)$
$\dfrac{W}{2}$	$\dfrac{W}{2}$	$\dfrac{W}{384EI}(8L^3 - 4Lb^2 + b^3)$ (when $b = L - 2a$)

Table 11.9 (contd)

Structure and loading	Bending moment at A	Bending moment at B
UDL W over span L	$-\dfrac{WL}{12}$	$-\dfrac{WL}{12}$
Triangular load (max at B)	$-\dfrac{WL}{15}$	$-\dfrac{WL}{10}$
Triangular load (max at centre)	$-\dfrac{5WL}{48}$	$-\dfrac{5WL}{48}$
Two triangular loads $W/2$ each	$-\dfrac{WL}{16}$	$-\dfrac{WL}{16}$
Point load P at distance a from A, b from B	$-\dfrac{Pab^2}{L^2}$	$-\dfrac{Pba^2}{L^2}$
Moment M applied at distance a from A	$\dfrac{Mb}{L^2}(3a - L)$	$-\dfrac{Ma}{L^2}(3b - L)$
Two triangular loads $W/2$ each over length a at ends	$-\dfrac{Wa}{12L}(4L - 3a)$	$-\dfrac{Wa}{12L}(4L - 3a)$
UDL W over central length $L - 2a$	$-\dfrac{W}{12L}(L^2 + 2aL - 2a^2)$	$-\dfrac{W}{12L}(L^2 + 2aL - 2a^2)$

Fixed-end beams

Shear force at A	Shear force at B	Maximum deflection
$\dfrac{W}{2}$	$\dfrac{W}{2}$	$\dfrac{WL^3}{384EI}$
$0.3W$	$0.7W$	$\dfrac{WL^3}{382EI}$
$\dfrac{W}{2}$	$\dfrac{W}{2}$	$\dfrac{1.4WL^3}{384EI}$
$\dfrac{W}{2}$	$\dfrac{W}{2}$	$\dfrac{0.6WL^3}{384EI}$
$P\left(\dfrac{b}{L}\right)^2\left(1+2\dfrac{a}{L}\right)$	$P\left(\dfrac{a}{L}\right)^2\left(1+2\dfrac{b}{L}\right)$	$\dfrac{2Pa^2b^3}{3EI(3L-2a)^2}$
Slope of moment diagram	Slope of moment diagram	–
$\dfrac{W}{2}$	$\dfrac{W}{2}$	$\dfrac{Wa^2}{480EI}(15L-16a)$
$\dfrac{W}{2}$	$\dfrac{W}{2}$	$\dfrac{W}{384EI}(L^3+2L^2a+4La^2-8a^3)$

Table 11.9 (contd)

Structure and loading	Bending moment at A	Positive bending moment in span
Propped cantilever, UDL W over span L	$-\dfrac{WL}{8}$	$\dfrac{9WL}{128}$
Propped cantilever, triangular load increasing to B	$-\dfrac{7WL}{60}$	$0.0846\,WL$
Propped cantilever, triangular load peak at midspan	$-\dfrac{5WL}{32}$	$0.0948\,WL$
Propped cantilever, two triangular loads $W/2$ each	$-\dfrac{3WL}{32}$	$0.0454\,WL$
Propped cantilever, point load P at distance a from A, b from B	$-\dfrac{Pb(L^2-b^2)}{2L^2}$	$\dfrac{Pb}{2}\left(2-\dfrac{3b}{L}+\dfrac{b^3}{L^3}\right)$
Propped cantilever with overhang a beyond B, point load P at tip	$-\dfrac{Pa}{2}$	Pa

Propped cantilever beams

Shear force at A	Shear force at B	Maximum deflection
$\dfrac{5W}{8}$	$\dfrac{3W}{8}$	$\dfrac{WL^3}{185EI}$
$\dfrac{9W}{20}$	$\dfrac{11W}{20}$	$\dfrac{0.0061\,WL^3}{EI}$
$\dfrac{21W}{32}$	$\dfrac{11W}{32}$	$0.00727\,\dfrac{WL^3}{EI}$
$\dfrac{19W}{32}$	$\dfrac{13W}{32}$	$0.0037\,\dfrac{WL^3}{EI}$
$P\left(1-\dfrac{a^2}{2L^3}(b+2L)\right)$	$\left(\dfrac{Pa^2}{2L^3}(b+2L)\right)$	$0.00932\,\dfrac{PL^3}{EI}$
$-\dfrac{3Pa}{2L}$	$P\left(1+\dfrac{3a}{2L}\right)$	$-\dfrac{PaL^2}{27EI}$

Table 11.10 Fire protection methods for steelwork.

Method of fire protection	Fire rating	Advantages	Limitations
Sprayed protection Thickness dependent on section factor (H_p/A) and fire rating	Up to 4 hours	Low cost Rapid application Cover complicated details Applied to non-primed steelwork Some products suitable for external use	Undesirable visual impact Overspray masking required Compatible primer required
Board protection Thickness dependent on section factor (H_p/A) and fire rating	Up to 4 hours	Visually acceptable boxed appearance Dry fixing methods Factory manufactured with tight tolerance Applied to non-primed steelwork Some products suitable for external use	Possibly more expensive than sprayed protection Difficult to fix around complicated details Slower to complete than spray protection
Intumescent coatings Applied by spray, brush or rollers Thickness dependent on section factor (H_p/A) and fire rating	Up to 2 hours	Possible to have a decorative finish Rapid application Cover complicated details Easier to fix post protection ancillaries	Possibly more expensive than sprayed protection Suitable for internal environments generally Generally requires blast cleaned surfaces and compatible primer
Block filled columns Use autoclaved aerated concrete blocks inside flanges of unprotected UC	Up to 30 minutes	Low cost Columns look slender and occupy less floor space Higher impact damage resistance	Limited fire rating Unwise application if the blockwork is also a partition wall
Concrete filled hollow columns	Up to 2 hours	Steel acts as permanent shuttering to concrete which carries the load in case of fire Columns look slender and occupy less floor space Higher impact damage resistance	Minimum size of column depends on the effective compaction of concrete with any reinforcement inside the hollow section
Composite slabs with profiled metal deck and unfilled voids	Up to 2 hours	Saving of time and cost due to unfilled voids With dovetail profile deck, void filling is unnecessary	For fire ratings over 90 minutes, voids must be filled where trapezoidal profile is used

For more detailed information see the following publications:
(1) *Fire Protection of Structural Steel in Buildings*, published jointly by ASFPCM and the Steel Construction Institute
(2) *BRE Digest 317*, Building Research Advisory Service
(3) BS 5950: Part 8, The British Standards Institution
(4) *Fire Resistance of Composite Beams*, Steel Construction Institute

Index

additional moment, 388, 396
aesthetic impact, trusses, 160
altitude factor, 86
amplitude of force of excitation, 66
amplitude of motion, 64
anchor plates, 451
angle of rotation, 14
angle of twist, 321
antimetric loading, 45, 46
anti-sag systems, 151
arbitrary time-dependent loading, 71
arches, 39
    elastic centre, 44
area moment theorems, 31
availability, trusses, 160
average load factor, 178
axial compression, columns, 157
axial strength modification factor, 394

back marks, 544, 545
barrier coats, 502
base plates
    analysis of, 447
    biaxial bending, 446
    capacity, 499
    empirical design, 443
    with stiffeners, 449
    trial size, 454, 496
base shear, 74, 126
base stiffness, 161
basic tension field strength, 231
basic wind speed, 86
batten systems, 218, 225
battens, 182, 219
beam end plates, 407, 422
beam flange bearing capacity, 410
beam splice, 486
beam stiffness, 165
beams, 227
    bending stress, 1
    haunched, 414, 421
    plastic behaviour, 79, 80, 81
    in pure bending, 1
    shear stress, 8
    strain energy, 14
    subject to torsion, 317
    torsional shear stress, 11
bearing resistance, 399
bearing stiffeners, 247, 252, 255, 262, 289, 295
bending compression, columns, 157

bending moment, 28
    cantilever beams, 546
    fixed-end beams, 550
    propped cantilever beam, 552
    simply supported beams, 548
bending strength, 245
bending stress, 1, 6
biaxial bending, 158, 264, 388
bolt boxes, 450
bolt load distribution, 408
bolt shear capacity reduction, 398
bolted column splice, 491
bolted connections, geometry, 397
bolted splice, 421, 485, 491
bolts
    capacities, 397, 398, 455, 459, 466, 470, 508–9
    clearance hole, 508
    connected ply, 398
    distribution of loads in, 409
    effective area, 397
    long slot, dimensions, 508
    maximum shear, 459
    modes of failure, 402
    potential maximum resistance, 480
    potential resistance, 474, 476
    resultant maximum shear, 455, 456
    resultant shear, 466
    shear capacity, 398, 455
    shank area, 508
    short slot, dimensions, 508
    spacing, 397
    tensile stress area, 508
    types of, 397, 508–9
    yield-line failure, 405
bolts in sleeves, minimum spacing, 452
boundary conditions, 156
box sections, 177, 180
bracing systems, 149
breadth of concrete rib, 358
brittle fracture, 505
buckling mode, 156
    of slender web, 288
    of web, 230
buckling parameter, 236
buckling resistance moment, 176, 245
buckling shear strength, 231
butt weld preparations, 432
butt welds, 429
    design, 434
    shear capacity, 509
    tension or compression capacity, 509

556 Index

canning, 143
cantilever beams
    bending moments, shear forces and deflections, 546–7
    influence lines, 46
carry-over factors, 37
Castigliano's theorem, 27, 43
cast-in bolts, 453
Charpy impact value, 505
circular frequency, 124
circular hollow sections, 177, 185
circular yielding, 475
classification of section, 174, 184, 186, 217, 225
classification of web, 338
coating, types of, 502
coating thickness, 502
coefficient of linear expansion, 505
column base, 443, 496
column compression stiffeners, 414, 483, 485
column end plate, 422
column flange backing plates, 416
column stiffness, 165
column web, 411
    bearing resistance, 411
    buckling resistance, 412
    effective width, 440
    load distribution, 411
    local shear, 412
column web panel shear, 412, 413
columns
    axial capacity, 155
    bending and axial compression, 157
    design basis, 156
    types of failure, 155
combined shear and tension, 399, 400
compact sections, 228
complementary shear stress, 8
composite beams, 329, 365
composite columns, 383, 390, 392
    encased, 384
composite construction, 144
composite sections, 338
    moment of inertia, 368
    plastic moment capacity, 336
composite stage, 376, 381
compound columns, 200, 201
compound sections, 305
compound tension members, 211
compound trusses, method of section, 23
compression flanges
    lateral torsional buckling, 227
    local buckling, 227
    stability, 373
compression resistance, 185, 188, 190, 205
compressive resistance, 176
compressive strength, 156, 174, 176, 184, 187, 190, 198
conceptual design, 142
concrete flange, resistance, 336
concrete resistance, 372
concrete stress block, 497
connected ply, bearing capacity, 455
connection
    beam-to-beam, 442, 465, 469
    beam-to-end plate, 485
    of bearing stiffeners, 264
    bolted, 397
    column compression stiffeners, 485
    column-to-foundation, 444
    column-to-roof truss, 454
    column/beam welded, 438
    eaves haunch, 473
    flange to web, 266
    moment capacity, 409
    moment of resistance, 479
    shear capacity, 410, 481
    of stiffeners, 264
    types of, 442
    using double angle web cleats, 465
    welded beam-to-column, 436
connection design, 139
connection diagrams, simple tension members, 215
construction stage, 371, 381
continuous beams, 35, 55, 95
    influence coefficients, 506–7
continuous composite beam, 347
continuous construction, 159
continuous frames, 57, 58
corner snipe, 483
corner yielding, 475
corrosion process, 501
corrosion protection, 143, 144, 160, 501
corrosion protection system, 502
crabbing, 303, 316
crabbing action, 268
cracked section, second moment of area, 333
creep, 330, 344
critical buckling load, 124
critical shear strength, 231, 255
critically damped structures, 63
cross bracing, 164
curvature, 344
cyclic frequency, 124

damping, 125
damping coefficient, 62, 73
damping force, 62
deflection, 208, 273, 278, 313, 343, 382
    cantilever beams, 547
    composite continuous beams, 363
    fixed-end beams, 551
    propped cantilever beam, 553
    pin-jointed structures, 27
    simply supported beams, 549
deformation matrix, 56, 104
degrees of freedom, 57
design stage check list, 142
design strength, 174, 187, 232, 270

destabilising forces, 245
destabilising load, 240
detailed design, 143
detailing, 388, 458, 465, 470, 473, 530–4
diagonal matrix, 54
diagonal shear stiffeners, 419, 420
direct stress, 132
direction factor, 86
displacement matrix, 54
displacement method, 52, 55, 57
distribution coefficients, 101, 106
distribution factor, 37
double intersection lacing, 164
ductility, 79, 505
dye penetrant inspection, 432
dynamic augmentation factor, 86
dynamic equilibrium, 63
dynamic magnification factor, 66, 67
dynamic model, 63
dynamic pressure, 86

eccentric connection, 161, 211, 221
eccentric vertical bracing, 151
economy of connections, trusses, 160
economy of materials, trusses, 160
edge distance, 397
effective area, 211, 215, 216, 221, 224
effective bracing, 166
effective breadth, 330, 346, 366
effective height, 86
effective length, 156, 158, 161, 162, 166, 184, 186, 191, 197, 203, 241, 242, 277
effective length ratio, 167, 168
effective moment of inertia, 415
effective payload, 153, 154
effective slenderness, 387
effective throat thickness, 428
effective width of flange, 344
elastic centres, 44, 116
elastic critical load, 156, 158
elastic critical load factor, 148
elastic critical moment, 176, 230, 245
elastic critical shear stress, 230
elastic modulus, 236
elastic neutral axis, 306
elastic section modulus, 334
electric arc welding, 427
encased columns, 384, 391
end distance, 397
end plates, 469
  bolt groups, 470
  shear capacity, 471
end posts, 261
end returns, 429
end stiffeners, 260
equation of motion, 63–64, 65, 124
equivalent applied moment, 410
equivalent bending moment, 408, 474
equivalent moment factor, 244
equivalent slenderness, 245

equivalent slenderness ratio, 176, 188, 230
equivalent uniform moment factor, 180, 212
erection, 144
Euler strength, 175, 187
Euler theory, 156
external force matrix, 58, 110, 111

failure of bolts by yielding, 403
failure of flange by yielding, 403
failure surfaces, 361
failure of web in tension, 407
fatigue checks, 269
fatigue of gantry girders, 315
fillet welds, 428, 429, 458
  capacities, 508
  design, 434
  design strength, 434
  effective length, 429
  throat thickness, 508
fin plates, 458
  moment capacity, 463
  shear capacity, 463
  welds, 464
finishes, 502
fire protection
  methods, 554
  ratings, 554
fitted compression stiffener, 473
fixed base, 445
fixed-end beams, 35
  bending moments, shear forces and deflections, 550–1
  plastic analysis, 83
fixed-end moments, 58, 101, 103, 106
flange-dependent shear strength, 232, 255
flange of beam, local bearing check, 412
flange cover plates, 422, 486, 488, 493
flange plates, 422, 425
flange plate–UB connections, 314
flange–web connection, 300
flexural buckling, 155
flexural buckling failure, 158
flexural shear stress, 325
forced vibration, 61, 65, 67
forcing function, 63
fracture toughness, 505
free-body diagram, 78, 119
free vibration, 61, 68, 71
frequency of excitation, 66
friction grip bolts, 142, 399, 488, 489
  tension capacity, 399, 508
full penetration butt welds, 429, 430
full shear connection, 338, 372
full tension-field action, 289
fully threaded bolts, 144
funicular polygon, 24, 91, 93

galvanising, 502
gantry girders, 245, 265, 266, 267, 269, 300
gas metal arc welding, 428

generalised matrix, 53
geometry (statics) matrix, 53, 54, 56, 59, 103, 110
gross area of stiffener, 415
gross outstand of the stiffener, 415
gross vehicle weight, 153, 154
group of bolts
   in-plane loading, 400
   out-of-plane bending, 407
   out-of-plane loading, 401
   polar moment of inertia, 455
gusset plates, 455, 456

Hamilton's principle, 121, 122
handling, trusses, 160
Hardy-Cross method, 102
harmonic excitation, 65
haunched connections, 143
haunches, 133
   local bearing of compression flange, 482
heavy bracings, 160
heavy columns, 160
high-cost connections, 142
high-strength friction grip (HSFG) bolts, 142, 488, 489
   proof load, 509
   slip resistance, 509
hingeless arches, 114
hingeless symmetrical arches, 43
holding-down (HD) bolts, 451, 497
   anchor plates, 453
   capacity, 452
holes, sizes of, 397, 508
hollow circular section, concrete filled, 389
horizontal surge, 268
   load, 303
Hp/A values
   equal angles, 539
   joists, 538
   parallel flange channels, 538
   structural tees, 541–3
   unequal angles, 540
   universal beams, 535–6
   universal columns, 537
hyperstatic reaction, 100

incomplete penetration, 431
industrial buildings, 160, 200
inertia ratio, 506, 507
inertial force, 62
influence coefficients
   bending moment, 506–7
   deflection, 507
   reaction, 506
influence diagrams, three-hinged arches, 50
influence lines, 46
interaction formula, 213, 390, 396, 425, 494
intermediate stiffeners, 295
intermediate transverse web stiffeners, 292

intermittent fillet welds, 143, 429
internal forces, 21
   trusses, 22

joint displacement matrix, 61
joint restraint coefficient, 165
joint rotation matrix, 57

K-bracing, 150
K type stiffeners, 420

lacing bars, 181
lacing systems, 181, 208, 218
lack of fusion, 431
lap joints, 429
large grip joints, 398
large trusses, 160
lateral bracing, 149
lateral loads, 148
lateral torsional buckling, 158, 211, 217, 224, 229, 240, 277, 282, 290, 309, 351
latticed girders, 142, 152, 161, 222
light bracings, 160
light columns, 160
light trusses, 160
lightweight concrete, 346
limited frame method, 161, 191
limiting equivalent slenderness ratio, 176, 188, 230
limits of deflection, 265
load combinations, 82, 146, 159
load-time history, 67
local bearing stress, 265, 311
local bending moments, 161, 184, 186
local buckling, 155
local capacity, 158, 188, 190, 193, 227
local capacity checks, 157
local failure, 156
local section capacity, 177
local shear capacity, 464, 468
local stresses, 265, 314
locked-in stress, 382
long joints, 399
longitudinal anchor force, 257
longitudinal shear, 359, 380
   resistance to, 361, 380
   transfer, 342
low-cost connections, 141
lumped masses, 62

magnetic particle inspection, 432
manufacturing, 143
manufacturing process, 138
material grade, 138, 505
material properties, 505–54
   bolts, 508–9
   Hp/A values, 535–44
   influence coefficients, 506–7
   steel sections, 510–29
   welds, 508–9

matrix method, 52, 103, 110
maximum weight, 154
Maxwell diagram, 24, 25, 91, 93
medium columns, 160
medium-cost connections, 142
method of section, 25, 91, 93
 compound trusses, 23
minor axis slenderness ratio, 375
modular ratio, 331, 346, 366
modulus of elasticity, 2, 505
modulus of rigidity, 11
moment capacity, 238, 271
moment connections, 140, 407, 408
moment distribution, 37, 101, 105, 106
moment of inertia, 4
motion, equation of, 63–64, 65, 124
motor vehicle regulations, 153
motor vehicles, special types, 154
multi-planar trusses, 142
multistorey buildings, 152, 161, 195, 273
multistorey frames, 152, 160
 braced, 160
 unbraced, 160

natural frequency, 63, 73, 364
negative moment capacity, 339
net area, 214, 224
net bearing resistance, 415
neutral axis, 3, 236, 332, 338
nominal moments, 176, 198
non-destructive testing, 431
normalized warping constant, 319
notched beams, 440
 block shear capacity, 461
 block shear check, 441
 failure pattern, 440
 plain shear check, 441
 shear capacity, 461
notches, moment capacity, 462
notional horizontal force, 146

open-web girders, connections, 141
ordinary articulated vehicles, 153
overall buckling, 157, 178, 180, 189, 190, 194
overall buckling check, 157
overdamped structures, 63

packing plates, 422
patterned loading, 506
painting specification, 503
parabolic arches, 114
parallel shank bolts, 399
partial penetration butt welds, 429, 430, 508
partial shear connection, 338, 342, 359, 363, 372
partial sway bracing, 168
partial tension-field action, 289
patterned loading, 147, 363, 506
percentage of redistribution, 341
permission to move, 153

Perry coefficient, 175, 176, 187, 245
phase angle, 64
piles, 517–18
pinned base, 445
plastic analysis, 82
 fixed-end beams, 83
 portal frames, 83, 127
plastic behaviour, 79
plastic design, 152
plastic hinges, 80, 128, 129
plastic modulus, 239
 of section, 80
plastic moment capacity, 158, 176, 372
 reduction of, 342, 350
plastic moment of resistance, 80, 119, 129, 130, 135, 230, 386, 387, 392, 394
plastic neutral axis, 306, 330, 336, 384, 386, 392, 393
plastic sections, 228
 properties, 335, 336, 337
plastic strain, 79
plate girders, 232, 266, 284, 297
plates, 74
 bending moment, 75
 flexural rigidity, 75
 pure bending, 75
 shear force, 76
 torsional moment, 76
 yield-line analysis, 76
Poisson's ratio, 75, 505
polar moment of inertia, 400
polygon of forces, 19, 91, 93
porosity, 431
portal apex connection, 422
portal frames, 152, 190, 473
 haunched, 133
 plastic analysis, 83, 127
positive plastic moment capacities, 338
potential energy, 122
potential resistance, 476
pre-cambering, 364
pressure coefficient, 87
primers, 502
principal axes, 4, 7
principal stress, 132
probability factor, 86
product inertia, 4, 5, 6
profile steel sheeting, 345, 357, 362, 365
propped cantilever beams: bending moments, shear forces and deflections, 552–3
propped construction, 363
protective system, 501
pseudo-spectral velocity, 73
punching shear perimeter, 451

radiography, 432
radius of curvature, 2
radius of gyration, 176, 236
rafter bracing, 149
rafters, 85

rectanglular hollow sections, 177, 189
rectangular box sections, 181
redistribution of support moments, 330, 340
reduced local section moment capacity, 213
reduction factor, 357, 390, 395
relative stiffness, 166, 168
reserve of strength, 136
resonance, 66, 74
response spectra, 73, 125
restraint system, 136
rigid design, 144
rigid frames, 105, 110, 148
rigid-jointed frames, 166, 167
Robertson constant, 175, 187
roof trusses, 183, 185, 220
  welded connections, 457

saddles, 143
seasonal factor, 86
second moment of area, 333, 395
second moment of inertia, 236
section classification, 185, 189, 234, 271
section properties, 287, 307, 334
  for positive moments, 367
section type, 138
sectional properties, 235
seismic accelerations, 72
seismic analysis, 120
seismic excitation, 125
seismic response, 72, 151
semi-compact sections, 229
semi-rigid design, 144
settlement of support, 34
shake-down effects, 363
shape factor, 80
shear buckling resistance, 350
  of web, 231, 255
shear capacity, 177, 237, 350, 373, 509
shear centre, 9, 10, 319
shear connection, 341, 347, 354, 379
  resistance, 337
shear connectors, 329, 341, 354, 375
  capacity of, 336
  spacing, 355, 356, 357
shear failure plane, 362
shear force, 28
  cantilever beams, 547
  fixed-end beams, 551
  propped cantilever beams, 553
  simply supported beams, 549
shear keys, 453
shear modulus, 317
shear stress, 8, 132
  torsional, 11
shear transfer, 453
shear walls, 152
shear zone, 410
shielded metal arc welding, 428
shrinkage, 330, 344
side yielding, 475

sidesway, 167
sign conventions, 111
simple connections, 140
simple construction, 159, 206, 273
simple design, 144
simple tension members, 211, 215, 221
simply supported beams, 31
  bending moments, shear forces and
    deflections, 548–9
  free vibration, 68
  influence lines, 47
  modes of vibration, 70
single degree of freedom (SDOF) systems
    62, 121
single diagonal bracing, 164
single fillet welds, 429
single intersection lacing, 164
single-storey buildings, 163, 164
site welding, 143
site wind speed, 86
slag inclusions, 431
slender sections, 229
slender webs, 240, 290
slenderness correction factor, 245, 375, 376
slenderness factor, 244, 388, 395
slenderness ratios, 169, 174, 184, 189
  limiting values, 181
  maximum, 186, 192
slip resistance, 399
slope–deflection equations, 29, 33, 56, 60
snow load, 89
spectral acceleration, 74
spectral displacement, 74
spectral velocity function, 73
speed limits, 153
splices, 422
  distribution of forces, 423
squashing, 155
staggered holes, 143, 214
standard effective wind speed, 86
standard pressure coefficient, 86
statics (geometry) matrix, 53, 54, 56, 59, 103,
    110
steel beams, resistance of
  beam, 372
  flange, 336
  slender, 336
  slender web, 336
steel sections
  detailing dimensions, 530–4
  dimensions and properties, 510–29
  Hp/A values, 535–44
  universal bearing piles, 517–18
stiffened girder, 237
stiffeners, 248
  area of, 415
  buckling resistance, 415
  connection, 300
  effective area of, 415
  effective length, 415

K type, 420
Morris, 420
N type, 420
stiffness matrix, 54, 60, 61, 112
stocky circular hollow steel section, 389
strain, 2
strain energy, 27
  arches, 40
  in axial load, 14
  in bending, 15
  due to shear, 16
  in torsion, 17
strength reduction factors, 174, 187, 193, 204
strength reserve, 132
stress concentration, 211
stress/strain curves, 505
structural dynamics, 61
structural hollow sections, 142, 213
structural stability, 148
strut action, 421, 423, 487, 493
  minor axis bending, 424
struts, 155
stud capacity, reduction of, 357
stud shear connectors, 379
studs
  dimension of, 359
  spacing of, 359
submerged arc welding, 428
superimposed fillet weld, 430, 508
supplementary column web plate, 416
surface preparation, 502
sway frames, 148, 152
sway index, 148

tall cantilever structures, 120
tall chimneys, 120
temporary lifting points, 144
tensile load capacity, 391, 509
tensile reinforcement, resistance of, 337
tension capacity, 217
tension-field action, 231, 238, 255, 258, 289, 297
tension stiffeners, 417, 439
tension zone, 410
thermal stresses
  hingeless symmetrical arches, 44
  two-hinged arches, 42
thin webs, buckling, 238
three-hinged arches
  influence diagrams, 50
  influence lines, 48, 49
three moment theorem, 35, 98
throat thickness, 428
ties, 149, 211
time-history, 74
time lag, 64
time period of oscillation, 124
torsion
  circular shaft, 11
  strain energy, 17

thin open sections, 13
thin rectangular members, 12
torsion constant, 236
torsion stiffeners, 263, 299
torsional buckling, 155
torsional constraint, 317
torsional index, 179, 236
torsional resistance, 317
torsional restraint, 151, 241, 312, 351, 377
torsional shear stress, 11, 326
torsional stiffness, 12
transformed area, 367
transformed section, 331
transport, trusses, 160
transportation, 143
transverse web stiffeners, 250, 251
triangular impulse, 67
trusses, 22, 85
  connections, 141
  determinate, 22
  Fink, 85
  Howe, 85
  indeterminate, 22
  influence lines, 51
  local bending moments, 161
  Mansard, 85
  method of section, 23
  Pratt, 85
  Warren, 85
two-hinged arches, 40
  thermal stresses, 42

ultimate axial tension, 214
ultimate unit resistance, 119, 120
ultrasonic flaw detection, 432
undamped free vibration, 65
undamped SDOF systems, 65, 67
undercut, 431
uniform moment factor, 159
uniform torsion, 317, 318
unit load analysis, 370
unity checks, 177, 188, 190, 198, 206, 217, 225
unpropped construction, 363, 366
unrestricted sidesway, 167
unstiffened webs, 246

V-bracing, 150
vector diagrams, 64
vertical bracing, 150
very short impulse, 71
vibration, 364, 382
vibration response, 64
virtual work, 78
viscous damping, 62, 65
viscous dashpot, 62
Von Mises criteria, 152
Von Mises yield criterion, 81

waisted shank friction grip bolts, 399
warping bending stress, 320

warping constant, 317
warping moments, 319
warping shear stress, 319
warping torsion, 317, 318
web angle cleat, 422
web bearing capacity, 410
web bearings, 247, 273, 283, 311, 411, 481
web buckling, 246, 272, 282, 309, 310, 412, 482
   capacity, 410
web cleats
   bolt arrangements, 467
   shear capacity, 467
web cover plates, 486, 490, 495
web panel shear capacity, 410
web plates, 422, 426
   distribution of loads, 427
web resistance, 372
web shear capacity, 482
web slenderness factor, 231
web stiffeners, 249
   intermediate transverse, 248
web stress ratio, 348
web tension failure, 404, 476
web thickness, minimum, 237

webs
   composite beams, 347
   with openings, 263
weld group, 435
   moment of inertia, 435
   polar moment of inertia, 435
welded connection, 314
welds and welding, 427
   capacities, 508–9
   defects, 431
   resultant shear stress, 436
   see also butt welds; fillet welds
wind girders, 149
wind loads, 87
   external pressure, 87
   internal pressure, 87

X-bracing, 150

yield lines, 78, 119, 475
yield-line analysis, 76, 118
yield-line patterns, 77

Z-bracing, 150
zero period acceleration, 125